**AIR
POLLUTION
CONTROL
THEORY**

McGRAW-HILL
BOOK COMPANY
New York
St. Louis
San Francisco
Auckland
Düsseldorf
Johannesburg
Kuala Lumpur
London
Mexico
Montreal
New Delhi
Panama
Paris
São Paulo
Singapore
Sydney
Tokyo
Toronto

MARTIN CRAWFORD

Professor of Engineering
University of Alabama in Birmingham

Air
Pollution
Control
Theory

This book was set in Times Roman.
The editors were B. J. Clark and M. E. Margolies;
the cover was designed by J. E. O'Connor;
the production supervisor was Leroy A. Young.
The drawings were done by ANCO Technical Services.
R. R. Donnelley & Sons Company was printer and binder.

Library of Congress Cataloging in Publication Data

Crawford, Martin.
    Air pollution control theory.

    Includes bibliographical references and index.
    1. Air—Pollution. 2. Gases—Cleaning. I. Title.
TD883.1.C7   628.5'3   75-20023
ISBN 0-07-013490-1

**AIR
POLLUTION
CONTROL
THEORY**

1234567890DODO79876

To
CAROLYN,
THOMAS,
and LILA

CONTENTS

Air pollution is a topic which, among many others, has been at the forefront of social concern for the past several years. Most of this concern, as far as the general public is concerned, has been directed toward the health, distributional, and regulatory aspects of the problem. To all appearances, the public has given little thought to the technological aspects of reducing emissions into the atmosphere. This may be due in large part to the fact that there has been no publication available which provides insight into the mode of operation of the various devices for removing pollutants prior to emission into the atmosphere.

The term *air pollution control* is capable of two interpretations. To the public at large it means probably the limitation or prohibition of emissions by force of law. Inherent in this interpretation is the determination of which substances should be limited and to what extent they should be limited, which requires determination of the effects of each substance on health, damage to property, and esthetic values. The interaction of different pollutant substances must also be considered. These areas of air pollution control have been extensively explored in the past several years.

In the second meaning of air pollution control, which is the one employed in this book, the word *control* is used in the sense of prevention. What means are available to prevent air pollution from occurring? Aside from shutting down all polluters, which would mean disabling our economy and disrupting our way of life, there are means available or potentially available to remove all pollutants to the extent necessary to prevent serious atmospheric contamination. These means take the form of certain devices which are subject to engineering analysis, and which can be designed, built, installed, and operated to achieve the desired result, but not without problems. In dealing with such problems, we need engineers who are trained to understand these devices; this book is designed to help provide such training.

A well-established branch of chemical engineering is that of gas cleaning. So how does air pollution control, in the sense used in this book, differ from gas cleaning? In truth, air pollution control lies within the discipline of gas cleaning. Nevertheless, the problems encountered in the design of control devices tend to be much more difficult than those associated with the usual gas-cleaning equipment. In processes where a solute is to be separated from a gas, the initial solute fracton is normally large and 90 to 99 percent removal is generally considered good practice. In air pollution control work, the initial concentration is usually quite low and the same 90 to 99 percent removal is required. Thus, in a sense, air pollution control takes up where ordinary gas cleaning leaves off.

This book is intended first of all as a textbook for use in a course in air pollution control theory. Such a course can be designed for senior engineering students or for graduate engineers. It is intended that the student gain familiarity in the following areas: (1) qualitative description and quantitative evaluation of the problems associated with control of air pollution; (2) understanding of the mode of operation of the various control devices; (3) evaluation of the performance of the various control devices; (4) evaluation of the performance of the total pollution control system, which may include several control devices, as well as the associated ductwork, hoods, and fans; (5) the methodology of gaining better insight into the operation of the devices and improving them; and (6) the methodology of design of air pollution control devices. The book attempts to deal with each of these areas of study.

It is hoped that this book will be of use to practicing engineers in the air pollution field. Its potential usefulness to such engineers is more seminal than remedial; that is, it should provide insight and inspiration toward arriving at new designs more than toward correcting defects in existing designs. The intent is to provide instruction in bringing theoretical insights into practical design and analysis of the various devices involved. It is in no way a handbook or compendium of practical solutions to practical problems.

The purpose of this text is to apply certain fields of engineering analysis, notably fluid mechanics, thermodynamics, heat transfer, dynamics, and electrodynamics, along with certain phases of chemistry and physics, to the rudimentary analysis and design of air pollution control devices. In this sense, it functions in the same way as texts on internal combustion engines, turbomachinery, or steam power plants which are used in senior mechanical engineering courses on these subjects, to mention three examples. Such books seek to bridge the gap between pure theory, as it is taught on the undergraduate level, and engineering practice in the field. Although such books are generally anchored equally well on the theory and practical sides of the gap, they carry the reader only so far into the practical field. To the author's knowledge, this is the first textbook which attempts to bridge the gap between theory and practice in air pollution control. Even a casual perusal shows that this book is anchored much more firmly on the theoretical side, which is a consequence of the newness of the approach, time limitations in developing the material, and the fact that much practical data in the field are still hoarded as proprietary information by the various companies active in developing air pollution control devices. The author hopes that future books will appear, including perhaps a later edition of this one, which will provide significantly greater emphasis on the practical aspects of applied theory.

The author believes that air pollution control theory is not the same thing as the combined total of the theory of operation of each of the available control devices; that is, in this sense, the whole is greater than the sum of its parts. This viewpoint can be defended: First, there are aspects of air pollution control which are common to all control devices and which can be treated separately from any particular device. Second, there exists the possibility that new devices may be invented from time to time and take their place in the theory. Third, there are synergistic effects involving the interaction of various control devices and other elements in any air pollution control system. And fourth, there is that branch of knowledge dealing with the selection of a particular control device to use in the first place.

There exists or can exist a field of air pollution control theory, which can take its place among the many specialized fields of engineering. Such a field is as difficult, challenging, and important to an industrial society as any other field. Whether or not this field is allowed and encouraged to develop depends upon the priorities and expectations which society places on the engineering profession and upon the resources allotted to it.

The material included in this text should be adequate for a course sequence totaling six quarter-hours credit. With some supplementation it should suffice to handle a course sequence of six semester hours. A course of three semester-hours credit can cover perhaps two-thirds of the book, say, to Chapter 9 or 10.

The author wishes to express his gratitude to the School of Engineering, University of Alabama in Birmingham, for its support during the preparation of the manuscript for this work. He is also grateful to his former students for their many helpful comments and criticisms. Particular thanks are due to Dr. E. R. Greene, Jr., for a number of illuminating discussions regarding technical matters, and to my father, Guy Crawford, for reading the manuscript.

MARTIN CRAWFORD

1

INTRODUCTION

With this chapter begins a study of many of the theoretical aspects of the science of reducing the level of pollution of our atmosphere. Predominantly, what is studied is the removal of pollutants from various effluent streams discharging into the atmosphere, a process which can be referred to as *depollution*. The vastly more desirable method of replacing a polluting process with a nonpolluting one is not treated here to any extent; this can be handled best in a detailed study of the particular process involved. The current chapter deals with the subject of air pollution control in a general way; subsequent chapters become more specific and more mathematical.

1-1 ROLE OF AIR POLLUTION CONTROL IN MODERN SOCIETY

Industrialization has provided humanity with many material and social benefits. At the same time it has brought in its wake many material and social problems. One of these is pollution of the environment. Most ecologists recognize pollution as a serious threat to the quality of our life and possibly to its very existence. Interest in environmental concerns has become intense in recent years, with much pressure being placed on public and corporate officials to alter policies and practices which in the past have resulted in degradation of the environment. Court actions and legislation have required industrial concerns to be more careful of pollution. It appears hopeful at least that in a few years our environment will be cleaned up to a considerable degree. Even so, much remains to be done, and many polluting practices continue to exist. There is a negative side to this concern for the environment. Environmentalists have succeeded in blocking many projects whose need to society is evident. These actions may be meritorious in most cases in that proper concern for the environment was not given in the planning of the project or that alternative plans should have been considered. Nevertheless, a better decision-making procedure needs to be developed, which will give proper concern to the effect on the environment but not allow a small group to block construction once a decision has been made.

Since pollution is concomitant with most industrial activity, methods of controlling it must be employed. The best method of control is to avoid pollution in the first place by replacing the present process with one which does not pollute. Usually this is not feasible, and so some method of control must be installed. Generally, the pollution control method adds to the cost of the process, and, for the most part, the added cost is passed on to the consuming public. In a few lucky instances, material of value will be recovered in the pollution control process, so that a net saving will result. In most such cases, however, material recovery will already be practiced and the degree of collection efficiency that is most economical for material recovery will be completely inadequate for effective pollution control.

When the social cost of pollution is considered—damage to health of the people in the community, damage to health of workers in the plant, damage to property, and damage to aesthetic values—a high degree of pollution control is nearly always indicated. Since the public pays the cost of pollution or its control, one way or the other, it seems reasonable that the public should insist that pollution control measures be installed where needed. This can best be done by requiring those who pollute to bear the full social cost of their pollution.

The following types of pollution are generally recognized. Some overlap exists between the different types.

Water pollution   The presence, in concentrations higher than normal, in natural waterways (lakes, streams, rivers, and oceans) of dissolved or suspended foreign material, such as silt, chemicals, fecal matter, metallic elements, organic material, or nutrients.

Air pollution   The presence in the atmosphere of solid particles, liquid droplets, or gaseous compounds which are not normally present or which are present in a concentration substantially greater than normal.

Waste pollution   The presence on land or water of solid material, organic or inorganic, which has no beneficial qualities.

Chemical pollution   The presence in plants and animal tissue, including feed and foodstuffs, of adulterant chemicals which have no beneficial effect.

Noise pollution   The presence in the open atmosphere or in a confined space of a noise generally considered undesirable, except possibly by the person responsible for it. In the latter case, noise pollution does not exist in the space immediately surrounding the person.

Thermal pollution   The discharge into the environment of a stream of air or water which is at a different temperature from that of the environment at the point of discharge or downstream of this point.

There is room for disagreement as to what constitutes pollution, especially in the case of chemical and noise pollution. For example, some food additives are regarded as beneficial by the manufacturer but as adulterants by some consumers. And a particular noise can be regarded as beautiful music by one person and as a discordant annoyance by another.

The concern of this book is with air pollution, which is described more completely in Sec. 1-3. At this point we shall consider the best results that air pollution control can achieve. It must be recognized that natural processes continually purify the air. Thus if all human sources of pollution were to stop today, the air would eventually be purified to its natural state. It must also be realized that natural processes pollute the air, and that natural air contains a certain level of what would otherwise be regarded as pollutants. Since such a level of pollution is natural and presumably organisms are adapted to it, we shall define polluted air as air which contains any substance in an amount significantly greater than that which appears in natural air. If the substance does not appear at all in natural air, then its presence constitutes pollution. Since water can appear in the atmosphere in all concentrations ranging from almost none to that required for saturation, we shall not regard water as a pollutant. This distinction has important practical consequences inasmuch as the effluent from many stacks discharging into the atmosphere may consist of almost pure water mixed with air and may be mistaken as a source of pollution.

It is possible to achieve virtually complete removal of any polluting substance from any airstream discharging into the atmosphere. Such complete pollution control is too expensive to be seriously considered in most cases, and, in fact, is unnecessary. However, there are situations where pollution control must be virtually complete. Nuclear power plants are a case in point, where radioactive materials must not be allowed to escape in quantities large enough to cause damage. Biological weapons research laboratories are another example where the escape of even the smallest amount of the wrong substances could cause complete catastrophe. In most cases, however, a certain amount of impurities can be allowed to pass into the atmosphere without serious harm. Natural purification processes will then keep the atmosphere reasonably pure. Since the cost of purifying an airstream increases drastically as complete purification is approached, a reasonable balance must be found.

In general, the cost of cleaning a gas stream increases with increasing gas-flow rate but at a rate somewhat less than proportional to the gas-flow rate. For a given amount of collected material, the cost is much less for a concentrated gas stream than for a dilute one; thus it is better to clean the stream before diluting it with additional air if this is possible. It is also more economical to clean a continuous gas stream than an intermittent one, particularly one which is used only occasionally. For these reasons, it is better to require 99 percent control on the effluent from a steel mill than to demand 60 percent control from 30,000 backyard barbecue grills. To be sure, backyard barbecue grills could be proscribed, and life would go on without them. But would the extra cleanliness of our air be worth that much deprivation to that many people? Choices such as this will have to be made. In any event, it is probably impractical to require as complete control over small sources of pollution as over large sources even though the small individual sources are far more numerous than the large ones and their total effluent may equal or exceed that from the large sources.

The question that concerns us most at this point is: To what extent can we reduce the total emissions into the atmosphere if we make the maximum cost-effective use of existing air pollution control technology? By cost effective is meant that degree of control which is beneficial to society as a whole, balancing the social cost of pollution against the cost of the control device. As explained previously, pollution control is seldom cost-effective on any other basis. To answer our question, at least qualitatively, let us assume that one-third of the total emissions is from large factories for which 98 to 99 percent control is feasible, one-third is from medium-size operations where 90 to 95 percent control can be economically obtained, and the remaining one-third is from small plants and individual operations where an average of 60 percent control can be obtained from control measures and elimination of certain unessential operations. Thus, the total reduction of the atmospheric pollution load is estimated to be about 84 percent, which appears to be acceptable as a current goal. It also appears to be obtainable, except possibly in two problem areas: The first problem relates to gaseous effluents from steam-generation plants, whether for power generation or for process and heating steam; the second problem relates to the automobile and other internal-combustion engines. In both cases, the pollutants involved are the oxides of sulfur and nitrogen. Much research is currently in progress on ways of controlling the emissions of these pollutants. Until this research bears fruit, realization of the above goal will not be entirely possible.

What about the future? Assuming that within the next few years we realize a reduction to 15 percent of current emission levels, will we be able to hold to that level indefinitely? To answer this question, one must make assumptions concerning the growth of industrial output, or at least that portion of industrial output which contributes to pollution, and concerning the advance of air pollution control technology. Most industrialists and economists like to predict a doubling of industrial capacity every 10 to 20 years. For example, utility executives frequently indicate that they expect the demand for electric power to double every 10 years for the foreseeable future. Of course, industrial capacity as a whole might not advance quite that fast, but a doubling every 15 years seems reasonable. However, we have now entered what has been called the *postindustrial age*. Much future increase in production may be oriented toward services, which are largely nonpolluting; computer software is a case in point. Also, many future production increases may take place in relatively nonpolluting activities, such as sophisticated electronic gear. Except for obvious pockets of poverty, there is reason to question whether the average American citizen really needs any increase in production. For these reasons, a much lower productivity increase rate may become the order of things. It is unlikely, however, that per capita production will decline as long as our resources hold out. Thus a lower limit to our future growth in industrial capacity is the population growth.

With these factors in mind, we might reasonably project a 4 percent annual growth rate in uncontrolled pollution. This figure corresponds to doubling the amount of pollutants before control every 18 years. To maintain a level after control equal to 15 percent of current emissions will require an average control of 92 percent after 18 years and 96 percent after 36 years. To achieve the latter requirement more advanced pollution control equipment must be developed or polluting activities must be restricted to large factories, where 98 to 99 percent control can be attained with present-day technology.

1-2 SCOPE OF AIR POLLUTION CONTROL THEORY

What is meant by air pollution control theory? And how does it differ from the study of air pollution control devices? Many devices are available for removing contaminants from air. These work on a variety of different principles, and in most cases the devices used today are similar to ones developed many years ago. Their designs have evolved over the years as experience was gained with their use and as theoretical insights were arrived at. Some of the more common devices include the settling chamber, the cyclone collector, inertial collectors of various types, filters, electrostatic precipitators, scrubbers, absorbers, adsorbers, combustion chambers, and condensers. Much air pollution control theory is indeed devoted to the study of how these devices operate and how they can best be designed.

The status of air pollution control theory, if limited to the detailed study of the operation and design of various control devices, can be described as one of adolescence; that is, the study of such devices has proceeded beyond the stage of infancy but has not yet reached maturity. By contrast, consider the gas turbine for aircraft application, or, in other words, the jet engine. A tremendous amount of research work has been conducted over the last 40 years to precisely describe the flow paths, boundary-layer development, and heat-transfer rates throughout the various components of the system. A complete bibliography in this area would run to several thousand entries. To be sure, the study of the jet engine is not complete by any means, and research will undoubtedly continue for many years to come. Consider, now, the quantity of research devoted to understanding the various pollution control devices. Some devices have been studied more extensively than others; the electrostatic precipitator has probably received the most attention. Even so, the number of research papers devoted to this one type of control device is only a small percentage of those devoted to the jet engine. Thus, the detailed operation of the electrostatic precipitator is not well understood. Most other air pollution control devices are even less well understood. This pessimistic view of our current knowledge should not obscure the fact that much is known about the operation of these devices.

Air pollution control theory, as envisioned in this book, encompasses the theory of each separate control device, but it is more comprehensive. The various control devices have much in common as to their net effect on the airstream. That is, their performance can be described in terms of a collection efficiency, and each device induces a certain pressure drop in the fluid stream. Therefore, the effect of a control device can be studied without knowing which control device is being used. Several control devices can be used in combination in a system, along with other elements. The interaction of various control devices, of the same type or of different types, with each other and with the other elements in the system can be studied in detail. Studies of this type represent a large portion of air pollution control theory. Studies of the entire system or comparisons of various different systems as to performance and economics fall under the scope of our theory. Finally, any systematic or theoretical means of selecting particular types of pollution control devices is a part of a comprehensive air pollution control theory. This book treats the theory of operation, design, and economic selection of the more common devices and also delves into the broader aspects of air pollution control theory as outlined in this section.

1-3 NATURE OF AIR POLLUTANTS

Polluted air was defined in Sec. 1-1 as air plus one or more constituents not normally present in air or present in greater than normal concentrations. The polluting constituents, or pollutants, can be in the form of solids, liquids, gases, or vapors. Solid and liquid pollutants are referred to as particulates, and these pollutant particles appear quite distinct from the gaseous phase when viewed under the microscope. Gaseous pollutants, on the other hand, exist as individual molecules diffused throughout the air; gaseous pollutant molecules cannot be distinguished from air molecules under the microscope.

The following classifications of polluted air masses are in common use:

Dust A mixture in air of irregular-shaped mineral particles in the size range from 1 to 200 μm formed by crushing, chipping, grinding, or like operations or by natural disintegration of rock and soil. Lint and particles of organic matter are also classified as dust.

Smoke A mixture in air of very fine particles formed by combustion or other chemical processes in the size range from 0.01 to 1 μm. The particles may be irregular in shape if formed of solid material, or they may be spherical if formed by condensation.

Mist A mixture in air of liquid droplets in the size range from 5 to 100 μm in diameter.

Fog   A mist in which the liquid is water.

Fume   A mixture in air of small particles in the size range from 0.1 to 1 μm. The particles are formed in industrial processes of various kinds. A fume is similar to a smoke, except that the particles can have a much wider range of composition.

Sulfurous smog   A mixture in air of sulfur dioxide and other sulfur compounds, along with particulates, including water droplets, and the reaction products which occur under these conditions.

Photochemical smog   A mixture in air of the products of very complex photochemical reactions involving various hydrocarbons and oxides of nitrogen in the presence of sunlight. Usually both particulates and gaseous pollutants are present, but photochemical smog can also exist with gaseous pollutants only.

Each of the preceding classifications of polluted air masses involves primarily particulate pollutants, except photochemical smog, which may have gaseous pollutants only.

When referring to the atmosphere in a particular locality, a large number of pollutants, both particulates and gases, will likely be present at a given time. In this case, the particular classification applied is governed by the condition which dominates at that time. Where no clear-cut distinction is possible, the classification used would be smog. In dealing with more limited air masses, such as exhaust gases flowing through a duct or stack or ventilation air in a workspace, the number of pollutants present is usually quite limited. The classification scheme then is quite useful. Another classification is for air polluted with one or a few gaseous contaminants; no specific term is used for this mixture.

An air mass containing gaseous pollutants can be described easily; an air mass containing particulates, however, is much more difficult to describe. The chemical composition of the pollutant must be known, of course. In addition, the size range of the particles is important, as is the relative number of particles of each size in the range. The total number of particles (or total mass) per unit volume of air must be known. The shape of the particles, or average shape and deviation from this shape, must be known. Also, the physical and chemical nature of the particle surface can be important.

Because of the tremendous number of particles present and the range of variation in size and shape, a series of weighted averages is used to characterize the particulate distribution. The weighting function used depends on the purpose for which the characterization is being made. This complex subject will be explored in greater detail in Chap. 4. For the present, it can be said that particulates in the size range of about 0.01 to 100 μm diameter are of significance to air pollution control theory. Particulates that are smaller than 0.01 μm are not easily distinguishable from large molecules and can be treated as though they were gaseous pollutants. Particulates that are larger than 100 μm settle out quickly and can be removed easily before discharge into the atmosphere. Moreover, if discharged into the atmosphere, these particulates settle out quickly, causing problems, perhaps, but not contributing to any general air pollution condition.

Despite the extensive amount of detailed information required to adequately characterize a particulate distribution, the most valuable parameter and the one usually reported is the particulate loading C_{mv} of the air, which is defined as the total mass of suspended particulate material per unit volume of mixture. Typical units for particulate loading are micrograms per cubic meter ($\mu g/m^3$) for atmospheric air and grams per cubic meter (g/m^3) for a polluted airstream. The strict International System of Units (see Sec. 1-5) are kilograms per cubic meter (kg/m^3). English units include grains per cubic foot (g/ft^3), pound mass per cubic foot (lb_m/ft^3), and ounces per cubic foot (oz/ft^3). Ideally atmospheric air should have a particulate loading in the range of 50 to 75 $\mu g/m^3$, and conditions are extremely hazardous with loadings about 1,500 $\mu g/m^3$. Typical dust loadings for polluted airstreams are in the range of 1 to 20 g/m^3.

The amount of gaseous pollutants contained in an air mass is given by the concentration. Concentration may be defined in a number of ways, and these are treated in considerable detail in Chap. 4. For the present, we shall define concentration in parts per million (C_{ppm}) as the ratio of the partial volume of pollutant gas to the partial volume of the air. The proper maximum allowable concentrations of various gases in air is subject to debate, and allowable limits have been promulgated for the protection of people living and working in various environments. These limits have been modified over the years, usually downward, to reflect increased knowledge of the long-range toxic effects of such substances. For workers in industrial environments, standards have been published by the American Conference of Governmental Industrial Hygienists [1], and their recommendations are given in Table 1-1 for a selected list of pollutants. These data can be used in the design of industrial ventilation systems.

Table 1-1 THRESHOLD-LIMIT VALUES FOR TOXIC POLLUTANTS

Substance	C_{ppm}	C_{mv}, mg/m^3
Acetaldehyde	100	180
Acetic acid	10	25
Acetic anhydride	5	20
Acetone	1000	2400
Acetonitrile	40	70
Acetylene tetrabromide	1	14
Acrolein	0.1	0.25
Acrylontrile	20	45
Allyl alcohol	2	5
Allyl chloride	1	3
Ammonia	25	18
n-Amyl acetate	100	525
sec-Amyl acetate	125	650
Aniline	5	19

(Continued)

Table 1-1 **THRESHOLD LIMIT VALUES FOR TOXIC POLLUTANTS** (*Continued*)

Substance	C_{ppm}	C_{mv}, mg/m^3
Arsine	0.05	0.2
Benzene	10	30
Benzyl chloride	1	5
Boron tribromide	1	10
Boron trifluoride	1	3
Bromine	0.1	0.7
Bromine pentafluoride	0.1	0.7
Bromoform	0.5	5
1,3-Butadiene	1000	2200
Butane	600	1450
2-Butanone	200	590
n-Butyl acetate	150	710
Butyl alcohol	100	300
Butyl mercaptan	0.5	1.5
Carbon disulfide	20	60
Carbon monoxide	50	55
Carbon tetrachloride	10	65
Chlorine	1	3
Chlorine trifluoride	0.1	0.4
Chlorobenzene	75	350
Chlorobromomethane	200	1050
Chloroform	25	120
Chloroprene	25	90
Copper fume		0.2
Cresol	5	22
Crotonaldehyde	2	6
Cyanogen	10	
Cyclohexane	300	1050
Cyclohexanol	50	200
Cyclohexanone	50	200
Cyclohexene	300	1015
Cyclopentadiene	75	200
Diacetone alcohol	50	240
Diazomethane	0.2	0.4
Diborane	0.1	0.1
1,2-Dibromoethane	20	145
Dibutyl phosphate	1	5
Dichlorodifluoromethane	1000	4950
1,1-Dichloroethane	200	320
1,2-Dichloroethane	50	200
1,2-Dichloroethylene	200	790
Dichloroethyl ether	5	30
Dichloromonofluoromethane	1000	4200
Dichlorotetrafluoroethane	1000	7000
Diethylamine	25	75
Difluorodibromomethane	100	860
Diglycidyl ether	0.5	2.8
Dimethylamine	10	18
Dioxane	50	
Diphenyl	0.2	1

(*Continued*)

Table 1-1   THRESHOLD LIMIT VALUES FOR TOXIC POLLUTANTS (*Continued*)

Substance	C_{ppm}	C_{mv}, mg/m^3
Ethyl acetate	400	1400
Ethyl alcohol	1000	1900
Ethylamine	10	18
Ethyl benzene	100	435
Ethyl bromide	200	890
Ethyl butyl ketone	50	230
Ethyl chloride	1000	2600
Ethylenediamine	10	25
Ethylene oxide	50	90
Ethyl ether	400	1200
Ethyl formate	100	300
Ethyl mercaptan	0.5	1
Fluorine	1	2
Fluorotrichloromethane	1000	5600
Formaldehyde	2	3
Formic acid	5	9
n-Heptane	400	1600
n-Hexane	100	360
2-Hexanone	100	410
Hydrogen bromide	3	10
Hydrogen chloride	5	7
Hydrogen cyanide	10	11
Hydrogen fluoride	3	2
Hydrogen sulfide	10	15
Iodine	0.1	1
Iron oxide fume		5
Isoamyl acetate	100	525
Isoamyl alcohol	100	360
Isobutyl acetate	150	700
Isobutyl alcohol	50	150
Isopropyl acetate	250	950
Isopropyl alcohol	400	980
Isopropylamine	5	12
Isopropylether	250	1050
Lead fumes and dusts		0.15
Methyl acetate	200	610
Methyl acrylonitrile	1	3
Methyl alcohol	200	260
Methylamine	10	12
Methyl bromide	15	60
Methyl chloride	100	210
Methyl chloroform	350	1900
Methyl formate	100	250
Methyl mercaptan	0.5	1
Methylene chloride	100	360
Naphthalene	10	50
Nitric oxide	25	30
Nitrobenzene	1	5
Nitroethane	100	310
Nitromethane	100	250
Nitrotoluene	5	30

(*Continued*)

Table 1-1 **THRESHOLD LIMIT VALUES FOR TOXIC POLLUTANTS** (*Continued*)

Substances	C_{ppm}	C_{mv}, mg/m^3
Octane	300	1450
Ozone	0.1	0.2
Pentane	600	1800
2-Pentanone	200	700
Perchloroethylene	100	670
Phenol	5	19
Phenylphosphine	0.05	0.25
Phosgene	0.05	0.2
Phosphorus trichloride	0.5	3
n-Propyl acetate	200	840
Propyl alcohol	200	500
n-Propyl nitrate	25	110
Propylene dichloride	75	350
Propylene imine	2	5
Propylene oxide	100	240
Pyridine	5	15
Selenium hexafluoride	0.05	0.4
Styrene	100	420
Sulfur dioxide	5	13
Sulfur hexafluoride	1000	6000
Sulfuric acid		1
Sulfur monochloride	1	6
1,1,2,2-Tetrachloroethane	5	35
Toluene	100	375
Trichloroethylene	100	535
1,2,3-Trichloropropane	50	300
Triethylamine	25	100
Trifluoromonobromomethane	1000	6100
Trimethyl benzene	25	120
Turpentine	100	560
Vinyl acetate	10	30
Vinyl bromide	250	1100
Vinyl toluene	100	480
Xylene	100	435
Xylidine	5	25
Zinc chloride fume		1
Zinc oxide fume		5

SOURCE: Excerpted from "TLVs, Threshold Limit Values for Chemical Substances and Physical Agents in the Workroom Environment with Intended Changes for 1974," American Conference of Governmental Industrial Hygienists, P.O. Box 1937, Cincinnati, Ohio 45201, 1974. Used by permission. These excerpts should not be taken as definitive; for definitive values, for examples of the use of these values, and for guidance in the interpretation of the values reference should be made to the preceding publication and to "Documentation of the Threshold Limit Values for Substances in Workroom Air," American Conference of Governmental Industrial Hygienists, Cincinnati, Ohio, 1974.

For atmospheric environments, definitive standards have not been published yet. However, in most cases the maximum limits of pollutant concentration in atmospheric air will certainly be considerably less than the values listed in Table 1-1. There are several reasons for this: First, the allowable limits for industrial workers are based primarily on the long-term toxic effects of different concentrations and are the highest concentrations for which no detrimental effects are believed to occur. For the population as a whole, however, the nuisance value of pollutants is as significant in setting the limits as is the health value, at least if the nuisance value results in a lower limiting value. Second, members of the general population are exposed to pollutants for a much longer period of time than industrial workers in a normal work week. Third, some members of the general population are in a much worse state of health or are more delicate as are children, than are most industrial workers, and for them exposure to very high concentrations of certain pollutants could be a direct cause of death or serious illness. Fourth, different substances present in a polluted atmosphere react with each other in the presence of sunlight to form highly complex compounds, many of which may have harmful toxic effects that the original pollutants might not have. These reactions can take place with very low concentrations, and so for this reason also the allowable concentrations in the atmosphere will probably be quite low.

1-4   AIR POLLUTION MEASUREMENT

One of the most crucial foundations of air pollution control theory as a discipline is an ability to define and measure pollution. Air pollution was defined in Sec. 1-1, and that definition is adequate here. We must still be able to identify which pollutants are present in a given air mass. We must also be able to determine how much of each pollutant is present; that is, we must be able to measure the concentration of a gaseous pollutant or the particulate loading of a particulate pollutant. Moreover, in the case of a particulate, we must also be able to determine many other properties of the distribution, such as size distribution and surface area. Measurements of this nature serve a variety of purposes.

Measurements of the atmosphere may be made to determine the effect of air pollution on an entire community or region or, more specifically, the effect of a single pollution source on its immediate environment. For example, measurements in a discharge stack or in the ductwork leading to the stack will determine how much of the pollutant or of some property of the pollutant is being discharged into the atmosphere from that source. These measurements will also determine the necessary characteristics of a pollution control device to be installed at that point. Measurements taken before and after installation of a pollution control device will determine its efficiency. Measurements of pollutant concentration in a workspace will determine its effect on the health of workers. In this section, we shall discuss some general aspects of air pollution measurement, but we shall not deal to any great extent with specific devices or the details of their design and operation.

The particular quantities which need to be measured will depend on the specific application being considered. In general, the quantities of an airstream which occasionally may need to be measured include the following: air-flow rate; air pressure, temperature, velocity, and humidity; identity and concentration of the various gaseous pollutants; identity and particulate loading of solid or liquid pollutants; size distribution of the particles; specific surface of the particles; and light absorption and scattering characteristics of the distribution.

The process of measurement of the characteristics of an airstream involves the following six steps:

1 Collection of a sample of the air pollutant mixture
2 Treatment of the sample
3 Separation of the pollutant (or pollutants) from the air in the sample
4 Measurement of the desired properties of the pollutants
5 Adoption of a mechanism for controlling and measuring the flow rate of the sample
6 Disposal of the sample gas after measurement is complete

In a particular case, not all the preceding steps are necessary or separately identifiable. In some cases, the measurement is made directly on the original gas stream, so that no sampling procedure is necessary. Let us examine each of these steps in some detail.

Although some measurements can be made directly on the airstream, in most cases it is necessary to draw off a representative sample of the original stream for analysis. The size of the sample to be drawn off can range from essentially zero to a considerable fraction of the original stream. A pitot tube for velocity measurement is an example of a case where essentially zero amount of gas needs to be sampled.

Several different types of sampling probes are shown in Fig. 1-1. Figure 1-1*a* shows an impact probe directed upstream. Figure 1-1*b* shows a static probe located at the duct wall and directed so that the axis of the nozzle is normal to the main flow direction. Figure 1-1*c* shows a side-port probe where the nozzle is in the side of the probe, with its axis normal to the main flow direction. Figure 1-1*d* and *e* show front- and rear-port probes, respectively, where the nozzle axes are parallel to the main flow direction. The end-port probe of Fig. 1-1*f* is similar to the static probe except that the probe extends into the duct. The nozzles shown in Fig. 1-1 are blunt orifices; other types of nozzles are often used in sampling probes. All the probes shown in Fig. 1-1 can be used for either continuous or intermittent sampling.

The treatment of the sample prior to separation of the pollutants from it usually involves measures for preserving the sample until separation can be effected. For example, it is often necessary to maintain the sample temperature above a certain value to avoid condensation. In many cases, the separation

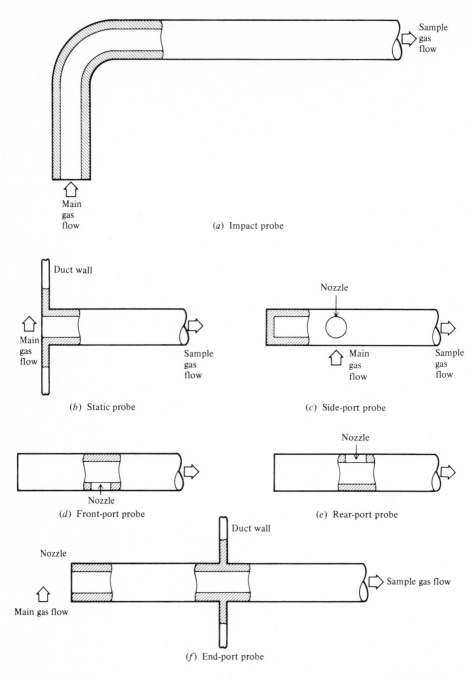

(a) Impact probe

(b) Static probe

(c) Side-port probe

(d) Front-port probe

(e) Rear-port probe

(f) End-port probe

FIGURE 1-1
Sampling probes.

mechanism is located inside the duct with the main gas stream so that no protection is necessary. Where this is not possible, the sample can be protected by insulating the tubing connecting the sampling probe or by electrically heating the tubing. In some cases, the sample needs to be heated or cooled after collection, which can be accomplished by suitable heat exchangers.

Separation of pollutants from the air in the sample is generally accomplished by devices which operate on the same principles as large-scale pollution control devices. For particulates, the favored devices are inertial impactors, cyclones, electrostatic precipitators, and filters. For gaseous pollutants, the devices include absorbers, adsorbers, condensers, and freeze traps. An Orsat analyzer is one device for separation and measurement of gases, gas concentration being measured; it operates on the principle of absorption of various gaseous components in sequence, with more sophisticated versions incorporating adsorption and combustion processes as well.

Measurement of the desired properties of the pollutants is often by far the most difficult part of the entire measuring process. Sometimes, however, this step can be rather simple, depending on what properties are to be measured. For gaseous pollutants, where only the properties of identity and concentration are required to be measured, gas-analysis equipment is employed. The Orsat analyzer, mentioned previously, is one gas-analysis device used primarily to analyze the gaseous products of combustion; it determines both the identity and the fraction by volume of the various gaseous constituents of the mixture. For adsorbing equipment and freeze traps, the weight fraction of the various constituents is measured by weighing the device before and after the sample is taken. All such equipment has the drawback that if unexpected constituents are present in the mixture, they will probably introduce errors into the various other readings.

Turning now to particulates, the simplest property that can be measured is the total weight, which is readily found by weighing the separating device before and after the test and taking the difference in weights. When this weight difference is divided by the total volume of the sample, the particulate loading C_{mv} is found. If a weight distribution by particle size is desired, the collection device can be arranged in stages, each stage collecting most of the particles in a given size range. When the different stages are weighed, a size-weight distribution is obtained, the degree of refinement of this distribution being determined by the number of separate stages employed. Where the total number of particles is the desired property, the particles can be collected on filter paper and the number of particles per unit area counted under the microscope. A size-number distribution can be obtained by counting under the microscope the number of particles per unit area in a given size range. A photomicrograph can be studied with the aid of templates to carry out this process more easily. Electron-microscope scanning techniques, using a computer, are also available for studies of this sort. One drawback of counting methods is that odd-shaped particles tend to align themselves on the filter in certain preferred directions, which can bias the statistical count of particle sizes.

Other techniques are available for particle measurement: Sedimentation and elutriation methods measure the falling velocity of the particles in air or other fluid. Measurements of porosity or capillarity of the collected particle mass give information as to the surface area of the particles. Additional methods use the change of value of certain properties, such as light transmittance or thermal conductivity, of a suspension of particles in a fluid.

A clear precaution for all measurements involving particulates, for all properties except total weight, is to avoid agglomeration of the particles during the separation or measuring processes. Such agglomeration will alter the size and number properties of the sample. This problem is especially acute with liquid particles.

An important step in the measuring process is to accurately control and measure the flow rate or total quantity of the sample. Usually the flow rate is measured by means of an orifice or flow nozzle either before or after the pollutants are separated from the air. The total amount of air inducted for the sample is the important consideration in most cases, and so either the flow rate must be held constant during the process or else integrating flow meters must be incorporated into the sampling device.

It is also very important to accurately control the flow rate at which the sample is being drawn into the sampling nozzle. This procedure is quite independent from measuring the flow rate and involves a subject known as isokinetic sampling, or isokinesis. Figure 1-2 shows three samples involving different flow rates. In Fig. 1-2a the flow rate is the same as the isokinetic rate; in Fig. 1-2b the flow rate is less than the isokinetic rate; and in Fig. 1-2c the flow rate is greater than the isokinetic rate. When the flow rate is the same as the isokinetic rate, the streamline pattern is smooth, with minimal disturbance by the presence of the probe. When the flow rate is less than the isokinetic rate, the streamlines tend to diverge outward near the edge of the probe, causing large particles to diverge into the probe and small particles to follow the streamlines; the result is a sample that contains too many large particles. When the flow rate is greater than the isokinetic rate, the streamlines diverge inward, large particles are flung outward, and the sample contains too few large particles. Isokinetic sampling is vital for large particles; in the normal range of particle densities, isokinetic sampling is unimportant for particles less than 5 μm in diameter. Obviously, it is of no consequence for gaseous pollutants.

In most sampling devices, the sample is drawn in by means of a small motor-driven blower or pump. To obtain isokinetic sampling, the flow rate can be controlled either by varying the speed of the blower or by operating a valve on the sampling line. Figure 1-3 shows a special type of sampling nozzle designed to facilitate isokinetic sampling. With proper design of the nozzle, isokinetic flow will be closely approximated when the pressure inside the tube equals the pressure inside the duct; these pressures are measured by the static tubes a and b, respectively. When these pressures are equal, as indicated by a manometer placed between them, isokinetic sampling results. Actually, equality of static

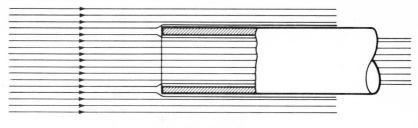

(*a*) Isokinetic sampling

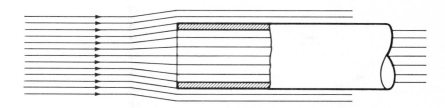

(*b*) Flow rate less than isokinetic rate

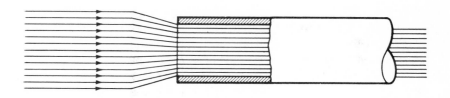

(*c*) Flow rate greater than isokinetic rate

FIGURE 1-2
Isokinetic sampling.

pressures at *a* and *b* only approximates the isokinetic condition. Because the velocity in the tube should equal that outside the tube for true isokinetic sampling, the static pressures would be exactly equal if friction could be neglected. With the same velocity, the pressure drop per unit length is greater inside the tube than inside the duct. There will also be a frictional loss due to the developing flow in the entrance of the probe tube. Because of these different rates of frictional loss, the locations of the pressure holes inside and outside the tube must be carefully adjusted if exact pressure balance is to indicate isokinesis.

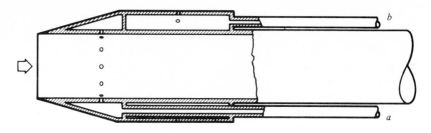

FIGURE 1-3
Nozzle for isokinetic sampling.

Since the location of the pressure holes is fixed for a given nozzle, we must conclude that the pressures at *a* and *b* are only approximately in balance for isokinetic sampling, except for one duct velocity. The approximation is usually adequate, and, besides, the nozzle can be calibrated so that the pressure differential between *a* and *b* is known for a given flow rate at the isokinetic condition. A more thorough treatment of gas sampling is given in Ref. [2].

Disposal of the sample gas after sampling is complete is the last step in the measurement of the characteristics of an airstream. At first glance, this step would not seem to be a problem, and in most cases it is not. However, if the sample gas contains poisonous constituents or radioactive elements, then the residual gas must be disposed of carefully. This is true even if the poisonous constituents are removed in the analysis since abortive samples must be allowed for. The most obvious way to dispose of residual gas is to vent it back into the main stream from which the sample was originally drawn. In many cases, the gas may be discharged directly into the atmosphere, either in the space or vented to outside. In some cases, the gas must be collected in special waste containers.

1-5 UNITS

The units used in this text are those incorporated in the International System of Units (SI). There are several reasons for choosing this particular set of units. First, there is no choice of units which offers a clear-cut advantage to an American author. Most current textbooks of engineering written in the United States, except electrical engineering works, use one of the English systems of units. However, it appears likely that the English systems are on the way out. Second, this book has special problems with respect to choice of units that most books do not have. The primary purpose of this book is to incorporate material from a variety of different engineering disciplines in one volume. These various disciplines use many different unit systems, although very few use the SI system. Most disciplines use either the English systems or the cgs metric system; many other metric unit systems are also widely used.

Thus, for the sake of consistency, the author has decided to adopt the SI system. We shall discuss this system in detail here and give an extensive list of conversion factors between various unit systems and the SI system. It is strongly recommended that students read this text in the SI units and that they solve the problems in these units. Where problem data are given in other units, students should first convert these to SI units and then solve the problem. Should the answer be required in other units, the answer in SI units can be converted back to the required units.

The SI system of units consists of a small number of basic units, a set of prefixes to extend the range of numbers that can be handled conveniently with the units used, and a set of derived units. All the derived units can be expressed in terms of the basic units, though use of the symbol for the derived unit saves time and space. The prefixes may be attached to either the basic units or to the derived units. The International System of Units is discussed in further detail in Ref. [3].

The basic SI units which are of concern to us are listed in Table 1-2. Each time the quantities shown in Table 1-2 appear, they should be expressed in the basic units or with a prefix attached. The standard prefixes are listed in Table 1-3. The symbol for the prefix is attached to the symbol for the unit with no space between them. Only one prefix should be used for a given unit. Thus, millimeter (mm) and micrometer (μm) are acceptable, but millimicrometer (mμm) is not advised. In this case, the proper usage is nanometer (nm), which is 10^{-9} meter (m). A value such as 10^{-4} m, which could be expressed as decimillimeter (dmm) were it not for the single-prefix rule, must be expressed as 0.1 mm or 100 μm. The unit micron (μ), which is equivalent to micrometer, will be used on occasion in expressing data but is not a consistent SI unit and should be replaced by micrometer (μm) in calculation.

A list of derived units is given in Table 1-4; this list is not exhaustive but should contain most of the derived units used in air pollution work. It will be observed that not all the quantities have special names for the appropriate unit. For instance, thermal conductivity is listed as $W/m \cdot K$ or $J/s \cdot m \cdot K$ and is equivalent to $kg \cdot m/s^3 \cdot K$. Note that a space is left between different units unless these are separated by a slash or other symbol. In certain cases, the same symbol is used for both a unit and a prefix, and it is very important not to

Table 1-2 BASIC SI UNITS

Dimension	Unit	Symbol
Length	meter	m
Mass	kilogram	kg
Time	second	s
Electric current	ampere	A
Thermodynamic temperature	degree Kelvin	K

Table 1-3 STANDARD PREFIXES

Factor	Symbol	Prefix
10^{-18}	a	atto
10^{-15}	f	femto
10^{-12}	p	pico
10^{-9}	n	nano
10^{-6}	μ	micro
10^{-3}	m	milli
10^{-2}	c	centi
10^{-1}	d	deci
10^{1}	da	deca or deka
10^{2}	h	hecto
10^{3}	k	kilo
10^{6}	M	mega
10^{9}	G	giga
10^{12}	T	tera

Table 1-4 DERIVED SI UNITS

Quantity	Name of unit	Unit symbol or abbreviation	In terms of basic units
Area	square meter		m^2
Volume	cubic meter		m^3
Frequency	hertz, cycle per second	Hz	s^{-1}
Density	kilogram per cubic meter		kg/m^3
Velocity	meter per second		m/s
Angular velocity	radian per second		rad/s
Acceleration	meter per second squared		m/s^2
Angular acceleration	radian per second squared		rad/s^2
Volumetric flow rate	cubic meter per second		m^3/s
Force	newton	N	$kg \cdot m/s^2$
Surface tension	newton per meter	N/m	kg/s^2
Pressure	newton per square meter	N/m^2	$kg/m \cdot s^2$
Dynamic viscosity	newton-second per square meter	$N \cdot s/m^2$	$kg/m \cdot s$
Kinematic viscosity	meter squared per second		m^2/s
Work, torque, energy	joule, newton-meter, watt-second	J, N $\cdot$ m, W $\cdot$ s	$kg \cdot m^2/s^2$
Power	watt, joule per second	W, J/s	$kg \cdot m^2/s^3$
Specific heat, gas constant	joule per kilogram degree	J/kg $\cdot$ K	$m^2/s^2 \cdot$ K
Enthalpy	joules per kilogram	J/kg	m^2/s^2
Entropy	joules per kilogram degree	J/kg $\cdot$ K	$m^2/s^2 \cdot$ K
Thermal conductivity	watt per meter degree	W/m $\cdot$ K	$kg \cdot m/s^3 \cdot$ K
Mass transfer coefficient	meter per second		m/s
Quantity of electricity	coulomb	C	$A \cdot s$
Electromotive force	volt	V	$kg \cdot m^2/A \cdot s^3$
Electric field strength	volt per meter	V/m	$kg \cdot m/A \cdot s^3$
Electric resistance	ohm	Ω	$kg \cdot m^2/A^2 \cdot s^2$
Electric conductivity	ampere per volt meter	A/V $\cdot$ m	$A^2 \cdot s^3/kg \cdot m^3$
Electric capacitance	farad	F	$A^2 \cdot s^4/kg \cdot m^2$

confuse the two. The chance of confusion can be reduced if different unit symbols are separated by a symbol, such as a dot or cross, for example, $N \cdot m$ or $N \times m$.

The use of a prefix with a derived unit which does not have a single symbol to represent it causes some difficulty. Thus, kinematic viscosity could be written as $m(m^2/s)$, but this is obviously awkward. A better choice would be to include the prefix with one term of the derived unit; thus, $1.0 \ m(m^2/s)$ is equal to $10 \ cm^2/s$, or $1.0 \ m^2/ks$. The use of prefixes in this way often leads to errors, and so before substitution into an equation for evaluation, it is usually best to reduce each quantity to the standard unit times a power of 10; for example, $1.0 \ m(m^2/s)$ is equal to $1.0 \times 10^{-3} \ m^2/s$. A further word of caution: When a derived unit contains a unit with a prefix that is squared, the prefixed unit is squared, not the unit itself; thus, cm^3/s means $(cm)^3/s$, not $c(m^3)/s$.

Frequently it will be necessary to convert a quantity from one unit to another. Table 1-5 is a table of conversion factors, provided for convenience in performing this operation. This table is arranged in the form of a set of identities involving various units. Any term in an equation can be multiplied or divided by unity without changing its value; and if the units are written down, particular units can be canceled to leave the desired set of units. To illustrate, suppose that the dynamic viscosity of a substance is given as $2.0 \times 10^{-5} \ lb_m/ft \cdot s$ and the conversion identity is

$$1 = 1.49 \ \frac{kg/m \cdot s}{lb_m/ft \cdot s}$$

Then the viscosity in SI units is

$$\mu = 2.0 \times 10^{-5} \ lb_m/ft \cdot s \times 1.49 \ \frac{kg/m \cdot s}{lb_m/ft \cdot s}$$

$$= 2.98 \times 10^{-5} \ kg/m \cdot s$$

Table 1-5 contains a number of conversion identities, each of which may be regarded as an expression of unity. These identities are arranged in the table first by categories and then within each category by quantity. The first category lists various units for the basic SI quantities, that is, those quantities for which the basic SI units are defined. No attempt is made to list the conversion identities for all possible choices of units. It may be necessary to use two or more conversion identities simultaneously to effect a given transformation, these identities being listed in either the same or different categories. One category of identities not listed in Table 1-5 is that relating quantities with different prefixes. For example, to convert from millimeters to meters, we may use the identity

$$1 = 10^{-3} \ m/mm$$

Similar relations hold for all units and prefixes.

Table 1-5 CONVERSION FACTORS TO SI UNITS

(a) Conversion identities involving units of quantities basic to the SI system

Length
$$0.01 \frac{m}{cm} \qquad 10^{-6} \frac{m}{\mu} \qquad 10^{-10} \frac{m}{\mathring{A}}$$
$$0.3048 \frac{m}{ft} \qquad 0.0254 \frac{m}{in} \qquad 0.9144 \frac{m}{yd}$$
$$1609.3 \frac{m}{mi}$$

Mass
$$10^{-3} \frac{kg}{g} \qquad 0.4536 \frac{kg}{lb_m} \qquad 6.48 \times 10^{-5} \frac{kg}{gr}$$
$$0.02835 \frac{kg}{oz\ (avdp)} \qquad 907.2 \frac{kg}{ton} \qquad 14.59 \frac{kg}{slug}$$

Time
$$60 \frac{s}{min} \qquad 3600 \frac{s}{h} \qquad 86,400 \frac{s}{day}$$
$$3.156 \times 10^7 \frac{s}{year}$$

Temperature
$$0.5555 \frac{K}{^\circ R} \qquad 0.5555 \frac{K}{^\circ F*} \qquad 1.0 \frac{K}{^\circ C*}$$

Electric current
$$10 \frac{A}{abampere} \qquad 3.3356 \times 10^{-10} \frac{A}{statampere}$$

(b) Geometric quantities

Area
$$10^{-4} \frac{m^2}{cm^2} \qquad 10^{-12} \frac{m^2}{\mu m^2} \qquad 0.0929 \frac{m^2}{ft^2}$$
$$6.452 \times 10^{-4} \frac{m^2}{in^2} \qquad 0.8361 \frac{m^2}{yd^2} \qquad 4047 \frac{m^2}{acre}$$
$$2.59 \times 10^6 \frac{m^2}{mi^2}$$

Volume
$$10^{-6} \frac{m^3}{cm^3} \qquad 10^{-3} \frac{m^3}{l} \qquad 10^{-18} \frac{m^3}{\mu m^3}$$
$$0.02832 \frac{m^3}{ft^3} \qquad 1.639 \times 10^{-5} \frac{m^3}{in^3} \qquad 3.785 \times 10^{-3} \frac{m^3}{gal\ (U.S.)}$$

Volumetric flow rate
$$10^{-6} \frac{m^3/s}{cm^3/s} \qquad 0.02832 \frac{m^3/s}{cfs} \qquad 1.639 \times 10^{-5} \frac{m^3/s}{in^3/s}$$
$$4.72 \times 10^{-4} \frac{m^3/s}{cfm} \qquad 7.87 \times 10^{-6} \frac{m^3/s}{cf\,h} \qquad 3.785 \times 10^{-3} \frac{m^3/s}{gal/s}$$
$$6.308 \times 10^{-5} \frac{m^3/s}{gpm} \qquad 1.051 \times 10^{-6} \frac{m^3/s}{gph}$$

Specific volume
$$10^{-3} \frac{m^3/kg}{cm^3/g} \qquad 10^{-15} \frac{m^3/kg}{\mu m^3/g} \qquad 0.0624 \frac{m^3/kg}{ft^3/lb_m}$$

* Temperature difference. (*Continued*)

Table 1-5 CONVERSION FACTORS TO SI UNITS (*Continued*)

Specific surface	$0.1 \dfrac{m^2/kg}{cm^2/g}$	$2.205 \times 10^{-12} \dfrac{m^2/kg}{\mu m^2/lb_m}$	$0.2048 \dfrac{m^2/kg}{ft^2/lb_m}$

(c) Kinematic quantities

Velocity	$0.01 \dfrac{m/s}{cm/s}$	$2.78 \times 10^{-4} \dfrac{m/s}{m/h}$	$0.278 \dfrac{m/s}{km/h}$
	$0.3048 \dfrac{m/s}{ft/s}$	$0.00508 \dfrac{m/s}{ft/min}$	$0.447 \dfrac{m/s}{mi/h}$
Acceleration	$0.01 \dfrac{m/s^2}{cm/s^2}$	$7.716 \times 10^{-8} \dfrac{m/s^2}{m/h^2}$	$0.3048 \dfrac{m/s^2}{ft/s^2}$
	$8.47 \times 10^{-5} \dfrac{m/s^2}{ft/min^2}$	$2.35 \times 10^{-8} \dfrac{m/s^2}{ft/h^2}$	
Momentum	$10^{-5} \dfrac{kg\,m/s}{g\,cm/s}$	$0.1383 \dfrac{kg\,m/s}{lb_m \cdot ft/s}$	$2.30 \times 10^{-3} \dfrac{kg\,m/s}{lb_m \cdot ft/min}$
Angular velocity	$0.01667 \dfrac{rad/s}{rad/min}$	$2.78 \times 10^{-4} \dfrac{rad/s}{rad/h}$	$0.1047 \dfrac{rad/s}{r/min}$
Angular acceleration	$2.78 \times 10^{-4} \dfrac{rad/s^2}{rad/min^2}$	$7.72 \times 10^{8} \dfrac{rad/s^2}{rad/h^2}$	$1.74 \times 10^{-3} \dfrac{rad/s^2}{r/min^2}$
Angular momentum	$10^{-7} \dfrac{kg \cdot m^2/s}{g \cdot cm^2/s}$	$0.04215 \dfrac{kg \cdot m^2/s}{lb_m \cdot ft^2/s}$	$7.02 \times 10^{-4} \dfrac{kg \cdot m^2/s}{lb_m \cdot ft^2/min}$
Area moment of inertia	$10^{-8} \dfrac{m^4}{cm^4}$	$4.16 \times 10^{-7} \dfrac{m^4}{in^4}$	$0.00863 \dfrac{m^4}{ft^4}$
Mass moment of inertia	$10^{-7} \dfrac{kg \cdot m^2}{g \cdot cm^2}$	$0.04214 \dfrac{kg \cdot m^2}{lb_m \cdot ft^2}$	$1.355 \dfrac{kg \cdot m^2}{lb_f \cdot ft\,s^2}$
	$2.93 \times 10^{-4} \dfrac{kg \cdot m^2}{lb_m\,in^2}$	$0.113 \dfrac{kg \cdot m^2}{lb_f\,in/s^2}$	
Momentum flow rate	$10^{-5} \dfrac{kg \cdot m/s^2}{g \cdot cm/s^2}$	$0.1383 \dfrac{kg \cdot m/s^2}{lb_m \cdot ft/s^2}$	$3.84 \times 10^{-5} \dfrac{kg \cdot m/s^2}{lb_m \cdot ft/min^2}$

(d) Quantities involving mass and mass flow

Mass flow rate	$10^{-3} \dfrac{kg/s}{g/s}$	$2.78 \times 10^{-4} \dfrac{kg/s}{kg/h}$	$0.4536 \dfrac{kg/s}{lb_m/s}$
	$0.00756 \dfrac{kg/s}{lb_m/min}$	$1.26 \times 10^{-4} \dfrac{kg/s}{lb_m/h}$	
Mass flux	$10 \dfrac{kg/m^2 \cdot s}{g/cm^2 \cdot s}$	$0.01667 \dfrac{kg/m^2 \cdot s}{g/m^2 \cdot min}$	$2.78 \times 10^{-7} \dfrac{kg/m^2 \cdot s}{g/m^2 \cdot h}$
	$4.883 \dfrac{kg/m^2\,s}{lb_m/ft^2 \cdot s}$	$0.0814 \dfrac{kg/m^2\,s}{lb_m/ft^2 \cdot min}$	$1.356 \times 10^{-3} \dfrac{kg/m^2 \cdot s}{lb_m/ft^2 \cdot h}$

(*Continued*)

Table 1-5 CONVERSION FACTORS TO SI UNITS (*Continued*)

Mass per unit area	$10 \dfrac{\text{kg/m}^2}{\text{g/cm}^2}$	$4.883 \dfrac{\text{kg/m}^2}{\text{lb}_m/\text{ft}^2}$	$703 \dfrac{\text{kg/m}^2}{\text{lb}_m/\text{in}^2}$
	$3.50 \times 10^{-4} \dfrac{\text{kg/m}^2}{\text{ton/mi}^2}$		

Density	$1000 \dfrac{\text{kg/m}^3}{\text{g/cm}^3}$	$16.02 \dfrac{\text{kg/m}^3}{\text{lb}_m/\text{ft}^3}$	$119.8 \dfrac{\text{kg/m}^3}{\text{lb}_m/\text{gal}}$
	$27{,}700 \dfrac{\text{kg/m}^3}{\text{lb}_m/\text{in}^3}$	$0.002289 \dfrac{\text{kg/m}^3}{\text{gr/ft}^3}$	

Mass release rate per unit volume	$1000 \dfrac{\text{kg/m}^3 \cdot \text{s}}{\text{g/cm}^3 \cdot \text{s}}$	$16.67 \dfrac{\text{kg/m}^3 \cdot \text{s}}{\text{g/cm}^3 \cdot \text{min}}$	$0.2778 \dfrac{\text{kg/m}^3 \cdot \text{s}}{\text{g/cm}^3 \cdot \text{h}}$
	$16.02 \dfrac{\text{kg/m}^3 \cdot \text{s}}{\text{lb}_m/\text{ft}^3 \cdot \text{s}}$	$0.267 \dfrac{\text{kg/m}^3 \cdot \text{s}}{\text{lb}_m/\text{ft}^3 \cdot \text{min}}$	$0.00445 \dfrac{\text{kg/m}^3 \cdot \text{s}}{\text{lb}_m/\text{ft}^3 \cdot \text{h}}$

Mass transfer coefficient†	$9.869 \times 10^{-5} \dfrac{\text{kg/N} \cdot \text{s}}{\text{g/cm}^2 \cdot \text{atm} \cdot \text{s}}$		$1.339 \times 10^{-8} \dfrac{\text{kg/N} \cdot \text{s}}{\text{lb}_m/\text{ft}^2 \cdot \text{atm} \cdot \text{h}}$

Specific volume	$10^{-3} \dfrac{\text{m}^3/\text{kg}}{\text{cm}^3/\text{g}}$	$0.06243 \dfrac{\text{m}^3/\text{kg}}{\text{ft}^3/\text{lb}_m}$	

Specific weight	$10 \dfrac{\text{N/m}^3}{\text{dyn/cm}^3}$	$157.1 \dfrac{\text{N/m}^3}{\text{lb}_f/\text{ft}^3}$	

(e) Dynamic quantities (quantities involving force)

Force	$10^{-5} \dfrac{\text{N}}{\text{dyn}}$	$1 \dfrac{\text{N}}{\text{kg} \cdot \text{m/s}^2}$	$9.8067 \dfrac{\text{N}}{\text{kg}_f}$
	$0.009807 \dfrac{\text{N}}{\text{g}_f}$	$0.1383 \dfrac{\text{N}}{\text{pdl}}$	$4.448 \dfrac{\text{N}}{\text{lb}_f}$
	$8896 \dfrac{\text{N}}{\text{ton}_f}$	$4448 \dfrac{\text{N}}{\text{kip}}$	

Surface tension	$10^{-3} \dfrac{\text{N/m}}{\text{dyn/cm}}$	$14.6 \dfrac{\text{N/m}}{\text{lb}_f/\text{ft}}$	$175 \dfrac{\text{N/m}}{\text{lb}_f/\text{in}}$

Pressure, stress	$0.1 \dfrac{\text{N/m}^2}{\text{dyn/cm}^2}$	$9.8067 \dfrac{\text{N/m}^2}{\text{kg}_f/\text{m}^2}$	$10^5 \dfrac{\text{N/m}^2}{\text{bar}}$
	$1.0133 \times 10^5 \dfrac{\text{N/m}^2}{\text{std atm}}$	$1.489 \dfrac{\text{N/m}^2}{\text{pdl/ft}^2}$	$47.88 \dfrac{\text{N/m}^2}{\text{lb}_f/\text{ft}^2}$
	$6894 \dfrac{\text{N/m}^2}{\text{lb}_f/\text{in}^2}$	$1.38 \times 10^7 \dfrac{\text{N/m}^2}{\text{ton}_f/\text{in}^2}$	$249.1 \dfrac{\text{N/m}^3}{\text{inH}_2\text{O}}$
	$2989 \dfrac{\text{N/m}^2}{\text{ftH}_2\text{O}}$	$133.3 \dfrac{\text{N/m}^2}{\text{mmHg}}$	$3386 \dfrac{\text{N/m}^2}{\text{inHg}}$

Body force	$10 \dfrac{\text{N/m}^3}{\text{dyn/cm}^3}$	$9.807 \times 10^6 \dfrac{\text{N/m}^3}{\text{kg}_f/\text{cm}^3}$	$157.1 \dfrac{\text{N/m}^3}{\text{lb}_f/\text{ft}^3}$
	$2.71 \times 10^5 \dfrac{\text{N/m}^3}{\text{lb}_f/\text{in}^3}$	$3.14 \times 10^5 \dfrac{\text{N/m}^3}{\text{ton}_f/\text{ft}^3}$	

† Based on pressure driving force.

(*Continued*)

Table 1-5 CONVERSION FACTORS TO SI UNITS (*Continued*)

Force per unit mass	$0.01 \dfrac{\text{N/kg}}{\text{dyn/g}}$	$9.807 \dfrac{\text{N/kg}}{\text{kg}_f/\text{kg}}$	$9.807 \dfrac{\text{N/kg}}{\text{lb}_f/\text{lb}_m}$
	$0.3049 \dfrac{\text{N/kg}}{\text{lb}_f/\text{slug}}$		

Torque	$10^{-7} \dfrac{\text{N} \cdot \text{m}}{\text{dyn} \cdot \text{cm}}$	$1.356 \dfrac{\text{N} \cdot \text{m}}{\text{lb}_f \cdot \text{ft}}$	$0.0421 \dfrac{\text{N} \cdot \text{m}}{\text{pdl} \cdot \text{ft}}$
	$2.989 \dfrac{\text{N} \cdot \text{m}}{\text{kg}_f \cdot \text{ft}}$		

Dynamic viscosity	$1 \dfrac{\text{kg/m} \cdot \text{s}}{\text{N} \cdot \text{s/m}^2}$	$1 \dfrac{\text{g/cm} \cdot \text{s}}{\text{P}}$	$0.1 \dfrac{\text{kg/m} \cdot \text{s}}{\text{P}}$
	$0.001 \dfrac{\text{kg/m} \cdot \text{s}}{\text{cP}}$	$2.78 \times 10^{-4} \dfrac{\text{kg/m} \cdot \text{s}}{\text{kg/m} \cdot \text{h}}$	$1.488 \dfrac{\text{kg/m} \cdot \text{s}}{\text{lb}_m/\text{ft} \cdot \text{s}}$
	$4.134 \times 10^{-4} \dfrac{\text{kg/m} \cdot \text{s}}{\text{lb}_m/\text{ft} \cdot \text{h}}$	$47.91 \dfrac{\text{kg/m} \cdot \text{s}}{\text{lb}_f \cdot \text{s/ft}^2}$	

Kinematic viscosity	$1 \dfrac{\text{cm}^2/\text{s}}{\text{St}}$	$10^{-4} \dfrac{\text{m}^2/\text{s}}{\text{St}}$	$2.778 \times 10^{-4} \dfrac{\text{m}^2/\text{s}}{\text{m}^2/\text{h}}$
	$0.0929 \dfrac{\text{m}^2/\text{s}}{\text{ft}^2/\text{s}}$	$2.581 \times 10^{-5} \dfrac{\text{m}^2/\text{s}}{\text{ft}^2/\text{h}}$	

Henry's law constant	$101{,}326 \dfrac{\text{N/m}^2}{\text{atm}}$	$133.3 \dfrac{\text{N/m}^2}{\text{mmHg}}$	$6893 \dfrac{\text{N/m}^2}{\text{lb}_f/\text{in}^2}$
	$47.89 \dfrac{\text{N/m}^2}{\text{lb}_f/\text{ft}^2}$		

Diffusion coefficient	$10^{-4} \dfrac{\text{m}^2/\text{s}}{\text{cm}^2/\text{s}}$	$2.78 \times 10^{-4} \dfrac{\text{m}^2/\text{s}}{\text{m}^2/\text{h}}$	$0.0929 \dfrac{\text{m}^2/\text{s}}{\text{ft}^2/\text{s}}$
	$2.58 \times 10^{-5} \dfrac{\text{m}^2/\text{s}}{\text{ft}^2/\text{h}}$		

(f) Quantities involving energy

Energy	$3.6 \times 10^6 \dfrac{\text{J}}{\text{kWh}}$	$4.187 \dfrac{\text{J}}{\text{cal}}$	$4187 \dfrac{\text{J}}{\text{kcal}}$
	$10^{-7} \dfrac{\text{J}}{\text{erg}}$	$1.356 \dfrac{\text{J}}{\text{ft} \cdot \text{lb}_f}$	$1055 \dfrac{\text{J}}{\text{Btu}}$
	$0.04214 \dfrac{\text{J}}{\text{ft} \cdot \text{pdl}}$	$2.685 \times 10^6 \dfrac{\text{J}}{\text{hp} \cdot \text{h}}$	$1.055 \times 10^8 \dfrac{\text{J}}{\text{therm}}$
	$0.1130 \dfrac{\text{J}}{\text{in} \cdot \text{lb}_f}$	$4.48 \times 10^4 \dfrac{\text{J}}{\text{hp} \cdot \text{min}}$	$745.8 \dfrac{\text{J}}{\text{hp} \cdot \text{s}}$

Power	$4.187 \dfrac{\text{W}}{\text{cal/s}}$	$4187 \dfrac{\text{W}}{\text{kcal/s}}$	$10^{-7} \dfrac{\text{W}}{\text{erg/s}}$
	$1.356 \dfrac{\text{W}}{\text{ft} \cdot \text{lb}_f/\text{s}}$	$0.293 \dfrac{\text{W}}{\text{Btu/h}}$	$0.04214 \dfrac{\text{W}}{\text{ft} \cdot \text{pdl/s}}$
	$1055 \dfrac{\text{W}}{\text{Btu/s}}$	$745.8 \dfrac{\text{W}}{\text{hp}}$	$0.1130 \dfrac{\text{W}}{\text{in} \cdot \text{lb}_f/\text{s}}$
	$3517 \dfrac{\text{W}}{\text{ton of refrigeration}}$	$17.6 \dfrac{\text{W}}{\text{Btu/min}}$	

(*Continued*)

Table 1-5 CONVERSION FACTORS TO SI UNITS (*Continued*)

Energy per unit length	$418.7 \dfrac{\text{J/m}}{\text{cal/cm}}$	$4.187 \times 10^5 \dfrac{\text{J/m}}{\text{kcal/cm}}$	$10^{-5} \dfrac{\text{J/m}}{\text{erg/cm}}$
	$4.449 \dfrac{\text{J/m}}{\text{ft} \cdot \text{lb}_f/\text{ft}}$	$3461 \dfrac{\text{J/m}}{\text{Btu/ft}}$	$8.81 \times 10^6 \dfrac{\text{J/m}}{\text{hp} \cdot \text{h/ft}}$
	$1.18 \times 10^7 \dfrac{\text{J/m}}{\text{kWh/ft}}$		
Energy per unit area	$41{,}868 \dfrac{\text{J/m}^2}{\text{cal/cm}^2}$	$4.187 \times 10^7 \dfrac{\text{J/m}^2}{\text{kcal/cm}^2}$	$0.001 \dfrac{\text{J/m}^2}{\text{erg/cm}^2}$
	$14.60 \dfrac{\text{J/m}^2}{\text{ft} \cdot \text{lb}_f/\text{ft}^2}$	$11{,}360 \dfrac{\text{J/m}^2}{\text{Btu/ft}^2}$	$2.89 \times 10^7 \dfrac{\text{J/m}^2}{\text{hp} \cdot \text{h/ft}^2}$
	$3.87 \times 10^7 \dfrac{\text{J/m}^2}{\text{kWh/ft}^2}$		
Energy per unit volume	$3.6 \times 10^6 \dfrac{\text{J/m}^3}{\text{kWh/m}^3}$	$4.187 \times 10^6 \dfrac{\text{J/m}^3}{\text{cal/cm}^3}$	$4.187 \times 10^9 \dfrac{\text{J/m}^3}{\text{kcal/cm}^3}$
	$0.1 \dfrac{\text{J/m}^3}{\text{erg/cm}^3}$	$47.9 \dfrac{\text{J/m}^3}{\text{ft} \cdot \text{lb}_f/\text{ft}^3}$	$3.73 \times 10^4 \dfrac{\text{J/m}^3}{\text{Btu/ft}^3}$
	$1.271 \times 10^8 \dfrac{\text{J/m}^3}{\text{kWh/ft}^3}$	$9.48 \times 10^7 \dfrac{\text{J/m}^3}{\text{hp} \cdot \text{h/ft}^3}$	
Power per unit length	$418.7 \dfrac{\text{W/m}}{\text{cal/s} \cdot \text{cm}}$	$4.187 \times 10^5 \dfrac{\text{W/m}}{\text{kcal/s} \cdot \text{cm}}$	$10^{-5} \dfrac{\text{W/m}}{\text{erg/s} \cdot \text{cm}}$
	$4.449 \dfrac{\text{W/m}}{\text{ft} \cdot \text{lb}_f/\text{s} \cdot \text{ft}}$	$3461 \dfrac{\text{W/m}}{\text{Btu/s} \cdot \text{ft}}$	$0.961 \dfrac{\text{W/m}}{\text{Btu/h} \cdot \text{ft}}$
	$2447 \dfrac{\text{W/m}}{\text{hp/ft}}$		
Energy flux (power per unit area)	$41{,}868 \dfrac{\text{W/m}^2}{\text{cal/s} \cdot \text{cm}^2}$	$4.187 \times 10^7 \dfrac{\text{W/m}^2}{\text{kcal/s} \cdot \text{cm}^2}$	$0.001 \dfrac{\text{W/m}^2}{\text{erg/s} \cdot \text{cm}^2}$
	$14.60 \dfrac{\text{W/m}^2}{\text{ft} \cdot \text{lb}_f/\text{sec ft}^2}$	$11{,}360 \dfrac{\text{W/m}^2}{\text{Btu/s} \cdot \text{ft}^2}$	$3.156 \dfrac{\text{W/m}^2}{\text{Btu/h} \cdot \text{ft}^2}$
	$8028 \dfrac{\text{W/m}^2}{\text{hp/ft}^2}$	$1.072 \times 10^4 \dfrac{\text{W/m}^2}{\text{kW/ft}^2}$	
Energy release rate (power per unit volume)	$4.187 \times 10^6 \dfrac{\text{W/m}^3}{\text{cal/s} \cdot \text{cm}^3}$	$4.187 \times 10^9 \dfrac{\text{W/m}^3}{\text{kcal/s} \cdot \text{cm}^3}$	$0.1 \dfrac{\text{W/m}^3}{\text{erg/s} \cdot \text{cm}^3}$
	$47.9 \dfrac{\text{W/m}^3}{\text{ft} \cdot \text{lb}_f/\text{s} \cdot \text{ft}^3}$	$3.73 \times 10^4 \dfrac{\text{W/m}^3}{\text{Btu/s} \cdot \text{ft}^3}$	$10.36 \dfrac{\text{W/m}^3}{\text{Btu/h} \cdot \text{ft}^3}$
	$3.53 \times 10^4 \dfrac{\text{W/m}^3}{\text{kW/ft}^3}$	$2.63 \times 10^4 \dfrac{\text{W/m}^3}{\text{hp/ft}^3}$	
Specific energy (energy per unit mass)	$1 \dfrac{\text{J/kg}}{\text{m}^2/\text{s}^2}$	$4187 \dfrac{\text{J/kg}}{\text{cal/g}}$	$4.187 \times 10^6 \dfrac{\text{J/kg}}{\text{kcal/g}}$
	$2.99 \dfrac{\text{J/kg}}{\text{ft} \cdot \text{lb}_f/\text{lb}_m}$	$2326 \dfrac{\text{J/kg}}{\text{Btu/lb}_m}$	$5.92 \times 10^6 \dfrac{\text{J/kg}}{\text{hp} \cdot \text{h/lb}_m}$
	$7.94 \times 10^6 \dfrac{\text{J/kg}}{\text{kWh/lb}_m}$		

(Continued)

Specific heat and gas constant	$1\dfrac{J/kg \cdot K}{m^2/s^2 \cdot K}$	$4187\dfrac{J/kg \cdot K}{cal/g \cdot °C}$	$10^{-4}\dfrac{J/kg \cdot K}{erg/g \cdot °C}$
	$4187\dfrac{J/kg \cdot K}{Btu/lb_m \cdot °F}$	$5.38\dfrac{J/kg \cdot K}{ft \cdot lb_f/lb_m \cdot °F}$	
Thermal conductivity	$418.7\dfrac{W/m \cdot K}{cal/s \cdot cm \cdot °C}$	$1.163\dfrac{W/m \cdot K}{kcal/h \cdot m \cdot °C}$	$10^{-5}\dfrac{W/m \cdot K}{erg/s \cdot cm \cdot °C}$
	$1.731\dfrac{W/m \cdot K}{Btu/h \cdot ft \cdot °F}$	$0.1442\dfrac{W/m \cdot K}{Btu \cdot in/h \cdot ft^2 \cdot °F}$	$0.00222\dfrac{W/m \cdot K}{ft \cdot lb_f/h \cdot ft \cdot °F}$
Heat-transfer coefficient	$41,868\dfrac{W/m^2 \cdot K}{cal/s \cdot cm^2 \cdot °C}$	$1.163\dfrac{W/m^2 \cdot K}{kcal/h \cdot m^2 \cdot °C}$	$0.001\dfrac{W/m^2 \cdot K}{erg/s \cdot cm^2 \cdot °C}$
	$5.679\dfrac{W/m^2 \cdot K}{Btu/h \cdot ft^2 \cdot °F}$	$12.52\dfrac{W/m^2 \cdot K}{kcal/h \cdot ft^2 \cdot °C}$	

(g) Electrical quantities

Quantity of electricity	$1\dfrac{C}{A \cdot s}$	$10\dfrac{C}{abcoulomb}$	$3.336 \times 10^{-10}\dfrac{C}{statcoulomb}$
Electromotive force (voltage)	$\dfrac{V}{kg \cdot m^2/A \cdot s^3}$ $1\dfrac{V}{W/A}$	$10^{-8}\dfrac{V}{abvolt}$	$299.8\dfrac{V}{statvolt}$
Electric resistance	$1\dfrac{\Omega}{kg \cdot m^2/A^2 \cdot s^3}$ $1\dfrac{\Omega}{V/A}$	$10^{-9}\dfrac{\Omega}{abohm}$	
	$8.988 \times 10^{11}\dfrac{\Omega}{statohm}$ $10^{-9}\dfrac{\Omega}{abohm}$		
Electrical resistivity	$1\dfrac{V \cdot m/A}{kg \cdot m^3/A^2 \cdot s^3}$ $1\dfrac{\Omega \cdot m}{kg \cdot m^5/A^2 \cdot s^3}$	$10^{-9}\dfrac{\Omega \cdot m}{abohm \cdot m}$	
	$8.988 \times 10^{11}\dfrac{\Omega \cdot m}{statohm \cdot m}$		
Electric capacitance	$1\dfrac{F}{A^2 \cdot s^4/kg \cdot m^2}$ $1\dfrac{F}{A \cdot s/V}$	$10^9\dfrac{F}{abfarad}$	
	$1.113 \times 10^{-12}\dfrac{F}{statfarad}$ $3.28\dfrac{V/m}{V/ft}$		
Electric field strength	$1\dfrac{V/m}{kg \cdot m/A\, s^3}$ $100\dfrac{V/m}{V/cm}$	$10^{-8}\dfrac{V/m}{abvolt/m}$	
	$299.8\dfrac{V/m}{statvolt/m}$ $39.4\dfrac{V/m}{V/in}$		
Magnetic flux	$1\dfrac{Wb}{kg \cdot m^2/A \cdot s^2}$ $1\dfrac{Wb}{V \cdot s}$		
Inductance	$1\dfrac{H}{kg \cdot m^2/A^2 \cdot s^2}$ $1\dfrac{H}{V \cdot s/A}$	$10^{-9}\dfrac{H}{abhenry}$	
	$8.988 \times 10^{11}\dfrac{H}{stathenry}$		

(h) Additional abbreviations
   gr, grain; P, poise; r, revolution; St, stoke

1-6 AIR POLLUTION CONTROL LITERATURE

There are no textbooks currently in print which attempt to cover the theoretical aspects of air pollution control, with particular reference to the operation of various control devices, in a manner similar to that attempted here. There are, however, a number of treatises available relating to various aspects of the subject. Probably the most comprehensive of these is the work by Stern [2, 4]. The "Air Pollution Engineering Manual" by Danielson [5] is also very comprehensive from the point of view of engineering practice. "Industrial Gas Cleaning" by Strauss [6] gives a good overall view of engineering practice in the area. A more recent work by Strauss [7], "Air Pollution Control," is a collection of several articles on specialized aspects of air pollution control. Other general works in the area include those of Nonhebel [8], Magill, Holden, and Ackley [9], Lund [10], the American Industrial Hygiene Association [11], the Research and Education Association [12], Hesketh [13], Williamson [14], Ross [15], Shaheen [16], Perkins [17], and Painter [18].

Several works are available which pertain to specialized types of air pollution control equipment or processes. Reference [1] and the books of Alden [19], Hemeon [20], and Baturin [21] cover the principles of industrial ventilation in detail. The works of White [22], Rose and Wood [23], and Oglesby and Nichols [24] are devoted to detailed analysis and discussion of engineering practice of the electrostatic precipitator. The cyclone collector is handled by Bradley [25], while filters are treated by Davies [26]. Some works relating to particle technology, which is closely related to air pollution control as well as other fields, include those of Orr [27], Fuchs [28], and Sanders [29]. The atmospheric dispersion of pollutants is handled by Meetham [30], Scorer [31], and Smith [32]. Other works are available and will be referred to later in this book.

Currently there are a number of periodicals devoted to air pollution, pollution in general, or various processes central to air pollution studies. The following are the principal periodicals in this category:

> *Atmospheric Environment*
> *Environmental Information Access*
> *Environmental Pollution*
> *Environmental Research*
> *Environmental Science and Technology*
> *Filtration and Separation*
> *International Journal of Air and Water Pollution*
> *Journal of Aerosol Science*
> *Journal of the Air Pollution Control Association*
> *Pollution Abstracts*
> *Staub, Reinhaltung der Luft*, English ed.

In addition, articles about air pollution control theory appear from time to time in a number of other engineering and scientific periodicals. Many articles in the field appear in the proceedings to various conferences, which is unfortunate in that such conference proceedings are not as available as periodicals. Even less widely available are the individual reports and bulletins published by various universities, laboratories, and other organizations, which were quite common up until a few years ago. The *Engineering Index*, the *Industrial Arts Index*, and the *Applied Mechanics Reviews* provide references to articles on air pollution. References [5], [8], and [9] also contain excellent bibliographies of journal articles and other publications.

1-7 DESIGN OF POLLUTION CONTROL EQUIPMENT

The purpose of this section is to discuss the basic philosophy and some general principles underlying the design of pollution control equipment. Much of what is said here pertains as well to the design of industrial equipment in general. In any design situation, the successful completion of a finished design must follow the steps indicated here:

1 List the functions which the completed control device must serve and the criteria it must fulfill.
2 Carefully delineate the constraints which the design must meet.
3 Explore the various modes of operation of the device, inventing new ones if necessary and possible.
4 Select a particular mode of operation for the device.
5 Marshall all the available theory, empirical results, and other information about devices which operate according to the mode selected. Develop additional theory and empirical results if necessary.
6 Work out a preliminary design of the control device.
7 Optimize the preliminary design so that it meets all constraints and will perform in the most economical manner. Determine what performance to expect if the device is operated at other than the design condition.
8 At this point it may or may not be appropriate to construct a model, or pilot plant, or even a full-scale model based on the preliminary design. If so, then this model should be tested to see if its performance meets the requirements in an economical manner.
9 Proceed now to the final design of the control device.
10 Verify that the final design meets the performance requirements, conforms to the stated constraints, and operates approximately at the optimum conditions.

As an example of this design process, let us apply the preceding steps to a particular problem. We shall not perform any calculations, but we shall outline what calculations need to be made as the design proceeds.

1 The pollution control device must remove 98 percent by mass of the particulates with a diameter exceeding 5 μm in a stream of 10,000 ft^3/min of air which also contains a corrosive gas. The density of the particles is given.

2   The device must fit within a space 10 ft wide by 15 ft long by 12 ft high. Its noise level must not exceed a certain decibel specification. It must last for at least 10 years under the corrosive action of the gases. It must connect into the inlet and outlet ducts at certain points, but let us assume that this requirement can be easily met by attaching additional ductwork. The disposal method must conform to that in use in the factory. (Note that these specifications are given in English units; we would convert these numbers to SI units for calculation.)

3   An exploration of the available modes of operation which reasonably might be successful reveals that we could use a filter, an electrostatic precipitator, a multiple cyclone, a straight-through cyclone, or a venturi scrubber. We conclude that one of these devices will surely work and that we do not have the resources on hand to attempt to invent new devices.

4   Upon examination of these various alternatives, we decide immediately that the electrostatic precipitator is too expensive in such small sizes, that the multiple cyclone has too high a pressure drop, that the straight-through cyclone only concentrates the dust stream, and that we have no provision for effecting a final separation. Elimination of these alternatives leaves for consideration a filter and a venturi scrubber. Upon further study, we decide that replacement of the filter element would be prohibitively expensive in the corrosive atmosphere present and that a filter would be too big to fit in the available space. This leaves the venturi scrubber; we believe it can be fitted into the available space, and the effluent in the form of a slurry poses no problems since our factory has other venturi scrubbers in operation and we can combine our effluent with the rest. (You must recognize that this reasoning illustrates the design procedure and may not be valid in your real-life situation.)

5   We study the theory of operation of the venturi scrubber, obtain performance data, some of it proprietary information perhaps, and see if we have enough information to proceed to the next step. If need be, we can develop more theory or obtain additional experimental or performance data.

6   Next we work out a preliminary design, which gives the size and proportions of the venturi scrubber and the collector which goes with it.

7   We optimize the preliminary design, so that it fits in the given space and so that the total overall cost, based on preliminary cost estimates, is minimized. Factors involved in estimating costs include the cost of material, the cost of labor in fabricating and installing the device, and the operating cost, which includes maintenance and power costs.

8   In this case, we decide to construct a plastic model and test it in a rig which we keep for such cases. We run the tests and find that the design-efficiency requirement is met with the model.

9   We now proceed to a final design of the device. First we establish

the primary dimensions of the various components. Then we apply stress analysis to determine the thickness of the various metal parts. We design the structural supports of the various components, piping and duct layouts, control systems, and any required power supplies, gages and indicators, and other necessary items. Working drawings are made.

10   Verification of the final design may be very difficult to achieve until the unit is actually constructed and tested in service. Prior to construction of the device, it may be advisable to examine the plans with an eye on the requirements and to do efficiency calculations and other measures to try to estimate the performance that the completed device will achieve. If care has been exercised in the previous steps of the design process, it is unlikely that such calculations will reveal anything new. They will merely constitute a final check before investing in the actual fabrication of the equipment. After the device is constructed, it may be wise to test it in simulated service at the manufacturing facility before shipping it to the site where it is to be installed. The final testing will be done after it is installed. After final adjustments and alterations are made so that the device meets specifications, it is unlikely that it will operate at the exact optimum point of minimum total cost. With reasonably good fortune, the increase in operating cost over the optimum value will not be great enough to displease the purchaser, if, indeed, the difference is detected at all.

The preceding example oversimplifies the design procedure since in reality hundreds or even thousands of engineering man-hours are devoted to the design of a pollution control device. The main idea here has been to give the reader a preliminary idea of what is involved in the design process.

The design criteria which must be met in a given situation usually relate to the required separation efficiency for the specific pollutants and their sizes (in the case of particulates). The flow rate, temperature, pressure, and humidity of the polluted air stream as well as the corrosive effect of the pollutants are all constraints which the collection device must meet satisfactorily. Additional constraints include limitations as to size and shape; limitations as to fluids, chemicals, and electric power which must be supplied to the device; limitations as to the allowable pressure drop across the device since otherwise additional pumps and blowers must be added to the system; and limitations as to the form in which the collected material can be discharged from the device. Other limitations concern the degree of control required; the extent to which operation should be automated or manually controlled; the time interval and conditions under which the device can be cleaned; whether removal of material should be continuous or intermittent; the time interval between shutdowns for maintenance; and synchronization of the operation of the pollution control device with the operation of other devices in the factory.

These different constraints have varying degrees of rigidity in a given application. Usually adjustments in the constraints can be made if no alternative can be found which will still satisfy the primary requirements. Thus, if more space is required for a given pollution control device than is available, more space can usually be found by knocking out a wall, locating the device somewhere else, or making other alternative arrangements. Similarly, if the only available device discharges the collected material in a form which is not workable with present disposal facilities, then additional disposal facilities can be added to the factory. Thus, the constraints are usually not completely rigid but can be adjusted if no solution is found which meets all the constraints specified initially.

The final criterion considered in the design of pollution control equipment is cost. Regardless of the truth of the accusation that if typical industrialists had their way, the only factor of any importance would be minimizing cost so as to maximize profit, social controls do dictate that other factors besides cost must be met. Once all such factors, as dictated by law, custom, union contract provisions, fear of legal action, customer's specifications, or simple humanitarian considerations, have been met, industrialists are then free to consider minimum cost as a criterion. This is reasonable because in design there is generally a wide latitude in choice of dimensions, flow rates, materials, and so forth, which meets the various constraints. So, within these constraints, the choice would be made which leads to the least total cost for the device. Let us now consider some of the factors which go into an analysis of cost.

The cost factors which must be considered in evaluating a design consist of the cost of materials in the manufacture and installation of the device; labor and indirect costs associated with the manufacture of the device; transportation costs; labor and indirect costs in installing the device; and operating costs, which include direct power costs, power costs due to increased friction losses in the flow induced by the device, labor costs for operation, labor and material costs for maintenance, and various indirect costs incurred by the presence of the device. The costs of manufacturing, transporting, and installing the device plus the expense necessary to put the device initially in good working order and certify its acceptance are usually regarded as capital outlay. Capital expenditures are generally recovered over a period of years, and they can be considered equivalent to a certain percentage of the original amount charged each year as an operating expense. This percentage rate depends on the current and future anticipated interest rate, life of the equipment, tax laws regarding write-off for depreciation, and any subsidies and other tax advantages available. In those cases where installation of the equipment is expected to lower manufacturing costs and thus increase profits, the corporate income tax becomes an important factor since the benefit to the company is reduced by the fraction of profits which must be paid as tax. Thus, the fraction of the original investment which must be charged each year as operating expenses is effectively increased.

In most installations of pollution control equipment, there is no question of profit; instead, both the annual capitalization and the actual operating expense are simply added in as part of the cost of doing business. Thus, comparison of two alternative pollution control devices, each of which fully meets the requirements for the job, is made by comparing the total annual costs, including both actual operating costs and capitalization charges. Thus, if device A has a large capitalization charge but small operating expense while device B has a much smaller initial capitalization charge but considerably greater actual operating expense, then the device having the smaller total expense would normally be chosen.

If tax matters are not considered, the annual capitalization rate is determined by the interest rate and the expected life of the equipment. This capitalization rate is equal to the amount of an annuity which the original investment will purchase over the life of the equipment at the current interest rate. These amounts can be found in annuity tables [33]. For example, if the interest rate is 6 percent and the life of the equipment is 20 years, the capitalization rate is found to be 8.72 percent/year. More elaborate formulas are available where special tax advantages must be considered or where maintenance or operating costs vary (usually upward) over the life of the equipment.

REFERENCES

1   American Conference of Governmental Industrial Hygienists, Committee on Industrial Ventilation: "Industrial Ventilation, a Manual of Recommended Practice," Lansing, Mich., 1974.
2   Stern, A. C., ed.: "Air Pollution," 2d ed., vol. II, "Analysis, Monitoring, and Surveying," Academic Press, Inc., New York, 1968.
3   The International System of Units, *Int. J. Heat Mass Transfer*, vol. 9, pp. 837–844, September 1966.
4   Stern, A. C., ed.: "Air Pollution," 2d ed., vol. I, "Air Pollution and Its Effects"; vol. II, "Analysis, Monitoring, and Surveying"; vol. III, "Sources of Air Pollution and Their Control," Academic Press, Inc., New York, 1968.
5   Danielson, J. A., ed.: "Air Pollution Engineering Manual," *Public Health Service Publ.* 999-AP-40, National Center for Air Pollution Control, Cincinnati, Ohio, 1967. (Available from the U.S. Government Printing Office, Washington, D.C.)
6   Strauss, W.: "Industrial Gas Cleaning," Pergamon Press, London, 1966.
7   Strauss, W., ed.: "Air Pollution Control," parts I and II, Interscience Publishers, New York, 1971, 1972.
8   Nonhebel, G., ed.: "Processes for Air Pollution Control," CRC Press, Cleveland, Ohio, 1972.
9   Magill, P. L., F. R. Holden, and C. Ackley, eds.: "Air Pollution Handbook," McGraw-Hill Book Company, New York, 1956.
10   Lund, H. F., ed.: "Industrial Pollution Control Handbook," McGraw-Hill Book Company, New York, 1971.
11   "Air Pollution Manual," part II, "Control Equipment," American Industrial Hygiene Association, Detroit, Mich., 1968.
12   "Pollution Control Technology," Research and Educational Association, New York, 1973.
13   Hesketh, H. E.: "Understanding and Controlling Air Pollution," Ann Arbor Science Publishers, Inc., Ann Arbor, Mich., 1972.
14   Williamson, S. J.: "Fundamentals of Air Pollution," Addison-Wesley Publishing Company, Inc., Reading, Mass., 1973.
15   Ross, R. D.: "Air Pollution and Industry," Van Nostrand Reinhold, New York, 1972.

16 Shaheen, E. I.: "Environmental Pollution: Awareness and Control," Engineering Technology Incorporated, Mahomet, Ill., 1974.

17 Perkins, H. C.: "Air Pollution," McGraw-Hill Book Company, New York, 1974.

18 Painter, D. E.: "Air Pollution Technology," Reston Publishing Company, Inc., Reston, Va., 1974.

19 Alden, J. L.: "Design of Industrial Exhaust Systems," 3d ed., Industrial Press, Inc., New York, 1959.

20 Hemeon, W. C. L.: "Plant and Process Ventilation," 2d ed., Industrial Press, Inc., New York, 1963.

21 Baturin, V. V.: "Fundamentals of Industrial Ventilation," 3d ed., transl. by O. M. Blunn, Pergamon Press, New York, 1972.

22 White, H. J.: "Industrial Electrostatic Precipitation," Addison-Wesley Publishing Company, Inc., Reading, Mass., 1963.

23 Rose, H. E., and A. J. Wood: "An Introduction to Electrostatic Precipitation in Theory and Practice," Constable & Co., Ltd., London, 1966.

24 Oglesby, S., and G. B. Nichols: "A Manual of Electrostatic Precipitator Technology," part I "Fundamentals"; part II, "Application Areas," Southern Research Institute, Birmingham, Ala., 1970.

25 Bradley, W. F.: "Hydrocyclone," Pergamon Press, New York, 1965.

26 Davies, C. N.: "Air Filtration," Academic Press, Inc., New York, 1973.

27 Orr, C., Jr.: "Particulate Technology," The Macmillian Company, New York, 1966.

28 Fuchs, N. A.: "The Mechanics of Aerosols," Pergamon Press, London, 1964.

29 Sanders, P. A.: "Principles of Aerosol Technology," Van Nostrand Reinhold, New York, 1970.

30 Meetham, A. R.: "Atmospheric Pollution, its Origins and Prevention," The Macmillan Company, New York, 1964.

31 Scorer, R. S.: "Air Pollution," Pergamon Press, London, 1968.

32 Smith, M., ed.: "Recommended Guide for the Prediction of the Dispersion of Airborne Effluents," American Society of Mechanical Engineers, New York, 1968.

33 Burington, R. S.: "Handbook of Mathematical Tables and Formulas," 4th ed., McGraw-Hill Book Company, New York, 1965.

PROBLEMS

1-1 Convert the following quantities to SI units:

(a) 661.3 ft; (b) 39.5 μm; (c) 30 yd; (d) 5.5 in; (e) 22 lb_m; (f) 5.25 gr/ft^3; (g) 65 oz/ft^2; (h) 3000 lb_m/h; (i) 14.7 lb_f/in^2; (j) 3.2 kip; (k) 36 ft^3; (l) 9000 ft^2/lb_m; (m) 60 mi/h; (n) 100 hp; (o) 10,000 Btu/h; (p) 0.1 cP; (q) 1.5×10^{-5} $lb_m/ft \cdot s$; (r) 3.5×10^{-4} ft^2/s; (s) 0.24 $Btu/lb_m \cdot °F$; (t) 1.5 $Btu/h \cdot ft^2 \cdot °F$; (u) 1.0 $Btu/lb_m \cdot h$; (v) 6.5 inH_2O; (w) 3000 statvolts; (x) 6000 statvolts/cm; (y) 10^6 abohms/mi; (z) 540 $lb_f \cdot ft \cdot s^2$.

1-2 Convert the following quantities given in SI units to the units specified:

(a) 3 mm to ft; (b) 1.0 GW to hp; (c) 10 am to Å; (d) 1.5 m^2/s to ft^2/h; (e) 50 N · cm to $lb_f \cdot ft$; (f) 5000 J to Btu; (g) 300 J/kg to Btu/lb_m; (h) 100°C to °R; (i) 90 kg/s to lb_m/s; (j) 5 m^3/s to ft^3/s.

1-3 Obtain the values of the following properties at the conditions indicated in SI units and in English units:

(a) Viscosity of water at 60°C

(b) Thermal conductivity of air at atmospheric pressure and 20°C

(c) Area moment of inertia about its centroidal axes of a square 2 m by 2 m
(d) Density of air at 0.101 MN/m^2 and 500 K
(e) Electrical resistivity of copper at 50°C
(f) Latent heat of vaporization of steam at 150°C
(g) Critical pressure and temperature of steam

1-4 A cyclone collector and an electrostatic precipitator are being considered for installation to cure a particular air pollution problem. The cyclone collector costs $25,000 installed, has annual maintenance charges of $2000, and costs $5/h for power consumption while operating. The electrostatic precipitator costs $100,000 initially, has annual maintenance costs of $1000, and costs $.50/h for power while operating. Capitalization charges are 9 percent of the original investment; the plant operates 6 h/day and 220 days/year. Compute the equivalent annual total operating costs of each device. Which device is preferable from a cost standpoint?

1-5 In economic studies, the present value of a future amount of money (whether the future amount will be spent, saved, or earned) is an important consideration. The present value is always less than the future amount because of the sum earned from compound interest. Although most tables and formulas assume that interest is compounded at fixed intervals, such as annually, it is possible to derive formulas based on continuous compounding of interest. Derive such a formula for the present value of a future sum.

If $S(t)$ represents the sum at any time t, and r is the interest rate, then

$$\frac{dS}{dt} = rS$$

is a differential equation satisfied by the sum S. This is subject to the boundary condition that when $t = t_f$, $S = S_f$, where S_f is the future amount at time t_f. Solve the differential equation, apply the boundary condition, and in this way derive the formula asked for.

1-6 For a sum of money S_0 currently in hand, an annuity can be purchased that will pay an amount cS_0 during each unit time interval (such as a year) for the duration t_f of the annuity. In our context, S_0 is the amount we are willing to spend now in order to save the amount cS_0 each year in operating costs. Assuming an interest rate r and also assuming continuous compounding of interest, derive an equation for c as a function of r and t_f.

To do this, consider a fund of amount $S(t)$, such that at time zero S is equal to S_0 and at time t_f, S is zero. The differential equation for S is

$$\frac{dS}{dt} = -cS_0 + rS$$

Solve this differential equation, then evaluate the constants from the boundary conditions, and finally solve for c. Evaluate the equation for the case where $r = 0.06$ and $t_f = 20$ years, and then compare with the value of c mentioned in the text.

1-7 Comparisons among alternative choices of equipment can also be made on the basis of the sum of the initial investment plus the present value of all future operating costs. Use the results of Prob. 1-5 to derive an equation for the present value of operating costs. Assume a uniform operating cost per unit time o extending over the period of time t_f.

2

PRINCIPLES OF FLUID FLOW

A study of air pollution control theory requires as a background a fairly considerable knowledge of fluid mechanics. Knowledge of heat transfer, mass transfer, chemistry, and physics is also required, especially for certain phases of the subject. Future chapters in this text are written on the assumption that the reader has completed freshman and sophomore chemistry and physics and a good basic course in fluid mechanics. It will be assumed that the reader has had nothing more than a brief introduction to heat and mass transfer, although without a more extensive background in these areas the results of certain sections of the book may have to be accepted without being able to follow the derivations closely.

Since fluid mechanics forms the principal foundation of air pollution control theory, the present chapter is devoted to a presentation of those aspects of fluid mechanics of greatest use. It may serve as a review or as a source of new material, depending on the scope of the reader's previous courses in the area.

2-1 BASIC DEFINITIONS

A fluid is defined as a substance which will not sustain a shear stress without continuous deformation for as long as the shear stress is applied. Fluids may be either liquids or gases and are subject to various other classifications as well. A liquid, under the action of gravity, will settle to the bottom of any container in which it is placed, forming a level surface, or meniscus, with vaporization taking place across the meniscus into the space above the liquid. A gas can be contained only in a closed container; otherwise it will escape into the surrounding atmosphere. (This is not always true since a more dense gas will tend to remain in a container that is open at the top if a less dense gas is above it.) Liquids are almost incompressible, their volumes changing only minutely with increases in pressure and only slightly with changes in temperature. Gases are very compressible since their volumes can change very substantially with changes in pressure and temperature.

A homogeneous fluid is one with fluid properties that do not vary from point to point in the flow field. A heterogeneous fluid is one with fluid properties that do vary in different regions of the flow field. By fluid property we mean a property of the fluid itself, not of the way in which it is flowing. Thus fluid properties are measured with the fluid stationary at some standard pressure and temperature. A mixture of particles in air is, strictly speaking, a heterogeneous fluid since the air and particles are separate phases and the region occupied by the particles has different properties from that occupied by the air. A mixture of air and gaseous pollutants, on the other hand, is a homogeneous fluid. Actually, an airstream polluted with particulates is often treated as a homogeneous mixture, especially if the particles are small, since the particles may behave in very much the same way as large molecules. However, a particulate airstream may never be treated as a homogeneous fluid in analyzing the internal flow pattern in a separation device since, by its very nature, the particles must behave differently from the gas if any separation is to be achieved.

The fluid properties of principal interest, that is, those properties with a numerical value that depends on the fluid but not on its condition of flow, are density, specific volume, specific weight, viscosity, and surface tension. Density is defined as the mass of a body of fluid divided by its volume. Since density may vary with position, this definition makes sense only in the limit as the fluid volume becomes very small. On the other hand, all matter consists of atoms and molecules which have finite, though small, dimensions, so that if the volume becomes too small, it will contain only a few molecules or perhaps none at all. In fact, since the molecules are in motion, a very small volume will contain a variable number of molecules and, defined in this way, density will be a highly variable quantity. For this reason, the volume used in defining density and other properties as well is small but large enough to contain many thousands of molecules. Call this volume V_0. The definition of density is then expressed mathematically as

$$\rho = \lim_{V \to V_0} \frac{m}{V} \quad kg/m^3 \qquad (2\text{-}1)$$

The specific volume is the reciprocal of density and is defined as

$$v = \lim_{V \to V_0} \frac{V}{m} = \frac{1}{\rho} \quad m^3/kg \qquad (2\text{-}2)$$

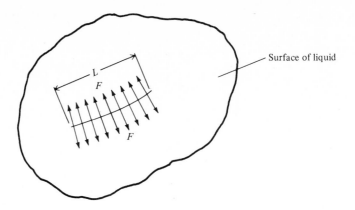

FIGURE 2-1
Surface tension in a liquid.

The specific weight is defined as the weight of material per unit volume and is

$$\gamma = \lim_{V \to V_0} \frac{w}{V} = \lim_{V \to V_0} \frac{mg}{V} = \rho g \qquad \text{N/m}^3 \qquad (2\text{-}3)$$

Viscosity is a fluid property in that it has the same value for all flow situations involving a specific fluid, but this value cannot be determined unless the fluid is flowing. It refers to the rate at which momentum is being transferred by molecular action in a flowing fluid having nonuniform velocity. Specifically, viscosity is the ratio of shear stress to a combination of the various velocity gradients in the flowing fluid. Its SI units are kilograms per meter-second (kg/m · s).

Surface tension is a measure of the internal forces generated between molecules by virtue of their position in the surface of the fluid or at the interface between two fluids. This force lies in the plane of the surface and is normal to any line or curve drawn in the surface. Surface tension is defined as the force per unit length of line, as shown in Fig. 2-1, and is expressed mathematically as

$$\sigma = \lim_{L \to L_0} \frac{F}{L} \qquad \text{N/m} \qquad (2\text{-}4)$$

Fluids are of most interest when they are flowing. Flows are depicted graphically by streamlines, as shown in Fig. 2-2. These are drawn so that at any point on the streamline the velocity of the fluid at that point is always tangent to the streamline. Mathematically, this condition requires that

$$\frac{dy}{dx} = \frac{v}{u} \qquad (2\text{-}5)$$

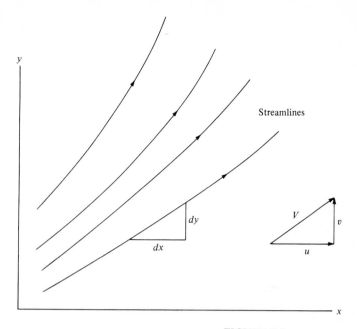

FIGURE 2-2
Streamlines in a fluid flow.

A streamtube is a collection of streamlines passing through successive points on a closed curve so as to form a closed conduit for flow. Since no flow can cross a streamline, a streamtube surface separates the flow inside the streamtube from the flow outside. Solid boundaries of the flow are the surfaces of streamtubes, but, of course, most streamtubes that can be drawn lie inside the flow field. In steady flow, the streamlines and streamtubes are fixed in position; in unsteady flow, they are continually shifting in position.

A flow is steady if the velocity at every point in the flow field does not vary with time. In this case, velocity is a function of position; that is, $V = V(x, y, z)$. In unsteady flow, velocity is a function of time, at least at some points in the flow field, and may or may not be a function of position. In uniform flow, velocity is not a function of position. Thus, general flow can be expressed mathematically as $V = V(x, y, z, t)$, unsteady uniform flow as $V = V(t)$, and steady uniform flow as $V = $ constant. Uniform flow in a conduit has a slightly different meaning. Here, the flow is uniform if the velocity profile normal to the axis of the duct does not change along the duct axis and if all properties except pressure are constant throughout the duct, as shown in Fig. 2-3; furthermore, the pressure must change linearly with distance along the tube.

Another important classification of flow is between laminar and turbulent flow. Flow is laminar if the only movement normal to the streamline is due to molecular motion. This action is illustrated in Fig. 2-4, in which a thin stream

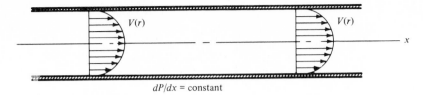

FIGURE 2-3
Uniform flow in a duct.

of dye is injected into the fluid at point *A*. A streak is formed which persists far downstream while gradually being diffused by molecular action. The curve formed by the dye streak, or streakline, follows the streamline in the case of steady flow. Although in laminar flow the gross fluid motion is along the streamlines, momentum is transported normal to the streamlines by molecular diffusion because the molecules of the fluid have a random motion superimposed on the orderly overall motion of the fluid. This random molecular motion carries individual molecules across the streamlines.

Turbulent flow does not possess the orderly motion of laminar flow. In turbulent flow, fluid moves across the streamlines in both molecular and large-scale motion; that is, the instantaneous streamlines are highly variable with time but tend to fluctuate around some average position. The average positions of the streamlines in turbulent flow are taken to be the locations of the streamlines. A dye stream injected into the fluid will break up rapidly as large-scale diffusion of the dye molecules takes place by turbulent action. This action is illustrated schematically in Fig. 2-5*a*, while Fig. 2-5*b* shows the positions of the velocity vectors at different times at a fixed point; Fig. 2-5*c* shows the instantaneous variation of the speed of the fluid at a point. The mean velocity is obtained by averaging the instantaneous velocity over a period of time that is sufficiently long to damp out the effect of the turbulent fluctuations but not so long that variations in the overall flow pattern are masked. The mean velocity is usually referred to as the velocity in turbulent flow.

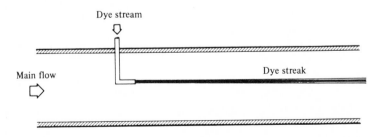

FIGURE 2-4
Injection of dye stream into laminar flow.

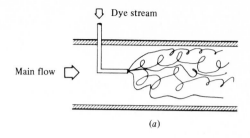

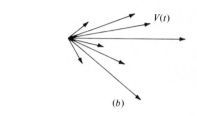

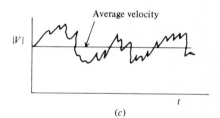

FIGURE 2-5
Turbulent fluctuations and instantaneous
velocities.

As noted previously, a fluid can be either compressible or incompressible. Gases are compressible fluids; liquids are nearly always essentially incompressible. The classification of compressible versus incompressible can also be applied to a particular fluid flow. However, classification with respect to flow is not the same as with respect to the fluid itself. Certainly an incompressible fluid can flow only in an incompressible manner, but a compressible fluid can nonetheless flow incompressibly. An incompressible flow is one in which the fluid density varies by only a negligible amount throughout the flow field. Many flow situations involving compressible fluids such as air are such that the fluid density does not change greatly in the flow field.

The usual criterion for deciding whether a compressible fluid is flowing in incompressible flow is the local Mach number, defined as the ratio of the local speed of the flow to the local speed of sound

$$M = \frac{V}{c} \qquad (2\text{-}6)$$

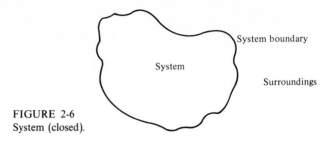

FIGURE 2-6
System (closed).

in which the speed of sound c is given in terms of absolute temperature T by

$$c = \sqrt{kRT} \qquad \text{m/s} \qquad (2\text{-}7)$$

If the maximum value of the local Mach number is limited to about 0.25, the flow can usually be considered as incompressible with acceptable accuracy for most engineering calculations. Most flow involved in air pollution control theory falls in this range. In a few devices, such as the venturi scrubber, the velocities become high enough to call this assumption into question, but even here this assumption of incompressibility is usually made since air pollution control theory is not yet sufficiently developed to warrant extremely accurate calculations.

Problems involving fluid flow and thermodynamics are usually analyzed using systems and control volumes as tools. Ordinarily it is wise to carefully draw either a system or a control volume in the beginning phases of an analysis. The system is defined as a region of space separated from its surroundings by an envelope called the *system boundary*. We shall regard the system as a closed system in which no mass crosses the system boundary during a process. The system can change shape during a process and can move about as the mass it contains flows through a device or region of space. Such a system is illustrated in Fig. 2-6. A control volume is a region fixed in space with rigid boundaries across which mass can flow. In most processes, mass will flow into the control volume over some portion of the boundary and will flow out over other portions of the boundary. The bounding surface of a control volume is called a *control surface*. A control volume is shown in Fig. 2-7.

2-2 FLUID PROPERTIES

The fluid properties of concern to air pollution control studies include density, specific heat, dynamic viscosity, and kinematic viscosity. We need to be able to evaluate these properties for air and for air in which either gaseous pollutants or particulates are mixed in small quantities. All these properties vary significantly with temperature; however, only the variation of the properties with temperature for air will be presented here. Presumably, for the small pollutant concentrations

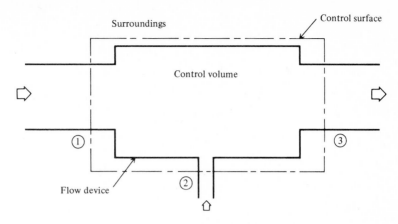

FIGURE 2-7
Control volume applied to a flow device.

found in most polluted air, only a small correction of the property values of air need be made to account for the effect of the pollutant; indeed, this correction can usually be neglected. It will be sufficiently accurate in virtually all cases to use room-temperature property values for pollutants in correcting the values for air to account for pollutant concentrations. With regard to the effect of pressure on fluid property values, only density and kinematic viscosity are noticeably affected by variations of pressure in the range encountered in air pollution control. These two properties can be readily evaluated at any pressure from the equations given in this section.

For all purposes encountered in this book, air may be regarded as a perfect gas. For a perfect gas or a mixture of perfect gases, the density may be obtained from the perfect-gas law

$$Pv = RT \quad \text{or} \quad \rho = \frac{P}{RT} \quad \text{kg/m}^3 \qquad (2\text{-}8)$$

The density of pure air at atmospheric pressure is plotted in Fig. 2-8 as a function of temperature. The constant R in Eq. (2-8), the specific gas constant, is related to the universal gas constant $R_u = 8314$ J/kg-mol $\cdot$ K by

$$R = \frac{R_u}{M} \quad \text{J/kg} \cdot \text{K} \qquad (2\text{-}9)$$

where M is the molecular weight of the gas.

The specific heats for air vary with temperature as shown in Fig. 2-9. The specific heat at constant pressure c_p and the specific heat at constant volume c_v are related by the gas constant for a perfect gas

$$c_p - c_v = R \qquad (2\text{-}10)$$

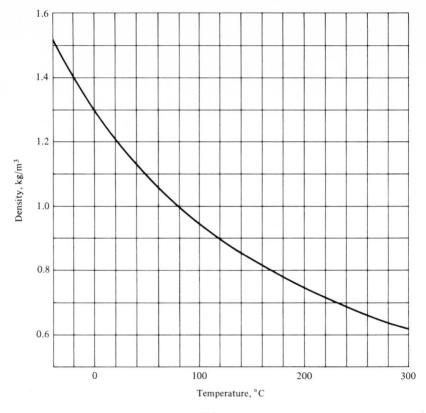

FIGURE 2-8
Density of pure air as a function of temperature.

The ratio of specific heats $k = c_p/c_v$ is a frequently used quantity. It, too, is plotted in Fig. 2-9 for air. Specific heats can be related to k and R by

$$c_p = \frac{kR}{k-1} \qquad (2\text{-}11)$$

$$c_v = \frac{R}{k-1} \qquad (2\text{-}12)$$

The dynamic viscosity, or viscosity, is plotted as a function of temperature in Fig. 2-10 for air. The kinematic viscosity $v = \mu/\rho$ is a strong function of pressure owing to the variation of density with pressure and also is plotted in Fig. 2-10 as a function of temperature for atmospheric pressure. The shear stress in a fluid flowing nonuniformly is related to the pressure gradient by Newton's law of viscosity as

$$\tau = \mu \frac{du}{dy} \qquad \text{N/m}^2 \qquad (2\text{-}13)$$

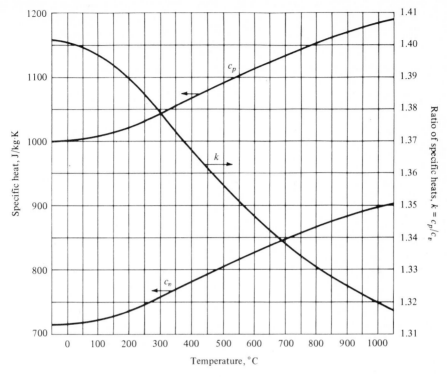

FIGURE 2-9
Specific heats of pure air as functions of temperature.

This equation applies to the one-dimensional flow shown in Fig. 2-11.

Next, consider mixtures of gaseous pollutants and air. Although the concentrations of gaseous pollutants may be quite large from the standpoint of air pollution, the influence on the properties of the mixture will be rather small at the concentration levels involved. The effect of the pollutants on the property values of the mixture will be of the nature of a small correction of the property values for pure air. A first-order approximation to the value of such a correction will be adequate for most purposes. We shall treat the case of n pollutant gases mixed with air in such a way that the greatest concentration of any of them is much smaller than the concentration of the air itself in the mixture.

The average molecular weight of a mixture is a weighted average of the molecular weights of the individual constituents of the mixture, where the weighting function is the mole fractions of the constituents. Using C_v to represent volumetric fraction and hence mole fraction for perfect-gas mixtures, the average molecular weight is given by

$$M = C_{v_a} M_a + \sum_{i=1}^{n} C_{v_i} M_i \qquad (2\text{-}14)$$

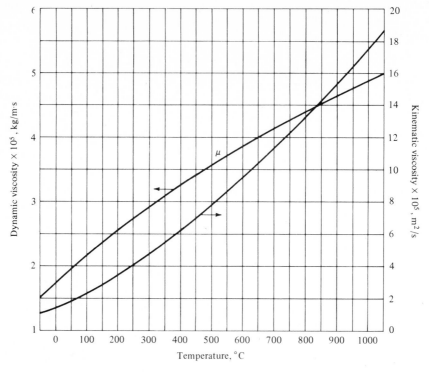

FIGURE 2-10
Dynamic and kinematic viscosities of pure air as functions of temperature.

The average gas constant for the mixture is a weighted average of the individual gas constants based on mass fractions, designated by C_m:

$$R = C_{m_a} R_a + \sum_{i=1}^{n} C_{m_i} R_i \quad \text{J/kg} \cdot \text{K} \quad (2\text{-}15)$$

The density is given as $\rho = P/RT$, which can be expressed approximately as

$$\rho = \frac{P}{C_{m_a} R_a T} \left(1 - \frac{\sum_{i=1}^{n} C_{m_i} R_i}{C_{m_a} R_a}\right) \quad \text{kg/m}^3 \quad (2\text{-}16)$$

The specific heats c_p and c_v are averaged in the same way as the gas constant R; thus

$$c_p = C_{m_a} c_{p_a} + \sum_{i=1}^{n} C_{m_i} c_{p_i} \quad (2\text{-}17)$$

$$c_v = C_{m_a} c_{v_a} + \sum_{i=1}^{n} C_{m_i} c_{v_i} \quad (2\text{-}18)$$

The dynamic viscosity of the mixture is evaluated by a method taken from

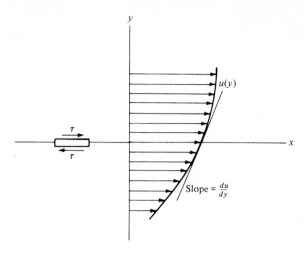

FIGURE 2-11
Shear stress related to the slope of the velocity profile.

Reid and Sherwood [1]. By an approximation to kinetic-theory results, the following equation holds:

$$\mu = \sum_{i=1}^{n} \frac{\mu_i}{1 + \sum_{\substack{j=1 \\ (j \neq i)}}^{n} \phi_{ij} C_{v_j}/C_{v_i}} \quad (2\text{-}19)$$

in which ϕ_{ij} is a defined function which can be approximated using the Wilke estimation method as

$$\phi_{ij} = \frac{[1 + \sqrt{\mu_i/\mu_j}(M_j/M_i)^{1/4}]^2}{\sqrt{8(1 + M_i/M_j)}} \quad (2\text{-}20)$$

When Eq. (2-20) is substituted into Eq. (2-19), there results

$$\mu = \sum_{i=1}^{n} \frac{\mu_i}{1 + \sum_{\substack{j=1 \\ (j \neq i)}}^{n} (C_{v_j}/C_{v_i})[1 + \sqrt{\mu_i/\mu_j}(M_j/M_i)^{1/4}]^2/\sqrt{8(1 + M_i/M_j)}} \quad (2\text{-}21)$$

We now simplify this result to allow for the small concentrations of pollutants. First replace the n in Eq. (2-21) by $n + 1$ and allow the $(n + 1)$st constituent to be the pure air, designated by subscript a. Then since $C_{v_i} \ll C_{v_a}$, Eq. (2-21) simplifies to

$$\mu = \mu_a + \frac{\sqrt{8}}{C_{v_a}} \sum_{i=1}^{n} \frac{\sqrt{1 + M_i/M_a}\, C_{v_i}\mu_i}{[1 + \sqrt{\mu_i/\mu_a}(M_a/M_i)^{1/4}]^2} \quad (2\text{-}22)$$

For a mixture of a single pollutant in air, Eq. (2-22) reduces to

$$\mu = \mu_a + \frac{\sqrt{8}}{C_{v_a}} \frac{\sqrt{1 + M_p/M_a}\, C_{v_p} \mu_p}{[1 + \sqrt{\mu_p/\mu_a}\,(M_a/M_p)^{1/4}]^2} \qquad (2\text{-}23)$$

Once the dynamic viscosity and density have been evaluated, the kinematic viscosity is obtained from $v = \mu/\rho$.

Values of the molecular weight, gas constant, specific heats, k, and dynamic viscosity are tabulated in Table 2-1 for a number of pollutant gases. The values of specific heats and viscosity given in Table 2-1 are taken from a wide variety of sources, including values calculated from various approximation formulas in many cases, so that the degree of accuracy of the various values is unknown. The values are given for room temperature (about 25°C) and low pressure. As explained previously, these values should be accurate enough to predict the effects of these pollutants on the property values of the mixture.

The treatment of a mixture of air and particulate pollutants parallels that of gaseous pollutants, with some differences, particularly in the case of viscosity. The density of the mixture of air and particulates is given by the following equation, which also holds for gaseous pollutants:

$$\rho = C_{v_a}\rho_a + \sum_{i=1}^{n} C_{v_i}\rho_i \qquad (2\text{-}24)$$

where the volumetric concentration of the air is related to those of the particulates by

$$C_{v_a} = 1 - \sum_{i=1}^{n} C_{v_i} \qquad (2\text{-}25)$$

and where ρ_i is the density at the total pressure of the mixture.

The specific heats are given by Eqs. (2-17) and (2-18), which can also be used for mixtures of air, gaseous pollutants, and particulates. The mass concentrations and volumetric concentrations are related by the equation

$$C_{m_i} = \frac{\rho_i C_{v_i}}{\rho_a C_{v_a} + \sum_{j=1}^{n} \rho_j C_{v_j}} \qquad i = a, 1, 2, \ldots, n \qquad (2\text{-}27)$$

To solve for C_v in terms of C_m we must invert this relation, giving

$$C_{v_i} = C_{m_i}\left(C_{v_a}\frac{\rho_a}{\rho_i} + \sum_{j=1}^{n} C_{v_j}\frac{\rho_j}{\rho_i}\right) \qquad i = a, 1, 2, \ldots, n \qquad (2\text{-}28)$$

Equation (2-28) is really $n + 1$ equations (for $i = a, 1, 2, \ldots, n$) which must be solved simultaneously for C_{v_i}. Since the pollutant concentrations are usually quite small, a reasonable approximation in most cases is

$$C_{v_i} = \frac{\rho_a}{\rho_i} C_{m_i} \qquad (2\text{-}29)$$

Table 2-1 PROPERTIES OF SOME POLLUTANT GASES AT STANDARD CONDITIONS

Gas	Formula	M	R, J/kg·K	c_p, J/kg·K	c_v, J/kg·K	k	$\mu \times 10^5$, kg/m·s
Acetaldehyde	C_2H_4O	44.1	188.5	1253	1065	1.18	0.89*
Acetic acid	$C_2H_4O_2$	60.1	138.3	1115	977	1.14	0.76
Acetic anhydride	$C_4H_6O_3$	102.1	81.4	1382*	1301*	1.06*	0.85*
Acetone	C_3H_6O	58.1	143.1	1453	1310	1.11	0.78
Acetonitrile	C_2H_3N	41.05	202.5	1275	1072	1.19	0.71*
Acetylene	C_2H_2	26.0	319.8	1604	1284	1.25	0.94
Acetylene tetrabromide	$C_2H_2Br_4$	345.7	24.1	289*	265*	1.09*	1.22*
Acrolein	C_3H_4O	56.1	148.2	1172*	1024*	1.14*	0.89*
Acrylic acid	$C_3H_4O_2$	72.1	115.3	1080*	965*	1.12*	0.83*
Acrylonitrile	C_3H_3N	53.1	156.6	1202	1045	1.15	0.28*
Allyl alcohol	C_3H_6O	58.1	143.1	1528*	1385*	1.10*	0.82*
Allyl chloride	C_3H_5Cl	76.52	108.7	1165*	1056*	1.10*	0.86*
Ammonia	NH_3	17.0	489.1	2191	1702	1.29	1.09
Amyl acetate	$C_7H_{14}O_2$	130.2	63.9	1347	1283	1.05	0.70*
n-Amyl alcohol	$C_5H_{12}O$	88.1	94.4	1701	1607	1.06	0.73*
Arsine	AsH_3	78.0	106.6	494	387	1.28	1.59
Benzene	C_6H_6	78.1	106.5	1250	1144	1.09	0.80
Benzyl chloride	C_7H_7Cl	126.6	65.7	919*	853*	1.08*	0.35*
Bromine	Br_2	159.8	52.0	230	178	1.29	1.56
n-Butane	C_4H_{10}	58.1	143.1	1686	1543	1.09	0.87
n-Butyl acetate	$C_6H_{12}O_2$	116.2	71.6	1301*	1229*	1.06*	0.34*
Butyl alcohol	$C_4H_{10}O$	74.1	112.2	1497	1385	1.08	0.75*
n-Butyl chloride	C_4H_9Cl	92.6	89.8	1198	1108	1.08	0.82*
Butylene	C_4H_8	56.1	148.2	1528	1380	1.11	0.77
n-Butyl mercaptan	$C_4H_{10}S$	90.2	92.2	1309	1217	1.08	0.77*
1-Butyne	C_4H_6	54.1	153.7	1509	1355	1.11	0.82*
n-Butyraldehyde	C_4H_8O	72.1	115.3	1771*	1656*	1.07*	0.81*
n-Butyric acid	$C_4H_8O_2$	88.1	94.4	1302*	1208*	1.08*	0.75*
n-Caprylic acid	$C_8H_{16}O_2$	144.2	57.7	1466*	1408*	1.04*	0.62*
Carbon dioxide	CO_2	44.01	189.9	832	642	1.30	1.62
Carbon disulfide	CS_2	76.1	109.3	657	548	1.20	1.00
Carbon monoxide	CO	28.01	296.8	1038	741	1.40	1.86
Carbon tetrachloride	CCl_4	153.8	54.1	560	506	1.11	1.01
Chlorine	Cl_2	70.9	117.3	482	365	1.32	1.48
Chloroform	$CHCl_3$	119.4	69.6	607	537	1.13	1.05
m-Cresol	C_7H_8O	108.1	76.9	1154	1077	1.07	0.68*
α-Crotonaldehyde	C_4H_6O	70.1	118.6	1278*	1159*	1.10*	0.79*
Cyanogen	C_2N_2	52.0	159.9	1715	1555	1.10	1.08
Cyclohexane	C_6H_{12}	84.1	98.9	1273	1174	1.08	0.70
Cyclohexanol	$C_6H_{12}O$	100.2	83.0	1345*	1262*	1.07*	0.70*
Cyclohexanone	$C_6H_{10}O$	98.1	84.8	1161*	1076*	1.08*	0.75*
Cyclohexene	C_6H_{10}	82.1	101.3	1289	1188	1.09	5.60
1,3-Cyclopentadiene	C_5H_6	66.1	125.8	1267*	1141*	1.11*	0.83*
n-Decane	$C_{10}H_{22}$	142.3	58.4	1658	1600	1.04	0.58*
Diacetone alcohol	$C_6H_{12}O_2$	116.2	71.6	1474*	1402*	1.05*	0.72*
1,1-Dichloroethane	$C_2H_4Cl_2$	99.0	84.0	770	686	1.12	0.92*
1,1-Dichloroethylene	$C_2H_2Cl_2$	96.9	85.8	670	584	1.15	0.98*
Dichloroethyl ether	$C_4H_8OCl_2$	143.0	58.1	978*	920*	1.06*	0.80*
Diethylamine	$C_4H_{11}N$	73.1	113.7	1546	1432	1.08	0.92
Diethyl ketone	$C_5H_{10}O$	86.1	96.6	1454*	1357*	1.07*	0.72*
Dimethylamine	C_2H_7N	45.1	184.4	1539	1355	1.14	0.82*
Dimethyl sulfide	C_2H_6S	62.1	133.9	1180	1046	1.13	0.88*
1,3-Dioxane	$C_4H_8O_2$	88.1	94.4	1046*	952*	1.10*	0.81*

(Continued)

Table 2-1   PROPERTIES OF SOME POLLUTANT GASES AT STANDARD CONDITIONS (*Continued*)

Gas	Formula	M	R, J/kg · K	c_p, J/kg · K	c_v, J/kg · K	k	$\mu \times 10^5$, kg/m · s
Ethane	C_2H_6	30.1	276.2	1760	1484	1.19	0.92
Ethyl acetate	$C_4H_8O_2$	88.1	94.4	992	898	1.11	0.75
Ethyl alcohol	C_2H_6O	46.1	180.4	1426	1246	1.14	0.91
Ethylamine	C_2H_7N	45.1	184.4	1634	1450	1.13	0.81*
Ethyl benzene	C_8H_{10}	106.2	78.3	1218	1140	1.07	0.66*
n-Ethyl bromide	C_2H_5Br	109.0	76.3	674	598	1.13	1.07*
n-Ethyl butyl ketone	$C_7H_{14}O$	114.2	72.8	1943*	1870*	1.04*	0.67*
Ethyl chloride	C_2H_5Cl	64.5	128.9	1005	876	1.15	1.06
Ethylene	C_2H_4	28.0	296.9	1503	1206	1.25	1.10
Ethylene bromide	$C_2H_4Br_2$	188.0	44.2	445*	401*	1.11*	1.17*
Ethylene chloride	$C_2H_4Cl_2$	99.0	84.0	770	686	1.12	0.97*
Ethyl ether	$C_4H_{10}O$	74.1	112.2	1863	1751	1.06	0.75
Ethyl formate	$C_3H_6O_2$	74.1	112.2	1237*	1125*	1.10*	0.92
Ethyl mercaptan	C_2H_6S	62.1	133.9	1176	1042	1.13	0.82*
Ethyl sulfide	$C_4H_{10}S$	90.2	92.2	1300	1208	1.08	0.69*
Fluorine	Fl	19.0	437.6	1199	761	1.57	0.83
Formaldehyde	CH_2O	30.0	277.1	1181	904	1.31	1.03
Formic acid	CH_2O_2	46.0	180.7	983	802	1.23	1.01*
Freon 11	CCl_3F	137.4	60.5	570	509	1.12	1.11
Freon 12	CCl_2F_2	120.9	68.8	603	534	1.13	1.25
Freon 21	$CHCl_2F$	102.9	80.8	594	513	1.16	1.13
Freon 22	$CHClF_2$	86.5	96.1	649	553	1.17	1.29
Freon 113	$C_2Cl_3F_3$	187.4	44.4	677	632	1.07	1.05
Helium	He	4.003	2077	5225	3148	1.66	2.01
n-Heptane	C_7H_{16}	100.2	83.0	1659	1576	1.05	0.62*
Hexane	C_6H_{14}	86.2	96.5	1669	1573	1.06	0.69
2-Hexanone	$C_6H_{12}O$	100.2	83.0	1492*	1409*	1.06*	0.71*
Hydrogen	H_2	2.016	4124	14,194	10,070	1.41	0.94
Hydrogen bromide	HBr	80.9	102.8	343	240	1.43	1.86
Hydrogen chloride	HCl	36.5	227.8	812	584	1.39	1.58
Hydrogen cyanide	HCN	27.1	306.8	1330	1023	1.30	1.00
Hydrogen fluoride	HF	20.0	415.7	1457	1041	1.40	0.52*
Hydrogen iodide	HI	127.9	65.0	228	163	1.40	1.89
Hydrogen sulfide	H_2S	34.1	243.8	1060	816	1.30	1.27
Iodine	I_2	253.8	32.8	145	112	1.30	1.37*
prim-Isoamyl alcohol	$C_5H_{12}O$	88.1	94.4	1659*	1565*	1.06*	0.76
Isobutyl acetate	$C_6H_{12}O_2$	116.2	71.6	1319*	1247*	1.06*	0.83
Isobutyl alcohol	$C_4H_{10}O$	74.1	112.2	1503	1391	1.08	0.77*
Isopropyl acetate	$C_5H_{10}O_2$	102.1	81.4	1210*	1129*	1.07*	0.77*
Isopropyl alcohol	C_3H_8O	60.1	138.3	1393	1255	1.11	0.76
Isopropyl chloride	C_3H_7Cl	78.5	105.9	1115	1009	1.10	0.89*
Isopropyl ether	$C_6H_{14}O$	102.2	81.4	1516	1435	1.06	0.69*
Methane	CH_4	16.04	518.3	2212	1694	1.31	1.21
Methyl acetate	$C_3H_6O_2$	74.1	112.2	914	802	1.14	0.71
Methyl alcohol	CH_4O	32.0	259.8	1374	1114	1.23	0.98
Methyl bromide	CH_3Br	95.0	87.5	445	358	1.24	1.04
Methyl chloride	CH_3Cl	50.5	164.6	807	642	1.26	1.17
Methyl chloroform	$C_2H_3Cl_3$	133.4	62.3	697	635	1.10	0.95*
Methyl ether	C_2H_6O	46.1	180.4	1435	1255	1.14	1.02
Methyl ethyl ketone	C_4H_8O	72.1	115.3	1434	1319	1.09	0.79*
Methyl formate	$C_2H_4O_2$	60.1	138.3	1142	1004	1.14	0.95
Methyl mercaptan	CH_4S	48.1	172.9	1049	876	1.20	1.02*
Methylene chloride	CH_2Cl_2	84.9	97.9	602	504	1.19	1.00

(*Continued*)

Table 2-1 PROPERTIES OF SOME POLLUTANT GASES AT STANDARD CONDITIONS (*Continued*)

Gas	Formula	M	R, J/kg · K	c_p, J/kg · K	c_v, J/kg · K	k	$\mu \times 10^5$, kg/m · s
Naphthalene	$C_{10}H_8$	128.2	64.9	1048	983	1.07	0.31*
Nitric oxide	NO	30.0	277.1	975	698	1.40	1.88
Nitrobenzene	$C_6H_5NO_2$	123.1	67.5	877*	810*	1.08*	0.79*
Nitroethane	$C_2H_5NO_2$	75.0	110.9	1089*	978*	1.11*	0.92*
Nitromethane	CH_3NO_2	61.0	136.3	940	804	1.17	0.94*
n-Nonane	C_9H_{20}	128.2	64.9	1653	1588	1.04	0.52
n-Octane	C_8H_{18}	114.2	72.8	1589	1516	1.05	0.55
Ozone	O_3	48.0	173.2	820	647	1.27	0.78*
n-Pentane	C_5H_{12}	72.1	115.3	1667	1552	1.07	0.69
2-Pentanone	$C_5H_{10}O$	86.1	96.6	1454*	1357*	1.07*	0.75*
Perchloroethylene	C_2Cl_4	165.9	50.1	580	530	1.09	1.00*
Phenol	C_6H_6O	94.1	88.4	1103	1015	1.09	0.78*
Phosgene	$COCl_2$	98.9	84.1	584	500	1.17	0.60*
Propane	C_3H_8	44.1	188.5	1677	1489	1.13	0.82
Propionic acid	$C_3H_6O_2$	74.1	112.2	1226*	1114*	1.10*	1.18
n-Propyl acetate	$C_5H_{10}O_2$	102.1	81.4	1247*	1166*	1.07*	0.78
n-Propyl alcohol	C_3H_8O	60.1	138.3	1456	1318	1.10	0.93
Propylene	C_3H_6	42.1	197.5	1522	1324	1.15	0.86
1,2-Propylene oxide	C_3H_6O	58.1	143.1	1247	1104	1.13	0.90*
n-Propyl mercaptan	C_3H_8S	76.2	109.1	1250	1141	1.10	0.83*
n-Propyl nitrate	$C_3H_7NO_3$	105.1	79.1	1131*	1052*	1.08*	0.97*
Pyridine	C_5H_5N	79.1	105.1	995	890	1.12	0.73*
Styrene	C_8H_8	104.1	79.9	1180	1100	1.07	0.72*
Sulfur dioxide	SO_2	64.1	129.7	635	505	1.26	1.27
Sulfur monochloride	S_2Cl_2	135.0	61.6	540	478	1.13	1.12*
Sulfur trioxide	SO_3	80.1	103.8	632	528	1.20	0.68*
Sulfuric acid	H_2SO_4	98.1	84.8	824	739	1.11	1.02*
Tetrachloroethane	$C_2H_2Cl_4$	167.9	49.5	601	552	1.09	0.98*
Toluene	C_7H_8	92.1	90.3	1134	1044	1.09	0.70
Trichloroethylene	C_2HCl_3	131.4	63.3	605	542	1.12	1.05*
1,2,3-Trichloropropane	$C_3H_5Cl_3$	147.4	56.4	758*	702*	1.08*	0.89*
Triethylamine	$C_6H_{15}N$	101.2	82.2	1564	1482	1.06	0.66*
n-Valeric acid	$C_5H_{10}O_2$	102.1	81.4	1361*	1280*	1.06*	0.72*
Water vapor	H_2O	18.02	460.0	1860	1400	1.33	0.862
o-Xylene	C_8H_{10}	106.2	78.3	1258	1180	1.07	0.65*

* These values were obtained from approximate calculations.

The typical mixture of particulate pollutants in air can be treated as a homogeneous mixture, which is discussed in some detail for a single such pollutant by Wallis [2]. It is reasonable to assume that the particles are so small as to be in instantaneous thermal equilibrium with the air. The influence of the particles on the viscosity of the mixture will be assumed to be unaffected by whether or not the particles have the same velocity as the air. Although this assumption may require refinement in the future, it will be adequate for this text.

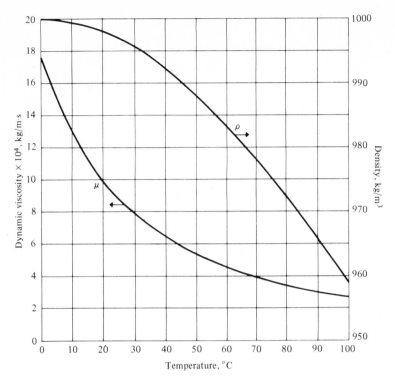

FIGURE 2-12
Density and dynamic viscosity of liquid water as functions of temperature.

The viscosity of a fluid containing liquid spheres moving at the same velocity is given by an equation developed by Taylor [3]:

$$\mu = \mu_a\left(1 + 2.5C_{v_p}\frac{1 + 0.4\mu_a/\mu_p}{1 + \mu_a/\mu_p}\right) \qquad (2\text{-}30)$$

For solid spheres, this equation reduces to

$$\mu = \mu_a(1 + 2.5C_{v_p}) \qquad (2\text{-}31)$$

In the case of liquid droplets, the viscosity of the liquid greatly exceeds that of the air, so that Eq. (2-31) is a good approximation in this case also. For example, with water droplets in air at room temperature, Eq. (2-30) becomes

$$\mu = \mu_a(1 + 2.46C_{v_p})$$

There remains the problem of generalizing Eq. (2-31) to the case of a mixture of several particulate pollutants. Since the viscosity of the particles makes little difference in their influence on the viscosity of the mixture, it seems

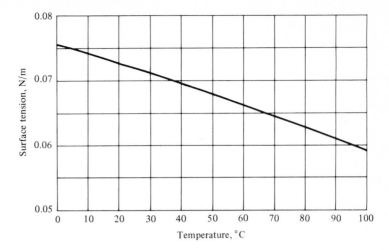

FIGURE 2-13
Surface tension of water as a function of temperature.

reasonable to replace C_{v_p} in Eq. (2-31) by the total volumetric concentration of all the pollutants present, thus giving

$$\mu = \mu_a\left(1 + 2.5\sum_{i=1}^{n} C_{v_i}\right) \qquad (2-32)$$

In any event, the typical value of C_{v_p} for a particulate pollutant in an industrial gas stream is of the order of 10^{-4} to 10^{-5}, so that the influence of the particles on the viscosity of the mixture is usually negligible.

Some properties of water are shown in Figs. 2-12 and 2-13. The density and dynamic viscosity are shown for saturated liquid in Fig. 2-12. The specific heats are not shown, but for liquid water c_p and c_v are virtually equal and have the approximate value of 4187 J/kg · K, accurate to about 1 percent over the range 0 to 100°C. Figure 2-13 shows the surface tension of liquid water in contact with air over the temperature range 0 to 100°C.

2-3 FLOW IN CONDUITS

As indicated in Sec. 2-1, flow in a duct can be either laminar or turbulent. Most duct flow of air in practical engineering applications is in the turbulent regime, though both types of flow need to be studied. Laminar flow occurs at low velocities in a duct; after a certain critical velocity is reached, the flow ceases to be laminar and assumes the characteristics of turbulent flow. There is a range of velocities immediately above the critical velocity in which the flow is neither laminar nor fully developed turbulent flow; this regime of flow is known as transition flow.

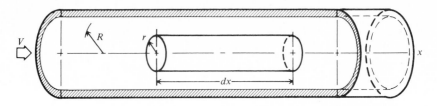

FIGURE 2-14
Flow in a circular duct.

The critical velocity at which laminar flow ceases is determined from the critical value of the Reynolds number Re. The Reynolds number is defined as follows for a circular duct:

$$\text{Re} = \frac{VD\rho}{\mu} = \frac{VD}{v} \qquad (2\text{-}33)$$

For a given pipe diameter and given fluid properties, the Reynolds number is proportional to the velocity. Thus, laminar flow corresponds to low Reynolds numbers, turbulent flow to high Reynolds numbers, and transition flow to intermediate Reynolds numbers. The value of the Reynolds number at the cessation of laminar flow, the critical Reynolds number, is about 2300 for flow in a duct. The upper end of the transition region, where flow becomes fully developed turbulent flow, occurs at a Reynolds number of about 10,000.

Consider, now, a steady flow, either laminar or turbulent, in a circular duct. Figure 2-14 shows the duct configuration, along with a concentric cylindrical element of fluid. The pressure variation in the plane normal to the duct axis is that due to hydrostatic forces only; for gases this pressure variation is negligible, and for liquids it is the same at all axial positions in a straight duct. In our analysis, then, we assume that pressure is a function only of the axial coordinate x. If the flow in the duct is steady and uniform, the pressure varies linearly with x so that dP/dx is constant along the pipe. The fluid element in Fig. 2-14 is shown as a free-body diagram in Fig. 2-15 with the pressure and shear forces illustrated. A force balance on the element yields the following equation:

$$\frac{dP}{dx} = -2\frac{\tau(r)}{r} \qquad (2\text{-}34)$$

For laminar flow, Eq. (2-13) applies and can be written in the notation of Fig. 2-15 as $\tau = -\mu\, du/dr$. When this equation is substituted into Eq. (2-34) and the resultant equation integrated, applying the boundary condition that u is zero at the pipe wall where $r = R$, the result is

$$u = -\frac{1}{4\mu}\frac{dP}{dx}(R^2 - r^2) \qquad (2\text{-}35)$$

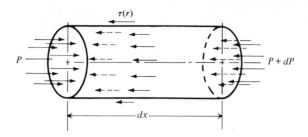

FIGURE 2-15
Cylindrical element of flow in a circular duct.

The average velocity is obtained from the definition that $V = \int_A u \, dA/A$, or

$$V = \frac{1}{\pi R^2} \int_0^R u 2\pi r \, dr = -\frac{1}{8\mu} \frac{dP}{dx} R^2 \qquad (2\text{-}36)$$

When Eq. (2-36) is substituted into Eq. (2-35), the result is

$$u = 2V\left(1 - \frac{r^2}{R^2}\right) = V_{\text{max}}\left(1 - \frac{r^2}{R^2}\right) \qquad (2\text{-}37)$$

Equation (2-37) is a parabolic velocity profile, as shown in Fig. 2-16, characteristic of laminar flow. The shear stress at the wall is given by

$$\tau_w = \frac{4\mu V}{R} \qquad (2\text{-}38)$$

Substituting Eq. (2-38) into Eq. (2-34) gives

$$\frac{dP}{dx} = -\frac{2\tau_w}{R} = -\frac{8\mu V}{R^2} \qquad (2\text{-}39)$$

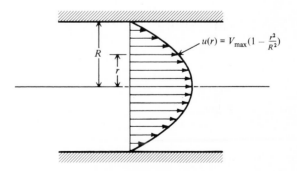

FIGURE 2-16
Velocity profile in laminar flow through a circular duct.

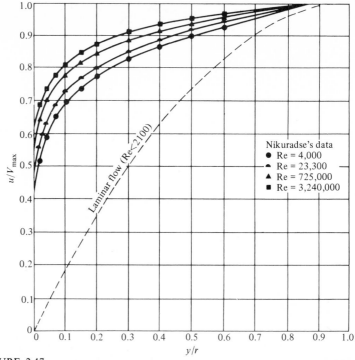

FIGURE 2-17
Velocity profiles in turbulent pipe flow. (*From J. G. Knudsen and D. L. Katz,*
"Fluid Dynamics and Heat Transfer," p. 152, McGraw-Hill Book Company, New
York, 1958. Used by permission.)

Since dP/dx is constant, Eq. (2-39) can be readily integrated to give

$$P_1 - P_2 = \frac{8\mu V L}{R^2} = \frac{2\mu V L}{D^2} \qquad (2\text{-}40)$$

in which L is the length of pipe between sections 1 and 2, where the respective
pressures are P_1 and P_2. A friction factor f is defined by the relation

$$f = \frac{(P_1 - P_2)D}{\rho V^2 L/2} \qquad (2\text{-}41)$$

Combining Eqs. (2-40) and (2-41) leads to the relation

$$f = \frac{64\mu}{VD} = \frac{64}{\text{Re}} \qquad (2\text{-}42)$$

which is valid for laminar flow in a circular duct.

If the flow in the duct is turbulent, the parabolic velocity profile developed
for laminar flow is no longer valid. Likewise, Eq. (2-42) for the friction factor

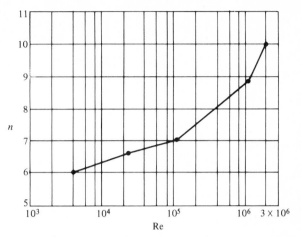

FIGURE 2-18
Exponent n in the equation $u/V_{max} = (y/R)^{1/n}$.

does not hold in the turbulent-flow regime. The velocity-profile characteristic of turbulent flow is flatter near the center and steeper near the wall of the duct than the corresponding laminar-flow velocity profile. Also, the turbulent-flow velocity profile depends on the Reynolds number of the flow, unlike the laminar-flow profile. Owing to the steeper profile near the wall, the friction factor in turbulent flow greatly exceeds that in laminar flow. Although wall roughness has no effect in laminar flow, it is a very important factor in turbulent flow. Figure 2-17, taken from Ref. [4], shows the turbulent-flow velocity profiles for several values of Reynolds number, together with the laminar-flow profile for comparison. In this plot, y is the distance from the wall toward the centerline, given by $y = R - r$. An approximate expression for the turbulent-flow velocity profile, valid except near the wall and centerline, is

$$\frac{u}{V_{max}} = \left(\frac{y}{R}\right)^{1/n} \qquad (2\text{-}43)$$

in which n varies with Reynolds number, as shown in Fig. 2-18.

The friction factor for flow in ducts is a function of the Reynolds number, and in the case of turbulent flow, it is also a function of the relative roughness e/D, where e is the mean height of the roughness projections and D is the pipe diameter. Figure 2-19, which is taken from Ref. [5], gives the relation between the friction factor f, Reynolds number Re, and relative roughness e/D in graphical form. Table 2-2 gives data for the roughness height e.

The friction-factor curve for turbulent flow in smooth pipes can be represented by

$$\frac{1}{\sqrt{f}} = 0.87 \ln (\text{Re} \sqrt{f}) - 0.80 \qquad (2\text{-}44)$$

58

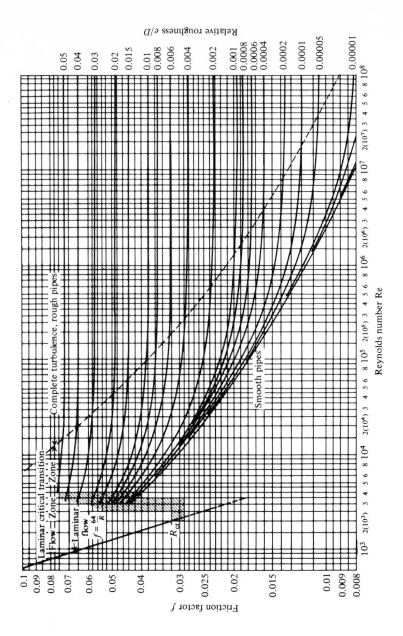

FIGURE 2-19
Friction-factor chart. *(From L. F. Moody, Friction Factors for Pipe Flow, Trans. ASME, vol. 66, pp. 671–684. Used by permission.)*

which may be approximated by

$$f = 0.0056 + 0.5(\text{Re})^{-0.32} \qquad (2\text{-}45)$$

The pressure drop over a length L of a uniform duct is given by the following equation, obtained from Eq. (2-41):

$$P_1 - P_2 = \frac{f L \rho V_{\text{ave}}^2}{2D} \qquad (2\text{-}46)$$

Flow in Noncircular Ducts

An exact treatment of the pressure drop in noncircular ducts poses great difficulties in general; a numerical solution is required, which in itself is not completely exact. Fortunately, for the majority of cases of noncircular ducts, a simple approximate treatment gives very good results. In this treatment, the cross-sectional shape of the duct is characterized by a single length known as the hydraulic diameter, which is defined as (see Fig. 2-20)

$$D_h = \frac{4A}{p} \qquad (2\text{-}47)$$

Table 2-2   ROUGHNESS-PROJECTION HEIGHTS FOR PIPE WALLS

	Values of e, cm	
Kind of pipe or lining (new)	Range	Design value
Brass	0.00015	0.00015
Copper	0.00015	0.00015
Concrete	0.03–0.3	0.12
Cast iron, uncoated	0.012–0.061	0.024
Cast iron, asphalt-dipped	0.0061–0.018	0.012
Cast iron, cement-lined	0.00024	0.00024
Cast iron, bituminous-lined	0.00024	0.00024
Cast iron, centrifugally spun	0.00030	0.00030
Galvanized iron	0.0061–0.024	0.015
Wrought iron	0.0030–0.0091	0.0061
Commercial and welded steel	0.0030–0.0091	0.0061
Riveted steel	0.091–0.91	0.18
Transite	0.00024	0.00024
Wood stave	0.018–0.091	0.061

SOURCE: R. V. Giles, "Theory and Problems of Fluid Mechanics and Hydraulics," Schaum's Outline Series, McGraw-Hill Book Company, New York, 1962, p. 257. Used by permission.

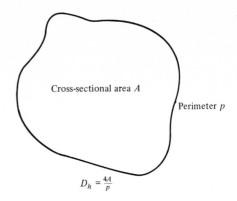

FIGURE 2-20
Hydraulic diameter for a duct of arbitrary
cross section.

in which D_h is the hydraulic diameter, A is the cross-sectional area of the duct, and p is the perimeter of duct that is in contact with the fluid (the so-called "wetted perimeter"). Where the flow completely fills the duct, as will invariably be the case in air pollution control study, the perimeter of the duct is used. For a circular duct, the hydraulic diameter reduces to the diameter; for a square duct of side w, the hydraulic diameter equals w; and for an annulus of inner and outer diameters D_i and D_o, respectively, $D_h = D_o - D_i$. The hydraulic diameter D_h is used in place of the diameter D in the formulas in this section: in Eq. (2-33), defining the Reynolds number; in Eqs. (2-41) and (2-42) for f; in Eq. (2-46) for $P_1 - P_2$; and in Fig. 2-19.

In assessing the accuracy of the hydraulic diameter approximation, it should be remembered that the method is very accurate for shapes which are nearly circular but becomes more and more inaccurate as a shape departs from that of a circle. A good measure of the validity of the approximation is based on the ratio D_h/D, where D is the diameter of a circle of equal area. This ratio is given by the equation

$$\frac{D_h}{D} = \frac{4A}{pD} = \frac{2\sqrt{\pi A}}{p} = 3.54 \frac{\sqrt{A}}{p} \qquad (2\text{-}48)$$

The ratio D_h/D for several configurations is given in the following table:

Configuration	D_h/D
Circle	1.000
Square	0.885
2 : 1 rectangle	0.835
4 : 1 rectangle	0.708
10 : 1 rectangle	0.509
Annulus, $D_o/D_i = 2$	0.578
Triangle, altitude/base $= 10$	0.378

As a general guide, the hydraulic diameter approximation is acceptable in engineering practice if D_h/D exceeds 0.5. Since no alternative exists that is easy to apply, the temptation to use this method for smaller values of D_h/D is very strong; if this is done, then the resulting answer must be applied cautiously.

The mass flow rate $\dot{m}$ and volume flow rate Q may be related to the velocity of flow and the properties of the duct and fluid, as follows: For a cross section of area dA, over which the velocity may be taken as uniform, the flow rates are

$$dQ = V\,dA \qquad d\dot{m} = \rho V\,dA$$

When integrated over the entire cross section of the duct, the results are

$$Q = \int_A V\,dA \qquad (2\text{-}49)$$

$$\dot{m} = \int_A \rho V\,dA \qquad (2\text{-}50)$$

If the velocity is uniform over the entire cross section, then Eq. (2-49) becomes $Q = AV$; and if, in addition, the density is constant over the cross section, as is usually assumed, then $\dot{m} = A\rho V$. In the real case, the velocity is not uniform over the cross section, and Eqs. (2-49) and (2-50) are simplified by using Eq. (2-49) as the basis for defining the average velocity over the cross section; that is, the average velocity is defined as

$$V_{ave} = \frac{1}{A}\int_A V\,dA \qquad (2\text{-}51)$$

Then Eq. (2-49) becomes

$$Q = AV_{ave} \qquad (2\text{-}52)$$

and since the density is constant over the cross section, Eq. (2-50) becomes

$$\dot{m} = A\rho V_{ave} \qquad (2\text{-}53)$$

For steady flow, the mass flow rate $\dot{m}$ will be the same at all sections of the duct at all times. Even for unsteady flow, the value of $\dot{m}$ will not change with axial position in the duct, though it may change with time. If the flow is incompressible, the volumetric flow rate Q will be constant along the duct and constant with time if the flow is also steady.

Minor Losses in Ducts

In addition to the pressure drop in ducts, on occasion it is also necessary to compute the pressure drop in duct fittings or components, such as bends and branch connections. There are two common methods of treating this problem in the literature: (1) The equivalent length of the straight duct is computed, and this length is added to the length of the straight duct in the system; and (2) the pressure drop in the fitting is computed from tabulated coefficients, and this pressure drop is added to that of the straight duct to obtain the total pressure drop of the system. We shall follow the latter approach here. The pressure drop across the fitting is related to the velocity head by a loss coefficient K_L, as

$$\Delta P = \frac{K_L \rho V_{ave}^{2}}{2} \qquad (2\text{-}54)$$

Some representative values of K_L for standard ductwork components are presented in Table 2-3, which is taken from Ref. [7].

Table 2-3   LOSS COEFFICIENTS FOR DUCT FITTINGS

Type of fitting	K_L
Tee connection, right-angle side outlet	2.0*
Branches from duct	
15° angle	0.10*
30° angle	0.20*
45° angle	0.25*
Tee connection, right-angle side inlet	1.0*
Branches into duct†	
30° angle	0.18*
45° angle	0.28*
60° angle	0.44*
Elbow, right-angle, rectangular duct with aspect ratio near unity	1.15
Elbow, right-angle, round duct	0.87
Elbow, radius/diameter 0.75, round duct	0.4
Elbow, radius/diameter 1.5, round duct	0.2
Entry to pipe from chamber	0.85
Entry to pipe from chamber, coned inlet	0.20
Pipe enlargement	
Abrupt or chamber inlet	0.8–1.0
Coned, 10 percent slope	0.25
Pipe entering chamber, coned 10 percent slope	0.50
Grilles, net area equal to duct area	1.25
Entrance loss	
Intake louvers and induction of outside air	1.5
Intake louvers without acceleration of inlet air	0.5

SOURCE: Taken primarily from B. H. Jennings, "Environmental Engineering, Analysis and Practice," International Textbook Company, Scranton, Pa., 1970, p. 424. Used by permission.
* Based on velocity head in branch.
† Taken from C. Strock and R. L. Koral, eds., "Handbook of Air Conditioning, Heating, and Ventilating," The Industrial Press, New York, 1965, pp. 5–14.

2-4 BOUNDARY-LAYER THEORY

In treating problems of fluid flow in regions bounded by solid surfaces, the effect of viscosity cannot be neglected. When viscosity is included in the governing differential equations of motion, these equations become very difficult to solve except in relatively simple cases. In order to obtain solutions which give reasonable approximations to the actual flow pattern, various simplifications have been employed.

The essence of the engineering approach to solving complex problems is to devise workable approximations and distinguish good approximations from bad ones. Much effort has been expended over the years by a great many engineers and researchers to arrive at good approximation methods for use in solving fluid-flow problems. In attempting to circumvent the difficulties imposed by the inclusion of viscosity into the equations of motion, it was realized that the effect of viscosity on the velocity profile is confined to a narrow region immediately adjacent to any solid surfaces present. This region is called the *boundary layer*. Outside the boundary layer, the viscosity of the fluid has no direct effect on the velocity profile, although it has an indirect effect in that the presence of the boundary layers does alter the geometrical boundaries of the flow somewhat, thus changing the velocity profile throughout the flow region.

The boundary-layer concept is utilized in the mathematical solution of a flow problem by neglecting the effect of viscosity in the equations of motion applicable outside the boundary layers and by applying other simplifying assumptions to the equations applicable within the boundary layers. These assumptions applicable inside the boundary layer are that the velocity component normal to the boundary is small compared with that parallel to the boundary, that the velocity components change much more rapidly in the direction normal to the boundary than in the direction parallel to the boundary, and that the pressure does not change appreciably across the boundary layer in the direction normal to the surface.

Boundary Layer on a Flat Plate

The classic example of boundary-layer development is that of a flat plate oriented parallel to the direction of main-stream flow. Usually, the main stream is assumed to have uniform velocity, called the *free-stream velocity*. Such a situation is shown in Fig. 2-21. If the leading edge of the plate is sharp, the boundary layer starts with zero thickness at the leading edge. The velocity profile inside the boundary layer at a typical section is shown in Fig. 2-22. It will be observed that the velocity inside the boundary layer approaches the free-stream velocity asymptotically at the edge of the boundary layer, so that the exact location of the edge of the boundary layer is somewhat nebulous. In order to separate the boundary-layer region from the free-stream region, the edge of the boundary layer is arbitrarily defined as that point where the velocity is 99 percent of the free-stream value.

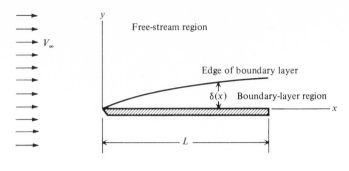

FIGURE 2-21
Boundary layer on a flat plate.

The concept of the boundary layer makes sense only if the flow outside the boundary layer in the free-stream region is turbulent. The flow inside the boundary layer may be either laminar or turbulent. Near the leading edge of the plate, the flow is laminar inside the boundary layer; farther back along the plate, assuming it extends far enough, a transition occurs and the flow becomes turbulent inside the boundary layer. Even in the turbulent boundary layer, however, a very thin region exists adjacent to the surfaces in which the flow has some of the characteristics of laminar flow; this region is referred to as the *laminar sublayer*.

Exact solutions exist for the laminar boundary layer on a flat plate and on other fairly simple geometries. For the turbulent boundary layers, exact solutions are impossible to obtain and approximate solutions to the boundary-layer equations must be used. Such approximate solutions may be obtained by expressing the velocity profile with a simple mathematical expression, such as a cubic function or a sine function. We shall not discuss the details of the solution of boundary-layer equations here, but we shall present the results of approximate solutions to the laminar and turbulent boundary layers over a flat plate.

The approximate solution for laminar flow over a flat plate is taken from Ref. [8], although it is available in many textbooks of fluid mechanics or heat transfer. The velocity profile in the boundary layer is approximated by

$$u = V_\infty \left[1.5 \frac{y}{\delta} - 0.5 \left(\frac{y}{\delta} \right)^3 \right] \qquad (2\text{-}55)$$

The boundary-layer thickness is given by

$$\delta = 4.64 \sqrt{\frac{vx}{V_\infty}} \qquad (2\text{-}56)$$

The friction factor f, analogous to the friction factor for flow in a duct, is given by

$$f = \frac{4\tau_w}{\rho V_\infty^2 / 2} = \frac{2.6}{\sqrt{V_\infty x / v}} = \frac{2.6}{\sqrt{\mathrm{Re}_x}} \qquad (2\text{-}57)$$

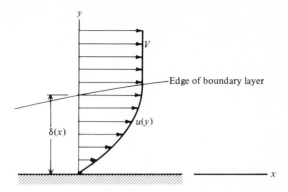

FIGURE 2-22
Velocity profile near a solid surface.

in which τ_w is the shear stress at the surface of the plate and Re_x is the Reynolds number based on distance along the plate from the leading edge, $Re_x = V_\infty x/v$. The friction factor defined by Eq. (2-57) varies along the plate and, in fact, is infinite at the leading edge. However, the boundary-layer solution described here is not valid for a very short distance immediately downstream from the leading edge, so that an infinite friction factor at the leading edge does not conform with reality at that point.

The average friction factor from the leading edge to any point x may be obtained by averaging the local value of f between the limits of 0 and x, giving

$$\bar{f} = \frac{1}{x}\int_0^x f\,dx = \frac{5.2}{\sqrt{V_\infty x/v}} = \frac{5.2}{\sqrt{Re_x}} \qquad (2\text{-}58)$$

Note that the average friction factor from the leading edge to a given point is just twice the local value at that point. For the total length of the plate, where $x = L$, the average friction factor is

$$\bar{f}_L = \frac{5.2}{\sqrt{Re_L}} \qquad (2\text{-}59)$$

where $Re_L = V_\infty L/v$.

The transition from a laminar boundary layer to a turbulent boundary layer occurs at a distance downstream from the leading edge of the plate at which the Reynolds number is equal to a certain critical value. The critical value generally used is $Re_c = 300{,}000$.

For the turbulent boundary layer, the velocity profile can be approximated everywhere except in the immediate vicinity of the wall by a relation similar to Eq. (2-43), with $n = 7$:

$$\frac{u}{V_\infty} = \left(\frac{y}{\delta}\right)^{1/7} \qquad (2\text{-}60)$$

However, Eq. (2-60) cannot be used to obtain the friction factor. An empirical equation for the friction factor is

$$f = 0.23 \, \text{Re}_x^{-0.2} \qquad (2\text{-}61)$$

Using Eqs. (2-60) and (2-61), a solution for the boundary-layer thickness is obtained, based on the assumption that the boundary layer is turbulent from its inception:

$$\delta = 0.37x \, \text{Re}_x^{-0.2} \qquad (2\text{-}62)$$

The assumption that the boundary layer is turbulent from the leading edge leads to errors in Eq. (2-62) near the transition point, of course, but there will be errors there anyway due to the neglect of the transition region. If most of the boundary layer over the plate is turbulent, then this assumption is a reasonably good one.

Boundary-Layer Flow around an Axisymmetric Body

The flow of a fluid around a body poses additional complexities to the boundary-layer problem . Usually, the body has a blunt nose on which a stagnation point develops. At a stagnation point, the free-stream velocity is zero and the boundary-layer thickness is finite. The free-stream velocity varies along the body according to a rather complex law, in general, and its evaluation requires application of potential-flow theory to the region outside the boundary layer. We shall not consider the problem of evaluating the free-stream velocity at this point, but we shall give some results of the boundary-layer solutions for known free-stream-velocity distributions.

Solutions are available for boundary-layer flow around two-dimensional bodies and bodies of revolution, as shown in Fig. 2-23. Actually, a two-dimensional body is a special case of a body of revolution, with the radius of the body equal to infinity. In Fig. 2-23, the coordinates used to describe a point in the boundary layer are the distance x from the stagnation point around the surface to the location in question and the distance y from the surface of the body to the point in question. Solutions based on the approximate technique are given in Ref. [9] for both laminar and turbulent boundary layers, with the turbulent boundary layer assumed to start at the stagnation point also. These solutions are for boundary-layer thickness δ: For laminar flow the solution is

$$\delta = 5.05 \frac{v^{0.5}}{RV_\infty^3} \left(\int_0^x V_\infty^5 R^2 \, dx \right)^{0.5} \qquad (2\text{-}63)$$

and for turbulent flow the solution is

$$\delta = 0.38 \frac{v^{0.2}}{RV_\infty^{3.29}} \left(\int_0^x V_\infty^{3.86} R^{1.25} \, dx \right)^{0.8} \qquad (2\text{-}64)$$

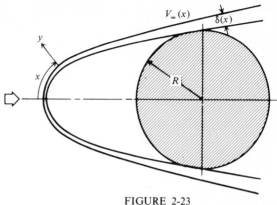

FIGURE 2-23
Flow around a body of revolution.

For two-dimensional bodies, R is assumed constant and equal to infinity. In such cases, R can be canceled out between the numerator and denominator of Eqs. (2-63) and (2-64). The shear stress for the laminar-flow case of Eq. (2-63) requires further analysis, and the reader should consult Ref. [9]. For the turbulent case of Eq. (2-64), the shear stress is given by Eq. (2-61).

When a fluid is flowing around a body, one factor which complicates the flow pattern tremendously is the formation of a wake behind the body. A wake will form in all cases except for very low velocities and streamlined bodies under suitable conditions. The formation of the wake is associated with a phenomenon known as *boundary-layer separation*. This phenomenon tends to occur when the flow outside the boundary layer is decelerating, which corresponds to a case of increasing pressure in the flow direction.

Boundary-layer separation and wake formation are shown in Fig. 2-24. The point at which separation occurs is marked by the point at which the velocity profile has zero slope at the wall; that is, $du/dy = 0$ at the wall ($y = 0$). In Fig. 2-24, transition from a laminar to a turbulent boundary layer takes place prior to the separation point. However, this is not necessarily the case, and in many situations separation occurs while the boundary layer is still laminar. The wake region is at a substantially lower pressure than the region outside the boundary layer, so that a large drag force is associated with the presence of a wake. Skin friction at the surface is less for the laminar boundary layer than for the turbulent boundary layer due to the reduced slope at the wall in the laminar case. On the other hand, a turbulent boundary layer can penetrate much farther downstream against an increasing pressure than can a laminar boundary layer, thus delaying the point of separation and reducing the size of the wake. The reduction in drag due to the wake more than offsets the increase in drag due to substituting a turbulent boundary layer for a laminar one, so that the situation shown in Fig. 2-24 results in substantially less drag force overall. A thorough discussion of wake formation is given by Schlichting [10].

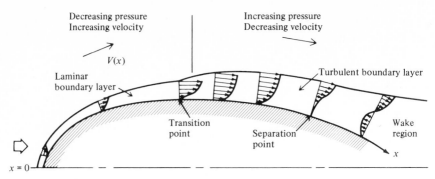

FIGURE 2-24
Boundary-layer separation and wake formation.

2-5 FLOW AROUND BODIES

Of paramount importance in the study of particulate air pollution control is
the pattern of flow of air past small particles. In many cases, such particles
can be idealized as spheres or cylinders, although in other cases, allowance
must be made for departure from these simple shapes. In most instances, the
particle will flow with the fluid, with some small deviation between the path of
the particle and that of the adjoining fluid molecules. If a coordinate system is
attached to the particle, the fluid velocity usually will be rather small in this
coordinate system, which represents the relative velocity between the fluid and the
particle. For small particles, particularly, the relative velocity will be quite small.

The pattern of flow past a particle is strongly dependent on the relative
velocity between the fluid and the particle and on the particle size. Several flow
regimes exist, and these are distinguished from each other by critical, or transition,
values of the Reynolds number, defined in this case by

$$\mathrm{Re} = \frac{V_\infty d}{\nu} \qquad (2\text{-}65)$$

in which d is the particle diameter and V_∞ is the relative velocity of the fluid
measured at a point several diameters away from the particle surface. The flow
regime most amenable to mathematical treatment is the creeping motion which
occurs for $\mathrm{Re} \le 0.6$. This regime is known as *Stokes' regime*, and the flow is
referred to as *Stokes' flow*. Since air pollution control problems involving the
smallest particles are the toughest ones to handle, the Stokes' flow regime will
receive the greatest attention here. The higher-velocity regimes will also be
considered but from a more empirical point of view.

Because it is so important in the study of particle motion, we shall now
present the equation for the drag force acting on a sphere in creeping motion
of a fluid. Owing to the complexity of the derivation, the details are worked
out in Appendix 1, with the final result given here. Thus, the drag force acting

on a sphere of diameter d placed in a stream of fluid having velocity V_∞, density ρ, and viscosity μ, as taken from Eq. (A1-31), is

$$F_D = 3\pi\mu V_\infty d \qquad (2\text{-}66)$$

This equation can be conveniently presented as a drag coefficient, defined as

$$C_D = \frac{F_D}{A_p \rho V_\infty^2/2} \qquad (2\text{-}67)$$

in which A_p is the projected frontal area of the body, in this case, the area of a circle of diameter d. Substituting Eq. (2-66) into Eq. (2-67) and replacing A_p by $\pi d^2/4$, we obtain

$$C_D = \frac{24\mu}{d\rho V_\infty} \qquad (2\text{-}68)$$

In terms of the Reynolds number, Eq. (2-68) becomes

$$C_D = \frac{24}{\text{Re}} \qquad (2\text{-}69)$$

Equation (2-69) and the equations of Appendix 1 are valid for Reynolds numbers up to about 0.6. Above this value, the effect of the neglected inertia terms in the equations of motion becomes significant. An improved solution due to Oseen, presented by Schlichting [10, pp. 97–98], leads to the following equation for drag coefficient:

$$C_D = \frac{24}{\text{Re}}\left(1 + \frac{3}{16}\,\text{Re}\right) \qquad (2\text{-}70)$$

which is valid up to Re = 2. This solution takes partial account of the inertia terms by linearizing them.

In the case of a long circular cylinder, a solution analogous to Stokes' solution for the sphere leads to invalid results owing to the fact that the inertia terms cannot be neglected. However, Oseen's improvement does lead to valid results for small Reynolds numbers. This solution is presented by Lamb [11], with the following result for the drag coefficient:

$$C_D = \frac{\pi(8/\text{Re})}{\ln\,(8/\text{Re}) - 0.07722} \qquad (2\text{-}71)$$

in which the Reynolds number is defined in terms of the cylinder diameter Re = $V_\infty\,d\rho/\mu$. Equation (2-71) is valid up to a Reynolds number of about 0.7. The drag force on a cylinder of length L is then

$$F_D = \frac{4\pi\mu V_\infty L}{\ln\,(8/\text{Re}) - 0.07722} \qquad (2\text{-}72)$$

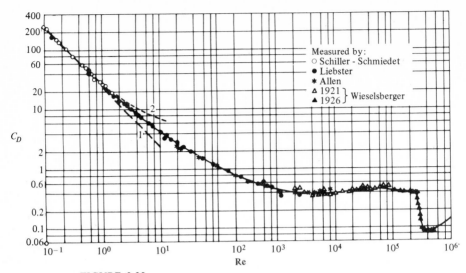

FIGURE 2-25

Drag coefficient for flow around a sphere. (*From H. Schlichting, "Boundary Layer Theory," 4th ed., p. 16, McGraw-Hill Book Company, New York, 1960. Used by permission.*)

As the velocity and hence the Reynolds number increases above the values for which Stokes' solution and Oseen's improvement are valid, a wake begins to form downstream of the body. As the velocity increases, the area occupied by the wake increases, with the result that the drag coefficient does not drop as rapidly as Stokes' solution would suggest if extrapolated to this region. As shown in Fig. 2-25, the drag coefficient levels off, eventually becoming almost constant. This region of constant drag coefficient extends from the Reynolds numbers 1000 to 300,000, at which point a sharp drop is experienced. This sharp drop in drag coefficient is due to the sudden onset of transition in the boundary layer, which enables the boundary layer to remain attached for a greater distance around the object, as explained in Sec. 2-4. Figure 2-26 shows the drag coefficient for a cylinder; Figs. 2-25 and 2-26 are taken from Schlichting [10, p. 16].

The curve of Fig. 2-25 is describable over its entire range only by a very complicated equation if, indeed, any single equation could describe it. For later analysis of the motion of particles, it is necessary to have available approximate equations for this curve; it is satisfactory to break up the total range of Reynolds numbers into several more limited ranges and to present approximate equations suitable for each range. At small Reynolds numbers, Stokes' equation or Oseen's improvement are satisfactory representations of the curve. At higher Reynolds numbers, a transition region exists requiring more complex representation of the curve. At still higher Reynolds numbers, the curve becomes even more complex but can be satisfactorily approximated by a constant value of C_D up to a Reynolds number of 270,000. Actually, several representations of the C_D curve are possible; each is a satisfactory approximation offering certain advantages in calculation.

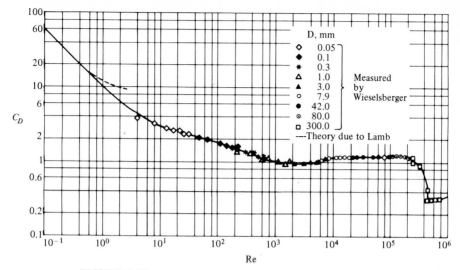

FIGURE 2-26
Drag coefficient for flow around a cylinder. (*From H. Schlichting, "Boundary Layer Theory," 4th ed., p. 16, McGraw-Hill Book Company, New York, 1960. Used by permission.*)

The choice among them will be made mainly on their ease of calculation for a particular purpose.

One representation of the drag curve for flow around a sphere is as follows:

$$C_D = \frac{24}{Re} = \frac{24\mu}{\rho V_\infty d} \qquad Re < 2 \qquad (2\text{-}73)$$

$$C_D = \frac{18.5}{Re^{0.6}} = \frac{18.5\mu^{0.6}}{\rho^{0.6}V_\infty^{0.6}d^{0.6}} \qquad 10 < Re < 700 \qquad (2\text{-}74)$$

$$C_D = 0.44 \qquad Re > 500 \qquad (2\text{-}75)$$

Another representation is

$$C_D = \frac{24}{Re} + 2 \qquad Re < 10 \qquad (2\text{-}76)$$

$$C_D = 0.344 + \frac{67.3}{Re} - \frac{287.4}{Re^2} \qquad 10 < Re < 700 \qquad (2\text{-}77)$$

$$C_D = 0.44 \qquad Re > 700 \qquad (2\text{-}78)$$

A third possibility is to use Eqs. (2-76) and (2-78) but to replace Eq. (2-77) with

$$C_D = \frac{15.6}{Re^{0.57}} \qquad 10 < Re < 700 \qquad (2\text{-}79)$$

We shall use one or another of these representations for C_D in our future developments.

2-6   ENERGY TRANSFER IN FLUID FLOW

In this section we shall consider certain energy effects as they relate to flow in a conduit. The starting point in this development is the first law of thermodynamics, which states that energy can neither be created nor destroyed. To apply this law to flow in a duct, let us consider the flow rate of various properties past a section of the duct. Consider a velocity distribution across the section, as shown in Fig. 2-27. For laminar flow the distribution is parabolic; for turbulent flow it is flatter in the center and steeper at the wall, as shown in Fig. 2-27. A laminar-flow distribution is shown in Fig. 2-16.

The mass flow rate past the section is given by Eq. (2-50); the average velocity past the section is defined in such a way that $\dot{m} = \rho A V_{ave}$, as given by Eq. (2-53). The momentum flow rate is given by the integral

$$\dot{M} = \int_A \rho V^2 \, dA \qquad (2\text{-}80)$$

which can be written as

$$\dot{M} = \beta \rho A V_{ave}^{\ 2} \qquad (2\text{-}81)$$

in which

$$\beta = \frac{1}{A V_{ave}^{\ 2}} \int_A V^2 \, dA \qquad (2\text{-}82)$$

For laminar flow, β has a value of 1.33; for turbulent flow, β has a value of about 1.05 and is usually assumed equal to 1.0.

Of more immediate importance to our present development are the energy flows past the section of Fig. 2-27. An energy flow is characterized by the fact that the energy in question is associated with each particle of mass which flows across the section. As this mass flows across the section from left to right, the energy goes along too. Thus, energy is carried into or out of a control volume, where the section is one portion of the control surface. The energies involved are kinetic energy, potential energy, internal energy, and flow-pressure energy.

Kinetic energy is that energy directly associated with the velocity of the fluid, and its flow rate is given by

$$\dot{E}_k = \tfrac{1}{2} \int_A \rho V^3 \, dA \qquad (2\text{-}83)$$

Kinetic energy is written in terms of the average velocity by

$$\dot{E}_k = \frac{\alpha \rho A V_{ave}^{\ 3}}{2} = \frac{\alpha \dot{m} V_{ave}^{\ 2}}{2} \qquad (2\text{-}84)$$

where

$$\alpha = \frac{1}{A V_{ave}^{\ 3}} \int_A V^3 \, dA \qquad (2\text{-}85)$$

Here, α is equal to 2.0 for laminar flow and to about 1.1 for turbulent flow. Frequently, α is assumed equal to 1.0, although the error involved in this assump-

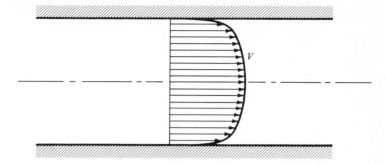

FIGURE 2-27
Velocity profile in a duct.

tion is noticeable. For flow emerging from a nozzle, α is very close to 1.0 (as is β, too, for that matter), which is the most common situation where kinetic energy is significant compared with other forms of energy.

Potential energy is due to the elevation of the fluid above some specified datum, and its flow rate is given by

$$\dot{E}_p = g \int_A \rho V z \, dA \qquad (2\text{-}86)$$

The average elevation is at the centroid of the duct cross section, so that this value is used to represent the elevation of the duct, giving for the potential-energy flow rate

$$\dot{E}_p = g\rho z A V_{\text{ave}} = \dot{m}gz \qquad (2\text{-}87)$$

Internal energy is characterized by temperature and is related to temperature by the specific heat of the fluid

$$\dot{U} = \int_A \rho c_v T V \, dA \qquad (2\text{-}88)$$

The average temperature is defined in terms of the internal energy as

$$T_{\text{ave}} = \frac{1}{A V_{\text{ave}}} \int_A T V \, dA \qquad (2\text{-}89)$$

so that internal energy is related to temperature by

$$\dot{U} = \rho c_v A V_{\text{ave}} T_{\text{ave}} = \dot{m} c_v T_{\text{ave}} \qquad (2\text{-}90)$$

For a liquid, c_v and c_p are identical and can be designated by c. For a gas, the specific heat to use in Eqs. (2-88) to (2-90) is c_v, or the specific heat at constant volume.

Flow-pressure energy, frequently referred to as *flow energy*, or *flow work*, and sometimes as *pressure energy*, is somewhat awkward to characterize. It is not a property of the fluid particle in the same way that kinetic, potential, and internal energies are. It is a property of the fluid particle only if it is flowing under pressure and is given as the product of pressure times volumetric flow rate:

$$\dot{E}_f = \dot{m}\frac{P}{\rho} = PQ = AV_{ave}P \qquad (2\text{-}91)$$

Here, the pressure is taken as the average pressure over the cross section, which will be the pressure at the centroid of the section.

The sum of the internal-energy flow rate and the flow-pressure energy rate, Eqs. (2-90) and (2-91), gives

$$\dot{U} + \dot{E}_f = \dot{m}c_v T_{ave} + \dot{m}Pv$$
$$= \dot{m}(u_{ave} + Pv)$$
$$= \dot{m}h_{ave}$$

where $h = u + Pv$. This becomes the enthalpy-flow rate $\dot{H}$, which for a liquid is given by

$$\dot{H} = \dot{m}\left(cT_{ave} + \frac{P}{\rho}\right) \qquad (2\text{-}92)$$

and for a gas is given by

$$\dot{H} = \dot{m}c_p T_{ave} = A\rho c_p V_{ave} T_{ave} \qquad (2\text{-}93)$$

The flow in a duct or other mechanical device can be analyzed by utilizing a control volume, which is a region fixed in space, with fluid and energy flows across its boundaries, selected to enable a thermodynamic analysis to be conveniently performed. We shall limit our consideration to control volumes which have one fluid stream flowing in and one stream flowing out. Also, we shall consider only steady flow, which requires that all flow rates, property values, and amount of fluid within the control volume be constant with time. These assumptions will give equations adequate for most air pollution control work. Such a control volume is illustrated in Fig. 2-28, which shows the inlet stream at state 1 and the exit stream at state 2, as well as heat and work transfers across the control surface. Heat is shown as positive if transferred to the control volume; work is positive if done by the control volume. These principles are in accord with the usual thermodynamic convention. However, where the direction of heat or work transfer is clear from the context of the analysis, these values are often treated as positive regardless of the sign convention.

The first law of thermodynamics, as applied to a steady-flow control volume, such as the one shown in Fig. 2-28, states that the rate of energy flow into the control volume equals the rate of energy flow leaving it. This law can be stated mathematically as

$$\dot{E}_{k_1} + \dot{E}_{p_1} + \dot{U}_1 + \dot{E}_{f_1} + \dot{Q} = \dot{E}_{k_2} + \dot{E}_{p_2} + \dot{U}_2 + \dot{E}_{f_2} + \dot{W} \qquad (2\text{-}94)$$

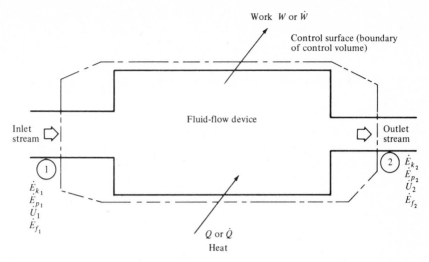

FIGURE 2-28
Control volume for one stream entering and one stream leaving.

Using Eqs. (2-84), (2-87), (2-90), and (2-91), Eq. (2-94) becomes

$$\dot{m}\left(\frac{\alpha V_1{}^2}{2} + gz_1 + c_v T_1 + P_1 v_1\right) + \dot{Q} = \dot{m}\left(\frac{\alpha V_2{}^2}{2} + gz_2 + c_v T_2 + P_2 v_2\right) + \dot{W}$$

$$(2\text{-}95)$$

As an approximation, let us assume that $\alpha = 1.0$, and also we replace $c_v T + Pv$ by enthalpy h. Then Eq. (2-95) becomes

$$\dot{m}\left(h_1 + \frac{V_1{}^2}{2} + gz_1\right) + \dot{Q} = \dot{m}\left(h_2 + \frac{V_2{}^2}{2} + gz_2\right) + \dot{W} \qquad (2\text{-}96)$$

If we divide by $\dot{m}$ and let $\dot{Q}/\dot{m} = Q/m$ and $\dot{W}/\dot{m} = W/m$, then

$$h_1 + \frac{V_1{}^2}{2} + gz_1 + \frac{Q}{m} = h_2 + \frac{V_2{}^2}{2} + gz_2 + \frac{W}{m} \qquad (2\text{-}97)$$

Returning to Eq. (2-95), let us divide by $\dot{m}$ and write it in the following way:

$$\frac{V_1{}^2}{2} + gz_1 + \frac{P_1}{\rho_1} = \frac{V_2{}^2}{2} + gz_2 + \frac{P_2}{\rho_2} + \frac{W}{m} + \left[-\frac{Q}{m} + c_v(T_2 - T_1)\right] \qquad (2\text{-}98)$$

Divide by g and note that $w = mg$ and $\gamma = \rho g$. Then

$$\frac{V_1{}^2}{2g} + z_1 + \frac{P_1}{\gamma_1} = \frac{V_2{}^2}{2g} + z_2 + \frac{P_2}{\gamma_2} + \frac{W}{w} + \left[-\frac{Q}{w} + \frac{c_v}{g}(T_2 - T_1)\right] \qquad (2\text{-}99)$$

We note that each term in Eq. (2-99) has units of length, that is, meters. Each term is referred to as a *head;* the velocity term is the velocity head, the elevation term the elevation head, and the pressure term the pressure head. The terms in brackets are collectively referred to as *head lost*, with the work term as the *head added*. Replacing the terms in brackets by h_L, we have*

$$\frac{V_1^2}{2g} + z_1 + \frac{P_1}{\gamma} = \frac{V_2^2}{2g} + z_2 + \frac{P_2}{\gamma} + \frac{W}{w} + h_{L_{1-2}} \qquad (2\text{-}100)$$

For flow processes in which heating or cooling is not a factor, the head-loss term represents conversion of mechanical energy to thermal energy. In Eqs. (2-95) to (2-100), the velocities V_1 and V_2 are the average velocities over the respective cross sections 1 and 2, the elevations z_1 and z_2 are those of the centroids of the sections, and the pressures are measured at the centroids of the sections.

The head-loss term in Eq. (2-100) can be evaluated for flow in a duct in terms of the friction factor by comparing Eq. (2-100) with Eq. (2-46). In Eq. (2-100) we omit the work term and set $V_1 = V_2$ and $z_1 = z_2$; the same conditions are obtained as apply to Eq. (2-46). Then, upon equating these two equations,

$$h_{L_{1-2}} = f \frac{L}{D} \frac{V_{\text{ave}}^2}{2g} \qquad (2\text{-}101)$$

On occasion, we shall need to compute the energy of the fluid flowing past a section or the power expended in pumping a fluid through a piece of machinery. The energy flowing past a section, per unit weight of fluid, is given by the left side of Eq. (2-99). To compute the power used in pumping a fluid through a device, Eq. (2-100) may be used. Usually, the head-loss term may be neglected, so that

$$\dot{W} = \dot{m} \left[\frac{V_1^2 - V_2^2}{2} + g(z_1 - z_2) + \frac{P_1}{\rho} - \frac{P_2}{\rho} \right] \qquad (2\text{-}102)$$

If the head-loss factor is important, as for a long pipe connected to a pump, the equation becomes

$$\dot{W} = \dot{m} \left[\frac{V_1^2 - V_2^2}{2} + g(z_1 - z_2) + \left(\frac{P_1}{\rho} - \frac{P_2}{\rho} \right) - gh_{L_{1-2}} \right] \qquad (2\text{-}103)$$

In Eqs. (2-102) and (2-103), if the work is done by the device, as for an engine or turbine, it will be positive; if the work is done on the device, as for a pump, blower, or compressor, it will be negative.

For heating or cooling a fluid in a duct without work transfer, Eq. (2-96) gives

$$\dot{Q} = \dot{m} \left[(h_2 - h_1) + \frac{V_2^2 - V_1^2}{2} + g(z_2 - z_1) \right] \qquad (2\text{-}104)$$

---

* The specific weight and density are made constant in Eqs. (2-100), (2-102), and (2-103), which restricts these equations to incompressible flow; the effect of compressibility is included in the definition of the head loss h_L.

For the case where changes in kinetic and potential energies are negligible, this equation becomes

$$\dot{Q} = \dot{m}(h_2 - h_1) \qquad (2\text{-}105)$$

For a gas which can be idealized as a perfect gas, $h_2 - h_1 = c_p(T_2 - T_1)$, which leads to

$$\dot{Q} = \dot{m}c_p(T_2 - T_1) \qquad (2\text{-}106)$$

For a liquid, the change in enthalpy is given by $c(T_2 - T_1) + (P_2 - P_1)/\rho$. In this case, the heat-transfer rate becomes

$$\dot{Q} = \dot{m}\left[c(T_2 - T_1) + \frac{P_2 - P_1}{\rho}\right] \qquad (2\text{-}107)$$

In most cases of heating or cooling, the pressure difference is negligible compared with the effect of the temperature change, so that Eq. (2-107) becomes

$$\dot{Q} = \dot{m}c(T_2 - T_1) \qquad (2\text{-}108)$$

For heating or cooling a fluid within a duct, the relations among the fluid temperature, flow parameters, duct-wall temperature, and duct length are important. For transfer of heat between a wall and a fluid flowing past it, a film coefficient h is defined such that

$$\dot{Q} = hA(T_w - T_{mb}) \qquad (2\text{-}109)$$

in which $\dot{Q}$ is the total heat-transfer rate to the fluid, A is the area of the wall surface in contact with the fluid, T_w is the wall temperature, and T_{mb} is the average temperature of the fluid. Let us consider a duct of length L, as shown in Fig. 2-29. The mean fluid temperature T_{mb}, which is referred to as the *mean-bulk temperature*, is defined as the average of the fluid temperatures entering and leaving the duct. Thus,

$$T_{mb} = \frac{T_1 + T_2}{2} \qquad (2\text{-}110)$$

Equations (2-109) and (2-110) can be used in a case of constant wall temperature if the temperature rise of the fluid $T_2 - T_1$ is small compared with the temperature difference between the wall and the fluid $T_w - T_{mb}$. If this is not the case, then a logarithmic mean temperature difference is a better approximation, given by

$$T_{mw} - T_{mb} = \frac{(T_{w_2} - T_2) - (T_{w_1} - T_1)}{\ln\left[(T_{w_2} - T_2)/(T_{w_1} - T_1)\right]} \qquad (2\text{-}111)$$

and the equation for the heat-transfer rate becomes

$$\dot{Q} = hA(T_{mw} - T_{mb}) \qquad (2\text{-}112)$$

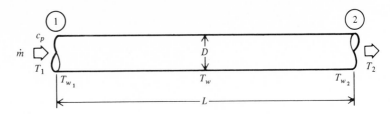

FIGURE 2-29
Heating or cooling of a fluid in a duct.

Determination of the film coefficient h is not easy in general; textbooks and treatises have been written on this subject. For laminar flow in a long duct having constant wall temperature, the film coefficient is given by

$$h = 3.658 \frac{k}{D} \qquad (2\text{-}113)$$

in which D is the hydraulic diameter of the duct and k is the thermal conductivity of the fluid. This equation is derived in Ref. [12], as is the analogous equation for the case of a long duct with constant heat-transfer rate per unit length, given by

$$h = 4.364 \frac{k}{D} \qquad (2\text{-}114)$$

For other conditions involving laminar flow, see the heat-transfer literature.

For turbulent flow in a duct, an empirical equation is recommended. The following equation, taken from Ref. [13], is suggested:

$$h = 0.023 \frac{k}{D} \, \text{Re}^{0.8} \, \text{Pr}^{1/3} \left(\frac{\mu_{mb}}{\mu_w} \right)^{0.14} \qquad (2\text{-}115)$$

where the Prandtl number Pr is defined by

$$\text{Pr} = \frac{c_p \mu}{k} \qquad (2\text{-}116)$$

Equation (2-115) is valid, to a fair degree of approximation, for Reynolds numbers greater than 10,000, for Prandtl numbers in the range 0.7 to 17,000, and for a length/diameter ratio $L/D > 60$. The viscosity μ_{mb} is the viscosity of the fluid measured at the mean-bulk temperature, while μ_w is measured at the wall temperature.

2-7   FLUID-FLOW MEASUREMENT

One of the more important aspects of the study of fluid flow in ducts is the determination of the flow rate. In an actual installation or in an experimental

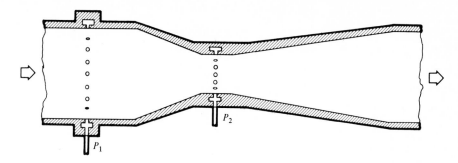

FIGURE 2-30
Venturi meter.

apparatus, this is accomplished by a flowmeter, which may be one of many types. In this section, we shall discuss a number of flowmeters which are useful in air pollution control studies. A good reference on this subject, from which greater detail may be obtained, is Holman [14].

The first category to be considered here consists of the displacement meters. Here a definite volume of the fluid or possibly a definite mass is allowed to pass through the instrument during each cycle of its operation. Some examples of this class of flowmeters include the nutating-disk meter, rotary-vane flowmeter (a wet version is used as a gas meter), lobed-impeller flowmeter, and piston flowmeter. The weighing tank and tipped-bucket flowmeters fall into this category, also, but can be used only for liquids. For a description of the mode of operation of certain of these devices, see Ref. [14]. Since these devices give an output directly in volume or mass flow, we shall not discuss the details of their operation here. At worst, it may be necessary to convert a mass taken over a measured time interval into an average flow rate, and it may be necessary to calibrate the instrument to obtain reliable readings.

The class of flowmeters most widely used is the obstruction meter. It includes principally the venturi meter, orifice flowmeter, and flow nozzle. Basically, the meter presents an obstacle to the flow, causing the velocity to increase and the pressure to decrease. Pressure taps are then inserted at points of different pressure, and the pressure difference between these two taps is measured and related to the flow rate. The venturi meter is illustrated in Fig. 2-30, the flow nozzle in Fig. 2-31, and the orifice meter in Fig. 2-32.

Let us first consider incompressible flow through an obstruction meter of the types indicated, neglecting friction during the analysis but adding a correction factor at the end to account for frictional effects. After neglecting the elevation heads, work, and head loss, Eq. (2-100) provides the basis for the analysis, giving

$$\frac{V_1^2}{2g} + \frac{P_1}{\gamma} = \frac{V_2^2}{2g} + \frac{P_2}{\gamma} \qquad (2\text{-}117)$$

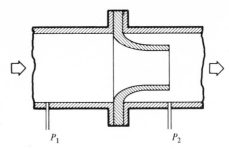

FIGURE 2-31
Flow nozzle.

P_1 P_2

Solving this equation for the velocity in the smallest portion of the meter V_2, using also the continuity equation $A_1 V_1 = A_2 V_2$, we have

$$V_2 = \frac{\sqrt{2(P_1 - P_2)/\rho}}{\sqrt{1 - (A_2/A_1)^2}} \qquad (2\text{-}118)$$

If we replace the area ratio by the ratio of diameters, we obtain

$$Q = \frac{A_2\sqrt{2(P_1 - P_2)/\rho}}{\sqrt{1 - (D_2/D_1)^4}} \qquad (2\text{-}119)$$

Next, we apply a discharge coefficient, defined as $c_d = Q_{\text{actual}}/Q_{\text{ideal}}$, where Q_{ideal} is given by Eq. (2-119), obtaining

$$Q = \frac{c_d A_2\sqrt{2(P_1 - P_2)/\rho}}{\sqrt{1 - (D_2/D_1)^4}} \qquad (2\text{-}120)$$

Equation (2-120) is valid for liquids and for gases if the pressure difference $P_1 - P_2$ is less than 10 percent of the inlet pressure P_1.

For the flow of gases where the pressure drop is too high to use the incompressible flow formula satisfactorily, the effect of compressibility must be included. We shall derive such a formula for the case of perfect gases with constant specific heats. Neglecting heat and work transfers and changes in potential energies, Eq. (2-97) gives

$$c_p T_1 + \frac{V_1^2}{2} = c_p T_2 + \frac{V_2^2}{2} \qquad (2\text{-}121)$$

We also use the perfect-gas law $P = \rho R T$, the relation $c_p = kR/(k - 1)$, the continuity equation $\dot{m} = A_1 \rho_1 V_1 = A_2 \rho_2 V_2$, and the following two equations derivable for an isentropic process of a perfect gas:

$$\frac{T_2}{T_1} = \left(\frac{P_2}{P_1}\right)^{(k-1)/k} \qquad (2\text{-}122)$$

$$\frac{\rho_2}{\rho_1} = \left(\frac{P_2}{P_1}\right)^{1/k} \qquad (2\text{-}123)$$

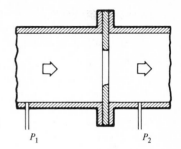

FIGURE 2-32
Orifice meter. P_1 P_2

Using these equations and solving for $V_2{}^2$ gives

$$V_2{}^2 = \frac{2kRT_1}{k-1}\frac{1-(P_2/P_1)^{(k-1)/k}}{1-(A_2/A_1)^2(P_2/P_1)^{2/k}} \qquad (2\text{-}124)$$

Upon further manipulation and solving for the mass flow rate $\dot{m}$,

$$\dot{m} = cA_2\sqrt{\frac{2k}{k-1}}\frac{P_1}{\sqrt{RT_1}}\frac{(P_2/P_1)^{1/k}\sqrt{1-(P_2/P_1)^{(k-1)/k}}}{\sqrt{1-(D_2/D_1)^4(P_2/P_1)^{2/k}}} \qquad (2\text{-}125)$$

where the coefficient c has been included to account for the effect of friction upon the flow. Frequently, the effect of compressibility is included in a factor Y, such that the mass flow rate can be written as

$$\dot{m} = YKA_2\sqrt{2\rho_1(P_1-P_2)} \qquad (2\text{-}126)$$

in which the factor K includes the coefficient c in Eq. (2-125) and also the effect of the area ratio A_2/A_1. For more specific details regarding the design, installation, and operation of these and other types of meters, it is suggested that Ref. [15] be consulted.

A device closely related to the obstruction meters, considered previously, yet used in a somewhat different way is the sonic nozzle. It is used only with compressible fluids. In the flow of such a fluid, the velocity at the throat or minimum section is limited to the local sonic velocity; and when this value is obtained, the flow through the converging portion of the device cannot be altered except by reducing the throat velocity. Thus the mass flow rate is fixed for the device and depends only on the geometry of the device. To achieve the sonic condition at the throat, the exit pressure must be sufficiently low so that the sonic nozzle requires a large pressure drop in its operation. The pressure required at the throat to achieve sonic velocity there is obtained from Eq. (2-124) by setting the velocity equal to the sonic velocity, given by Eq. (2-7), and using Eq. (2-122). After simplification, this equation becomes

$$\frac{P_2}{P_1} = \left(\frac{2}{k+1}\right)^{k/(k-1)} \qquad (2\text{-}127)$$

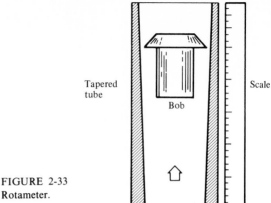

Tapered tube

Bob

Scale

FIGURE 2-33
Rotameter.

and, substituting into Eq. (2-125), the mass flow rate becomes

$$\dot{m} = cA_2 \frac{P_1}{\sqrt{RT_1}} \left(\frac{2}{k+1}\right)^{1/(k-1)} \sqrt{\frac{2k}{k+1}} \qquad (2\text{-}128)$$

In deriving Eqs. (2-127) and (2-128), the inlet velocity effect, as manifested by the area ratio A_2/A_1 or the diameter ratio D_2/D_1, is neglected. If a nozzle or orifice is used, such as in Fig. 2-31 or 2-32, the pressure P_2, as calculated by Eq. (2-127), must be maintained downstream of the nozzle to obtain full flow; that is, the pressure must not exceed the value P_2. If a venturi is used, such as in Fig. 2-30, the downstream pressure can be much nearer to the upstream pressure and still maintain sonic flow in the throat. Thus a venturi can be used with much less pressure loss.

Sonic nozzles can be used for very precise measurements, and by using a variable throat for the nozzle, flow measurements can be made over a continuous range. Sonic nozzles are most useful for controlling the rate of flow, however. For example, a sampling tube can be made to draw off a fixed rate of flow from the source by passing the flow through a sonic nozzle, which requires that a low pressure be maintained in the probe downstream of the nozzle.

The next category of flowmeters to be considered, the drag-effect meters, have a mode of operation that is based on the drag of an object in the stream. The rotameter, shown in Fig. 2-33, is one such meter. The area of flow between the bob and the tube increases as the bob moves upward due to the taper in the tube. This increase in area of flow allows a greater flow rate for the same drag force, so that at higher flow rates the bob must ride higher in the tube. The height of the bob is then a measure of the flow rate, and with proper design of the bob, the flow rate is a linear function of the height; also, small variations of the fluid density will not change the calibration. Rotameters are usually calibrated when supplied for a specific fluid, so that in practice they are direct-reading instruments.

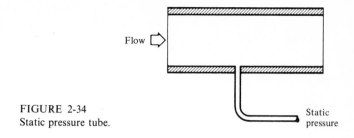

FIGURE 2-34
Static pressure tube.

The turbine flowmeter is also categorized as a drag-effect meter. It operates by indicating the rate at which a turbine wheel rotates when placed in the flow stream. It, too, is a direct-reading instrument subject to calibration. The propeller-type anemometer works in the same way, the main difference being that only a small portion of the total flow passes through the propeller of the anemometer, whereas all the flow passes through the turbine wheel in a turbine meter. The vane-type anemometer is also a drag-effect meter, operating on the principle that a vane is deflected when a stream of fluid impinges on it. The amount of the deflection is a measure of the flow velocity and hence of the flow rate.

Velocity probes are of importance in measuring flow rates through ducts and passages. The propeller-type anemometer and the deflecting-vane type anemometer may often be used to measure velocities at particular points in a flow, though the propeller type, particularly, covers too large an area to measure point velocities accurately for many cases of duct flow. Two other types of velocity probes are the pitot-static tube and the hot-wire anemometer. These devices can be inserted into a duct or other flow passage to measure velocities at several points in the flow. From such a velocity traverse, the flow rate and average velocity can be obtained.

A pitot-static tube consists of a pitot tube and a static tube located in a single probe. The static tube consists of an opening at right angles to the flow direction. This measures the static pressure of the stream at the location of the opening, that is, the ordinary pressure of the fluid. The pitot tube, on the other hand, is an opening pointed upstream so that the axis of the opening is parallel to the flow direction. The pressure inside the pitot tube is greater than that in the stream owing to the effect of impact of the stream on the stationary fluid in the opening. Figure 2-34 shows a static tube, Fig. 2-35 shows a pitot tube, and Fig. 2-36 shows a pitot-static combination. Figure 2-36 also shows a manometer attached to the pitot and static legs of the tube. Such a manometer reads the dynamic pressure, which is a function of the flow velocity. The output of the static tube is termed the *static pressure* and that of the pitot tube is called the *total pressure*, so that the *dynamic pressure* is the difference between the two.

Next, we consider how the velocity can be computed from the measured dynamic pressure from a pitot-static tube. The process involved occurs in the

Flow

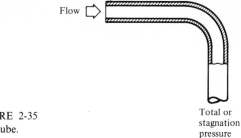

FIGURE 2-35
Pitot tube.

Total or
stagnation
pressure

region just ahead of the impact or pitot tube. Equation (2-100) may be applied in this region. We let point 1 be some distance upstream from the pitot opening and point 2 be just inside the opening. After equilibrium is reached, no flow will occur in the tube, so that the pressure at point 2 is the same as that read on the manometer. Equation (2-100) reduces to

$$\frac{V_1{}^2}{2g} + \frac{P_1}{\gamma} = \frac{P_2}{\gamma} \qquad (2\text{-}129)$$

Solving for V_1 gives

$$V_1 = \sqrt{\frac{2(P_2 - P_1)}{\rho}} \qquad (2\text{-}130)$$

Here, $P_2 - P_1$ is the dynamic pressure read by the manometer in Fig. 2-36. In the derivation of Eq. (2-130), the fluid is assumed to be incompressible, which limits the flow to low Mach numbers. For higher subsonic Mach numbers, the proper equation is

$$\frac{P_2}{P_1} = \left(1 + \frac{k-1}{2} M_1{}^2\right)^{k/(k-1)} \qquad (2\text{-}131)$$

which can be solved for the Mach number. The velocity can then be found if the temperature is known, so that the sonic velocity can be computed from Eq. (2-7).

A portable velocity probe for use in ducts and at air grills is shown in Fig. 2-37. The measuring principle is based on the rotameter. A probe is attached to the inlet opening on the upper left corner; the air leaves the instrument by the center opening. A small white plastic ball serves as the rotameter bob and indicates the velocity at the sampling point, which is related to the flow rate through the rotameter. The instrument has two ranges: a small hole in the side of the outlet tube allows the air to escape if the outlet opening is closed by the finger; the high range is read if the outlet opening at the top center is covered over. This instrument, which is quite economical, is described in Ref. [16].

The hot-wire anemometer and its companion, the hot-film anemometer, are very useful in laboratory situations for measuring velocity at a point, especially

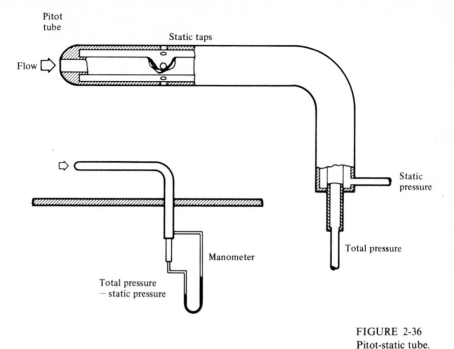

FIGURE 2-36
Pitot-static tube.

where the velocity is changing very rapidly with either distance or with time. It is useful in measuring turbulent fluctuations, for example. Some of the more rugged and consequently less sensitive versions of the hot-wire anemometer are also useful in industrial and field situations. In the hot-wire anemometer, a very fine wire is heated by the flow of an electric current through it. The wire is located in the fluid stream, whose velocity is being measured, and is cooled by heat transfer to the fluid stream. The wire soon reaches an equilibrium temperature such that the heat-transfer rate by convection from the wire to the stream exactly equals the heat-dissipation rate inside the wire due to resistance heating. The velocity measurement can be made either by measuring the wire temperature, which can be done by measuring the wire resistance and relating this to temperature, or by maintaining the wire temperature at a fixed value and measuring the electric current flowing through the wire. Either way, the measured quantity is a function of the temperature of the fluid, which must be measured separately, the fluid velocity, and the characteristics of the instrument. Since the characteristics of the instrument can be eliminated as a factor by calibration, the fluid velocity can then be determined.

The hot-film anemometer is similar to the hot-wire anemometer except that a very thin film of electrically conducting material is plated onto a nonconductive substrate, the film serving the same purpose as the wire. The film has the advantages of greater physical strength and quicker response, and also it can be

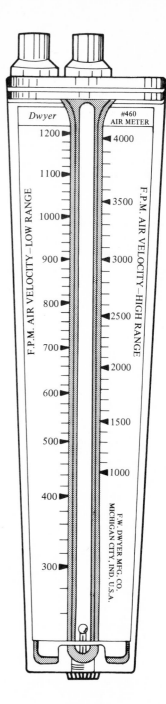

FIGURE 2-37
Portable anemometer using the rotameter
principle. (*From F. W. Dwyer Mfg. Co.,*
Michigan City, Ind. Bull. No. H-9, 1970.
Used by permission.)

used for measuring higher-temperature gases since the substrate can be cooled internally by flow of water or other cooling fluid. Quite complicated bridge circuits are available to produce the necessary readout and to compensate for the inherent nonlinearities of the system. These are described briefly in Ref. [17]. It is possible to arrange two or three sensors on mutually perpendicular axes to measure two- and three-dimensional velocity fluctuations; Ref. [17] also discusses how this can be accomplished.

Flow visualization techniques are important in studying flows in complex configurations or in studying turbulent effects. The simple shadowgraph is useful in locating turbulent regions. The Schlieren apparatus and the Mach-Zehnder interferometer are very useful in quantitatively studying flow patterns. These devices are discussed in Ref. [14]. Smoke methods are commonly used, both in the laboratory and in the field, to show flow patterns and to identify turbulent regions and regions of nonuniform flow. Cigarette smoke can be used, or titanium tetrachloride is frequently used. The latter reacts with moisture in the air to form titanium oxide and hydrochloric acid. Smoke guns are available using this chemical, with which smoke can be injected at any desired point in a flow field.

REFERENCES

1 Reid, R. C., and T. K. Sherwood: "The Properties of Gases and Liquids, Their Estimation and Correlation," 2d ed., pp. 420–421, McGraw-Hill Book Company, New York, 1966.
2 Wallis, G. B.: "One-Dimensional Two-Phase Flow," chap. 2, McGraw-Hill Book Company, New York, 1969.
3 Taylor, G. I.: The Viscosity of a Fluid Containing Small Drops of Another Fluid, *Proc. R. Soc. London, Ser. A*, vol. 138, pp. 41–48, 1932.
4 Knudsen, J. G., and D. L. Katz: "Fluid Dynamics and Heat Transfer," p. 152, McGraw-Hill Book Company, New York, 1958.
5 Moody, L. F.: Friction Factors for Pipe Flow, *Trans. ASME*, vol. 66, pp. 671–84, 1944.
6 Giles, R. V.: "Theory and Problems of Fluid Mechanics and Hydraulics," Schaum's Outline Series, McGraw-Hill Book Company, New York, 1962.
7 Jennings, B. H.: "Environmental Engineering, Analysis and Practice," p. 424, International Textbook Company, Scranton, Pa., 1970.
8 Parker, J. D., J. H. Boggs, and E. F. Blick: "Introduction to Fluid Mechanics and Heat Transfer," pp. 208–211, Addison-Wesley Publishing Company, Inc., Reading, Mass., 1969.
9 Kays, W. M.: "Convective Heat and Mass Transfer," pp. 90–92, 95–96, McGraw-Hill Book Company, New York, 1966.
10 Schlichting, H.: "Boundary Layer Theory," 4th ed., McGraw-Hill Book Company, New York, 1960.
11 Lamb, H.: "Hydrodynamics," 6th ed., pp. 610–616, Dover Publications, Inc., New York, 1945.

12 Kays, W. M.: "Convective Heat and Mass Transfer," pp. 107–110, McGraw-Hill Book Company, New York, 1966.

13 Parker, J. D., J. H. Boggs, and E. F. Blick: "Introduction to Fluid Mechanics and Heat Transfer," p. 245, Addison-Wesley Publishing Company, Inc., Reading, Mass., 1969.

14 Holman, J. P.: "Experimental Methods for Engineers," chap. 7, McGraw-Hill Book Company, New York, 1966.

15 Bean, H. S., ed.: "Fluid Meters, Their Theory and Application," 6th ed., American Society of Mechanical Engineers, New York, 1971.

16 F. W. Dwyer Mfg. Co., Michigan City, Ind., *Bull.* H-9, 1970.

17 Hot Film and Hot Wire Anemometry, *Bull.* TB15, Thermo-Systems Inc., St. Paul, Minn.

PROBLEMS

2-1 Derive Eq. (2-22) from Eq. (2-21), making the assumption indicated in the text.

2-2 Pure air flows through a duct at a pressure of 1.0 atm and a temperature of 52°C. Determine its density, specific heats, ratio of specific heats, and dynamic and kinematic viscosities.

2-3 Repeat Prob. 2-2 if the pressure is 80,000 N/m^2 and the temperature is 52°C.

2-4 Repeat Prob. 2-2 if the pressure is 14.7 psia and the temperature is 100°F. Express the answers in feet, pound mass, seconds, and British Thermal Units.

2-5 Carbon monoxide is mixed as a pollutant in pure air with a volumetric concentration of 0.001. For 1.0 atm pressure and 20°C temperature, determine the
(*a*) Mass concentration of the carbon monoxide in the mixture
(*b*) Density of the mixture
(*c*) Specific heats c_p and c_v of the mixture and also the ratio of specific heats k
(*d*) Dynamic and kinematic viscosities of the mixture

2-6 A mixture of pure air, methane ($C_v = 0.000025$), and ammonia ($C_v = 0.000045$) exists at 0.1 MN/m^2 pressure and 300 K temperature. Determine the density, specific heats k, and dynamic and kinematic viscosities of the mixture. What is C_v for the air?

2-7 A mixture of pure air and a series of gaseous pollutants exists at 1.0 atm pressure and 20°C temperature. The pollutants and their volumetric concentrations are:

Sulfur dioxide	400×10^{-6}
Hydrogen sulfide	65×10^{-6}
Ethyl acetate	175×10^{-6}
Chlorine	25×10^{-6}
Methyl mercaptan	660×10^{-6}

Determine ρ, c_p, c_v, k, μ, and v for this mixture.

2-8 A mixture of pure air and a solid particulate pollutant exists at 1.0 atm pressure and 20°C temperature. The particulate has a density of 2000 kg/m^3 and a specific heat of 600 $J/kg \cdot K$. The mass concentration of the particulate is 0.00002. Determine the
(*a*) Volumetric concentration of the particulate substance
(*b*) Density of the mixture
(*c*) Specific heats of the mixture
(*d*) Dynamic and kinematic viscosities of the mixture

2-9   An airstream at a pressure of 0.101 MN/m^2 and temperature of 300°K contains sulfur dioxide with $C_v = 0.0005$ and a particulate with $C_m = 0.0001$. The particulate density and specific heat are 3500 kg/m^3 and 750 J/kg · K, respectively. Determine ρ, c_p, c_v, k, μ, and v for this mixture.

2-10 Liquid water at 1.0 atm and 20°C is being sprayed into an airstream. Determine the density, dynamic and kinematic viscosities, and surface tension for the water. If the water is in the form of droplets of 0.1 mm diameter, how much surface energy is present per kilogram of water? (Surface energy is equal to the product of surface area times surface tension.)

2-11 Air at 45°C flows through a 20-cm diameter smooth duct at a velocity of 30 m/s. The pressure is 1.0 atm. Determine the Reynolds number, the friction factor, and whether the flow is laminar or turbulent. What will be the pressure drop in 50 m of the duct?

2-12 Water flows at 20°C and atmospheric pressure through a 3.0-cm-diameter pipe 600 m long. The pipe is made of commercial steel. If the water flow rate is 3.5 dm^3/s, determine the average velocity in the pipe, the Reynolds number, and the pressure drop along the pipe.

2-13 A duct is being designed to carry 10 m^3/s of air at 20°C and 300,000 N/m^2 pressure. What size duct, made of commercial steel, is required if the pressure drop over 100 m length is not to exceed 10,000 N/m^2?

2-14 Water flows through a capillary tube 1.0 mm in diameter, made of smooth plastic. For a flow rate of 1.0 cm^3/s, at 20°C, how long must the tube be if the pressure drop along it is to be 1.0 MN/m^2?

2-15 Air at 1.0 atm pressure and 260°C flows through a duct of dimensions 1.0 by 3.0 m. The flow rate is 25 m^3/s, the length of duct is 100 m, and the duct material is galvanized steel. Compute the average velocity in the duct, the Reynolds number, the friction factor, and the pressure drop in the duct.

2-16 Compute the hydraulic diameter and the ratio D_h/D for an annulus for diameter ratios D_o/D_i of 3.0, 1.5, 1.25, 1.10, 1.05, and 1.01.

2-17 Compute the hydraulic diameter and the ratio D_h/D for an annulus with diameter ratio D_o/D_i of 2.0, with the inner circle eccentrically located with respect to the outer circle, and with the centers of the circles displaced by $0.1D_o$.

2-18 Derive an expression for the velocity profile in laminar flow in a concentric annulus in terms of D_o, D_i, the properties of the fluid, and dP/dx. Compute the average velocity and the maximum velocity, and relate velocity to the average and maximum velocities. Find the diameter of a circular duct which gives the same pressure drop for equal flow. Compare this diameter with the hydraulic diameter of the annulus.

2-19 Determine the hydraulic diameter of a duct with a cross section that is an equilateral triangle 1.0 m on each side.

2-20 Determine the pressure drop in a 90° elbow in a rectangular duct having dimensions of 15 by 22 cm if air is flowing at a velocity of 15 m/s at standard atmospheric pressure and temperature.

2-21 Air at standard conditions flows past a flat plate with a free-stream velocity of 75 m/s. At what point along the plate will transition from laminar to turbulent boundary-layer flow occur? What will be the boundary-layer thickness at this point? What will be the boundary-layer thickness at a point 1.5 m down the plate from the leading edge?

2-22 Derive an expression for the boundary-layer thickness at the point of transition for flow over a flat plate in terms of kinematic viscosity and free-stream velocity. Use a transition Reynolds number of 300,000.

2-23 Consider the boundary-layer development on a sphere, as shown in Fig. 2-38. Relate R to x for use in Eq. (2-63). Assume the velocity V_∞ is given in terms of the angle θ by $3U(\sin \theta)/2$. Relate V_∞ to x. Evaluate δ as a function of x and of θ. What is the value of δ when $x = 0$?

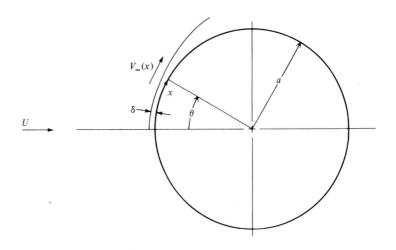

FIGURE 2-38

2-24 Derive Eq. (A1-7) (Appendix 1) from Eqs. (A1-5) and (A1-6).

2-25 Derive Eq. (A1-10).

2-26 Show that Eq. (A1-11) is equivalent to Eq. (A1-10).

2-27 Derive Eq. (A1-19).

2-28 Derive Oseen's improved solution [Eq. (2-70)] by replacing u_r and u_θ by the terms $V_\infty \cos \theta + u'_r$ and $-V_\infty \sin \theta + u'_\theta$, where the primed terms are small compared with V_∞. Hence the Navier-Stokes equations can be linearized by neglecting terms involving the product of two primed quantities.

2-29 A sphere of 5 μm diameter moves in a fluid at a velocity of 3.0 cm/s. If the fluid density is 2.0 g/cm^3 and its viscosity is 3.0×10^{-4} kg/m · s, what will be the drag coefficient and the drag force acting on the sphere?

2-30 Derive Eq. (2-71).

2-31 A cylinder 2.0 μm in diameter and 0.5 cm long moves in water at a velocity of 1.5 cm/s. Neglecting end effects, what drag force and drag coefficient can be expected? Assume the temperature of the water is 25°C.

2-32 Compute and plot C_D as a function of Reynolds number using Eqs. (2-73) to (2-75). Compare with the experimental values taken from Fig. 2-25.

2-33 Compute and plot C_D as a function of Re using Eqs. (2-76) to (2-78) and compare with experimental values. Also compute and plot the results of Eq. (2-79) in the range of $10 < \text{Re} < 700$.

2-34 Water at 20°C flows through a circular duct 10 cm in diameter with a volumetric flow rate of 0.1 m³/s. For laminar flow, determine the
(a) Maximum and average velocities
(b) Momentum flow rate past a section and β
(c) Kinetic energy flow rate past a section and α
(d) Reynolds number
Is laminar flow likely in this situation? For what maximum flow rate can laminar flow be expected?

2-35 Assume that the velocity profile for turbulent flow in a circular pipe can be adequately expressed by Eq. (2-60), written as

$$\frac{u}{V_{max}} = \left(1 - \frac{r}{R}\right)^{1/7}$$

Determine the values of α and β for this velocity profile.

2-36 An airstream of 10,000 m³/s at 30°C and 1.0 atm pressure is pumped to a pressure of 1.01 atm. Assuming negligible potential and kinetic energy changes, how much power is required? Assuming a combined motor and compressor efficiency of 0.65, how much electric power is required for the job?

2-37 Water at 20°C flows at 0.2 m³/s through a pipe. A pump increases the pressure of the water from 1.0 MN/m² to 3.0 MN/m². Assuming that no changes in kinetic or potential energies are involved, how much power is imparted to the water?

2-38 Water at 20°C and 5.0 MN/m² expands through a nozzle to 1.0 atm pressure. If the nozzle is horizontal, no heat is transferred, and the inlet kinetic energy is negligible, what exit velocity will be achieved, neglecting losses?

2-39 Water at 20°C flows through a concrete pipe of 1.0 m diameter at a flow rate of 3.0 m³/s. The pipe originates at an elevation of 1000 m above sea level at atmospheric pressure and terminates after a length of 300 km at an elevation of 100 m above sea level, also at atmospheric pressure. How much pumping power, if any, is necessary to produce this flow?

2-40 Air at 25°C and 1.0 atm pressure flows in a duct of 0.5 m diameter at an average velocity of 25 m/s. The duct is 100 m long, and the duct wall is maintained at a temperature of 300°C. At what temperature will the air leave the duct, and how much heat will be transferred per second?

2-41 Water at 60°C is cooled in a pipe to 50°C. The pipe is 20 cm in diameter and 20 m long, and the flow rate is 0.25 m³/s. If the pipe-wall surface temperature is a fixed amount less than the water temperature at each point along the pipe, what must be the temperature difference between the water and the pipe wall at each point?

2-42 Derive Eq. (2-119).

2-43 Derive Eqs. (2-122) and (2-123).

2-44 Derive Eq. (2-125).

2-45 Water at 20°C flows through a venturi meter having an inlet diameter of 4.0 cm and a throat diameter of 2.0 cm. The pressure difference between the entrance and the throat is 20,000 N/m². Determine the flow rate of water in cubic meters per second. Take $c_d = 0.96$.

2-46 Air at 1.0 atm pressure and 30°C temperature flows through an orifice of 0.3 m diameter in a 0.6-m-diameter pipe. The pressure drop across the orifice is 7500 N/m², and the discharge coefficient is taken as 0.90. What flow rate of air is indicated, and what will be its velocity in the pipe?

2-47 Air flows through a venturi meter having throat diameter of 0.1 m located in a pipe of 0.3 m diameter. The air pressure at the entrance is 1.0 MN/m^2, and that at the throat is 0.6 MN/m^2. The air temperature at the entrance is 300°C. For a discharge coefficient c of 0.95, determine the mass flow rate of the air.

2-48 A sonic nozzle is to be designed for use in a sampling tube. A nozzle somewhat similar to the one in Fig. 2-31 is to be used. The flow rate entering the sampling tube is to be 50 cm^3/s, and the diameter of the sampling tube is 2.0 cm. The fluid is air at 1.01 atm pressure and 200°C temperature. What should be the diameter of the nozzle throat, and what pressure must be maintained downstream of the nozzle to achieve sonic flow at the nozzle throat?

2-49 A pitot-static tube inserted into a duct carrying air at 1.0 atm pressure and 25°C temperature indicates a dynamic pressure reading of 15 cm of water. Compute the air velocity at that point.

3

DYNAMICS OF PARTICLES IN FLUIDS

A special branch of dynamics deals with the motion of particles in fluids. In addition to the momentum and inertia of a particle and the gravitational and other body forces acting on it, a drag force acts on the particle due to its motion relative to the fluid surrounding it. This drag force tends to be a rather complicated function of the relative velocity, thus rendering the mathematical treatment of this branch of dynamics a rather formidable task. We are concerned with two basic types of particle motion: In the first case, the particle falls under the action of gravity or some other body force with negligible initial velocity. In the second case, the particle is projected with a relatively high velocity, and its motion is studied under the assumption that gravitational and other body forces are negligible.

3-1 MOTION OF A SPHERICAL PARTICLE INITIALLY AT REST UNDER THE ACTION OF AN EXTERNAL FORCE

In Sec. 2-5 we studied the motion of a fluid around a particle, especially a spherical particle. In that section, the assumptions were made that the flow was steady and that the velocity of the particle was constant. In this section, we deal with the flow around a particle which is accelerating under the action of an external force. The velocity of the particle relative to the fluid is not necessarily constant. We shall assume that the particle is accelerating slowly enough that the results from Sec. 2-5 still apply at any particular particle velocity; that is, the time required for the flow field around the particle to adjust to a change in particle velocity is much less than the time required for the particle velocity to change appreciably. The particle and the forces acting upon it are shown in Fig. 3-1. The x, y, z coordinate system shown is attached to the fluid molecules and moves with the fluid, making a fixed angle with the fluid streamlines. Thus, the particle velocity of concern is the relative velocity between the particle and the fluid. The coordinate s is in the direction of the particle motion.

In the real case, the particle is assumed to follow the gas motion in the absence of the force $\mathbf{F}$, which causes a deflection of the particle in the direction of $\mathbf{F}$. Since the gas motion is not necessarily uniform, the particle will undergo an acceleration due to the gas motion, which is separate from its acceleration due to the force $\mathbf{F}$. One consequence of this is that the x, y, z coordinate system is not an inertial one, which may lead to appreciable error in the analysis. Errors may also be introduced from the assumption mentioned in the preceding paragraph and from the presence of induced turbulence in the fluid stream, which would not be present if the fluid were at rest.

Summing forces acting on the sphere and equating this sum to the rate of change of momentum, in accordance with Newton's second law of motion, gives

$$\mathbf{F} + \mathbf{F}_D = m\frac{d\mathbf{V}}{dt} = \rho_p V \frac{d\mathbf{V}}{dt} \qquad (3\text{-}1)$$

The drag force is given in terms of the projected frontal area A_p, the velocity, and the drag coefficient defined by Eq. (2-67) as

$$\mathbf{F}_D = -C_D \frac{\rho A_p}{2} V\mathbf{V} \qquad (3\text{-}2)$$

When Eq. (3-2) is substituted into Eq. (3-1), the result is

$$\rho_p V \frac{d\mathbf{V}}{dt} + \frac{1}{2} C_D \rho A_p V\mathbf{V} = \mathbf{F} \qquad (3\text{-}3)$$

In the preceding equations $\mathbf{V}$ is the vector velocity while V is the magnitude of this velocity, $V = |\mathbf{V}|$. In terms of the velocity components u, v, and w, and in terms of the applied-force components F_x, F_y, and F_z, the vector equation (3-3) can be written as three scalar equations:

$$\rho_p V \frac{du}{dt} + \frac{1}{2} C_D \rho A_p V u = F_x \qquad (3\text{-}4)$$

$$\rho_p V \frac{dv}{dt} + \frac{1}{2} C_D \rho A_p V v = F_y \qquad (3\text{-}5)$$

$$\rho_p V \frac{dw}{dt} + \frac{1}{2} C_D \rho A_p V w = F_z \qquad (3\text{-}6)$$

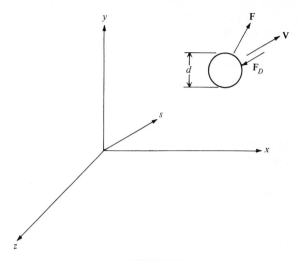

FIGURE 3-1
Acceleration of a particle in a fluid.

Now, let the s coordinate be in the direction of motion of the particle and assume that changes in direction of motion of the particle do not affect the equation of motion in the s direction. Then Eq. (3-4) can be written in terms of V as a function of s:

$$\frac{dV}{dt} = \frac{F}{\rho_p \overline{V}} - \frac{1}{2} \frac{C_D A_p \rho}{\overline{V} \rho_p} V^2 \qquad (3\text{-}7)$$

We may solve this equation for dt, and upon indicating the integration from $t = 0$ where $V = 0$, we obtain

$$t = \frac{\rho_p}{\rho} \int_0^V \frac{dV}{F/\rho \overline{V} - C_D A_p V^2 / 2\overline{V}} \qquad (3\text{-}8)$$

The distance s traveled by the particle from time zero, where s is zero, to time t is given by

$$s = \int_0^t V \, dt = \frac{\rho_p}{\rho} \int_0^V \frac{V \, dV}{F/\rho \overline{V} - C_D A_p V^2 / 2\overline{V}} \qquad (3\text{-}9)$$

If the force F inducing motion of the particle relative to the gas stream is constant, then the particle will approach asymptotically a condition of constant velocity, referred to as the *terminal velocity* V_t. When this condition obtains, dV/dt will be zero in Eq. (3-7), so that V_t may be readily solved for, giving

$$C_{D_t} V_t^2 = \frac{2F}{\rho A_p} \qquad (3\text{-}10)$$

Since C_D is in general a function of the velocity, it is not possible to solve Eq. (3-10) for velocity directly until the variation of C_D with velocity has been obtained. In Eqs. (3-7) to (3-9), V is the particle velocity while $\overline{V}$ is the volume of the particle. These equations have not yet been restricted to spherical particles and are good for any shape of particle as long as the appropriate expression for C_D is employed.

Equations Strictly for Spherical Particles

Let us now specialize the analysis so that it pertains to spherical particles only. For a sphere, the projected area is $A_p = \pi d^2/4$ while the volume of the sphere is $\overline{V} = \pi d^3/6$. Equations (3-4) to (3-6), the differential equations of motion in three dimensions, become

$$\frac{du}{dt} + \frac{3}{4}\frac{\rho C_D}{\rho_p d} V u = \frac{6}{\pi \rho_p d^3} F_x \qquad (3\text{-}11)$$

$$\frac{dv}{dt} + \frac{3}{4}\frac{\rho C_D}{\rho_p d} V v = \frac{6}{\pi \rho_p d^3} F_y \qquad (3\text{-}12)$$

$$\frac{dw}{dt} + \frac{3}{4}\frac{\rho C_D}{\rho_p d} V w = \frac{6}{\pi \rho_p d^3} F_z \qquad (3\text{-}13)$$

whereas, Eqs. (3-7) to (3-10) become

$$\frac{dV}{dt} = \frac{6}{\pi \rho_p d^3} F - \frac{3}{4}\frac{\rho C_D}{\rho_p d} V^2 \qquad (3\text{-}14)$$

$$t = \frac{\rho_p}{\rho}\int_0^V \frac{dV}{6F/\pi\rho d^3 - 3C_D V^2/4d} \qquad (3\text{-}15)$$

$$s = \frac{\rho_p}{\rho}\int_0^V \frac{V\,dV}{6F/\pi\rho d^3 - 3C_D V^2/4d} \qquad (3\text{-}16)$$

$$C_{D_t} V_t^2 = \frac{8F}{\mu\rho d^2} \qquad (3\text{-}17)$$

Next, let us consider the case of laminar flow around a sphere, for which Eq. (2-73) applies if the Reynolds number is less than about 2.0. For this case

$$C_D = \frac{24\mu}{\rho V d} \qquad (3\text{-}18)$$

Equation (3-17) gives the terminal velocity as

$$V_t = \frac{F}{3\pi d\mu} \qquad (3\text{-}19)$$

The differential equation of motion, Eq. (3-14), becomes

$$\frac{dV}{dt} + \frac{18\mu}{\rho_p d^2} V = \frac{6F}{\pi \rho_p d^3} \qquad (3\text{-}20)$$

Equation (3-15) becomes

$$t = \frac{\rho_p}{\rho} \int_0^V \frac{dV}{6F/\pi\rho d^3 - 18\mu V/\rho d^2}$$

which integrates to

$$t = -\frac{\rho_p d^2}{18\mu} \ln\left(1 - \frac{3\pi\mu dV}{F}\right) \qquad (3\text{-}21)$$

Using Eq. (3-19), Eq. (3-21) can be written as

$$t = -\frac{\rho_p d^2}{18\mu} \ln\left(1 - \frac{V}{V_t}\right) \qquad (3\text{-}22)$$

For the distance traveled by the particle, Eq. (3-16) gives

$$s = \frac{\rho_p}{\rho} \int_0^V \frac{V\, dV}{6F/\pi\rho d^3 - 18\mu V/\rho d^2}$$

which leads to

$$s = -\left[\frac{\rho_p d^2 V}{18\mu} + \frac{\rho_p F d}{54\pi\mu^2} \ln\left(1 - \frac{3\pi\mu dV}{F}\right)\right] \qquad (3\text{-}23)$$

or

$$s = -\left[\frac{\rho_p d^2 V}{18\mu} + \frac{\rho_p F d}{54\pi\mu^2} \ln\left(1 - \frac{V}{V_t}\right)\right] \qquad (3\text{-}24)$$

Equation (3-22) can be solved directly for velocity as a function of time, giving

$$V = V_t(1 - e^{-18\mu t/\rho_p d^2}) \qquad (3\text{-}25)$$

Substituting Eq. (3-25) into Eq. (3-24) gives for the distance traveled

$$s = V_t t - \frac{\rho_p d^2 V_t}{18\mu}(1 - e^{-18\mu t/\rho_p d^2}) \qquad (3\text{-}26)$$

Equations (3-18) to (3-26) are valid for Stokes' flow of a spherical particle, subject to the various restrictions mentioned previously; Stokes' flow occurs if

$$\text{Re}_t = \frac{V_t d\rho}{\mu} \le 2.0 \qquad (3\text{-}27)$$

For the most part, work involving particulates can be handled using the formulas just developed owing to the fact that most such particles are small enough so that the terminal velocity is reached at very small Reynolds numbers. In some cases, however, preliminary control measures involve removing very large particles for which the preceding equations do not apply. Also, in studying the action of scrubbers, where relatively large particles of water or other substance are injected into the air pollutant stream, the motion of the injected particles requires equations applicable to larger Reynolds numbers than those we have just presented. Finally, certain research problems may require analysis of the motion of particles appreciably larger than most pollutant particles. For these reasons, it is desirable to include in this section an analysis of the motion of large particles. Such an analysis is based on the drag-coefficient representation of Eqs. (2-76) to (2-78).

The analysis of the motion of large particles is rather lengthy, involving some very messy integration, the details of which we shall omit here. The procedure parallels that used in developing Eqs. (3-19), (3-22), and (3-24). Since three equations are required to present the drag coefficient C_D over the total range of Reynolds numbers, it is necessary to break the integration into three regions. This operation requires the definition of two transition values each of velocity, time, and distance; these values will be presented shortly. We shall summarize the results for the three regions as follows.

For the terminal velocity

$$V_t = \frac{2\mu}{\rho d}\left(\sqrt{9 + \frac{\rho F}{\pi\mu^2}} - 3\right) \qquad\qquad \mathrm{Re}_t \leq 10 \qquad (3\text{-}28)$$

$$V_t = \frac{4.8\mu}{\rho d}\left(\sqrt{447 + \frac{\rho F}{\pi\mu^2}} - 20.4\right) \qquad 10 \leq \mathrm{Re}_t \leq 700 \qquad (3\text{-}29)$$

$$V_t = \frac{2.4}{d}\sqrt{\frac{F}{\rho}} \qquad\qquad \mathrm{Re}_t \geq 700 \qquad (3\text{-}30)$$

The equations for time explicitly as a function of velocity are

$$t = \frac{\rho_p d^2}{6\mu\sqrt{9 + \rho F/\pi\mu^2}}\ln\frac{\rho F/\pi\mu^2 - (3 - \sqrt{9 + \rho F/\pi\mu^2})\rho dV/2\mu}{\rho F/\pi\mu^2 - (3 + \sqrt{9 + \rho F/\pi\mu^2})\rho dV/2\mu} \qquad \mathrm{Re} \leq 10$$

$$(3\text{-}31)$$

$$t = t_1 + \frac{\rho_p d^2}{2.49\mu\sqrt{447 + \rho F/\pi\mu^2}}$$

$$\times \ln\frac{(98 + 4.9\sqrt{447 + \rho F/\pi\mu^2} + \rho dV/\mu)(22 - \sqrt{447 + \rho F/\pi\mu^2})}{(98 - 4.9\sqrt{447 + \rho F/\pi\mu^2} + \rho dV/\mu)(22 + \sqrt{447 + \rho F/\pi\mu^2})}$$

$$10 \leq \mathrm{Re} \leq 70 \qquad (3\text{-}32)$$

$$t = t_2 + \frac{\rho_p d^2}{2.8\sqrt{\rho F/\pi}}\ln\frac{(0.235\sqrt{\rho F/\pi\mu^2} + \rho dV/\mu)(\sqrt{\rho F/\pi\mu^2} - 164)}{(0.235\sqrt{\rho F/\pi\mu^2} - \rho dV/\mu)(\sqrt{\rho F/\pi\mu^2} + 164)} \qquad \mathrm{Re} \geq 700$$

$$(3\text{-}33)$$

The equations for distance explicitly as a function of velocity are

$$s = -\frac{\rho_p d}{3\rho} \ln\left(1 - \frac{3\pi\mu dV}{F} - \frac{\pi d^2 V^2 \rho}{4F}\right)$$

$$-\frac{\rho_p d}{\rho\sqrt{9 + \rho F/\pi\mu^2}} \ln \frac{\rho F/\pi\mu^2 - (3 - \sqrt{9 + \rho F/\pi\mu^2})\rho dV/2\mu}{\rho F/\pi\mu^2 - (3 + \sqrt{9 + \rho F/\pi\mu^2})\rho dV/2\mu} \qquad \mathrm{Re} \leq 10$$

$$(3\text{-}34)$$

$$s = s_1 - \frac{\rho_p d}{0.516\rho} \ln \frac{\rho F/\pi\mu^2 + 36 - 8.4\rho dV/\mu - 0.043(\rho dV/\mu)^2}{\rho F/\pi\mu^2 - 52.5}$$

$$-\frac{39.31\rho_p d}{\rho\sqrt{447 + \rho F/\pi\mu^2}}$$

$$\times \ln \frac{(98 + 4.83\sqrt{447 + \rho F/\pi\mu^2} + \rho dV/\mu)(22.4 - \sqrt{447 + \rho F/\pi\mu^2})}{(98 - 4.83\sqrt{447 + \rho F/\pi\mu^2} + \rho dV/\mu)(22.4 + \sqrt{447 + \rho F/\pi\mu^2})}$$

$$10 \leq \mathrm{Re} \leq 700 \qquad (3\text{-}35)$$

$$s = s_2 - \frac{\rho_p d}{0.66\rho} \ln \frac{\rho F/\pi\mu^2 - 0.055\,(\rho dV/\mu)^2}{\rho F/\pi\mu^2 - 26,950} \qquad \mathrm{Re} \geq 700 \qquad (3\text{-}36)$$

In the preceding equations, Re_t is the Reynolds number based on terminal velocity, t_1 and s_1 are time and distance, respectively, at which Re_t is 10, and t_2 and s_2 are the values at which Re_t equals 700. These values of Re_t are the transition values in Eqs. (2-76) to (2-78). The values of t_1, t_2, s_1, and s_2 are as follows:

$$t_1 = \frac{\rho_p d^2}{6\mu\sqrt{9 + \rho F/\pi\mu^2}} \ln \frac{15 - \rho F/\pi\mu^2 - 5\sqrt{9 + \rho F/\pi\mu^2}}{15 - \rho F/\pi\mu^2 + 5\sqrt{9 + \rho F/\pi\mu^2}} \qquad (3\text{-}37)$$

$$t_2 = t_1 + \frac{\rho_p d^2}{2.49\sqrt{447\mu^2 + \rho F/\pi}} \ln \frac{3136 - \rho F/\pi\mu^2 - 157\sqrt{447 + \rho F/\pi\mu^2}}{3136 - \rho F/\pi\mu^2 + 157\sqrt{447 + \rho F/\pi\mu^2}} \qquad (3\text{-}38)$$

$$s_1 = -\frac{\rho_p d}{3\rho} \ln\left(1 - \frac{55\pi\mu^2}{\rho F}\right)$$

$$-\frac{\rho_p d}{\sqrt{9 + \rho F/\pi\mu^2}} \ln \frac{1.67 - \rho F/9\pi\mu^2 + 0.556\sqrt{9 + \rho F/\pi\mu^2}}{1.67 - \rho F/9\pi\mu^2 - 0.556\sqrt{9 + \rho F/\pi\mu^2}} \qquad (3\text{-}39)$$

$$s_2 = s_1 - \frac{\rho_d d}{0.516\rho} \ln \frac{\rho F/\pi\mu^2 - 26,926}{\rho F/\pi\mu^2 - 52.5}$$

$$-\frac{39.31\rho_p d}{\rho\sqrt{447 + \rho F/\pi\mu^2}} \ln \frac{3253 - \rho F/\pi\mu^2 - 143\sqrt{447 + \rho F/\pi\mu^2}}{3253 - \rho F/\pi\mu^2 + 143\sqrt{447 + \rho F/\pi\mu^2}} \qquad (3\text{-}40)$$

Motion in Standard Air at Low Velocity

If a particle is traveling in air at standard conditions, the equations for small velocity, that is, with Re less than 2.0, become: From Eq. (3-19),

$$V_t = 5766 \frac{F}{d} \qquad (3\text{-}41)$$

From Eq. (3-22),

$$t = -3019\rho_p d^2 \ln\left(1 - \frac{V}{V_t}\right) \qquad (3\text{-}42)$$

From Eq. (3-24),

$$s = -\left[3019\rho_p d^2 V + 174 \times 10^5 \rho_p F d \ln\left(1 - \frac{V}{V_t}\right)\right] \qquad (3\text{-}43)$$

From Eq. (3-25),

$$V = V_t(1 - e^{-3.31 \times 10^{-4} t/d^2 \rho_p}) \qquad (3\text{-}44)$$

From Eq. (3-26),

$$s = V_t t - 3019\, d^2 V_t(1 - e^{-3.31 \times 10^{-4} t/d^2 \rho_p}) \qquad (3\text{-}45)$$

And, finally, the Reynolds number becomes

$$\text{Re} = 64{,}516 V_t d \qquad (3\text{-}46)$$

Motion under the Influence of Gravity

So far, nothing has been said about the force F acting on a particle. Practical possibilities include the force of gravity and the force of electrostatic attraction or repulsion. Various combinations of forces can exist also. Although the electrostatic force cannot be evaluated without detailed knowledge of the specific situation, the force of gravity is known to a high degree of approximation and will be virtually the same in all cases inasmuch as all air pollution work in the foreseeable future will be applicable to the earth's surface.

The equations developed so far in this section can be made applicable to the free fall of a particle in a gravitational field by substituting for force F the value

$$F = mg = \frac{\pi \rho_p d^3 g}{6} \qquad (3\text{-}47)$$

Equation (3-47) applies only to spherical particles. Although this substitution can be made in all the equations in which F appears, we shall perform the substitution here only for the equations for low-velocity flow, Eqs. (3-19), (3-21), and (3-24). These become, respectively,

$$V_t = \frac{\rho_p d^2 g}{18\mu} \tag{3-48}$$

$$t = -\frac{\rho_p d^2}{18\mu} \ln\left(1 - \frac{18\mu V}{\rho_p d^2 g}\right) = -\frac{\rho_p d^2}{18\mu} \ln\left(1 - \frac{V}{V_t}\right) \tag{3-49}$$

$$s = -\left[\frac{\rho_p d^2 V}{18\mu} + \frac{\rho_p{}^2 d^4 g}{324\mu^2} \ln\left(1 - \frac{V}{V_t}\right)\right] \tag{3-50}$$

Motion under the Influence of Gravity in Standard Air

For spheres falling in standard air under the influence of gravitation only, Eqs. (3-47) to (3-50) become

$$F = 5.1348\rho_p d^3 \tag{3-51}$$

$$V_t = 29{,}609\rho_p d^2 \tag{3-52}$$

$$t = -3019\rho_p d^2 \ln\left(1 - \frac{V}{V_t}\right) \tag{3-53}$$

$$s = -\left[3019 d^2 \rho_p V + 8.94 \times 10^7 \rho_p{}^2 d^4 \ln\left(1 - \frac{V}{V_t}\right)\right] \tag{3-54}$$

Equations (3-44) to (3-46) remain the same.

EXAMPLE 3-1 A water droplet with a diameter of 0.8 mm falls through standard air under the influence of gravity. What will be its terminal velocity?

SOLUTION Equation (3-46) gives for the Reynolds number

$$\text{Re}_t = 64{,}516 V_t d = 64{,}516(0.8 \times 10^{-3})V_t = 51.6 V_t$$

Equation (3-51) gives for the force acting on the droplet

$$F = 5.135\rho_p d^3 = 5.135(1000)(0.8 \times 10^{-3})^3 = 2.63 \times 10^{-6} \text{ N}$$

Assume initially that the terminal Reynolds number exceeds 700, for which Eq. (3-30) gives

$$V_t = \frac{2.4}{d}\sqrt{\frac{F}{\rho}} = \frac{2.4}{0.8 \times 10^{-3}}\sqrt{\frac{2.63 \times 10^{-6}}{1.185}} = 4.47 \text{ m/s}$$

The terminal Reynolds number is then computed to be

$$\text{Re}_t = 51.6(4.47) = 231$$

Since this value does not meet the original assumption, assume next that the Reynolds number is between 10 and 700 and use Eq. (3-29):

$$V_t = \frac{4.8}{d}\left(\sqrt{447\mu^2 + \frac{\rho F}{\pi}} - 20.4\mu\right)$$

$$= \frac{4.8}{1.185(0.8 \times 10^{-3})}$$

$$\times \left[\sqrt{447(1.8 \times 10^{-5})^2 + \frac{1.185(2.63 \times 10^{-6})}{\pi}} - 20.4(1.8 \times 10^{-5})\right]$$

$$= 3.51 \text{ m/s} \qquad Ans.$$

The Reynolds number is now computed to be 181; thus the preceding value of V_t is in the correct range.   ////

3-2   CUNNINGHAM'S CORRECTION FACTOR

The analysis of Sec. 3-1 is valid, subject to the restrictions discussed in that section, for particles that are not too small. For very small particles, however, molecular slip can occur, resulting in a lower drag force than that calculated by the formulas developed in Sec. 3-1. A correction for this effect was developed by Cunningham [1] and is treated in this section. For particles moving in air, this correction is significant only if the diameter is of the order of 1.0 μm or less. For this reason, the correction need be applied only to flow at very low Reynolds numbers.

The magnitude of the Cunningham correction factor, as given by Strauss [2], is

$$C = 1 + 2\frac{\lambda}{d}(1.257 + 0.400e^{-0.55d/\lambda}) \qquad (3\text{-}55)$$

in which d is the particle diameter and λ is the mean free path of the gas molecules, given by

$$\lambda = \frac{\mu}{0.499\rho\bar{u}} \qquad (3\text{-}56)$$

In Eq. (3-56), $\bar{u}$ is the mean molecular velocity, given by

$$\bar{u} = \sqrt{\frac{8R_u T}{\pi M}} \qquad (3\text{-}57)$$

The correction factor, which is greater than unity, acts to decrease the resistance to particle motion. From Eqs. (3-2) and (3-18), the drag force for low-velocity flow is

$$F_D = \frac{C_D}{C}\frac{\rho A_p}{2} = \frac{\pi}{8}\frac{C_D}{C}\rho d^2 V^2 = \frac{3\pi\mu V d}{C} \qquad (3\text{-}58)$$

If we replace C_D/C in Eq. (3-58) by C'_D, we have

$$F_D = \frac{\pi}{8} C'_D \rho d^2 V^2 \qquad (3\text{-}59)$$

where, from Eq. (3-18), C'_D is given by

$$C'_D = \frac{24\mu}{\rho V dC} \qquad (3\text{-}60)$$

Replacing C_D by C'_D in Eqs. (3-15) to (3-17), using C'_D as given by Eq. (3-60), and proceeding to develop the equivalent forms of Eqs. (3-19) and (3-21) to (3-26), we obtain the following equations:

$$V_t = \frac{FC}{3\pi\mu d} \qquad (3\text{-}61)$$

$$t = -\frac{\rho_p d^2 C}{18\mu} \ln\left(1 - \frac{3\pi\mu dV}{FC}\right) \qquad (3\text{-}62)$$

$$t = -\frac{\rho_p d^2 C}{18\mu} \ln\left(1 - \frac{V}{V_t}\right) \qquad (3\text{-}63)$$

$$s = -\left[\frac{\rho_p d^2 VC}{18\mu} + \frac{\rho_p FdC^2}{54\pi\mu^2} \ln\left(1 - \frac{3\pi\mu d}{FC} V\right)\right] \qquad (3\text{-}64)$$

$$s = -\left[\frac{\rho_p d^2 VC}{18\mu} + \frac{\rho_p FdC^2}{54\pi\mu^2} \ln\left(1 - \frac{V}{V_t}\right)\right] \qquad (3\text{-}65)$$

$$V = V_t(1 - e^{-18\mu t/\rho_p d^2 C}) \qquad (3\text{-}66)$$

$$s = V_t t - \frac{\rho_p d^2 V_t C}{18\mu}(1 - e^{-18\mu t/\rho_p d^2 C}) \qquad (3\text{-}67)$$

The equation for Reynolds number is unaffected by the Cunningham correction factor and is given as Eq. (3-27).

Motion of the Particle in Standard Air

For the motion of a particle in standard air, Eqs. (3-61) to (3-67) become

$$V_t = 5766 \frac{FC}{d} \qquad (3\text{-}68)$$

$$t = -3019\rho_p d^2 C \ln\left(1 - \frac{V}{V_t}\right) \qquad (3\text{-}69)$$

$$s = -\left[3019\rho_p d^2 VC + 174 \times 10^5 \rho_p FdC^2 \ln\left(1 - \frac{V}{V_t}\right)\right] \qquad (3\text{-}70)$$

$$V = V_t(1 - e^{-3.31 \times 10^{-4} t/\rho_p d^2 C}) \qquad (3\text{-}71)$$

$$s = V_t t - 3019\rho_p d^2 V_t C(1 - e^{-3.31 \times 10^{-4} t/\rho_p d^2 C}) \qquad (3\text{-}72)$$

If the spherical particle falls from rest under the action of gravity, the force F is given by Eq. (3-47). Equations (3-61), (3-63), and (3-65) become

$$V_t = \frac{\rho_p d g C}{18\mu} \tag{3-73}$$

$$t = -\frac{\rho_p d^2 C}{18\mu} \ln\left(1 - \frac{V}{V_t}\right) \tag{3-74}$$

$$s = -\left[\frac{\rho_p d^2 V C}{18\mu} + \frac{\rho_p^2 d^4 g C^2}{324\mu^2} \ln\left(1 - \frac{V}{V_t}\right)\right] \tag{3-75}$$

If the particle falls through standard air under the action of gravity, then Eqs. (3-73) to (3-75) become

$$V_t = 29{,}609\rho_p d^2 C \tag{3-76}$$

$$t = -3019\rho_p d^2 C \ln\left(1 - \frac{V}{V_t}\right) \tag{3-77}$$

$$s = -\left[3019\rho_p d^2 V C + 8.94 \times 10^7 \rho_p^2 d^4 C^2 \ln\left(1 - \frac{V}{V_t}\right)\right] \tag{3-78}$$

For standard air, $\lambda = 6.67 \times 10^{-8}$ m $= 0.0667$ μm. Table 3-1 gives values of the Cunningham correction factor for various particle diameters moving in standard air (298 K, 1.0 atm).

Table 3-1   CUNNINGHAM   CORRECTION FACTOR FOR STANDARD AIR

d, μm	C	d, μm	C
0.001	221.6	0.1	2.911
0.002	111.1	0.2	1.890
0.003	74.25	0.3	1.574
0.004	55.83	0.4	1.424
0.005	44.78	0.5	1.337
0.006	37.41	0.6	1.280
0.007	32.15	0.7	1.240
0.008	28.20	0.8	1.210
0.009	25.14	0.9	1.186
0.01	22.68	1.0	1.168
0.02	11.65	2.0	1.084
0.03	7.978	3.0	1.056
0.04	6.151	4.0	1.042
0.05	5.060	5.0	1.034
0.06	4.337	6.0	1.028
0.07	3.823	7.0	1.024
0.08	3.441	8.0	1.021
0.09	3.145	9.0	1.019
		10.0	1.017

EXAMPLE 3-2 A spherical particle with a diameter of 0.25 μm and a specific gravity of 2.25 falls in standard air under the action of gravity. Compute its terminal velocity. How much time elapses before the particle reaches 90 percent of its terminal velocity?

SOLUTION For standard air, $\lambda = 0.0667$ μm. Equation (3-55) gives the Cunningham correction factor C. First we calculate $d/\lambda = 3.75$:

$$C = 1 + \frac{2}{3.75} [1.257 + 0.4e^{-0.55(3.75)}] = 1.70$$

Equation (3-76) gives the terminal velocity for a sufficiently small particle falling in standard air:

$$V_t = 29,609(2250)(0.25 \times 10^{-6})^2(1.70) = 7.07 \ \mu\text{m/s} \qquad Ans.$$

The time required to reach 90 percent of this velocity is given by Eq. (3-77):

$$t = -3019(2250)(0.25 \times 10^{-6})^2(1.70) \ln (1 - 0.9) = 1.66 \ \mu\text{s} \qquad Ans. \qquad ////$$

3-3 PROJECTION OF A SPHERICAL PARTICLE IN A FLUID

In many processes related to air pollution control, particles of varying size are projected through air or other fluid with high initial velocities. These particles are then slowed down due to air friction, eventually reaching the terminal settling velocity induced by gravity. If the air is moving and the gravitational settling velocity is negligible, the particles will move with the airstream once the initial projection velocity has been overcome by friction. Examples of such processes include the projection of particles by chipping, grinding, or other forming processes. Projection of particles can also occur in conveying and handling processes of dusts and powders. In certain air pollution control devices, droplets of water or other liquid are injected into an airstream. In studying various such devices and processes, it is necessary to evaluate the trajectory of particles which are projected into a fluid body with significant initial velocity V_0. With regard to formation of particles in industrial processes and projection of particles from conveying processes, the work by Hemeon [3] provides a great deal of discussion and analysis. He refers to the distance traveled by a particle from its initial projection until it stops relative to the gas as the *pulvation distance*.

In this section we shall treat the motion of a particle which is projected into a body of fluid with an initial velocity V_0. The particle is assumed to be acted upon only by the drag force between it and the fluid. The Cunningham correction factor will be neglected since this analysis is useful only for relatively large particles, where the correction would have negligible effect. We shall also neglect the effect of gravity and other forces in this analysis. The motion will be described by Eq. (2-73) for the drag coefficient if the initial velocity is low enough to meet the restriction that $Re_0 < 2.0$; if not, then Eqs. (2-76) to (2-78) will be used. In this analysis, the terminal velocity is zero, although, as it turns out, the time required to reach this velocity is infinite. The distance required to reach the terminal velocity of zero, or the total distance moved by the particle, is finite, as is to be expected.

Equations (3-8) and (3-9) may be used to begin the analysis, where the external force F is set equal to zero. This gives

$$t = -\frac{4}{3}d\frac{\rho_p}{\rho}\int_{V_0}^{V}\frac{dV}{C_D V^2} \qquad (3\text{-}79)$$

$$s = -\frac{4}{3}d\frac{\rho_p}{\rho}\int_{V_0}^{V}\frac{dV}{C_D V} \qquad (3\text{-}80)$$

in which the limitation to spherical particles has been imposed. The analysis is in four sections, according to the value of the initial Reynolds number, $Re_0 = \rho d V_0/\mu$. For $Re_0 < 2.0$, Eq. (2-73) is used, that is

$$C_D = \frac{24\mu}{\rho d V} \qquad Re_0 < 2.0 \qquad (3\text{-}81)$$

Equations (3-79) and (3-80) then lead to the following results:

$$t = \frac{\rho_p d^2}{18\mu}\ln\frac{V_0}{V} \qquad Re_0 < 2.0 \qquad (3\text{-}82)$$

$$s = \frac{\rho_p d^2}{18\mu}(V_0 - V) \qquad Re_0 < 2.0 \qquad (3\text{-}83)$$

The maximum distance reached by the particle when its velocity has been reduced to zero occurs at infinite time and is

$$s_{max} = \frac{\rho_p d^2 V_0}{18\mu} \qquad Re_0 < 2.0 \qquad (3\text{-}84)$$

If the fluid in which the particle travels is standard air, these equations become

$$Re_0 = 64{,}516 d V_0 \qquad Re_0 < 2.0 \qquad (3\text{-}85)$$

$$C_D = \frac{3.73 \times 10^{-4}}{dV} \qquad Re_0 < 2.0 \qquad (3\text{-}86)$$

$$t = 3019\rho_p d^2 \ln\frac{V_0}{V} \qquad Re_0 < 2.0 \qquad (3\text{-}87)$$

$$s = 3019\rho_p d^2(V_0 - V) \qquad Re_0 < 2.0 \qquad (3\text{-}88)$$

$$s_{max} = 3019\rho_p d^2 V_0 \qquad Re_0 < 2.0 \qquad (3\text{-}89)$$

If $Re_0 < 10$, then Eq. (2-76) gives for C_D

$$C_D = 2 + \frac{24\mu}{\rho d V} \qquad Re_0 < 10 \qquad (3\text{-}90)$$

The resulting equations for this case are

$$t = \frac{\rho_p d^2}{18\mu} \ln \frac{V_0}{V} \frac{\rho dV/\mu + 12}{\rho dV_0/\mu + 12} \qquad 0 < \text{Re} < \text{Re}_0 < 10 \qquad (3\text{-}91)$$

$$s = \frac{2}{3} \frac{\rho_p d}{\rho} \ln \frac{\rho dV_0/\mu + 12}{\rho dV/\mu + 12} \qquad 0 < \text{Re} < \text{Re}_0 < 10 \qquad (3\text{-}92)$$

$$s_{\text{max}} = \frac{2}{3} \frac{\rho_p d}{\rho} \ln \left(1 + \frac{\rho dV_0}{12\mu}\right) \qquad 0 < \text{Re} < \text{Re}_0 < 10 \qquad (3\text{-}93)$$

and for motion in standard air, these reduce to

$$t = 3019\rho_p d^2 \ln \frac{V_0}{V} \frac{\rho dV/\mu + 12}{\rho dV_0/\mu + 12} \qquad 0 < \text{Re} < \text{Re}_0 < 10 \qquad (3\text{-}94)$$

$$s = 0.5626\rho_p d \ln \frac{\rho dV_0/\mu + 12}{\rho dV/\mu + 12} \qquad 0 < \text{Re} < \text{Re}_0 < 10 \qquad (3\text{-}95)$$

$$s_{\text{max}} = 0.5626\rho_p d \ln \left(1 + \frac{\rho dV_0}{12\mu}\right) \qquad 0 < \text{Re} < \text{Re}_0 < 10 \qquad (3\text{-}96)$$

If $10 < \text{Re}_0 < 700$, then C_D is given by Eq. (2-77) for $\text{Re} > 10$ and by Eq. (3-90) for $\text{Re} < 10$. Equation (2-77) becomes

$$C_D = 0.344 + \frac{67.3\mu}{\rho dV} - \frac{287.4\mu^2}{\rho^2 d^2 V^2} \qquad 10 < \text{Re} < \text{Re}_0 < 700 \qquad (3\text{-}97)$$

There will be two expressions for t, according as Re is greater than or less than 10. These expressions are

$$t = - \frac{\rho_p d^2}{52.63\mu} \ln \frac{(\rho dV/\mu - 4.18)(\rho dV_0/\mu + 200)}{(\rho dV/\mu + 200)(\rho dV_0/\mu - 4.18)} \qquad 10 < \text{Re} < \text{Re}_0 < 700 \quad (3\text{-}98)$$

$$t = t_2 + \frac{\rho_p d^2}{18\mu} \ln 0.4545\left(1 + \frac{12\mu}{\rho dV}\right) \qquad \text{Re} < 10 \qquad (3\text{-}99)$$

The quantity t_2 appearing in Eq. (3-99) is the value of t which occurs when $\text{Re} = 10$. This value is

$$t_2 = - \frac{\rho_p d^2}{52.63\mu} \ln \frac{\rho dV_0/\mu + 200}{36.08(\rho dV_0/\mu - 4.18)} \qquad (3\text{-}100)$$

There are also two expressions for distance s:

$$s = 1.938 \frac{\rho_p d}{\rho} \ln \frac{(\rho dV_0/\mu)^2 + 195.6\rho dV_0/\mu - 835.5}{(\rho dV/\mu)^2 + 195.6\rho dV/\mu - 835.5}$$

$$+ 1.858 \frac{\rho_p d}{\rho} \ln \frac{(\rho dV/\mu - 4.18)(\rho dV_0/\mu + 200)}{(\rho dV/\mu + 200)(\rho dV_0/\mu - 4.18)} \qquad 10 < \text{Re} < \text{Re}_0 < 700$$

$$(3\text{-}101)$$

$$s = s_2 + \frac{2}{3}\frac{\rho_p d}{\rho} \ln \frac{22}{\rho dV/\mu + 12} \qquad \text{Re} < 10 \qquad (3\text{-}102)$$

The maximum distance traveled is

$$s_{max} = s_2 + 0.4041 \frac{\rho_p d}{\rho} \qquad (3\text{-}103)$$

The value of s_2, the distance traveled by the particle when Re = 10, is

$$s_2 = 1.938 \frac{\rho_p d}{\rho} \ln \frac{(\rho dV_0/\mu)^2 - 195.6\rho dV_0/\mu - 835.5}{1220.5}$$

$$+ 1.858 \frac{\rho_p d}{\rho} \ln \frac{0.02774(\rho dV_0/\mu + 200)}{\rho dV_0/\mu - 4.18} \qquad (3\text{-}104)$$

For motion in standard air, Eqs. (3-98) to (3-104) become, respectively,

$$t = -1032\rho_p d^2 \ln \frac{(dV - 0.0000649)(dV_0 + 0.003105)}{(dV + 0.003105)(dV_0 - 0.0000649)}$$

$$10 < \text{Re} < \text{Re}_0 < 700 \qquad (3\text{-}105)$$

$$t = t_2 - 3019\rho_p d^2 \ln \left(0.4545 + \frac{8.45 \times 10^{-5}}{dV}\right) \qquad \text{Re} < 10 \qquad (3\text{-}106)$$

$$t_2 = -1032\rho_p d^2 \ln \frac{dV_0 + 0.003105}{36.08(dV_0 - 0.0000649)} \qquad (3\text{-}107)$$

$$s = 1.635\rho_p d \ln \frac{(dV_0)^2 + 0.00303dV_0 - 2.01 \times 10^{-7}}{(dV)^2 + 0.00303dV - 2.01 \times 10^{-7}}$$

$$+ 1.568\rho_p d \ln \frac{(dV - 0.0000649)(dV_0 + 0.003105)}{(dV + 0.003105)(dV_0 - 0.0000649)}$$

$$10 < \text{Re} < \text{Re}_0 < 700 \qquad (3\text{-}108)$$

$$s = s_2 + 0.5626\rho_p d \ln \frac{0.000341}{dV + 0.000186} \qquad \text{Re} < 10 \qquad (3\text{-}109)$$

$$s_{max} = s_2 + 0.3410\rho_p d \qquad (3\text{-}110)$$

$$s_2 = 1.635\rho_p d \ln \frac{(dV_0)^2 + 0.00303dV_0 - 2.01 \times 10^{-7}}{2.93 \times 10^{-7}}$$

$$+ 1.568\rho_p d \ln \frac{0.02774(dV_0 + 0.003105)}{dV_0 - 0.0000649} \qquad (3\text{-}111)$$

Finally, for the case where the initial Reynolds number exceeds 700, $C_D = 0.44$, as given by Eq. (2-78). There are three equations for t and s, corresponding to the cases where the Reynolds number is greater than 700, between 700 and 10, and less than 10. There will be two transition values of t and s; these are t_1 and s_1 which occur when Re = 700, and t_2 and s_2 which occur when Re = 10. The expressions and values for t_2 and s_2 are not the same as those listed previously. The equations for time t are

$$t = 3.03 \frac{\rho_p d}{\rho}\left(\frac{1}{V} - \frac{1}{V_0}\right) \qquad 700 < \text{Re} < \text{Re}_0 \qquad (3\text{-}112)$$

$$t = t_1 - \frac{\rho_p d^2}{52.63\mu} \ln \frac{1.293(\rho dV/\mu - 4.18)}{\rho dV/\mu + 200} \qquad 10 < \text{Re} < 700 \qquad (3\text{-}113)$$

$$t = t_2 + \frac{\rho_p d^2}{18\mu} \ln \frac{0.4545(\rho dV/\mu + 12)}{\rho dV/\mu} \qquad \text{Re} < 10 \qquad (3\text{-}114)$$

The values of t_1 and t_2 are

$$t_1 = 3.03 \frac{\rho_p d}{\rho}\left(\frac{\rho d}{700\mu} - \frac{1}{V_0}\right) \qquad (3\text{-}115)$$

$$t_2 = t_1 + 0.06325 \frac{\rho_p d^2}{\mu} \qquad (3\text{-}116)$$

The equations for distance s are

$$s = 3.03 \frac{\rho_p d}{\rho} \ln \frac{V_0}{V} \qquad 700 < \text{Re} < \text{Re}_0 \qquad (3\text{-}117)$$

$$s = s_1 - 1.938 \frac{\rho_p d}{\rho} \ln \frac{(\rho dV/\mu)^2 + 195.6\rho dV/\mu - 835.5}{626,084}$$

$$+ 1.858 \frac{\rho_p d}{\rho} \ln \frac{1.293(\rho dV/\mu - 4.18)}{\rho dV/\mu + 200} \qquad 10 < \text{Re} < 700 \qquad (3\text{-}118)$$

$$s = s_1 + 5.909 \frac{\rho_p d}{\rho} + \frac{2}{3}\frac{\rho_p d}{\rho} \ln \frac{22}{\rho dV/\mu + 12} \qquad \text{Re} < 10 \qquad (3\text{-}119)$$

$$s_{\max} = s_2 + 0.4041 \frac{\rho_p d}{\rho} \qquad (3\text{-}120)$$

The values of s_1 and s_2 are

$$s_1 = 3.03 \frac{\rho_p d}{\rho} \ln \frac{\rho dV_0}{700\mu} \qquad (3\text{-}121)$$

$$s_2 = s_1 + 5.909 \frac{\rho_p d}{\rho} \qquad (3\text{-}122)$$

For motion in standard air, Eqs. (3-112) to (3-122) become, respectively,

$$t = 2.557\rho_p d\left(\frac{1}{V} - \frac{1}{V_0}\right) \qquad 700 < \text{Re} < \text{Re}_0 \tag{3-123}$$

$$t = t_1 - 1032\rho_p d^2 \ln\frac{1.293(dV - 0.0000649)}{dV + 0.003105} \qquad 10 < \text{Re} < 700 \tag{3-124}$$

$$t = t_2 + 3019\rho_p d^2 \ln\frac{0.4545(dV + 1.86 \times 10^{-4})}{dV} \qquad \text{Re} < 10 \tag{3-125}$$

$$t_1 = 2.557\rho_p d\left(92d - \frac{1}{V_0}\right) \tag{3-126}$$

$$t_2 = t_1 + 3438\rho_p d^2 \tag{3-127}$$

$$s = 2.557\rho_p d \ln\frac{V_0}{V} \qquad 700 < \text{Re} < \text{Re}_0 \tag{3-128}$$

$$s = s_1 - 1.635\rho_p d \ln\left(6648d^2V^2 + 20.16dV - 0.00133\right)$$
$$\quad - 1.568\rho_p d \ln\frac{1.293(dV - 0.0000648)}{dV + 0.00310} \qquad 10 < \text{Re} < 700 \tag{3-129}$$

$$s = s_1 + 4.986\rho_p d - 0.5621\rho_d\ d \ln\left(2933dV + 0.5455\right) \qquad \text{Re} < 10 \tag{3-130}$$

$$s_{\max} = s_2 + 0.3410\rho_p d \tag{3-131}$$

$$s_1 = 2.525\rho_p d \ln 92.2dV_0 \tag{3-132}$$

$$s_2 = s_1 + 4.986\rho_p d \tag{3-133}$$

In all the equations given in this chapter for motion in standard air for which values of ρ and μ suitable for air at standard conditions are substituted, the standard SI units must be used. Figure 3-2 shows in graphical form the sequence of equations which should be used in calculating the motion of the particle for various initial values of Reynolds number Re_0. The numbers represent the equation numbers applicable to the different ranges of Reynolds number; the general equations are indicated first, followed by those for standard air, which are labeled by an asterisk.

EXAMPLE 3-3   A water droplet 1.2 mm in diameter is projected in still air (at standard conditions) with an initial velocity of 50 m/s. Neglecting gravitational effects, how far will the drop travel before coming to rest?

SOLUTION   Equation (3-85) gives the initial Reynolds number:

$$\text{Re}_0 = 64{,}516(0.0012)(50) = 3871$$

The quantity $\rho_p d/\rho$ is computed as 1.013. Equations (3-121), (3-122), and (3-120) give s_1, s_2, and $s_{\max}$, respectively:

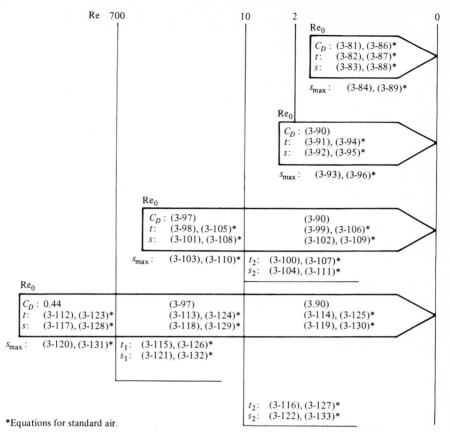

*Equations for standard air.

FIGURE 3-2

Sequence of equations for projection of particles for different initial values of Reynolds number.

$$s_1 = 3.03(1.013) \ln \frac{3871}{700} = 5.25 \text{ m}$$

$$s_2 = 5.25 + 5.91(1.013) = 11.2 \text{ m}$$

$$s_{\text{max}} = 11.2 + 0.404(1.013) = 11.6 \text{ m} \qquad Ans. \qquad ////$$

3-4 MOTION OF A SPHERICAL PARTICLE IN AN ACCELERATING FLUID STREAM WITH NO EXTERNAL FORCE

In the analyses performed heretofore in this chapter, the coordinate system was attached to the fluid stream so that the particle motion appeared as the deviation between the motion of the particles and that of the gas. The particle was acted upon by an external force which caused this deviation. Where the gas stream is accelerating, the particle will always lag behind in its motion. We have assumed that this lag, or the deviation in particle motion from fluid motion attributable to the acceleration of the fluid, is negligible compared with the deviation caused by the external force.

The assumption that the coordinate system is attached to the gas flow needs further clarification where the motion of the particle is in a nonuniform fluid motion. Again we assume that any lag in particle motion from fluid motion is negligible. The previous analysis may be applied if it is considered that the coordinate system is always attached to the fluid through which the particle is passing at the moment. A detailed discussion of more complex two- and three-dimensional problems of this sort is beyond the scope of our present discussion.

In this section, we wish to consider the motion of a particle in a fluid where no external force is acting on the particle other than the drag force of the fluid and where the fluid stream is accelerating rapidly such that the lag between particle and fluid motions is of primary significance. Such a problem is of prime concern in dealing with inertial separation mechanisms. We shall assume that the coordinate system is stationary in the following analysis. Figure 3-3 shows such a coordinate system, along with the streamlines of the fluid motion and the path of the particle motion. We designate by u_g, v_g, and w_g the velocity components of the fluid, and by u, v, and w the velocity components of the particle motion. The velocity $\mathbf{V}_g$, whose components are u_g, v_g, and w_g, is assumed to be a known function of position x, y, and z for steady flow and of time, in addition, for unsteady flow.

If the external force $\mathbf{F}$ is zero, Eq. (3-1) becomes

$$\mathbf{F}_D = \rho_p V \frac{d\mathbf{V}}{dt} \qquad (3\text{-}134)$$

The drag force is given by the difference between the particle velocity and the fluid velocity as

$$\mathbf{F}_D = -C_D \frac{\rho A_p}{2} (\mathbf{V} - \mathbf{V}_g)|\mathbf{V} - \mathbf{V}_g| \qquad (3\text{-}135)$$

Substituting Eq. (3-135) gives for Eq. (3-134)

$$\rho_p V \frac{d\mathbf{V}}{dt} + \frac{C_D \rho A_p}{2} |\mathbf{V} - \mathbf{V}_g|(\mathbf{V} - \mathbf{V}_g) = 0 \qquad (3\text{-}136)$$

If we limit our treatment to the case of small Reynolds numbers, based on the difference between particle and fluid velocities, the drag coefficient is given by

$$C_D = \frac{24\mu}{\rho d|\mathbf{V} - \mathbf{V}_g|C} \qquad (3\text{-}137)$$

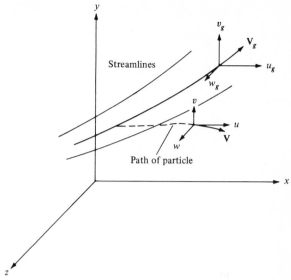

FIGURE 3-3
Motion of a particle in an accelerating fluid.

and Eq. (3-136) gives the vector equation of motion of the particle:

$$\frac{d\mathbf{V}}{dt} + \frac{18\mu}{\rho_p d^2 C} \mathbf{V} = \frac{18\mu}{\rho_p d^2 C} \mathbf{V}_g \qquad (3\text{-}138)$$

The vector velocity $\mathbf{V}$ can be written in terms of the velocity components u, v, and w by the relation

$$\mathbf{V} = \mathbf{i}u + \mathbf{j}v + \mathbf{k}w = \mathbf{i}\frac{dx}{dt} + \mathbf{j}\frac{dy}{dt} + \mathbf{k}\frac{dz}{dt} \qquad (3\text{-}139)$$

in which $\mathbf{i}$, $\mathbf{j}$, and $\mathbf{k}$ are unit vectors in the x, y, and z directions, respectively. Therefore Eq. (3-138) can be expanded into three scalar equations, one for each of the three coordinate directions:

$$\frac{d^2 x}{dt^2} + \frac{18\mu}{\rho_p d^2 C}\frac{dx}{dt} - \frac{18\mu}{\rho_p d^2 C} u_g = 0 \qquad (3\text{-}140)$$

$$\frac{d^2 y}{dt^2} + \frac{18\mu}{\rho_p d^2 C}\frac{dy}{dt} - \frac{18\mu}{\rho_p d^2 C} v_g = 0 \qquad (3\text{-}141)$$

$$\frac{d^2 z}{dt^2} + \frac{18\mu}{\rho_p d^2 C}\frac{dz}{dt} - \frac{18\mu}{\rho_p d^2 C} w_g = 0 \qquad (3\text{-}142)$$

In the preceding equations, the motion has been restricted to that of a spherical particle. The gas velocity components u_g, v_g, and w_g are functions of x, y, z, and t in general, or for steady flow they are functions of x, y, and z. Since these gas

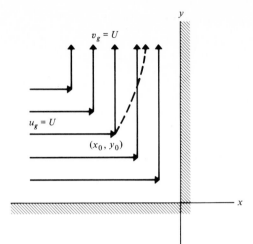

FIGURE 3-4
Motion of a particle for an abrupt change
in fluid velocity.

velocities are usually rather complex functions of these variables, the mathematical complexity of Eqs. (3-140) to (3-142) is extreme. For this reason, we shall make no attempt to solve these differential equations in a general way.

The initial conditions associated with Eqs. (3-140) to (3-142), which will be required to effect a solution, are that when $t = 0$:

$$x = x_0 \qquad y = y_0 \qquad z = z_0 \qquad (3\text{-}143)$$

$$u = u_0 \qquad v = v_0 \qquad w = w_0 \qquad (3\text{-}144)$$

In addition, the functions u_g, v_g, and w_g must be specified. If the problem can be reduced to a two-dimensional one, only Eqs. (3-140) and (3-141) are needed and only the first two parts of Eqs. (3-143) and (3-144) are relevant.

As an example of the method of evaluation of Eqs. (3-140) to (3-142), consider the gas motion shown in Fig. 3-4 in which the gas makes an abrupt right-angle turn, with no change in magnitude of velocity. The particle will start to diverge from the gas motion at the point where the abrupt turn is made. The path of the particle will shift to the right during its subsequent motion, eventually reaching a path coincident with another streamline but shifted a certain distance from the original streamline.

The motion is two-dimensional, so that Eqs. (3-140) and (3-141) may be written as follows for this particular situation:

$$\frac{d^2x}{dt^2} + \frac{18\mu}{\rho_p d^2 C} \frac{dx}{dt} = 0 \qquad (3\text{-}145)$$

$$\frac{d^2y}{dt^2} + \frac{18\mu}{\rho_p d^2 C} \frac{dy}{dt} = \frac{18\mu}{\rho_p d^2 C} U \qquad (3\text{-}146)$$

The appropriate initial conditions are

$$y_0 = -x_0 \qquad u_0 = U \qquad v_0 = 0 \qquad \text{when } t = 0 \qquad (3\text{-}147)$$

The differential equations are easily solved, giving

$$x = C_1 + C_2 e^{-at}$$
$$y = C_3 + C_4 e^{-at} + Ut$$

where $a = 18\mu/\rho_p d^2 C$. When the boundary conditions of Eq. (3-147) are applied and the resulting equations solved for C_1, C_2, C_3, and C_4, we have

$$x = x_0 + \frac{U}{a}(1 - e^{-at}) \qquad (3\text{-}148)$$

$$y = -x_0 - \frac{U}{a}(1 - e^{-at}) + Ut \qquad (3\text{-}149)$$

If Eq. (3-148) is solved for t and the result substituted into Eq. (3-149), we have the equation of the particle trajectory:

$$y = -x - \frac{U}{a} \ln\left(1 - \frac{x - x_0}{U/a}\right) \qquad (3\text{-}150)$$

At large time $(t \to \infty)$

$$x - x_0 \to \frac{U}{a} = \frac{\rho_p d^2 UC}{18\mu} \qquad (3\text{-}151)$$

$$y \to -x + Ut \qquad (3\text{-}152)$$

As a numerical example, suppose

$$\rho_d = 2.0 \text{ g/cm}^3 = 2000 \text{ kg/m}^3$$
$$d = 20 \ \mu\text{m}$$
$$U = 20 \text{ m/s}$$
$$\mu = 1.84 \times 10^{-5} \text{ kg/m} \cdot \text{s}$$
$$C = 1$$

Equation (3-151) gives

$$x - x_0 \to \frac{2000(20 \times 10^{-6})^2(20)(1.0)}{18(1.84 \times 10^{-5})} = 0.0483 \text{ m} = 4.83 \text{ cm}$$

On the other hand, if d is 2.0 μm in the preceding calculation, then

$$x - x_0 \to 0.483 \text{ mm}$$

For motion in standard air, Eq. (3-151) becomes

$$x - x_0 \to 3019\rho_p d^2 UC \qquad (3\text{-}153)$$

The preceding example is highly unrealistic physically since the flow pattern given would be quite difficult to achieve. Nevertheless, it is quite tractable mathematically and, in addition, serves as a limiting case useful in estimating behavior of real systems. The 5-cm separation distance achieved with a 20-μm particle is adequate to effect separation in various types of inertial separation devices, whereas the 2-μm particle would be difficult to separate in such a device. In fiber filters, a 0.5-mm separation distance is enough to achieve significant separation. Even though the numerical values of separation distances estimated by the preceding example are subject to considerable error, a great deal of insight into various separation mechanisms can be obtained from the use of these results.

REFERENCES

1   Cunningham, E.: On the Velocity of Steady Fall of Spherical Particles through Fluid Medium, *Proc. R. Soc. London, Ser. A*, vol. 83, pp. 357–365, 1910.
2   Strauss, W.: "Industrial Gas Cleaning," pp. 63 and 130, Pergamon Press, London, 1966.
3   Hemeon, W. C. L.: "Plant and Process Ventilation," 2d ed., chap. 3, The Industrial Press, New York, 1963.

PROBLEMS

3-1   Derive Eq. (3-23).
3-2   Derive Eqs. (3-28) to (3-30).
3-3   Derive Eq. (3-31).
3-4   Derive Eq. (3-32).
3-5   Derive Eq. (3-33).
3-6   Derive Eq. (3-34).
3-7   Derive Eq. (3-35).
3-8   Derive Eq. (3-36).
3-9   Derive Eqs. (3-37) to (3-40).
3-10 A particle of 10 μm diameter having a density of 2.5 g/cm^3 falls from rest under the action of gravity in air at 1.0 atm pressure and 20°C temperature. Compute the terminal velocity, the terminal Reynolds number, and the time and distance required for the particle to reach 95 percent of the terminal velocity.
3-11 Repeat Prob. 3-10 if the particle diameter is 100 μm.
3-12 Repeat Prob. 3-10 if the particle diameter is 1.0 mm.
3-13 A particle 0.5 mm in diameter and of 3.0 g/cm^3 density falls in water at 20°C. If the particle starts from rest, compute its terminal velocity and the Reynolds number when terminal velocity is reached. What velocity will the particle have attained after it has traveled for 10 cm? What will be its velocity after having fallen for 2.0 s from rest?
3-14 Compute the terminal velocity of a particle of 50 μm diameter falling freely from rest in standard air if its specific gravity is (*a*) 0.6; (*b*) 1.5; (*c*) 5.0; (*d*) 13.6.

3-15 A particle falls from rest in a stratified fluid consisting of two layers having different properties, with the properties being constant for each layer. Assume that the Reynolds number never exceeds 2.0. If the particle starts its fall from a point located s_0 above the interface of the two fluids, derive expressions for terminal velocity, time, and distance as the particle falls. Consider the motion of the particle in both fluids.

3-16 Repeat Prob. 3-10 for particle diameters of 1.0, 0.5, 0.1, 0.05, and 0.01 μm. Use Cunningham's correction factor.

3-17 Compute the terminal velocity of a particle of 10 μm diameter falling in air having a temperature of 25°C if its density is 1.5 g/cm^3 and the air pressure is (*a*) 1.0 atm; (*b*) 10,000 N/m^2; (*c*) 1000 N/m^2; (*d*) 100 N/m^2; and (*e*) 10 N/m^2.

3-18 A spherical particle whose diameter is 0.1 mm, made of a material having a density of 4.2 g/cm^3, is projected in standard air at a velocity of 30 m/s. How far will the particle travel before coming to rest; that is, what is the magnitude of the *pulvation distance?*

3-19 A water droplet of 1.0 mm diameter is projected into a body of standard air at an initial velocity of 100 m/s. What is its initial Reynolds number? How far will it travel before coming to rest?

3-20 A particle having a density of 1.75 g/cm^3 is projected into standard air at an initial velocity of 70 m/s. Calculate and plot pulvation distance as a function of particle diameter for a range of particle diameters from 5.0 to 100 μm.

3-21 A particle whose density is 1.25 g/cm^3 is projected with an initial velocity of 30 m/s into a body of air at 0.095 MN/m^2 and 375°C. Compute the distance traveled by the particle before coming to rest. Assume that the particle diameter is 50 μm.

3-22 A steel sphere of 5 mm diameter is projected into a body of water at 30°C with an initial velocity of 50 m/s. Neglecting the effect of gravity, how far will the sphere travel before coming to rest?

3-23 A spherical bullet 1.0 cm in diameter is projected from the muzzle of a gun at a velocity of 1000 m/s. The density of the bullet is 7.5 g/cm^3. Neglecting shock-wave and gravitational effects, how far will the bullet travel before coming to rest, assuming, of course, that it does not strike an object?

3-24 A spherical particle with a density of 2.0 g/cm^3 and a diameter of 30 μm is projected vertically downward with an initial velocity of 10 m/s. The fluid medium through which the particle travels is still air at standard conditions. Allowing for the effect of gravity, how much time must elapse before the particle reaches a point 0.1 m below the point of initial projection?

3-25 Rework the example of Sec. 3-4 if the deflection angle of the flow is an arbitrary angle α, with an abrupt change of flow direction along the bisector of the included angle $\pi - \alpha$ and with the magnitude of the flow velocity equal to U before and after the abrupt change in direction.

3-26 A water droplet 0.1 mm in diameter undergoes a motion similar to that treated in Sec. 3-4. If the gas is air, what separation distance is expected for an air velocity of 40 m/s?

3-27 For steady, irrotational, two-dimensional flow about a right-angled corner, as shown in Fig. 3-5 (p. 118), the velocity components are given as

$$u_g = bx \qquad v_g = -by$$

To avoid problems dealing with infinite distances, suppose a particle starts its motion at a point (x_0, y_0) near the top. Derive an equation for the particle trajectory and determine the point at which the particle strikes the wall, if it does.

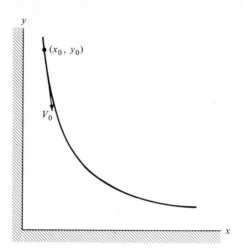

FIGURE 3-5

4

POLLUTANT DISTRIBUTIONS AND COLLECTION EFFICIENCIES

There are a number of properties of individual particles which are of significance in air pollution control work. Such quantities as mass, surface area, shape, linear size, or diameter, when summed over all the particles present in the collection, yield information as to the overall properties of the particle collection. Although gaseous pollutants have particles of uniform size, so that much of the material in this chapter does not apply to them or applies in a trivial way, collections of particulates usually contain particles having a wide range of sizes.

The overall properties of the particle collection depend, not only on the sum of the properties of individual particles, but also on the way in which these property values are distributed over the particles in the collection. For instance, a particle collection consisting of many large particles and few small ones might have the same mass as one consisting of fewer large particles and tremendously more small ones. Yet the two collections are significantly different, the first being much more amenable to separation by mechanical means. Often, the specification of two or more properties of the collection, each determined as the sum of a like property of each individual particle, will suffice to adequately specify the distributional aspect of the property in the collection. Sometimes, though, it is necessary to develop special properties of the collection for the express purpose of treating the distributional nature of the collection.

4-1   PROPERTIES AND COLLECTIONS OF PARTICLES

Part of the work of this chapter involves methods of summing property values of individual particles to yield a value for the collection as a whole. For this purpose, we shall introduce the idea of a generalized property P (having no relation to pressure), whose value represents the property in question for either a single particle or the collection. We usually characterize a single particle in terms of a diameter, whose exact specification will be discussed later on.

The generalized property for a single particle is given as $P(d)$, where the assumption is made that all particles having a particular value of diameter have the same value of the property. We also assume that the property $P(d)$ is continuously distributed, so that $P(d + d(d))$ varies only slightly from $P(d)$ if $d(d)$ is small. Note the somewhat unfortunate aspect of choosing the symbol d to represent particle diameter: It leads to an awkward combination with the differential symbol d. Thus $d(d)$ means a differential change in diameter. Despite its awkwardness, we shall retain the use of this symbol for its convenience in other respects. The symbol P without the diameter attached will represent the total value of the property for the entire collection.

We have indicated that $P(d)$ is the value of a property for a particle of diameter d and that this value does not change much over a range of diameter from d to $d + d(d)$ if $d(d)$ is small. We shall be dealing with collections of very large numbers of particles. It must be known how many particles are present having diameter d. Strictly speaking, if d is specified to sufficiently many decimal places, there will be no particles having diameter d. There will be a certain number whose diameter is very close to d, however. More specifically, there will be a certain number of particles whose diameter lies in the range from d to $d + d(d)$. Since $d(d)$ is arbitrary, though it must be small, the number of particles defined this way is an arbitrary number. To uniquely quantify the number distribution, let us denote by $N(d)$ the total number of particles whose diameters are less than or equal to d. Also, we denote by N the total number of particles in the collection, or $N = N(\infty)$. Now, since $N(d)$ is a continuous function of d, or at least almost so in a collection of very many particles, we may determine its derivative with respect to d, and this derivative is denoted by $n(d)$:

$$n(d) = \frac{dN(d)}{d(d)} \qquad (4\text{-}1)$$

It is this quantity $n(d)$ which is useful to characterize the particle distribution in a collection. The number of particles whose diameter is less than or equal to d is given by

$$N(d) = \int_0^d n(d)\, d(d) \qquad (4\text{-}2)$$

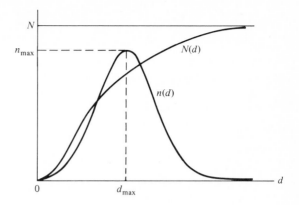

FIGURE 4-1
Typical particle-distribution curves.

and the total number of particles present is given by

$$N = \int_0^\infty n(d)\, d(d) \qquad (4\text{-}3)$$

The limit of ∞ in Eq. (4-3) causes no difficulty since $n(d)$ approaches zero as d becomes large. Figure 4-1 shows a typical distribution, indicating $n(d)$ and $N(d)$ as functions of d.

Having the generalized property value specified at diameter d, and having the number-distribution function $n(d)$ defined by Eq. (4-1), we can now determine the value of the generalized property P of the total collection. For the total number of particles lying in the range from d to $d + d(d)$, the value of property P is $P(d)n(d)\, d(d)$. Integrating this over the total range of diameter gives

$$P = \int_0^\infty P(d)n(d)\, d(d) \qquad (4\text{-}4)$$

If number of particles is the property of interest, then $P(d) = 1$ since each particle contributes 1 to the total number. Also, $P = N$. Then, Eq. (4-3) results.

The mass of the particle collection is obtained from Eq. (4-4) if $P(d)$ is the mass of a single particle:

$$m = \int_0^\infty m(d)n(d)\, d(d) \qquad (4\text{-}5)$$

For a spherical particle of density ρ_p the particle mass is given by $\pi\rho_p d^3/6$. Then

$$m = \frac{\pi\rho_p}{6}\int_0^\infty n(d)d^3\, d(d) \qquad \text{(spherical particles)} \qquad (4\text{-}6)$$

The total surface area of the particles in the collection can be obtained from Eq. (4-4) if $P(d)$ is the surface area of each particle, given by πd^2 for spherical particles:

$$A_s = \int_0^\infty A_s(d)n(d) \, d(d) \tag{4-7}$$

$$A_s = \pi \int_0^\infty n(d)d^2 \, d(d) \qquad \text{(spherical particles)} \tag{4-8}$$

The specific surface is defined as the surface area per unit mass of the particle collection. For a single spherical particle its value is

$$A_{ss} = \frac{6}{\rho_p d} \qquad \text{(spherical particles)} \tag{4-9}$$

For a collection of general particles, the specific surface is given by

$$A_{ss} = \frac{\int_0^\infty A_s(d)n(d) \, d(d)}{\int_0^\infty m(d)n(d) \, d(d)} \tag{4-10}$$

and for a collection of strictly spherical particles, this becomes

$$A_{ss} = \frac{6}{\rho_p} \frac{\int_0^\infty n(d)d^2 \, d(d)}{\int_0^\infty n(d)d^3 \, d(d)} \tag{4-11}$$

Values of specific surface range from 5 to over 10^5 m^2/kg in the case of fine carbon black. Specific surface is clearly of interest in processes involving chemical reaction or adsorption. The light-scattering ability of a particle collection is also related to this property.

The formulas developed to this point are expressed in terms of integrals whose integrands involve the terms $n(d)$ and d. The number-distribution function $n(d)$ will be discussed in some detail in Sec. 4-3. The diameter d is a measured quantity characteristic of the particles themselves and will be discussed at length here. If the particles are all spherical, then the diameter to use is obviously the diameter of the sphere. Particle distributions consisting of liquid droplets formed by condensation or by shearing action from a body of liquid are normally spherical. On the other hand, particles formed by fracture from a solid and particles formed by crystallization are usually nowhere near spherical in shape. It is for these particles that the concept of diameter requires modification. In order to facilitate the analysis, a *mean diameter* is defined.

There are several possible ways to define mean diameter of a particle or of a particle collection. These relate either to a particular technique of size measurement or to a derived equation for particle behavior, such as that for free fall of a particle. The particular choice of mean diameter is made on the basis of the use to which it will be put, that is, which properties of the particle

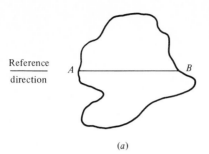 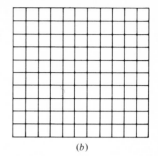

(a)   (b)

FIGURE 4-2
Measurement of mean diameter d_1.

collection will be calculated using the mean diameter. The measured diameters will be based on a large number of particles since there will be a considerable number of particles of varying shapes and volumes having about the same mean diameter. In measuring mean diameter, a representative sample of the original particle collection is taken and the measuring process is applied to it. A series of ranges in diameter is set up, either chosen arbitrarily or related to the output of the measuring instrument. The numbers of particles or weights of particles whose diameters fall in these respective ranges are tabulated. From these data, a distribution curve is plotted, and, if desired, a distribution function is fitted to this curve.

Next we shall define six mean diameters, referred to as d_1 to d_6. In general, all these mean diameters will have different values for the same collection of particles. The first definition of mean diameter is the measurement of length in a particular direction. To uniquely define length parallel to the reference direction and at the same time provide an easily workable process, the length in question will be measured through a point midway between the top- and bottom-most points of the particle as it is oriented relative to the reference direction. This is illustrated in Fig. 4-2a; Fig. 4-2b shows a transparent grid which will facilitate measurement. This particular mean diameter is called d_1 and is equal to distance AB in Fig. 4-2a. (The actual measurement of d_1 and d_2 will be made from a photomicrograph of a sample of the original particle collection. A scale equal to the magnification of the photomicrograph must be applied to the resultant mean diameter.)

The second definition of mean diameter, d_2, consists of measuring the length across the particle in several directions and taking the average of these. This measurement is illustrated in Fig. 4-3a, and a transparent grid that is useful in making the measurement is shown in Fig. 4-3b. The intersection point should be as near to the centroid of the particle as possible, although some error in this regard is not critical.

The third definition of mean diameter, d_3, relates to surface area and

(a)

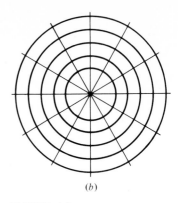

(b)

FIGURE 4-3
Measurement of mean diameter d_2.

states that a sphere having diameter d_3 will have the same surface area as the particle in question. Thus, since the surface area of a sphere is πd^2,

$$d_3 = \sqrt{\frac{A_s}{\pi}} \qquad (4\text{-}12)$$

The specific surface of this particle is

$$A_{ss} = \frac{\pi d_3{}^2}{m} = \frac{\pi d_3{}^2}{\rho_p \overline{V}} \qquad (4\text{-}13)$$

The fourth mean diameter d_4 is based on specific surface, stating that a sphere of diameter d_4 has the same specific surface as the particle in question. This leads to the following definition:

$$d_4 = \frac{6\overline{V}}{A_s} \qquad (4\text{-}14)$$

For d_5, the volumetric diameter, a sphere with diameter d_5 will have the same volume as the particle in question and will have the same mass if the densities are equal. This gives

$$d_5 = \left(\frac{6\overline{V}}{\pi}\right)^{1/3} \qquad (4\text{-}15)$$

The sixth definition for mean diameter d_6 is based on free fall of the particle in the laminar regime, or for Re < 2.0. A sphere having the diameter d_6 will have the same terminal velocity under the influence of gravity in standard air as will the particle in question. The measured property of the particle in this case will be the terminal settling velocity. Equation (3-52) forms the basis for the calculation of d_6 from terminal settling velocity V_t:

$$V_t = 29609 \rho_p \, d_6{}^2$$

from which

$$d_6 = 0.00581 \sqrt{\frac{V_t}{\rho_p}} \qquad (4\text{-}16)$$

In some instruments, settling velocity is measured in a fluid other than standard air. In such a case, Eq. (4-16) is replaced by

$$d_6 = 1.355 \sqrt{\frac{\mu V_t}{\rho_p}} \qquad (4\text{-}17)$$

It must be realized that in many cases particle samples will be biased in some way. For example, a sample exposed to view will most likely be collected on a filter with the long dimensions of the particles lying in the plane of the filter, or at least an unnaturally large fraction of the particles will be so oriented. This sample will give a particle size count that is much larger than it should be. Similar problems of bias exist with the other mean diameters. To obtain a true mean-diameter count using existing instrumentation is not easy. Best results are obtained if the diameter readings are taken from an instrument whose mode of operation most closely approximates the process for which the data will be used. Thus, if the behavior of the particles in a separator is to be studied, their mean diameter might be suitably determined from a free-fall apparatus.

4-2 POLLUTANT CONCENTRATIONS

The *particle loading* of the air is equal to the total mass of particulate pollutants in the air. This term obviously does not apply to gaseous pollutants; the term *concentration* serves the purpose instead. In fact, the various definitions of concentration apply equally to gaseous and particulate pollutants. We shall define four concentrations which are widely used either with gaseous or particulate pollutants. Each of these is defined as an amount of pollutant divided by the corresponding amount of air in which this pollutant is dispersed.

The first concentration to be defined here is mass concentration C_m, the mass of pollutant divided by the mass of air plus mass of pollutant:

$$C_m = \frac{m_p}{m_a + m_p} \qquad (4\text{-}18)$$

In this equation, m_p is the mass of pollutant substance and m_a is the mass of pure air in some volume of air pollutant mixture. The second concentration is the volumetric concentration C_v, the volume of pollutant divided by the volume of air and pollutant combined:

$$C_v = \frac{V_p}{V_a + V_p} \quad (4\text{-}19)$$

The third concentration is mass-volume concentration C_{mv}, the mass of pollutant divided by the volume of air and pollutant combined:

$$C_{mv} = \frac{m_p}{V_a + V_p} \quad (4\text{-}20)$$

Finally, the fourth concentration is volumetric concentration in parts per million C_{ppm}, generally used only for gaseous pollutants:

$$C_{ppm} = \frac{V_p}{V_a} \times 10^6 \quad (4\text{-}21)$$

Other concentrations will be defined in later chapters for specific purposes.

We can develop a number of relationships among these different concentrations. The basic relations needed for this are

$$m_p = \rho_p V_p \qquad m_a = \rho_a V_a$$

Upon solving Eq. (4-19) for V_p/V_a, we obtain

$$\frac{V_p}{V_a} = \frac{C_v}{1 - C_v} \quad (4\text{-}22)$$

When these equations are used, Eq. (4-18) relates C_m to C_v as

$$C_m = \frac{\rho_p C_v}{\rho_a(1 - C_v) + \rho_p C_v} \quad (4\text{-}23)$$

while Eq. (4-20) relates C_{mv} to C_v as

$$C_{mv} = \rho_p C_v \quad (4\text{-}24)$$

For the usual range of pollutant concentrations and densities, Eqs. (4-22) and (4-23) can be written to a good approximation as

$$\frac{V_p}{V_a} \approx C_v \quad (4\text{-}25)$$

$$C_m \approx \frac{\rho_p}{\rho_a} C_v \quad (4\text{-}26)$$

From Eqs. (4-23) and (4-24), additional relations may be obtained. These are written, along with their approximations, as

$$C_v = \frac{C_{mv}}{\rho_p} \tag{4-27}$$

$$C_v = \frac{\rho_a C_m}{\rho_p - (\rho_p - \rho_a)C_m} \approx \frac{\rho_a}{\rho_p} C_m \tag{4-28}$$

$$C_{mv} = \frac{\rho_a \rho_p C_m}{\rho_p - (\rho_p - \rho_a)C_m} \approx \rho_a C_m \tag{4-29}$$

$$C_m = \frac{\rho_p C_{mv}}{\rho_a \rho_p + (\rho_p - \rho_a)C_{mv}} \approx \frac{C_{mv}}{\rho_a} \tag{4-30}$$

We can relate C_v to C_{ppm} from Eqs. (4-21) and (4-22) as

$$C_v = \frac{C_{ppm}}{10^6 + C_{ppm}} \approx 10^{-6} C_{ppm} \tag{4-31}$$

The density of the air is given by the perfect-gas law

$$\rho_a = \frac{P}{RT} \tag{4-32}$$

in which R has the value of 287 J/kg · K for air. Here P is the pressure of the mixture. For gaseous and vapor pollutants at small concentrations, the perfect-gas law holds quite well also:

$$\rho_p = \frac{PM_p}{8314.3T} \tag{4-33}$$

where M_p is the molecular weight of the pollutant substance and P is the pressure of the mixture.

For particulate pollutants, the concentrations can be written in terms of the number distribution function $n(d)$, where $n(d)$ is now modified to represent number of particles per unit volume of mixture. We shall use the approximations inherent in small concentrations of pollutants. Equation (4-18) for C_m, together with Eq. (4-5), leads to

$$C_m = \frac{m_p}{m_a}$$

$$= \frac{1}{\rho_a} \int_0^\infty n(d)m_p(d)\,d(d)$$

$$= \frac{\rho_p}{\rho_a} \int_0^\infty \frac{\pi}{6} n(d_s)d_s{}^3\,d(d_s) \tag{4-34}$$

In Eq. (4-34) the second integral may be used if the diameter used is the mean diameter d_s based on mass; in any case, if m_p is the particle mass based on diameter d, any mean diameter may be used in the integration. Equation (4-19) gives C_v as $\overline{V}_p/\overline{V}_a$ approximately. From Eq. (4-4), using $P(d)$ as the volume of the particle and again assuming that $n(d)$ is the number of particles per unit volume,

$$C_v = \frac{\pi}{6} \int_0^\infty n(d_s) d_s{}^3 \, d(d_s) \qquad (4\text{-}35)$$

Equation (4-20) may be written

$$C_{mv} = \int_0^\infty n(d) m_p(d) \, d(d) = \frac{\pi}{6} \rho_p \int_0^\infty n(d_s) d_s{}^3 \, d(d_s) \qquad (4\text{-}36)$$

When a pollutant air mixture is flowing through a duct, the volumetric flow rate of mixture is denoted by Q, which represents the volume of mixture that crosses any fixed section of the duct in unit time. But other properties besides volume also flow past the fixed section. In fact, any property which can be identified with a quantity of fluid or mixture can be considered to flow past a section of a duct.

For example, the mass flow rate of pollutant past the section is given by

$$\dot{m}_p = C_{mv} Q \qquad (4\text{-}37)$$

Using $\dot{P}$ to represent the flow rate of any property, we have

$$\dot{P} = \frac{P}{V_a + V_p} Q \qquad (4\text{-}38)$$

in which P is the value of the property in question for a volume $V_a + V_p$ of the mixture. For surface area flowing past the section,

$$\dot{A}_s = \dot{m}_p A_{ss} = C_{mv} A_{ss} Q \qquad (4\text{-}39)$$

The volume rate of pollutant flow is given by

$$\dot{Q}_p = C_v Q \qquad (4\text{-}40)$$

The number of particles flowing past a section per unit time is given as

$$\dot{N} = NQ \qquad (4\text{-}41)$$

in which N is the total number of particles per unit volume of mixture.

EXAMPLE 4-1   Carbon tetrachloride (CCl_4, molecular weight $= 153.8$) is combined with air flowing in a duct at a concentration of 150 ppm. The air, at standard conditions, flows at the rate of 10 m^3/s.

(a) Compute C_m, C_v, and C_{mv} for the CCl_4 in the mixture.

(b) At what rate, in kilograms per day, does the CCl_4 flow through the duct?

SOLUTION   (a) From the problem statement, $C_{ppm} = 150$. The air density is 1.185 kg/m³, and the pollutant density is given by Eq. (4-33) as

$$\rho_p = \frac{PM_p}{8314.3T} = \frac{101326(153.8)}{8314.3(298)} = 6.29 \text{ kg/m}^3$$

Equations (4-31), (4-26), and (4-27) give the other concentrations:

$$C_v = 10^{-6}C_{ppm} = 0.00015 \qquad Ans.$$

$$C_m = \frac{\rho_p}{\rho_a}C_v = \frac{6.29}{1.185}\, 0.00015 = 0.000796 \qquad Ans.$$

$$C_{mv} = \rho_p C_v = 6.29(0.00015) = 0.000944 \text{ kg/m}^3 \qquad Ans.$$

(b) Equation (4-37) gives the mass flow rate of the pollutant:

$$\dot{m}_p = C_{mv}Q = 0.000944(10)3600(24) = 815 \text{ kg/day} \qquad Ans. \qquad ////$$

4-3 PARTICLE DISTRIBUTIONS

The number-distribution function $n(d)$ for a collection whose particles span a wide range of sizes has been defined in Eq. (4-1). The use of this function in various equations has been discussed somewhat in Sec. 4-1. A typical number-distribution curve is shown in Fig. 4-1. In this section we are concerned with formulating a suitable mathematical equation describing this function. With such an equation available, the integrals in the various equations of Sec. 4-1 can be evaluated, though a closed-form solution is sometimes difficult to obtain. In that case, a numerical quadrature may have to be resorted to.

The specific particle distribution to be expected in particulate collections is subject to considerable variation. The distribution having relatively simple mathematical form which best fits a wide variety of dusts and mists is the log-normal distribution, which will be discussed in some detail later in this section. Other mathematical forms are useful in specific situations also. In addition, it is worthwhile to consider how an empirically determined particle distribution can be treated. If the mathematical function describing the distribution curve is known with complete accuracy, then all properties of the distribution can be evaluated. In the statistical analysis of distributions, a few properties which have numerical values are used instead of trying to operate with functions directly. The properties in common use are certain measures of mean value, dispersion, and skewness. These will be defined and described.

Figure 4-4 shows a log-normal distribution which for the moment will serve as an illustration of a general distribution. Both the number distribution $n(d)$ and the cumulative distribution $N(d)$ are shown, each normalized by the total number of particles N. Also shown in Fig. 4-4 are the three mean diameters most commonly used to describe a distribution.

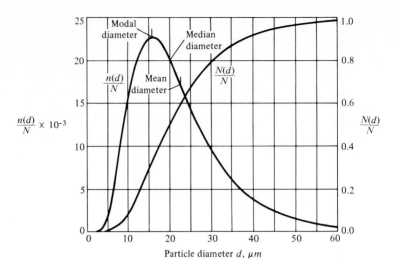

FIGURE 4-4
Log-normal particle distribution.

First in importance is the mean diameter $\bar{d}$, in this case, with a value of 22.7 μm, defined as

$$\bar{d} = \frac{\int_0^\infty n(d)d\,d(d)}{\int_0^\infty n(d)\,d(d)} = \frac{\int_0^\infty n(d)d\,d(d)}{N} \qquad (4\text{-}42)$$

The mean diameter is such that the centroid of the area under the curve for $n(d)$ lies on a vertical line passing through this diameter. It is the mean diameter most often used in describing the distribution. The second mean diameter is the median diameter d_{me}, defined as the diameter at which the cumulative distribution has a value of 0.5. That is,

$$N(d_{me}) = \frac{N}{2} \qquad (4\text{-}43)$$

The area under the curve for $n(d)$ is the same on both sides of a vertical line passing through the median diameter d_{me}. The third mean diameter is the modal diameter d_{mo}, defined as the diameter at which the greatest number of particles is clustered. This diameter is located at the maximum point of the curve for $n(d)$ and is defined mathematically as the point where

$$\frac{dn(d)}{d(d)} = 0 \qquad (4\text{-}44)$$

The measure of dispersion in common use is the standard deviation or its square, the variance, given by

$$\sigma^2 = \frac{1}{N}\int_0^\infty (d - \bar{d})^2 n(d)\,d(d) \qquad (4\text{-}45)$$

The standard deviation is the positive square root of the variance. The variance is taken about the mean, as defined by Eq. (4-42), and is equal to the moment of inertia of the area under the curve for $n(d)$ about the vertical line passing through the centroid. A complete description of dispersion would require specification of all higher even moments for which the exponent in Eq. (4-45) is 4, 6, 8, ... The second moment, however, usually suffices.

Skewness is defined in terms of the third moment about the centroid, given by

$$\beta^3 = \frac{1}{N} \int_0^\infty (d - \bar{d})^3 n(d)\, d(d) \qquad (4\text{-}46)$$

Higher-order odd moments are required for a complete description, but the third moment is usually enough for sufficient accuracy.

Normal Distribution Function

At this point, we may look at some particular number-distribution functions. The one we shall consider first is the normal distribution function, shown in Fig. 4-5, which is coincident with the normal, or gaussian, probability curve. For this type of curve, the mean diameter, median diameter, and modal diameter are all identical, equal to 50 μm in Fig. 4-5.

The equation for $n(d)$ is given by

$$n(d) = \frac{N}{\sqrt{2\pi}} \frac{1}{\sigma} e^{-(d - \bar{d})^2 / 2\sigma^2} \qquad (4\text{-}47)$$

in which N is the total number of particles, σ is the standard deviation for the distribution, and $\bar{d}$ is the mean particle diameter. The cumulative particle diameter curve $N(d)$ is obtained by integrating Eq. (4-47):

$$N(d) = \frac{N}{\sqrt{2\pi}} \frac{1}{\sigma} \int_0^d e^{-(d - \bar{d})^2 / 2\sigma^2}\, d(d) \qquad (4\text{-}48)$$

The integral in Eq. (4-48) is not capable of evaluation in closed form. It is a standard form, however, known as the *error function*, and its value can be obtained from any of a number of reference sources, such as Jahnke, Emde, and Lösch [1].

The error function is defined as

$$\text{erf } z = \frac{2}{\sqrt{\pi}} \int_0^z e^{-z^2}\, dz \qquad (4\text{-}49)$$

In Eq. (4-48), we may replace $(d - \bar{d})/\sqrt{2}\,\sigma$ by z, which gives

$$N(d) = \frac{N}{\sqrt{\pi}} \left(\int_{-\bar{d}/\sqrt{2}\sigma}^{0} e^{-z^2}\, dz + \int_0^{(d - \bar{d})/\sqrt{2}\sigma} e^{-z^2}\, dz \right)$$

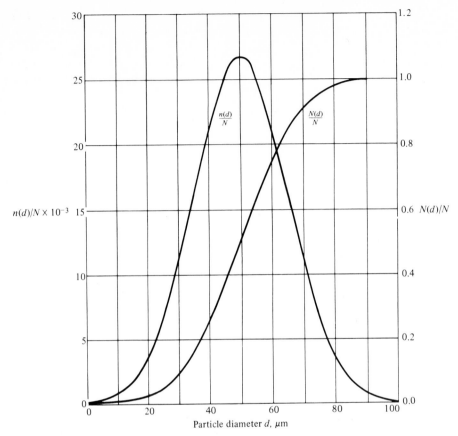

FIGURE 4-5
Normal particle distribution.

Now we may replace $-\bar{d}$ in the first integral by $-\infty$ without causing noticeable error since the normal distribution must not show any appreciable number of particles with negative diameter, else the normal distribution is completely inappropriate. We may replace z by $-z$ in the first integral, giving

$$N(d) = \frac{N}{\sqrt{\pi}} \int_0^\infty e^{-z^2}\, dz + \frac{N}{\sqrt{\pi}} \int_0^{(d-\bar{d})/\sqrt{2}\sigma} e^{-z^2}\, dz$$

Using Eq. (4-49) then gives

$$N(d) = \frac{N}{2} + \frac{N}{2}\operatorname{erf}\frac{d-\bar{d}}{\sqrt{2}\sigma} \qquad (4\text{-}50)$$

in which use is made of the fact that the error function erf z approaches unity as z approaches infinity.

Only rarely does a particle distribution follow the normal distribution function closely. This might be expected intuitively owing to the requirement of negative diameters in the function. Of course, in the example of Fig. 4-5, the error caused by this discrepancy is insignificant. Still, this discrepancy makes it clear that the normal function cannot satisfy all real particle distributions. In actual fact, it satisfies very few of them; it is included here because of its mathematical simplicity and because it is a useful preliminary for the log-normal distribution to follow.

The Log-Normal Distribution Function

White [2] discusses the applicability of the log-normal distribution function to real particulate collections. He cites Kolmogorov [3] in a theoretical study which leads to the conclusion that physical processes forming particulate distributions tend to produce a log-normal distribution. Although real distributions do deviate some from the log-normal pattern, the deviation is no greater than would be expected with any theoretical curve.

In a log-normal distribution we can define a diameter parameter u as the logarithm of the diameter:

$$u = \ln d \qquad (4\text{-}51)$$

If the number distribution function, written now as $n(u)$, is plotted as a function of u, the normal probability curve results, as shown in Fig. 4-6, in which the data of Fig. 4-4 is replotted in this form. In examining Figs. 4-4 and 4-6, note that diameter d is expressed in meters even though the label on the abscissa of Fig. 4-4 is in micrometers, so that the logarithm becomes a large negative number. Also, the large values of $n(d)$ come about from this fact and from the definition of $n(d)$ as $dN(d)/d(d)$. In the definition of u given in Eq. (4-51), the dimension of d must be ignored in taking the logarithm.

The relation between $n(u)$ and $n(d)$ is obtained from Eq. (4-51) and the definition of $n(d) = dN(d)/d(d)$. We have

$$n(d) = \frac{dN(d)}{d(d)} = \frac{dN(u)}{du}\frac{du}{d(d)}$$

From Eq. (4-51)

$$\frac{du}{d(d)} = \frac{1}{d} \qquad (4\text{-}52)$$

Then

$$n(d) = \frac{n(u)}{d}$$

or

$$n(u) = n(d)d \qquad (4\text{-}53)$$

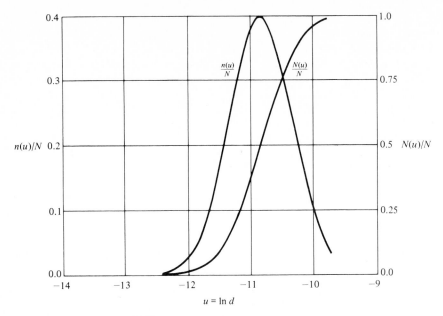

FIGURE 4-6
Log-normal particle distribution plotted with logarithmic abscissa.

As Fig. 4-6 shows, in the log-normal distribution, the curve for $n(u)$ is a normal probability curve in u, centered about a value $\bar{u}$ and with a standard deviation σ_u. The equation for this curve is

$$n(u) = \frac{N}{\sqrt{2\pi}\,\sigma_u}\, e^{-(u-\bar{u})^2/2\sigma_u^2} \qquad (4\text{-}54)$$

The cumulative size distribution function $N(u)$, which is equal to $N(d)$ if d and u are related by Eq. (4-51), is obtained by integrating Eq. (4-54):

$$N(u) = \frac{N}{\sqrt{2\pi}\,\sigma_u} \int_{-\infty}^{u} e^{-(u-\bar{u})^2/2\sigma_u^2}\, du$$

Letting $\eta = (u - \bar{u})/\sqrt{2}\,\sigma_u$ in the preceding integral, we obtain the integral form

$$N(u) = \frac{N}{\sqrt{2\pi}} \int_{-\infty}^{(u-\bar{u})/\sqrt{2}\sigma_u} e^{-\eta^2}\, d\eta$$

Then, proceeding with a derivation identical to that used in obtaining Eq. (4-50), the result is

$$N(u) = \frac{N}{2} + \frac{N}{2}\, \text{erf}\, \frac{u - \bar{u}}{\sqrt{2}\,\sigma_u} \qquad (4\text{-}55)$$

It is desirable to relate Eqs. (4-54) and (4-55) to the corresponding equations involving diameter d. Also, the properties of the distribution in diameter d should be considered. In transforming these equations, we shall need to define a new mean diameter d_m and a new standard deviation σ_m, which are related to the log-normal distribution as

$$\bar{u} = \ln d_m \qquad d_m = e^{\bar{u}} \qquad (4\text{-}56)$$

$$\sigma_u = \ln \sigma_m \qquad \sigma_m = e^{\sigma_u} \qquad (4\text{-}57)$$

It should be emphasized that the diameter d_m is not the same as the mean diameter $\bar{d}$ of the distribution, nor is the standard deviation σ_m the same as the standard deviation σ of the distribution.

Making use of Eqs. (4-51), (4-53), (4-56), and (4-57), the distribution equations (4-54) and (4-55) become

$$n(d) = \frac{N}{\sqrt{2\pi}\, d \ln \sigma_m} \exp\left[-\frac{(\ln d/d_m)^2}{2(\ln \sigma_m)^2} \right] \qquad (4\text{-}58)$$

$$N(d) = \frac{N}{2} + \frac{N}{2} \operatorname{erf} \frac{\ln d/d_m}{\sqrt{2} \ln \sigma_m} \qquad (4\text{-}59)$$

The mean diameter $\bar{d}$ is obtained from Eq. (4-42), using also Eqs. (4-52) to (4-54), giving

$$\bar{d} = \frac{1}{N} \int_0^\infty n(d) d \, d(d)$$

$$= \frac{1}{N} \int_0^\infty n(u) \frac{du}{du/d(d)}$$

$$= \frac{1}{\sqrt{2\pi}\, \sigma_u} \int_{-\infty}^\infty e^u e^{-(u-\bar{u})^2/2\sigma_u^2} \, du \qquad (4\text{-}60)$$

Upon expanding the integrand and simplifying, we have

$$\bar{d} = \frac{1}{\sqrt{2\pi}} e^{\bar{u}} e^{\sigma_u^2/2} \int_{-\infty}^\infty e^{-[u-(\bar{u}+\sigma_u^2)]^2/2\sigma_u^2} \frac{du}{\sigma_u}$$

Letting

$$t = \frac{u - (\bar{u} + \sigma_u^2)}{\sqrt{2}\, \sigma_u}$$

$$dt = \frac{du}{\sqrt{2}\, \sigma_u}$$

then the expression for $\bar{d}$ is

$$\bar{d} = e^{\bar{u}} e^{\sigma_u^2/2} \frac{2}{\sqrt{\pi}} \int_0^\infty e^{-t^2} \, dt$$

Since the error function of infinity is unity, we have

$$\bar{d} = e^{\bar{u}}e^{\sigma_u{}^2/2} = d_m e^{(\ln \sigma_m)^2/2} \qquad (4\text{-}61)$$

Next, we shall develop equations for the standard deviation and skewness of the log-normal distribution. Equations (4-45) and (4-46) provide the basic definitions; and when Eqs. (4-52) to (4-54) are used, these become

$$\sigma^2 = \frac{1}{N} \int_0^\infty (d - \bar{d})^2 n(d) \, d(d) = \frac{1}{\sqrt{2\pi}\,\sigma_u} \int_{-\infty}^\infty (e^u - \bar{d})^2 e^{-(u-\bar{u})^2/2\sigma_u{}^2} \, du \qquad (4\text{-}62)$$

$$\beta^3 = \frac{1}{N} \int_0^\infty (d - \bar{d})^3 n(d) \, d(d) = \frac{1}{\sqrt{2\pi}\,\sigma_u} \int_{-\infty}^\infty (e^u - \bar{d})^3 e^{-(u-\bar{u})^2/2\sigma_u{}^2} \, du \qquad (4\text{-}63)$$

These equations can be expanded to give

$$\sigma^2 = \frac{1}{\sqrt{2\pi}\,\sigma_u} \int_{-\infty}^\infty (e^{2u} - 2\bar{d}e^u + \bar{d}^2)e^{-(u-\bar{u})^2/2\sigma_u{}^2} \, du \qquad (4\text{-}64)$$

$$\beta^3 = \frac{1}{\sqrt{2\pi}\,\sigma_u} \int_{-\infty}^\infty (e^{3u} - 3\bar{d}e^{2u} + 3\bar{d}^2 e^u - \bar{d}^3)e^{-(u-\bar{u})^2/2\sigma_u{}^2} \, du \qquad (4\text{-}65)$$

With the use of the identity

$$\int_{-\infty}^\infty e^{ru}e^{-(u-\bar{u})^2/2\sigma_u{}^2} \, du = \sqrt{2\pi}\,\sigma_u\, e^{ru}e^{r^2\sigma_u{}^2/2} \qquad (4\text{-}66)$$

whose proof is not especially difficult, the integrals of Eqs. (4-64) and (4-65) are easily performed, giving

$$\sigma = (e^{2\bar{u}}e^{2\sigma_u{}^2} - 2\bar{d}e^{\bar{u}}e^{\sigma_u{}^2/2} + \bar{d}^2)^{1/2} \qquad (4\text{-}67)$$

$$\beta = (e^{3\bar{u}}e^{9\sigma_u{}^2/2} - 3\bar{d}e^{2\bar{u}}e^{2\sigma_u{}^2} + 3\bar{d}^2 e^{\bar{u}}e^{\sigma_u{}^2/2} - \bar{d}^3)^{1/3} \qquad (4\text{-}68)$$

When Eqs. (4-56) and (4-57) are substituted, Eqs. (4-67) and (4-68) may be written

$$\sigma = \bar{d}[e^{(\ln \sigma_m)^2} - 1]^{1/2} \qquad (4\text{-}69)$$

$$\beta = \bar{d}[e^{3(\ln \sigma_m)^2} - 3e^{(\ln \sigma_m)^2} + 2]^{1/3} \qquad (4\text{-}70)$$

When Eq. (4-69) is solved for σ_m, the result is

$$\sigma_m = \exp\left[\ln\left(1 + \frac{\sigma^2}{\bar{d}^2}\right)\right]^{1/2} \qquad (4\text{-}71)$$

Using this, Eq. (4-70) becomes

$$\beta = \sigma\left(\frac{3\sigma}{\bar{d}} + \frac{\sigma^3}{\bar{d}^3}\right)^{1/3} \qquad (4\text{-}72)$$

Application of the Log-Normal Distribution Function to an Actual Distribution

As indicated previously, the log-normal distribution function is the most convenient mathematical form which fits most data satisfactorily. As a consequence, in analyzing empirical data, it is well to first attempt to fit such a curve to the data. As a convenience in such an attempt, the log-normal distribution illustrated in Figs. 4-4 and 4-6 takes on a particularly convenient form if the cumulative-distribution curve is plotted on log-probability paper. This curve is shown in Fig. 4-7, in which the abscissa is plotted on a logarithmic coordinate while the ordinate is plotted on a probability scale. The cumulative-distribution curve becomes a straight line on these coordinates for the log-normal distribution. Real distributions frequently follow this curve quite closely in the midrange even if they deviate significantly from the curve at the ends. By plotting the empirical data on this type of graph paper, it can be seen immediately whether the log-normal distribution is a satisfactory mathematical model for the particular distribution.

Assuming that a curve such as the one in Fig. 4-7 has been plotted and found satisfactory, the parameters of the distribution can be obtained directly from this curve. Repeating Eq. (4-59),

$$\frac{N(d)}{N} = 0.5 + 0.5 \operatorname{erf} \frac{\ln d/d_m}{\sqrt{2} \ln \sigma_m} \qquad (4\text{-}73)$$

The diameter d_m, which is equal to the median diameter d_{me}, can be found immediately if we let $N(d)/N$ be equal to 0.5. Then

$$\operatorname{erf} \frac{\ln d/d_m}{\sqrt{2} \ln \sigma_m} = 0$$

which can be satisfied only if

$$d = d_m = d_{me} \qquad (4\text{-}74)$$

Thus, the diameter d_m is that diameter for which the cumulative distribution curve has a value of 0.5. Also, in Eq. (4-73), we can let

$$\ln \sigma_m = \ln \frac{d_\sigma}{d_m} \qquad (4\text{-}75)$$

where d_σ is the value of d for which this condition holds. Selecting this particular value of d provides two relations:

$$\sigma_m = \frac{d_\sigma}{d_m} \qquad (4\text{-}76)$$

and

$$\frac{N(d)}{N} = 0.5 + 0.5 \operatorname{erf} (0.707)$$

$$= 0.841 \qquad (4\text{-}77)$$

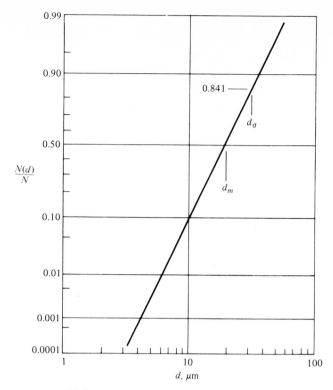

FIGURE 4-7
Cumulative distribution on log-probability coordinates.

Thus, σ_m is the ratio of the diameter for which the cumulative-distribution curve has the value of 0.841 to the median diameter. These values can be read directly off the graph. Equations (4-56) and (4-57) will then give σ_u and $\bar{u}$; Eq. (4-61) will give $\bar{d}$; and Eqs. (4-67) and (4-68) will give σ and β. All that is required is to have some log-probability paper on hand.

Utilization of Experimental Data

The experimental data from most sampling instruments is likely to be in the form of ΔN for a series of diameter ranges; that is, over a certain series of ranges in diameter, the number of particles in each range is presented. In addition, a pretty good guess as to the largest and smallest particle diameters is possible. The problem is to convert these data into values of $N(d)$ for a series of values of d. The simplest procedure is to let d_i represent the endpoints of the different ranges, with d_0 being the smallest particle diameter present. Then the respective values of ΔN are summed to obtain $N(d_i)$ at the various endpoints. This procedure can be expressed as

$$N(d_i) = \sum_{j=1}^{i} \Delta N_j \qquad (4\text{-}78)$$

Table 4-1 lists some ranges in d, along with values of ΔN for each range, from which the curve of Fig. 4-7 could be constructed. Since these data were deliberately selected as fitting the log-normal distribution of Fig. 4-7, the points calculated will lie on the curve exactly. Equation (4-78) is used to calculate the values of $N(d_i)$ shown in the table.

The foregoing analysis is based, of course, on the assumption that accurate data of the type assumed are indeed available. Measurement of particle distributions is very difficult and is as much an art as a science at present. To convert the data output of existing instruments to the form assumed in this section requires judgment and expertise, and even then the accuracy is not particularly good. Some attempts have been made to remove the judgmental factor by tailoring the analysis to fit the output of the instruments. Picknett [4] and Soole [5] have made such attempts to fit the results of a particular cascade impactor. Their analyses take into account the collection efficiencies of the separate stages of the impactor. Since we have not yet discussed collection efficiency, we shall not pursue this subject any further in the present chapter.

EXAMPLE 4-2 A collection of particles has $\bar{d} = 10$ μm with $\sigma = 5$ μm; the collection is assumed to follow a log-normal distribution. Compute d_m, d_{mo}, d_{me}, σ_m, σ_u, $\bar{u}$, and β.

SOLUTION   Equation (4-71) gives

$$\sigma_m = \exp\left[\ln\left(1 + \frac{\sigma^2}{\bar{d}^2}\right)\right]^{1/2} = \exp\left[\ln\left(1 + \frac{25}{100}\right)\right]^{1/2} = 1.604 \qquad Ans. \qquad (4\text{-}79)$$

Table 4-1   DATA FOR FIG. 4-7

Range of d, μm	i	d_i, μm	$\Delta N_i/N$	$N(d_i)/N$
0–2	0	2.0	0.0	0.0
2–8	1	8.0	0.0359	0.0359
8–14	2	14.0	0.1999	0.2358
14–20	3	20.0	0.2642	0.500
20–30	4	30.0	0.2940	0.647
30–45	5	45.0	0.1556	0.8718
45–60	6	60.0	0.0365	0.9861
60–80	7	80.0	0.0139	1.0

Note that σ_m has no units. Equation (4-72) gives

$$\beta = 5\left(3\frac{5}{10} + \frac{5^3}{10^3}\right)^{1/3} = 5.878 \ \mu m \qquad Ans.$$

Equation (4-61) gives d_m as

$$d_m = \bar{d}e^{-(\ln \sigma_m)^2/2} = 10e^{-(\ln 1.604)^2/2} = 8.944 \ \mu m \qquad Ans.$$

Equation (4-56) gives $\bar{u}$; in taking the logarithm of d_m, d_m must be expressed in m and the units suppressed:

$$\bar{u} = \ln d_m = \ln 8.944 \times 10^{-6} = -11.62 \qquad Ans.$$

Equation (4-57) gives σ_u, which, like $\bar{u}$, has no units:

$$\sigma_u = \ln \sigma_m = \ln 1.604 = 0.473 \qquad Ans.$$

The median diameter d_{me} is equal to d_m, or 8.944 μm. The modal diameter d_{mo} can be obtained from the following equation developed in Prob. 4-29 at the end of the chapter:

$$d_{mo} = \frac{d_m}{1 + \sigma^2/\bar{d}^2} = \frac{8.944}{1 + \left(\frac{5}{10}\right)^2} = 7.155 \ \mu m \qquad Ans. \qquad ////$$

4-4 COLLECTION EFFICIENCIES

Few, if any, collection devices are capable of removing all the pollutant substance from a gas stream. Some devices are more effective than others, and in the case of particulates, a given device will remove some of the particles more effectively than others. This effectiveness in removing the pollutant from the stream is measured by means of the collection efficiency. Collection efficiency is defined as the ratio of the amount collected to the total amount present. But amount of what? Several properties of the pollutant substance can be used in defining efficiency. In the case of particulates, this property can be at a single diameter or over a range of diameters. If the definition encompasses the total range of particle diameters, the efficiency is referred to as *overall efficiency*. If only particles at or near a single diameter are used, the efficiency is at a certain particle diameter.

Let us first write expressions for the collection efficiency in terms of a general property P, letting P_p be the value of this property in the entering stream, P_c the amount collected, and P_o the amount which passes through the collector. The overall efficiency is defined as

$$\eta = \frac{P_c}{P_p} \qquad (4\text{-}80)$$

The amount passing through, also called the *penetration*, is given by

$$P_o = P_p - P_c \qquad (4\text{-}81)$$

We may then write the alternative expressions for the efficiency:

$$\eta = \frac{P_c}{P_o + P_c} = \frac{1}{1 + P_o/P_c} \qquad (4\text{-}82)$$

$$\eta = \frac{P_p - P_o}{P_p} = 1 - \frac{P_o}{P_p} \qquad (4\text{-}83)$$

The preceding definitions of efficiency can also be applied to efficiencies at a single diameter $\eta(d)$, for example:

$$\eta(d) = \frac{P_c(d)}{P_p(d)} \qquad (4\text{-}84)$$

$$\eta(d) = \frac{P_c(d)}{P_o(d) + P_c(d)} = \frac{1}{1 + P_o(d)/P_c(d)} \qquad (4\text{-}85)$$

$$\eta(d) = \frac{P_p(d) - P_o(d)}{P_p(d)} = 1 - \frac{P_o(d)}{P_p(d)} \qquad (4\text{-}86)$$

in which $P_p(d)$, $P_c(d)$, and $P_o(d)$ are the property values initially, collected, and passed through, respectively, at the diameter d. The overall efficiency can be related to the efficiency at a single diameter by means of the equation

$$\eta = \frac{P_c}{P_p} = \frac{\int_0^\infty \eta(d) P_p(d)\, d(d)}{\int_0^\infty P_p(d)\, d(d)} \qquad (4\text{-}87)$$

Let us now look at some of the types of properties representable by the generalized property P. Most commonly used is mass, which can be readily specified for spherical particles or in terms of d_s for any shape of particle. Equation (4-84) becomes

$$\eta(d) = \frac{m_c(d)}{m_p(d)} \qquad (4\text{-}88)$$

Equations (4-85) and (4-86) become

$$\eta(d) = \frac{1}{1 + m_o(d)/m_c(d)} \qquad (4\text{-}89)$$

$$\eta(d) = 1 - \frac{m_o(d)}{m_p(d)} \qquad (4\text{-}90)$$

Equation (4-87) may be written using $P_p(d)$ equal to $n_p(d)m_p(d)$, where $n_p(d)$ is the distribution function for the original stream and $m_p(d)$ is the mass of a single particle of diameter d. Again using d_s,

$$\eta_m = \frac{m_c}{m_p} = \frac{\int_0^\infty \eta(d) n(d) m_p(d)\, d(d)}{\int_0^\infty n(d) m_p(d)\, d(d)} \qquad (4\text{-}91)$$

The mass of an individual particle is given by $\pi d_5^3/6$, so that Eq. (4-91) becomes

$$\eta_m = \frac{\int_0^\infty \eta(d_5)n(d_5)d_5^3\, d(d_5)}{\int_0^\infty n(d_5)d_5^3\, d(d_5)} \qquad (4\text{-}92)$$

If we write this for a log-normal particle distribution, making use of Eqs. (4-51) to (4-54), we have

$$\eta_m = \frac{\int_{-\infty}^\infty \eta(u)n(u)e^{3u}\, du}{\int_{-\infty}^\infty n(u)e^{3u}\, du} \qquad (4\text{-}93)$$

$$\eta_m = \frac{\int_{-\infty}^\infty \eta(u)e^{3u}e^{-(u-\bar{u})^2/2\sigma_u^2}\, du}{\int_{-\infty}^\infty e^{3u}e^{-(u-\bar{u})^2/2\sigma_u^2}\, du} \qquad (4\text{-}94)$$

in which $u = \ln d_5$ for nonspherical particles. Using Eq. (4-66) to evaluate the denominator, Eq. (4-94) becomes

$$\eta_m = \frac{\int_{-\infty}^\infty \eta(u)e^{3u}e^{-(u-\bar{u})^2/2\sigma_u^2}\, du}{\sqrt{2\pi}\,\sigma_u\, e^{3\bar{u}}e^{9\sigma_u^2/2}} \qquad (4\text{-}95)$$

The remaining integral can be evaluated if $\eta(u)$ is specified. We shall attempt to do this when specific collection devices are studied.

Next, let us look at the use of particle surface area as the property to define the efficiency. For a spherical particle, $P_p(d)$ becomes $n_p(d)\pi d^2$, and this same expression holds for a nonspherical particle if d is taken as d_3. Equations (4-88) to (4-90) are unchanged for the use of surface area. Equations (4-91) to (4-95) become, respectively,

$$\eta_A = \frac{A_{sc}}{A_{sp}} = \frac{\int_0^\infty \eta(d)n(d)A_{sp}(d)\, d(d)}{\int_0^\infty n(d)A_{sp}(d)\, d(d)} \qquad (4\text{-}96)$$

$$\eta_A = \frac{\int_0^\infty \eta(d_3)n(d_3)d_3^2\, d(d_3)}{\int_0^\infty n(d_3)d_3^2\, d(d_3)} \qquad (4\text{-}97)$$

$$\eta_A = \frac{\int_{-\infty}^\infty \eta(u)n(u)e^{2u}\, du}{\int_{-\infty}^\infty n(u)e^{2u}\, du} \qquad (4\text{-}98)$$

$$\eta_A = \frac{\int_{-\infty}^\infty \eta(u)e^{2u}e^{-(u-\bar{u})^2/2\sigma_u^2}\, du}{\int_{-\infty}^\infty e^{2u}e^{-(u-\bar{u})^2/2\sigma_u^2}\, du} \qquad (4\text{-}99)$$

$$\eta_A = \frac{\int_{-\infty}^\infty \eta(u)e^{2u}e^{-(u-\bar{u})^2/2\sigma_u^2}\, du}{\sqrt{2\pi}\,\sigma_u\, e^{2\bar{u}}e^{2\sigma_u^2}} \qquad (4\text{-}100)$$

Equations (4-98) to (4-100) apply to a log-normal distribution with $u = \ln d_3$ for nonspherical particles.

Next, if the number of particles is the property of interest in defining collection efficiency, $P_p(d)$ becomes $n_p(d)$ and the appropriate diameter to use is d_1 or d_2. Equations (4-91) to (4-95) become

$$\eta_n = \frac{N_c}{N_p} = \frac{\int_0^\infty \eta(d)n(d)\,d(d)}{\int_0^\infty n(d)\,d(d)} \qquad (4\text{-}101)$$

$$\eta_n = \frac{\int_{-\infty}^\infty \eta(n)n(u)\,du}{\int_{-\infty}^\infty n(u)\,du} \qquad (4\text{-}102)$$

$$\eta_n = \frac{\int_{-\infty}^\infty \eta(u)e^{-(u-\bar u)^2/2\sigma_u^2}\,du}{\int_{-\infty}^\infty e^{-(u-\bar u)^2/2\sigma_u^2}\,du}$$

$$\eta_n = \frac{\int_{-\infty}^\infty \eta(u)e^{-(u-\bar u)^2/2\sigma_u^2}\,du}{\sqrt{2\pi}\,\sigma_u} \qquad (4\text{-}103)$$

In the expressions for η_m, η_A, and η_n, the integral has not been evaluated owing to the need for detailed information about $\eta(u)$. Although such detailed information can be obtained only from a study of a particular collection device, we can at this time evaluate the integral for certain forms in which the information might appear. Thus, if $\eta(u)$ is given by an expression of the form

$$\eta(u) = 1 - \sum_{i=1}^\infty \alpha_i e^{-\beta_i u} \qquad d_l < d < \infty \qquad (4\text{-}104)$$

we can proceed to evaluate the integrals in a general way. We shall assume that the expression of Eq. (4-104) holds only for diameter d larger than a certain minimum value d_l. Below this value, $\eta(u)$ will not be specified at this time, although in practice its value will often be zero.

To evaluate the integrals, an identity similar to Eq. (4-66) but with finite limits will be needed. This can be shown to be

$$\int_{L_1}^{L_2} e^{ru}e^{-(u-\bar u)^2/2\sigma_u^2}\,du = \sqrt{\frac{\pi}{2}}\,\sigma_u e^{r\bar u}e^{r^2\sigma_u^2/2}\left[\text{erf}\frac{L_2-(\bar u+r\sigma_u^2)}{\sqrt 2\,\sigma_u} - \text{erf}\frac{L_1-(\bar u+r\sigma_u^2)}{\sqrt 2\,\sigma_u}\right]$$
$$(4\text{-}105)$$

Using this identity, Eq. (4-95) for η_m can be written

$$\eta_m = (\sqrt{2\pi}\,\sigma_u e^{3\bar u}e^{9\sigma_u^2/2})^{-1}\left[\int_{-\infty}^{u_l}\eta(u)e^{3u}e^{-(u-\bar u)^2/2\sigma_u^2}\,du\right.$$

$$\left. + \int_{u_l}^\infty e^{3u}e^{-(u-\bar u)^2/2\sigma_u^2}\,du - \sum_{i=1}^\infty \alpha_i\int_{u_l}^\infty e^{(3-\beta_i)u}e^{-(u-\bar u)^2/2\sigma_u^2}\,du\right]$$

$$\eta_m = \frac{\int_{-\infty}^{u_l}\eta(u)e^{3u}e^{-(u-\bar u)^2/2\sigma_u^2}\,du}{\sqrt{2\pi}\,\sigma_u e^{3\bar u}e^{9\sigma_u^2/2}} + \frac{1}{2} - \frac{1}{2}\text{erf}\frac{u_l-\bar u-3\sigma_u^2}{\sqrt 2\,\sigma_u}$$

$$- \frac{1}{2}\sum_{i=1}^\infty \alpha_i e^{-\beta_i\bar u}e^{\beta_i(\beta_i-6)\sigma_u^2/2}\left(1 - \text{erf}\frac{u_l-\bar u-3\sigma_u^2+\beta_i\sigma_u^2}{\sqrt 2\,\sigma_u}\right) \qquad (4\text{-}106)$$

In a similar way, η_A and η_n become

$$\eta_A = \frac{\int_{-\infty}^{u_l} \eta(u)e^{2u}e^{-(u-\bar{u})^2/2\sigma_u^2}\,du}{\sqrt{2\pi}\,\sigma_u\,e^{2\bar{u}}e^{2\sigma_u^2}} + \frac{1}{2} - \frac{1}{2}\,\mathrm{erf}\,\frac{u_l - \bar{u} - 2\sigma_u^2}{\sqrt{2}\,\sigma_u}$$
$$- \frac{1}{2}\sum_{i=1}^{\infty}\alpha_i\,e^{-\beta_i\bar{u}}e^{\beta_i(\beta_i-4)\sigma_u^2/2}\left(1 - \mathrm{erf}\,\frac{u_l - \bar{u} - 2\sigma_u^2 + \beta_i\sigma_u^2}{\sqrt{2}\,\sigma_u}\right) \quad (4\text{-}107)$$

$$\eta_n = \frac{\int_{-\infty}^{u_l} \eta(u)e^{-(u-\bar{u})^2/2\sigma_u^2}\,du}{\sqrt{2\pi}\,\sigma_u} + \frac{1}{2} - \frac{1}{2}\,\mathrm{erf}\,\frac{u_l - \bar{u}}{\sqrt{2}\,\sigma_u}$$
$$- \frac{1}{2}\sum_{i=1}^{\infty}\alpha_i\,e^{-\beta_i\bar{u}}e^{\beta_i^2\sigma_u^2/2}\left(1 - \mathrm{erf}\,\frac{u_l - \bar{u} + \beta_i\sigma_u^2}{\sqrt{2}\,\sigma_u}\right) \quad (4\text{-}108)$$

For future reference also, Eq. (4-104) can be related to diameter d, noting that $u = \ln d$, as

$$\eta(d) = 1 - \sum_{i=1}^{\infty}\frac{\alpha_i}{d^{\beta_i}} \quad (4\text{-}109)$$

In most cases, it is expected that where Eq. (4-109) is used to approximate an actual distribution for $\eta(d)$, the choice for β_i will be

$$\beta_i = i \quad (4\text{-}110)$$

We shall rewrite Eqs. (4-106) to (4-108) to reflect this additional expectation. Note, again, that we shall most often use the preceding equations with $\eta(u)$ equal to zero for $u < u_l$. With these two changes the equations become

$$\eta_m = \frac{1}{2}\,\mathrm{erfc}\,\frac{u_l - \bar{u} - 3\sigma_u^2}{\sqrt{2}\,\sigma_u} - \frac{1}{2}\sum_{i=1}^{\infty}\alpha_i\,e^{-i\bar{u}}e^{i(i-6)\sigma_u^2/2}\,\mathrm{erfc}\,\frac{u_l - \bar{u} - 3\sigma_u^2 + i\sigma_u^2}{\sqrt{2}\,\sigma_u}$$
$$(4\text{-}111)$$

$$\eta_A = \frac{1}{2}\,\mathrm{erfc}\,\frac{u_l - \bar{u} - 2\sigma_u^2}{\sqrt{2}\,\sigma_u} - \frac{1}{2}\sum_{i=1}^{\infty}\alpha_i\,e^{-i\bar{u}}e^{i(i-4)\sigma_u^2/2}\,\mathrm{erfc}\,\frac{u_l - \bar{u} - 2\sigma_u^2 + i\sigma_u^2}{\sqrt{2}\,\sigma_u}$$
$$(4\text{-}112)$$

$$\eta_n = \frac{1}{2}\,\mathrm{erfc}\,\frac{u_l - \bar{u}}{\sqrt{2}\,\sigma_u} - \frac{1}{2}\sum_{i=1}^{\infty}\alpha_i\,e^{-i\bar{u}}e^{i^2\sigma_u^2/2}\,\mathrm{erfc}\,\frac{u_l - \bar{u} + i\sigma_u^2}{\sqrt{2}\,\sigma_u} \quad (4\text{-}113)$$

in which the complementary error function, erfc (x), is used to replace the error function

$$\mathrm{erfc}\,(x) = 1 - \mathrm{erf}\,(x) \quad (4\text{-}114)$$

[A table of values of erf (x) and erfc (x) is provided in Appendix 2.]

In assessing the collection efficiency of a device, it must be kept in mind that $(1 - \eta)$ is the fraction of the original particles which get through the collector without being captured. Thus a collector with 0.99 efficiency allows 0.01 of the original particles to escape. A collector twice as effective could hardly capture 198 percent of the original particles, and so we must look elsewhere to define what this collector would do relative to the first collector. It makes more sense to say that the second collector allows only half as many particles to escape if it is twice as good, so that its efficiency would be 0.995. Thus, a collector is rated as to effectiveness by how small its value of $(1 - \eta)$ is.

EXAMPLE 4-3   A collector has a diameter efficiency for $d > 10$ μm given by

$$\eta(d) = 1 - \frac{10}{d} \qquad d, \mu m$$

For $d < 10$ μm, the efficiency is zero. The collector operates on a log-normal particle distribution for which $\bar{u}$ is -10.5 and σ_u is 0.3. Compute η_m, η_A, and η_n for this collector acting on this particle distribution.

SOLUTION   By comparison with Eq. (4-109), $\alpha_1 = 10^{-5}$. The minimum diameter for which particles can be collected d_l is 10 μm, so that $u_l = \ln d_l = \ln 10^{-5} = -11.51$. Equation (4-111) becomes

$$\eta_m = \frac{1}{2} \text{erfc} \frac{u_l - \bar{u} - 3\sigma_u^2}{\sqrt{2}\,\sigma_u} - \frac{10^{-5}}{2} e^{-\bar{u}} e^{-2.5\sigma_u^2} \text{erfc} \frac{u_l - \bar{u} - 2\sigma_u^2}{\sqrt{2}\,\sigma_u}$$

which reduces to

$$\eta_m = \tfrac{1}{2} \text{erfc}\,(-3.017) - 0.145\,\text{erfc}\,(-2.805) = 0.710 \qquad Ans.$$

Equation (4-112) for η_A and Eq. (4-113) for η_n are evaluated similarly giving

$$\eta_A = 0.683 \qquad Ans.$$
$$\eta_n = 0.653 \qquad Ans. \qquad ////$$

4-5   MULTIPLE COLLECTORS

One of the most seemingly fertile fields of exploration to improve efficiency of collection using existing technology is to combine two or more devices into a single unit. By this means it is hoped to achieve higher collection efficiencies overall than either of the devices can achieve by itself. Usually this hope is not realized in practice or often even in theory since for most collectors collection efficiency drops off at smaller diameters. In other words, the smaller particles that get through the first collector will probably get through the second collector also, thus rendering the combination of little greater usefulness. Although some slight improvement can usually be obtained in this way, more often it is cheaper to purchase a single more efficient collector than to try to obtain better efficiency by placing two cheaper collectors in series.

On the other hand, there are ways in which several collectors can be arranged to give substantially improved efficiency, at least in theory. There are also certain advantages to the use of multiple collectors other than improved overall efficiency. For instance, a concentrator (to be discussed later) can be used to clean an airstream by diverting a portion of it containing most of the dust to some central or convenient point for final collection. Another instance is where a collector that is efficient for large particles only is used ahead of a collector that is efficient for all particles. The first collector removes most of the mass, thus making it easier for the second collector to operate and possibly improving its efficiency for small particles. In this section we shall determine how to evaluate the efficiency of a combination of collectors if the efficiency of each collector is known as a function of particle diameter.

As a first example, consider the case of two collectors in series, as shown in Fig. 4-8. The partially cleaned gas from collector 1 is fed into collector 2, where it is then cleaned more thoroughly. The overall efficiency of two collectors is found by treating the combination as though it were a single collector. The amount collected from the combination is the total collected from each individual collector; the cleaned gas from the combination is that leaving the last individual collector. In some combinations, more than one stream of cleaned gas is possible; in such a case, the gas leaving will be the sum of these streams.

The amount of the generalized property P or in terms of diameter $P(d)$ which is collected by collector 1 is given as

$$P_{c1}(d) = \eta_1(d)P_{p1}(d) \qquad (4\text{-}115)$$

and the amount leaving the collector in the gas stream is

$$P_{o1}(d) = [1 - \eta_1(d)]P_{p1}(d) \qquad (4\text{-}116)$$

This is also the amount of P which enters collector 2. The amount collected by 2 and the amount leaving 2 in the gas stream are, respectively,

$$P_{c2}(d) = \eta_2(d)P_{p2}(d) = \eta_2(d)[1 - \eta_1(d)]P_{p1}(d) \qquad (4\text{-}117)$$

$$P_{o2}(d) = [1 - \eta_2(d)]P_{p2}(d) = [1 - \eta_2(d)][1 - \eta_1(d)]P_{p1}(d) \qquad (4\text{-}118)$$

The efficiency of the combination at diameter d is given by Eq. (4-86) as

$$\eta(d) = 1 - \frac{P_{o2}(d)}{P_{p1}(d)} \qquad (4\text{-}119)$$

and making use of Eq. (4-118), this becomes

$$\eta(d) = 1 - [1 - \eta_2(d)][1 - \eta_1(d)] = \eta_1(d) + \eta_2(d) - \eta_1(d)\eta_2(d) \qquad (4\text{-}120)$$

The overall value of P leaving collector 2 is obtained by integrating $P_{o2}(d)$ as

$$P_{o2} = \int_0^\infty P_{o2}(d)\,d(d) = \int_0^\infty [1 - \eta_1(d)][1 - \eta_2(d)]P_{p1}(d)\,d(d) \qquad (4\text{-}121)$$

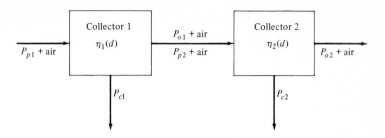

FIGURE 4-8
Combination of two collectors in series.

Then the overall efficiency of the combination is

$$\eta = 1 - \frac{P_{o2}}{P_{p1}} = 1 - \frac{\int_0^\infty [1 - \eta_1(d) - \eta_2(d) + \eta_1(d)\eta_2(d)]P_{p1}(d)\,d(d)}{\int_0^\infty P_{p1}(d)\,d(d)}$$

which reduces to

$$\eta = \frac{\int_0^\infty [\eta_1(d) + \eta_2(d) - \eta_1(d)\eta_2(d)]P_{p1}(d)\,d(d)}{\int_0^\infty P_{p1}(d)\,d(d)} \qquad (4\text{-}122)$$

Equation (4-122) can be written for the particular values of P—mass, surface area, and number of particles—in a development paralleling that of Eqs. (4-92), (4-97), and (4-101), using the appropriate diameter in each case. This gives

$$\eta_m = \frac{\int_0^\infty [\eta_1(d) + \eta_2(d) - \eta_1(d)\eta_2(d)]n(d)d^3\,d(d)}{\int_0^\infty n(d)d^3\,d(d)} \qquad (4\text{-}123)$$

$$\eta_A = \frac{\int_0^\infty [\eta_1(d) + \eta_2(d) - \eta_1(d)\eta_2(d)]n(d)d^2\,d(d)}{\int_0^\infty n(d)d^2\,d(d)} \qquad (4\text{-}124)$$

$$\eta_n = \frac{\int_0^\infty [\eta_1(d) + \eta_2(d) - \eta_1(d)\eta_2(d)]n(d)\,d(d)}{\int_0^\infty n(d)\,d(d)} \qquad (4\text{-}125)$$

These can also be written in terms of u using Eqs. (4-51) to (4-53), giving, respectively,

$$\eta_m = \frac{\int_{-\infty}^\infty [\eta_1(u) + \eta_2(u) - \eta_1(u)\eta_2(u)]n(u)e^{3u}\,du}{\int_{-\infty}^\infty n(u)e^{3u}\,du} \qquad (4\text{-}126)$$

$$\eta_A = \frac{\int_{-\infty}^\infty [\eta_1(u) + \eta_2(u) - \eta_1(u)\eta_2(u)]n(u)e^{2u}\,du}{\int_{-\infty}^\infty n(u)e^{2u}\,du} \qquad (4\text{-}127)$$

$$\eta_n = \frac{\int_{-\infty}^\infty [\eta_1(u) + \eta_2(u) - \eta_1(u)\eta_2(u)]n(u)\,du}{\int_{-\infty}^\infty n(u)\,du} \qquad (4\text{-}128)$$

If we now use Eq. (4-54) for the log-normal distribution, these equations become

$$\eta_m = \frac{\int_{-\infty}^{\infty} [\eta_1(u) + \eta_2(u) - \eta_1(u)\eta_2(u)]e^{3u}e^{-(u-\bar{u})^2/2\sigma_u^2}\,du}{\sqrt{2\pi}\,\sigma_u\,e^{3\bar{u}}e^{9\sigma_u^2/2}} \qquad (4\text{-}129)$$

$$\eta_A = \frac{\int_{-\infty}^{\infty} [\eta_1(u) + \eta_2(u) - \eta_1(u)\eta_2(u)]e^{2u}e^{-(u-\bar{u})^2/2\sigma_u^2}\,du}{\sqrt{2\pi}\,\sigma_u\,e^{2\bar{u}}e^{2\sigma_u^2}} \qquad (4\text{-}130)$$

$$\eta_n = \frac{\int_{-\infty}^{\infty} [\eta_1(u) + \eta_2(u) - \eta_1(u)\eta_2(u)]e^{-(u-\bar{u})^2/2\sigma_u^2}\,du}{\sqrt{2\pi}\,\sigma_u} \qquad (4\text{-}131)$$

Note that these equations are obtained from Eqs. (4-95), (4-100), and (4-103) by replacing $\eta(u)$ by $\eta_1(u) + \eta_2(u) - \eta_1(u)\eta_2(u)$, as given by Eq. (4-120) for this particular combination of two collectors in series.

Let us now use an expression similar to Eq. (4-104) to evaluate further Eqs. (4-129) to (4-131). Let us assume that $\eta_1(u)$ and $\eta_2(u)$ can be written as

$$\eta_1 = 1 - \sum_{i=1}^{\infty} \alpha_i e^{-iu} \qquad u > u_l \qquad (4\text{-}132)$$

$$\eta_2 = 1 - \sum_{j=1}^{\infty} \beta_j e^{-ju} \qquad u > u_l \qquad (4\text{-}133)$$

and each efficiency is zero for $u < u_l$. The combined efficiency expression becomes

$$\eta_1(u) + \eta_2(u) - \eta_1(u)\eta_2(u) = 1 - \sum_{i=1}^{\infty}\sum_{j=1}^{\infty} \alpha_i \beta_j e^{-(i+j)u} \qquad (4\text{-}134)$$

and Eqs. (4-111) to (4-113) become

$$\eta_m = \frac{1}{2}\,\mathrm{erfc}\,\frac{u_l - \bar{u} - 3\sigma_u^2}{\sqrt{2}\,\sigma_u}$$
$$-\frac{1}{2}\sum_{i=1}^{\infty}\sum_{j=1}^{\infty} \alpha_i \beta_j e^{-(i+j)\bar{u}}e^{(i+j)(i+j-6)\sigma_u^2/2}\,\mathrm{erfc}\frac{u_l - \bar{u} - (3 - i - j)\sigma_u^2}{\sqrt{2}\sigma_u} \qquad (4\text{-}135)$$

$$\eta_A = \frac{1}{2}\,\mathrm{erfc}\,\frac{u_l - \bar{u} - 2\sigma_u^2}{\sqrt{2}\,\sigma_u}$$
$$-\frac{1}{2}\sum_{i=1}^{\infty}\sum_{j=1}^{\infty} \alpha_i \beta_j e^{-(i+j)\bar{u}}e^{(i+j)(i+j-4)\sigma_u^2/2}\,\mathrm{erfc}\frac{u_l - \bar{u} - (2 - i - j)\sigma_u^2}{\sqrt{2}\,\sigma_u} \qquad (4\text{-}136)$$

$$\eta_n = \frac{1}{2}\,\mathrm{erfc}\!\left(\frac{u_l - \bar{u}}{\sqrt{2}\sigma_u}\right)$$
$$-\frac{1}{2}\sum_{i=1}^{\infty}\sum_{j=1}^{\infty} \alpha_i \beta_j e^{-(i+j)\bar{u}}e^{(i+j)^2\sigma_u^2/2}\,\mathrm{erfc}\frac{u_l - \bar{u} + (i + j)\sigma_u^2}{\sqrt{2}\,\sigma_u} \qquad (4\text{-}137)$$

The particular efficiency distribution used in Eqs. (4-132) to (4-137) has one serious drawback which renders these equations of questionable value in practice: The same lower limit u_l is used for each individual efficiency. In practice, the two collectors would have quite different characteristics, so that the lower limit would have different values for the two collectors. This factor can be taken into account at the expense of making the resulting equations very complicated. We shall not attempt this refinement at the present time.

A device similar to a collector but serving a somewhat different function is known as a *concentrator*. A concentrator is a device which takes in the stream of air and pollutant and discharges two streams. One stream is cleaned, so that the amount of pollutant contained in it is small compared with that in the original stream. The second stream is very dirty, containing most of the original pollutant. Normally, the dirty stream will contain less than 10 percent of the gas, so that the pollutant loading will be relatively high in that stream. A number of collector devices can be modified to become concentrators; the cyclone collector is a case in point. Fairly effective concentrators can be manufactured and installed very cheaply. Thus the use of concentrators has many advantages. However, no concentrator can be used by itself since a collector must be provided somewhere to do the final collection. It is obvious that in a large factory, where many sources of polluted air are spread over a wide area, concentrators could be located at convenient points, with the dirty streams from all these concentrators brought together to a central collector. We shall consider some examples of concentrators used in combination with collectors.

Consider, first, the case of a single concentrator used in series with a single collector, as shown in Fig. 4-9. The efficiency of a concentrator is defined in a manner similar to that of a collector:

$$\eta(d) = \frac{P_d(d)}{P_p(d)} \quad (4\text{-}138)$$

For P_{p2} we may write

$$P_{p2}(d) = \eta_1(d)P_{p1}(d)$$

Then P_{o2} is given by

$$P_{o2}(d) = [1 - \eta_2(d)]P_{p2}(d) = \eta_1(d)[1 - \eta_2(d)]P_{p1}(d)$$

so that

$$\frac{P_{o2}(d)}{P_{p1}(d)} = \eta_1(d)[1 - \eta_2(d)]$$

The efficiency of the combination is given by

$$\eta(d) = 1 - \frac{P_{o1}(d) + P_{o2}(d)}{P_{p1}(d)}$$

$$= 1 - \frac{P_{o1}(d)}{P_{p1}(d)} - \frac{P_{o2}(d)}{P_{p1}(d)}$$

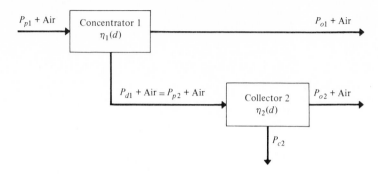

FIGURE 4-9
Concentrator and collector in series.

From the definition of efficiency of the concentrator,

$$\frac{P_{o1}(d)}{P_{p1}(d)} = 1 - \eta_1(d)$$

Then the overall efficiency of the combination at diameter d is

$$\eta(d) = 1 - [1 - \eta_1(d)] - \eta_1(d)[1 - \eta_2(d)] = \eta_1(d)\eta_2(d) \qquad (4\text{-}139)$$

This expression can be used in conjunction with Eqs. (4-95), (4-100), and (4-103) to evaluate the overall efficiency of the combination, based on a log-normal particle distribution and an appropriate efficiency distribution. The resulting equations are analogous to Eqs. (4-129) to (4-131):

$$\eta_m = \frac{\int_{-\infty}^{\infty} \eta_1(u)\eta_2(u)e^{3u}e^{-(u-\bar{u})^2/2\sigma_u^2}\,du}{\sqrt{2\pi}\,\sigma_u\,e^{3\bar{u}}e^{9\sigma_u^2/2}} \qquad (4\text{-}140)$$

$$\eta_A = \frac{\int_{-\infty}^{\infty} \eta_1(u)\eta_2(u)e^{2u}e^{-(u-\bar{u})^2/2\sigma_u^2}\,du}{\sqrt{2\pi}\,\sigma_u\,e^{2\bar{u}}e^{2\sigma_u^2}} \qquad (4\text{-}141)$$

$$\eta_n = \frac{\int_{-\infty}^{\infty} \eta_1(u)\eta_2(u)e^{-(u-\bar{u})^2/2\sigma_u^2}\,du}{\sqrt{2\pi}\,\sigma_u} \qquad (4\text{-}142)$$

Using the efficiency distributions given by Eqs. (4-132) and (4-133), we can evaluate the integrals in the preceding equations. Note that the product $\eta_1\eta_2$ is zero if either efficiency is zero, so that u_l can be the larger of the two actual cutoff diameters. The equations are

$$\eta_m = \frac{1}{2}\,\mathrm{erfc}\,\frac{u_l - \bar{u} - 3\sigma_u^2}{\sqrt{2}\,\sigma_u}$$

$$-\frac{1}{2}\sum_{i=1}^{\infty}(\alpha_i + \beta_i)e^{-i\bar{u}}e^{i(i-6)\sigma_u^2/2}\,\mathrm{erfc}\,\frac{u_l - \bar{u} - 3\sigma_u^2 + i\sigma_u^2}{\sqrt{2}\,\sigma_u}$$

$$+\frac{1}{2}\sum_{i=1}^{\infty}\sum_{j=1}^{\infty}\alpha_i\beta_j e^{-(i+j)\bar{u}}e^{(i+j)(i+j-6)\sigma_u^2/2}\,\mathrm{erfc}\,\frac{u_l - \bar{u} - (3 - i - j)\sigma_u^2}{\sqrt{2}\,\sigma_u} \qquad (4\text{-}143)$$

$$\eta_A = \frac{1}{2} \, \text{erfc} \, \frac{u_l - \bar{u} - 2\sigma_u^2}{\sqrt{2}\,\sigma_u}$$

$$-\frac{1}{2} \sum_{i=1}^{\infty} (\alpha_i + \beta_i) e^{-i\bar{u}} e^{i(i-4)\sigma_u^2/2} \, \text{erfc} \, \frac{u_l - \bar{u} - 2\sigma_u^2 + i\sigma_u^2}{\sqrt{2}\,\sigma_u}$$

$$+\frac{1}{2} \sum_{i=1}^{\infty} \sum_{j=1}^{\infty} \alpha_i \beta_j \, e^{-(i+j)\bar{u}} e^{(i+j)(i+j-4)\sigma_u^2/2} \, \text{erfc} \, \frac{u_l - \bar{u} - (2 - i - j)\sigma_u^2}{\sqrt{2}\,\sigma_u} \qquad (4\text{-}144)$$

$$\eta_n = \frac{1}{2} \, \text{erfc} \, \frac{u_l - \bar{u}}{\sqrt{2}\,\sigma_u}$$

$$-\frac{1}{2} \sum_{i=1}^{\infty} (\alpha_i + \beta_i) e^{-i\bar{u}} e^{i^2\sigma_u^2/2} \, \text{erfc} \, \frac{u_l - \bar{u} + i\sigma_u^2}{\sqrt{2}\,\sigma_u}$$

$$+\frac{1}{2} \sum_{i=1}^{\infty} \sum_{j=1}^{\infty} \alpha_i \beta_j \, e^{-(i+j)\bar{u}} e^{(i+j)^2\sigma_u^2/2} \, \text{erfc} \, \frac{u_l - \bar{u} + i\sigma_u^2 + j\sigma_u^2}{\sqrt{2}\,\sigma_u} \qquad (4\text{-}145)$$

The concentrator-collector combination of Fig. 4-9 can be modified as shown in Fig. 4-10 to feed the output from the collector back into the input to the concentrator. To determine the efficiency of the combination, note that $P_{o1} = (1 - \eta_1)P_{p1}$. Then

$$\eta = 1 - \frac{P_{o1}}{P_{p0}} = 1 - (1 - \eta_1)\frac{P_{p1}}{P_{p0}}$$

Next, we shall evaluate P_{p1}, which is equal to $P_{p0} + P_{o2}$. To evaluate P_{o2}, we start with

$$P_{d1} = \eta_1 P_{p1}$$

Then

$$P_{o2} = (1 - \eta_2)P_{d1} = \eta_1(1 - \eta_2)P_{p1} = \eta_1(1 - \eta_2)(P_{p0} + P_{o2})$$

Solving this for P_{o2}, we obtain

$$P_{o2} = \frac{\eta_1(1 - \eta_2)P_{p0}}{1 - \eta_1 + \eta_1\eta_2}$$

Then P_{p1} becomes

$$P_{p1} = P_{p0} + P_{o2} = P_{p0}\left[1 + \frac{\eta_1(1 - \eta_2)}{1 - \eta_1 + \eta_1\eta_2}\right] = \frac{P_{p0}}{1 - \eta_1 + \eta_1\eta_2}$$

Then the expression for efficiency becomes

$$\eta = 1 - \frac{1 - \eta_1}{1 - \eta_1 + \eta_1\eta_2} = \frac{\eta_1\eta_2}{1 - \eta_1 + \eta_1\eta_2}$$

This is the efficiency at a particular diameter of particle, so that we may write

$$\eta(d) = \frac{\eta_1(d)\eta_2(d)}{1 - \eta_1(d) + \eta_1(d)\eta_2(d)} \qquad (4\text{-}146)$$

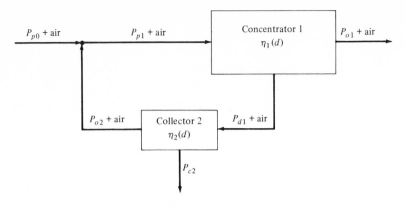

FIGURE 4-10
Concentrator and collector in series with feedback.

This equation is sufficiently complex that evaluation of the overall efficiency of the combination will need the use of a computer. Hence, we shall not substitute this relation into Eqs. (4-95), (4-100), and (4-103) for the log-normal distribution.

As a final example, let us consider the case of a single concentrator in series with two collectors, the output of the smaller collector being fed back to the supply to the concentrator, as shown in Fig. 4-11. To analyze this combination, we first write

$$\eta = 1 - \frac{P_{o2}}{P_{p0}}$$

Next, for P_{o2} we may write in succession

$$P_{o2} = (1 - \eta_2)P_{o1} = (1 - \eta_1)(1 - \eta_2)P_{p1} = (1 - \eta_1)(1 - \eta_2)(P_{p0} + P_{o3})$$

Now, for P_{o3} we may write successively

$$\begin{aligned}
P_{o3} &= (1 - \eta_3)P_{p3} \\
&= (1 - \eta_3)P_{d1} \\
&= (1 - \eta_3)\eta_1 P_{p1} \\
&= \eta_1(1 - \eta_3)(P_{p0} + P_{o3})
\end{aligned}$$

so that P_{o3} becomes

$$P_{o3} = \frac{\eta_1(1 - \eta_3)P_{p0}}{1 - \eta_1(1 - \eta_3)}$$

Then, substituting this into the expression for P_{o2} gives

$$P_{o2} = P_{p0}(1 - \eta_1)(1 - \eta_2)\left[1 + \frac{\eta_1(1 - \eta_3)}{1 - \eta_1(1 - \eta_3)}\right] = \frac{(1 - \eta_1)(1 - \eta_2)P_{p0}}{1 - \eta_1(1 - \eta_3)}$$

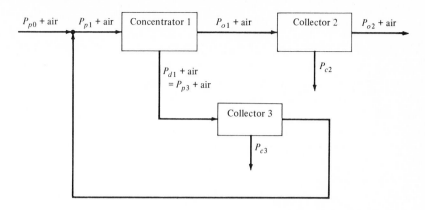

FIGURE 4-11
Concentrator and two collectors.

Finally, the equation for η becomes

$$\eta = 1 - \frac{(1 - \eta_1)(1 - \eta_2)}{1 - \eta_1(1 - \eta_3)} = \frac{\eta_2 + \eta_1\eta_3 - \eta_1\eta_2}{1 - \eta_1(1 - \eta_3)}$$

Writing this for the efficiency at a single diameter gives

$$\eta(d) = \frac{\eta_2(d) + \eta_1(d)\eta_3(d) - \eta_1(d)\eta_2(d)}{1 - \eta_1(d) + \eta_1(d)\eta_3(d)} \qquad (4\text{-}147)$$

This relation can be substituted into Eqs. (4-95), (4-100), or (4-103) to obtain expressions for η_m, η_A, or η_n, respectively. As in the previous case, a computer evaluation will be required.

We may rewrite Eq. (4-147) in the following forms:

$$\eta = \eta_1 + \frac{(1 - \eta_1)[\eta_2 - \eta_1(1 - \eta_3)]}{1 - \eta_1 + \eta_1\eta_3} \qquad (4\text{-}148)$$

$$\eta = \eta_2 + \frac{\eta_1\eta_3(1 - \eta_2)}{1 - \eta_1 + \eta_1\eta_3} \qquad (4\text{-}149)$$

We see that for reasonable values of efficiencies, η is greater than η_1, and that for any choice of values of the three efficiencies, η is greater than η_2. Calculating the overall efficiency for a few choices of η_1, η_2, and η_3, gives the following values:

η_1	η_2	η_3	η
0.9	0.9	0.9	0.989
0.99	0.99	0.99	0.999899
0.7	0.99	0.9	0.99677
0.5	0.99	0.5	0.99333
0.5	0.9	0.5	0.93333

We see that this arrangement can be very advantageous for cases where a highly efficient collector is used as collector 2, where a relatively cheap and only moderately efficient concentrator and collector 3 are used, and where it is desirable to remove the bulk of large particles from the stream before it enters the efficient collector. The efficiencies just calculated might apply to a particle with a fairly small diameter; for larger particles the efficiencies would be much higher.

EXAMPLE 4-4 The combination of concentrator and collector of Fig. 4-10 has individual efficiencies of $\eta_1 = 0.9$ and $\eta_2 = 0.6$. What is the efficiency of the combination?

SOLUTION Equation (4-146) gives

$$\eta = \frac{\eta_1 \eta_2}{1 - \eta_1 + \eta_1 \eta_2} = \frac{0.9(0.6)}{1 - 0.9 + 0.9(0.6)} = 0.844 \qquad Ans. \qquad ////$$

4-6 CONCENTRATION OF POLLUTANT IN ENCLOSURES WITH CONTROLLED AIR SUPPLY

Let us change our focus at this point to consider a problem associated with the maintenance of a conditioned space at suitable levels of pollution. Although the theory would be equally applicable if a high level of pollution concentration were desired, in practice a low level of concentration is sought. This can be achieved if air is withdrawn from the space, some of this air exchanged for outside air, and the mixture filtered to lower the contaminant level before reinjecting the air into the space. This process is illustrated in Fig. 4-12, which shows two filters, one active on the supply air and one treating the replacement air from outside. Although filters are used here, any type of conditioning device could be used.

We shall derive an expression for the concentration of pollutant in the conditioned space in terms of the other variables of the system. The concentration used can be either C_m, C_v, C_{mv}, or C_{ppm}. For our derivation we shall designate the concentration simply as C. For equilibrium, the rate at which makeup air is supplied must equal the rate at which air is discharged. The rate at which air is exhausted from the space equals the supply rate $Q_r = Q_s$. The recirculated air rate is $Q_{rec} = Q_r - Q_o$. The quantity of pollutant supplied to the space is $Q_s C_s$ unless C is taken as C_m, in which case a density factor should be included. For our purposes, density can be assumed constant, and so the same result is obtained for C_m, as well as for C_v, C_{mv}, or C_{ppm}. Thus

$$Q_s C_s = [Q_r C_r - Q_o C_r + Q_o C_o (1 - \eta_2)](1 - \eta_1) \qquad (4\text{-}150)$$

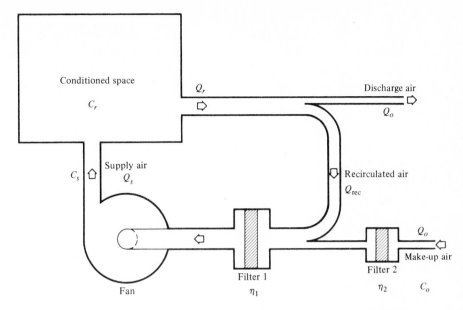

FIGURE 4-12
Air-filtering system.

There are two cases we should consider. In the first case, there will be substantially no production of pollutant in the conditioned space, so that $C_r = C_s$. For this condition

$$C_r Q_s = (1 - \eta_1)C_r Q_r - (1 - \eta_1)C_r Q_o + (1 - \eta_1)(1 - \eta_2)C_o Q_o$$

Solving this for C_r leads to

$$C_r = \frac{(1 - \eta_1)(1 - \eta_2)Q_o C_o}{\eta_1 Q_r + (1 - \eta_1)Q_o} \qquad (4\text{-}151)$$

If we define a make-up fraction f as

$$f = \frac{Q_o}{Q_s} = \frac{Q_o}{Q_r} \qquad (4\text{-}152)$$

then Eq. (4-151) becomes

$$C_r = \frac{(1 - \eta_1)(1 - \eta_2)f C_o}{\eta_1 + (1 - \eta_1)f} \qquad (4\text{-}153)$$

The second case of importance is that which will occur in industrial ventilation situations, where a pollutant is released into the air in the space and ventilation is provided to keep the room concentration level acceptably low. For this purpose, we let R be the rate at which the pollutant is being discharged into the room air. In the equations which follow, C_r and C_s must be restricted in form to either C_{mv}, in which case R is the mass release rate of pollutant, or to C_v, in which case R is the volume release rate:

$$R = Q_r(C_r - C_s) = Q_s(C_r - C_s) \qquad (4\text{-}154)$$

Then Eq. (4-150) becomes

$$C_r Q_s - R = (1 - \eta_1)Q_s C_r + (1 - \eta_1)[(1 - \eta_2)C_o - C_r]Q_o$$

Solving this for Q_s gives

$$Q_s = \frac{R + (1 - \eta_1)[(1 - \eta_2)C_o - C_r]Q_o}{\eta_1 C_r} \qquad (4\text{-}155)$$

which is the flow rate necessary to maintain a pollutant level not greater than C_r in the conditioned space.

Of course, local concentrations of the pollutant can exceed this level by far in certain parts of the space before dilution has taken place. If, as is usually the case, there is no pollutant being brought in with the outside air, then C_o is zero and

$$Q_o = f Q_s \qquad (4\text{-}156)$$

$$Q_s = \frac{R}{C_r[(1 - f)\eta_1 + f]} \qquad (4\text{-}157)$$

The concentration to which the supply must be reduced by a collector is given by Eq. (4-154) as

$$C_s = C_r - \frac{R}{Q_s} \qquad (4\text{-}158)$$

EXAMPLE 4-5   The allowable concentration of methylamine in a workspace is 10 ppm. If the air-filtering system of Fig. 4-12 is used, and if $f = 0.25$ and $\eta_1 = 0.75$, what flow rate Q_s of ventilation air must be supplied to overcome a volume release rate of 10^{-4} m^3/s of the pollutant into the space?

SOLUTION   Equation (4-157) gives the supply-air rate Q_s. For use in the equation, $R_v = 10^{-4}$ m^3/s, $C_{vr} = 10^{-6}C_{ppm} = 10^{-5}$, $f = 0.25$, and $\eta_1 = 0.75$. Then

$$Q_s = \frac{R_v}{C_{vr}[(1 - f)\eta_1 + f]}$$

$$= \frac{10^{-4}}{10^{-5}[0.75(0.75) + 0.25]} = 12.31 \text{ m}^3/\text{s} \qquad Ans. \qquad ////$$

REFERENCES

1   Jahnke, E., F. Emde, and F. Lösch: "Tables of Higher Functions," 6th ed., McGraw-Hill Book Company, New York, 1960.
2   White, H. J.: "Industrial Electrostatic Precipitation," pp. 58–63, Addison-Wesley Publishing Company, Inc., Reading, Mass., 1963.

3 Kolmogoroff, A. N.: Über das Logarithmisch Normale Verteilungsgesetz der Dimensionen der Teilchen bei Zerstückelung, *C. R. Dokl. Acad. Sci. URSS (Akad. Nauk USSR)*, vol. 31, no. 2, pp. 99–101, 1941.
4 Picknett, R. G.: A New Method of Determining Aerosol Size Distributions from Multistage Sampler Data, *J. Aerosol Sci.*, vol. 3, no. 3, pp. 185–98, May 1972.
5 Soole, B. W.: Concerning the Calibration Constants of Cascade Impactors, with Special Reference to the Casella MK. 2, *J. Aerosol Sci.*, vol. 2, no. 1, pp. 1–14, February 1971.

PROBLEMS

4-1 Compute the specific surface of a sphere of 1.0 μm diameter if its density is 800 kg/m^3.

4-2 What is the specific surface of a cube 10 μm on each side if its density is 2000 kg/m^3?

4-3 The surface energy of a body, that is, the energy due to surface-tension forces, is given by σA_s, where σ is the surface tension and A_s is the surface area. Relate surface energy to specific surface of a body. Evaluate the surface energy for 1.0 kg of water at 20°C, assuming spherical particles, for the following conditions. Also, determine the number of particles in each case.
  (a) 0.1 μm diameter; (b) 1.0 μm diameter; (c) 10 μm diameter; (d) 100 μm diameter; (e) 1.0 mm diameter; (f) a single particle

4-4 Derive a formula for the specific surface based on the measurements required in evaluating d_2, as shown in Fig. 4-3.

4-5 Write an equation for the specific surface for the case of d_5, as given by Eq. (4-15).

4-6 Derive an expression for mean diameter d_6 analogous to Eq. (4-17), but assume that the particle is large enough that its terminal velocity is in the turbulent regime Re > 700.

4-7 Repeat Prob. 4-6, except assume that the particle size is such that the terminal velocity lies in the range where the Reynolds number is between 10 and 700.

4-8 Derive Eq. (4-17) for the case where Cunningham's correction factor must be taken into account.

4-9 A particle collection has a size distribution function $n(d)$ given as

$$n(d) = 0 \qquad d < 2.0\ \mu m$$
$$n(d) = 10^8(d - 2) \qquad 2.0\ \mu m < d < 50\ \mu m$$
$$n(d) = 4.8 \times 10^9 \qquad 50\ \mu m < d < 100\ \mu m$$
$$n(d) = 0 \qquad d > 100\ \mu m$$

In the preceding equations, d is in micrometers and $n(d)$ is in reciprocal micrometers. The particle density is 950 kg/m^3. Determine the number of particles, mass of the particles in the collection, and specific surface, assuming spherical particles.

4-10 A sample from a particle collection is measured by the method used to generate d_1, with the following results:

Diameter range, μm	No. of particles
1.0–5.0	250
5.0–10.0	600
10.0–30.0	900
30.0–50.0	700
50.0–75.0	400
75.0–100.0	100

Compute and plot the distribution curve $n(d)$. Using numerical or graphical integration, compute the mass of the sample, assuming a density of 3000 kg/m^3.

4-11 Figure 4-13 shows 100 particles, magnified 2000 times. The particle density is 1500 kg/m^3. Determine $n(d)$ for this sample, and find the mass of the sample. Use mean diameter d_1 and assume that this diameter can be used to calculate the mass of each particle.

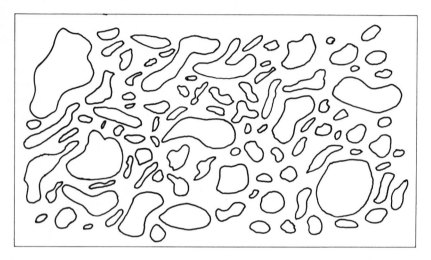

FIGURE 4-13

4-12 Repeat Prob. 4-11 using diameter d_2.

4-13 A device used to measure d_3 by measuring surface area and number of particles in certain size ranges gives the following data:

No. of particles	Surface area, mm^2
3000	0.038
6000	0.471
9000	2.828
7000	4.949
5000	6.285
3000	5.892
1000	2.827

From these data, determine the number distribution function $n(d_3)$, plot it, and estimate the specific surface of the particle collection. Assume $\rho_p = 3000$ kg/m^3.

4-14 A device used to measure d_4 gives the same data as in Prob. 4-13 but in addition gives the total mass of the particles in each size range. These mass quantities, in micrograms, are, respectively, for each line of the data in Prob. 4-13: 0.049; 1.111; 9.22; 19.31; 33.66; 42.11; and 25.72. The density of the material comprising the particles is 3000 kg/m^3. Determine and plot the number distribution $n(d_4)$, the total mass, and the specific surface of the collection. Compare with the results of Prob. 4-13.

4-15 A particle collection comprised of material having a density of 3000 kg/m^3 is separated by a classifier into various size ranges. The particles in each range are weighed and counted, resulting in the following data:

No. of particles	Mass, μg
3000	0.0378
6000	1.178
9000	14.14
7000	37.11
5000	62.84
3000	73.63
1000	42.41

From these data, determine the number distribution $n(d_5)$ based on the mass mean diameter d_5. Estimate the total surface area and specific surface of the particle collection.

4-16 A particle analyzer classifies particles according to their free-fall velocity in a fluid, giving mean diameter d_6. Data on numbers of particles and their free-fall velocities are:

No. of particles	Free-fall velocity, mm/s
3000	0.249
6000	1.557
9000	6.226
7000	14.01
5000	24.91
3000	38.92
1000	56.04

Determine the number distribution $n(d_6)$ from these data, and estimate the specific surface of the collection. The density of the particle material is 2000 kg/m^3.

4-17 Derive Eqs. (4-23) and (4-24).

4-18 Derive Eqs. (4-28) to (4-30).

4-19 Justify the approximations used in Eqs. (4-25) to (4-30). Assume C_{mv} is 30.0 g/m^3, ρ_a is 1.2 kg/m^3, and ρ_p is 5000 kg/m^3.

4-20 A gaseous pollutant is present in standard atmospheric air at a concentration C_{ppm} equal to 700. The molecular weight of the pollutant is 275. Compute C_m, C_v, and C_{mv}.

4-21 A particulate pollutant for which ρ_p is 1200 kg/m^3 in standard air has a loading C_{mv} of 19 g/m^3. Determine C_v and C_m for this mixture.

4-22 If the particle collection of Prob. 4-9 represents a sample of 1000 m^3 of mixture, calculate C_m, C_v, and C_{mv} for the mixture, assuming standard air. Assume d in Prob. 4-9 represents d_5.

4-23 For the data of Prob. 4-22, if the mixture flows in a duct at a rate of $40 \text{ m}^3/\text{s}$, determine the mass-flow rate of pollutant, the surface-area flow rate, and the rate of flow of number of particles.

4-24 Benzene is mixed with standard air at a concentration of 75 ppm. Determine C_m, C_v, and C_{mv}.

4-25 A dust having a specific surface of $1000 \text{ m}^2/\text{kg}$ has a volumetric concentration of 0.02 in an airstream (standard air). The flow rate of the mixture is $2.5 \text{ m}^3/\text{s}$. The density of the dust is 2500 kg/m^3. At what rate is surface area flowing past a section?

4-26 Derive Eq. (4-66).

4-27 Derive Eqs. (4-69) to (4-72).

4-28 Show that for a log-normal distribution the median diameter d_{me} is the same as the diameter d_m given by Eq. (4-56).

4-29 Derive the following expression for the modal diameter in a log-normal particle distribution:

$$d_{mo} = d_m\left(1 + \frac{\sigma^2}{\bar{d}^2}\right)^{-1} = \bar{d}\left(1 + \frac{\sigma^2}{\bar{d}^2}\right)^{-3/2}$$

4-30 Write expressions for $\bar{u}$ and σ_u in terms of $\bar{d}$ and σ.

4-31 Consider a log-normal particle distribution for which $\bar{u} = -9.7$ and $\sigma_u = 0.7$. Evaluate d_m, $\bar{d}$, d_{me}, d_{mo}, σ_m, and σ, and plot the curve of $n(d)/N$ as a function of d. Also, plot the curve of $N(d)/N$ as a function of d.

4-32 Derive an expression for the specific surface of a particle collection obeying the log-normal distribution. The diameter will be interpreted as d_4.

4-33 A log-normal particle collection has $\bar{d} = 30$ μm and $\sigma = 5$ μm. Determine d_m, d_{mo}, σ_m, $\bar{u}$, σ_u, and β.

4-34 A log-normal distribution has values of $\bar{u} = -10.0$ and $\sigma_u = 0.5$. If there are 10^9 particles having a density of 1500 kg/m^3/m^3 of air-particulate mixture, and the mixture flows at 10,000 m^3/s past a point in a duct, determine the mass flow rate of particles past the point.

4-35 Determine C_m, C_v, and C_{mv} for the particle distribution of Prob. 4-34.

4-36 Derive an expression for the mass of particles in a particle collection having a log-normal distribution function.

4-37 Derive an expression for C_m in Eq. (4-34), using the log-normal particle distribution.

4-38 Derive an expression for C_v in Eq. (4-35), using the log-normal particle distribution.

4-39 Repeat Prob. 4-32 using the normal probability distribution function.

4-40 Repeat Prob. 4-36 using the normal probability distribution function.

4-41 Repeat Prob. 4-37 using the normal probability distribution function.

4-42 Repeat Prob. 4-38 using the normal probability distribution function.

4-43 Repeat Prob. 4-34 using the distribution function of Prob. 4-33.

4-44 A particle collection, consisting of particles having a density of 2000 kg/m^3 and obeying a normal distribution function, for which $\bar{d}$ is 50 μm and σ is 15 μm, passes through a collector which removes 0.95 of the mass which enters it. There are 10^{12} particles per cubic meter, and the flow rate of mixture is 1000 m^3/s. How much mass is removed from the mixture per second, and how much is discharged into the atmosphere?

4-45 The following data were obtained from an impactor:

Diameter range, μm	$\Delta N_i/N$
1–2	0.013
2–3	0.186
3–4	0.250
4–5	0.250
5–6	0.130
6–7	0.080
7–8	0.050
8–9	0.020
9–10	0.010

The smallest diameter is assumed to be 1.0 μm, and the largest to be 12.0 μm. Compute and plot the cumulative-distribution curve from these data. Determine d_m and σ_m ; and from these compute $\bar{d}$, σ, and β.

4-46 Equation (4-76) is obtained by setting the argument of the error function in Eq. (4-73) equal to $1/\sqrt{2}$. Obtain a corresponding equation for σ_m if the argument of the error function is set equal to $\sqrt{2}$.

4-47 Derive Eqs. (4-89) and (4-90).

4-48 Derive or verify Eq. (4-105).

4-49 A particle collector has an efficiency as a function of particle diameter given by the following expression, where d is in micrometers:

$$\eta = 1 - e^{-(d-5)/15} \qquad d > 5\ \mu m$$
$$\eta = 0 \qquad d < 5\ \mu m$$

A collector having this efficiency receives dirty air having the particle distribution indicated in Prob. 4-34. Determine the overall efficiencies of the collector based on mass, surface area, and number of particles.

4-50 A particle collector has an efficiency as a function of particle size given by

$$\eta = 0 \qquad d < 10\ \mu m$$
$$\eta = 0.025(d - 10) \qquad 10 < d < 50\ \mu m$$
$$\eta = 1.0 \qquad d > 50\ \mu m$$

The collector receives air having a particle distribution given by

$$n_p(d) = 0 \qquad d < 1\ \mu m \text{ and } d > 100\ \mu m$$
$$n_p(d) = 0.001062(d - 1)N \qquad 1 < d < 20\ \mu m$$
$$n_p(d) = [0.0202 - 0.000252(d - 20)]N \qquad 20 < d < 100\ \mu m$$

In the preceding equations, d is given in micrometers. What is the overall efficiency of the collector based on mass? On surface area? On number of particles?

4-51 A particle collector operates on a particle collection for which $\bar{d}$ is 25.0 μm and σ is 10 μm in a log-normal particle distribution. The collection efficiency is given by the expression

$$\eta(d) = 1 - \frac{25}{d^2} \qquad d > 10\ \mu m$$

in which d is in micrometers. Using Eqs. (4-111) to (4-113), find the three collection efficiencies.

4-52 Derive equations analogous to Eqs. (4-95), (4-106), and (4-111) using specific surface as the property upon which efficiency is based. Refer to the efficiency as η_{ss}.

4-53 Derive Eqs. (4-135) to (4-137).

4-54 Derive Eqs. (4-143) to (4-145).

4-55 Derive Eqs. (4-148) to (4-149).

4-56 Is the efficiency given by Eq. (4-146) greater than or less than η_1? Than η_2 ?

4-57 Calculate the efficiency of the combination of Fig. 4-10 for the following sets of values of η_1 and η_2 : 0.99, 0.99; 0.9, 0.9; 0.99, 0.7; 0.9, 0.7; 0.99, 0.5; 0.9, 0.5; 0.7, 0.7; 0.7, 0.5; 0.5, 0.5.

4-58 Two concentrators and one collector are utilized in the combination shown in Fig. 4-14. Derive an expression for the efficiency at diameter d for the combination in terms of the individual efficiencies at this diameter.

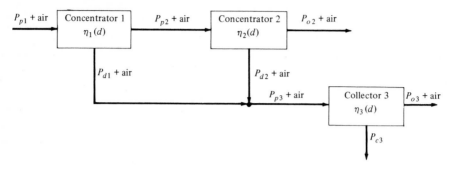

FIGURE 4-14

4-59 Repeat Prob. 4-58 if the output P_{o3} from the collector is returned to mix with the input to concentrator 1.
4-60 The combination of Prob. 4-58 has components whose efficiencies are $\eta_1 = 0.80$, $\eta_2 = 0.75$, and $\eta_3 = 0.60$. What is the efficiency of the combination?
4-61 Repeat Prob. 4-60 using the modification specified in Prob. 4-59.
4-62 Derive an expression for the diametral efficiency of the combination shown in Fig. 4-15 in terms of the individual diametral efficiencies. The bypass factor κ is arbitrary, except that it must lie between 0 and 1.

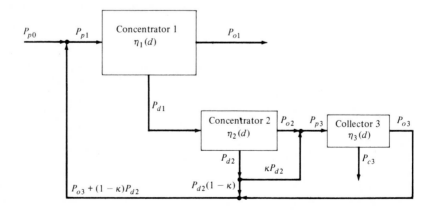

FIGURE 4-15

4-63 Evaluate $\eta(d)$ for the combination of Prob. 4-62 if $\eta_1(d) = 0.9$, $\eta_2(d) = 0.6$, $\eta_3(d) = 0.9$, and $\kappa = 0.5$.

4-64 A house is located in a polluted region where the pollutant count in the atmosphere is 2000 $\mu g/m^3$. A system such as that of Fig. 4-12 is to be used continuously to keep the pollutant level down inside. The supply rate is 1.0 m^3/s, and the makeup air fraction is 0.25. The efficiency of filter 1 is 0.60, and filter 2 is omitted. What equilibrium concentration of pollutant will exist in the house?

4-65 Suppose that the makeup air is supplied by infiltration through the doors, windows, and walls of the building enclosing the conditioned space, so that, in effect, filter 2 of Fig. 4-12 is absent. This makeup air will then be supplied continuously. Suppose that the conditioning system operates intermittently, with on and off periods of t_1 and t_2, respectively. Derive an expression for the concentration in the space as a function of time, and determine the maximum concentration.

4-66 A house in the country, where the only significant pollution is due tc pollen, is to be kept at a pollen level inside of 0.1 $\mu g/m^3$. If the pollen level outside is 30 $\mu g/m^3$, the efficiency of filter 2 in Fig. 4-12 is 0.7, and the makeup air fraction f is 0.3, what must be the efficiency of filter 1 to achieve the specified pollen level inside the house?

4-67 In an industrial situation, a pollutant is discharged into a conditioned space at the rate of 10 m^3/min. The allowable concentration of the pollutant in the space is 10 ppm. The filter efficiency is 0.9. If Q_o/Q_s is 0.5, what air-flow rate must be supplied, and what concentration must be maintained in the supply air? Assume that the outside air is free of the pollutant.

4-68 An industrial ventilation system makes use of the arrangement shown in Fig. 4-16. Derive an expression for the supply rate Q_s for a fixed pollutant level C_r leaving the room.

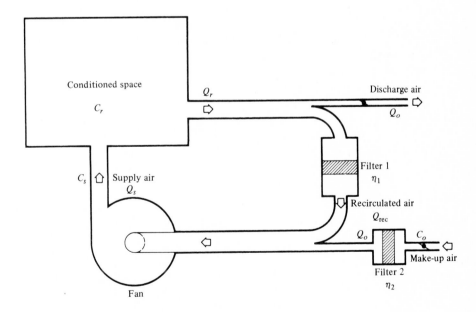

FIGURE 4-16

4-69 Repeat Prob. 4-68 if the pollutant is generated in the conditioned space at the rate R.

4-70 Repeat Prob. 4-66 using the equation derived in Prob. 4-68.

4-71 Repeat Prob. 4-67 using the formula derived in Prob. 4-69.

4-72 The solvent trichloroethylene is used in a space at the rate of 5 gal/h. If no recirculated air is present ($f = 1$), how much ventilation air is required (in cubic meters per second)? Repeat if $f = 0.2$ and the filter efficiency is $\eta_1 = 0.95$, using the results of Prob. 4-69.

4-73 Repeat Prob. 4-72 using carbon tetrachloride.

DESIGN OF INDUSTRIAL
VENTILATION SYSTEMS

In various industrial enterprises, hazardous or potentially hazardous conditions exist which can affect the health and safety of those who work there. Fumes and vapors are given off from storage tanks, processing vats, and other types of processing equipment. Dusts are given off from grinders, pulverizers, hammermills, conveying equipment, mixers, and many other types of equipment. Paint sprayers produce mists which can get into the workspace if uncontrolled. Indeed, most heavy industry has any number of situations in which maintenance of a clean working environment is essential.

5-1 INTRODUCTION

The problem of industrial ventilation is distinct from that of pollution control in that the usual method of ventilation is to remove the pollutants from the workspace by exhausting them outdoors. Pollution control, however, requires that the pollutants not be exhausted outdoors. These two opposing requirements can both be met by installing pollution control devices prior to exhausting the ventilation stream. Thus, industrial ventilation and pollution control often go hand in hand; indeed, the engineering principles involved in both subjects are rather similar.

It can be safely said that the most desirable method of pollution control is to avoid producing the pollutant in the first place, which also makes industrial ventilation a much simpler matter. The replacement of a polluting process by a nonpolluting one often is not practicable or acceptable to the industry involved. Occasionally it is possible, and the engineer should be alert to this possibility. More frequently, if a heavily polluting process is to be replaced by another process, it is because the new process is more amenable to the installation of effective pollution control devices.

Pollutants given off in an industrial environment vary over a wide range of quantity and toxicity. It may be assumed that a potential hazard exists when dust is formed, where combustion or high-temperature processing takes place, where chemicals are poured or stored in open containers, or where items are dipped into vats of chemicals. Four general methods to prevent exposure of workers to harmful levels of pollutants are:

1   Replace the hazardous process by a nonhazardous one, or at least substitute a less hazardous substance for the one currently in use.
2   Enclose the source of the pollutant; for example, use a closed container instead of an open vat for storage of a chemical, or perform an operation by remote control from behind an enclosure. It is assumed here that all workers will be outside the enclosure at all times when exposure is likely.
3   Dilute the concentration of the pollutant by providing adequate ventilation air to the region near the source of the pollutant.
4   Provide exhaust hoods which will ensure that the escaping substance will be sucked into the hood and not allowed to escape into the workspace.

A fifth possible solution, providing protective devices such as gas masks, is not considered to be a satisfactory one for everyday use, though it is valuable in emergencies and can be used as standard procedure in maintenance of enclosures and ventilation systems. In this chapter we shall be concerned with the third and fourth methods.

5-2   VENTILATION BY DILUTION

In this section we shall consider the requirements for maintaining a workspace at an acceptably low level of pollutant concentration by supplying an adequate volume of fresh air to the workspace. This problem has already been treated to a considerable extent in Sec. 4-6, and also Prob. 4-68 illustrates an alternative ventilation system. Therefore there remains relatively little of this subject to discuss in the present section.

The usual industrial ventilation system is simpler than that shown in Fig. 4-12; a typical system is shown in Fig. 5-1. We may analyze this system from

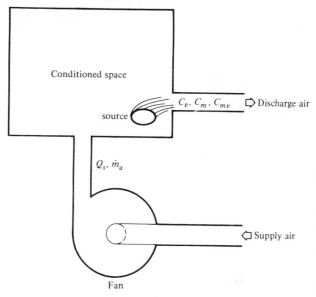

FIGURE 5-1
Simple industrial ventilation system.

first principles by noting that the mass flow rate leaving the conditioned space is $\dot{m}_a + \dot{m}_p$ and that this can be closely approximated as $\dot{m}_a$ since in most situations $\dot{m}_p$ is small; here $\dot{m}_p$ is the mass flow rate of pollutant. A mass balance on the conditioned space gives the result

$$(\dot{m}_a + \dot{m}_p)C_m = \dot{m}_p$$

or, making the approximation indicated previously and solving for $\dot{m}_a$ gives

$$\dot{m}_a = \frac{\dot{m}_p}{C_m} \qquad (5\text{-}1)$$

The volumetric flow rate of supply air Q_s is given by

$$Q_s = \frac{\dot{m}_p}{\rho C_m} = \frac{\dot{m}_p}{C_{mv}} = \frac{Q_p}{C_v} \qquad (5\text{-}2)$$

In all the preceding equations the concentrations are those of the pollutant substance in the stream leaving the conditioned space. The flow rates of pollutant are equal to those given off by the source or the total from all sources present if there are more than one.

In industrial practice there will often be several sources present, introducing a number of different pollutants into the same space. In dealing with more than one pollutant substance, ventilation requirements can be calculated for each pollutant in turn, using the appropriate allowable concentration of each; the most severe requirement will determine the ventilation air quantity required.

Ventilation by dilution is fraught with hazards if very toxic substances are employed or if care is not used in locating the ventilation source relative to the pollutant source and the area where workers will be positioned. The reason for this is that although overall concentrations of pollutant will be acceptably low when the ventilation system is properly designed, there will exist local regions near the source where pollutant levels will be dangerously high. It is imperative that workers not be located in such regions for appreciable lengths of time. The best skill of the designer may be required to avoid placing workers in a region where the pollutant concentration is too high.

In determining where the regions of high pollutant concentration will be, it is important to note that there are two effects which will disperse the pollutant into the air in the room: molecular diffusion and convective transport. Molecular diffusion can be neglected in any space which is ventilated at all since diffusion velocities are several orders of magnitude less than air velocities in any space other than a sealed room. However, diffusion is potentially significant when the pollutant source is located in a zone having little ventilation; the pollutant must leave the source somehow, and in such stagnant air conditions, high pollutant concentrations may build up, which would be hazardous to workers.

The main mechanism for removal of the pollutant from the source is air currents passing over or near the source. These currents carry the substance away and mix it with other air in the room. The pollutant source should be well ventilated, the source of fresh air should be behind the workers, the exhaust duct should be located near the source, and no workers should be located between the source and the exhaust duct. Usually this can be conveniently achieved by locating the exhaust ducts in the ceiling over the pollutant source or high on a wall near the source. If the pollutant source is small, as would be the case where a large number of sources are present and distributed throughout the space, these precautions may be unnecessary. It may be necessary only for workers to avoid approaching the source too closely.

Where a relatively small pollutant source is present at a location not immediately in front of an exhaust duct, it is possible to estimate the concentration near the source. To do this, an imaginary envelope is drawn around the source in a convenient shape and the space inside the envelope is regarded as an enclosure which receives ventilation air and in which a pollutant is released. Any convenient shape will do; the half-cylinder shown in Fig. 5-2 will be used in the analysis here. We assume that the velocity of the air in the vicinity of the pollutant source is known or can be estimated closely enough. Its value can be found using an anemometer in an existing workspace. The mass flow rate of air entering the envelope is given by $\dot{m}_a = \rho V \pi R^2 / 2$. Combining this with Eq. (5-1) and solving for C_m gives

$$C_m = \frac{2\dot{m}_p}{\rho V \pi R^2} \qquad (5\text{-}3)$$

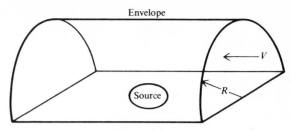

FIGURE 5-2
Average concentration near a small source.

and solving for R gives

$$R = \sqrt{\frac{2\dot{m}_p}{\pi \rho V C_m}} \qquad (5\text{-}4)$$

Equations (5-3) and (5-4) can be interpreted in the following way: If an envelope of radius R around the source can be kept free from exposure to workers, then C_m is regarded as the maximum incremental concentration to which anyone might be exposed who stays outside this envelope. If C_m is specified, then R is the radius of the space inside which serious exposure is liable to occur. There is, of course, no assurance that the pollutant will be evenly distributed in the airstream leaving the envelope; thus it is unwise for anyone to be continuously exposed directly downwind from the source if this is avoidable.

The analysis given in the preceding paragraphs is concerned with gross behavior of the air in the space only and does not treat small-scale effects very well. Since small local concentrations of pollutants could have disastrous effects if highly toxic substances are present, it is recommended that ventilation by dilution be employed only for substances of lesser toxicity. That is, if brief exposure to high concentrations is highly dangerous, then this method of ventilation should be avoided and a hood should be employed. We shall discuss the design and selection of hoods next.

5-3 HOOD SPECIFICATIONS

The use of hoods represents a strong step forward over ventilation by dilution. A hood is defined as any opening through which air is sucked in and later exhausted. The types of hoods in use include exhausted enclosures, booths, exterior hoods, and receiving hoods. Normally, the air collected by the hood is exhausted outside after appropriate treatment, but in special cases it can be exhausted back into the conditioned space after removal of most of the pollutant substances.

A few simple hood configurations are illustrated in Fig. 5-3. Figure 5-3a shows an opening fitted into the wall immediately above a workbench, which

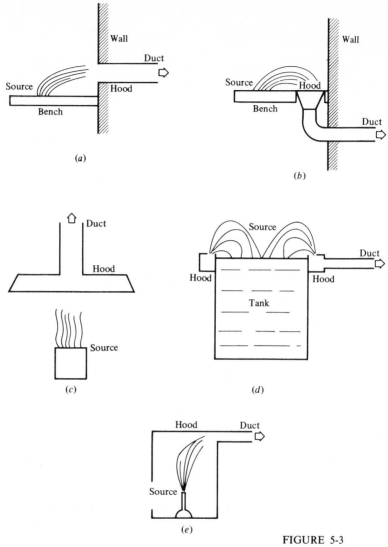

FIGURE 5-3
Several types of hoods.

draws in the fumes generated on the bench. Figure 5-3*b* is similar, except that the hood is located in the plane of the workbench at the rear near the wall. Figure 5-3*c* shows a canopy hood located above a pollutant source. Such hood arrangements are particularly valuable if the pollutant source consists of a heated gas. Figure 5-3*d* shows an arrangement for venting an open vat or tank; slots are provided along the top edge of the tank, which can be either circular or rectangular. An enclosed hood is shown in Fig. 5-3*e*; these are most useful where highly toxic substances are generated and accidental escape into the workspace

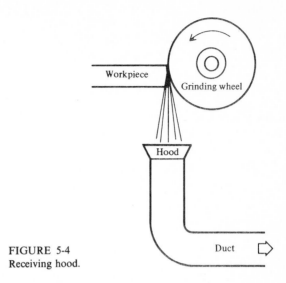

FIGURE 5-4
Receiving hood.

is to be avoided. A receiving hood is arranged so as to receive large particles which are projected outward from a source; these particles impinge directly into the hood. Figure 5-4 shows a receiving hood used in conjunction with a grinding wheel.

The shape of the hood opening is pretty much the choice of the designer. Most hood openings are either circular or rectangular, including long slots. In special cases other shapes may be used, but the ones we have mentioned are chosen most often for ease of fabrication.

Hood design is usually accomplished by considering a control volume and its accompanying control surface, as shown in Fig. 5-5. The control volume is drawn so as to include the source being controlled and to extend far enough beyond it to draw in any particles projected from the source in a direction opposite to the hood. The symbol X represents the distance in a direction normal to the plane of the hood face from the hood to the control surface. The air velocity coming in across the control surface is referred to as the *control velocity* and is customarily assumed to be uniform over the control surface. This assumption is closely realized in practice provided the value of X exceeds the smaller dimension of the hood opening by a factor of 2 or 3.

If the pollutant source involves the forcible projection of particles, as is the case for grinding, chipping, sawing, dumping of powder, and other operations, the particles projected from the source away from the hood will have to slow down to essentially zero velocity by air friction before they can be captured. The distance required for this reduction in velocity to occur has been termed *pulvation distance* by Hemeon [1]. The control surface must extend beyond the pollutant source by at least the pulvation distance P, as shown in Fig. 5-6. Thus $X \geq P + L$, as shown in Fig. 5-6; in fact, X must exceed $P + L$ if the source is not located on the centerline of the hood. The magnitude of the pulvation distance

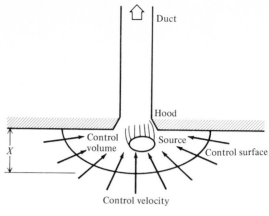

FIGURE 5-5
Control surface surrounding a hood.

is difficult to predict beforehand since the maximum size of particles formed or the force with which they will be projected is difficult to forecast. However, the pulvation distance may be easily measured once the process is in operation.

The value of the control velocity required to ensure capture of all the pollutant substance depends on a number of factors. The condition of air movement in the space, exclusive of the effect of the hood, is very important. In drafty spaces, much higher control velocities must be used to prevent continuous or intermittent spillover of the contaminated air when the hood is used. Where the contaminant is very toxic, higher control velocities must be used. Table 5-1, which has been adapted from Hemeon [2], can serve as a rough guide to the selection of control velocities but should not be a substitute for sound judgment in weighing the pertinent factors of a given application. Table 5-1 applies to cases where the contaminant is at room temperature. Where the contaminant is at an elevated temperature, other formulas and tabular data are available.

The velocity of the fluid entering the hood is not important from the standpoint of capture but is important in setting the power requirements of the

Table 5-1 RECOMMENDED CONTROL VELOCITIES

Condition	Pollutant of lesser toxicity, m/s	Pollutant of greater toxicity, m/s
Draftless or well baffled	0.2–0.3	0.25–0.35
Moderately drafty	0.25–0.35	0.3–0.4
Very drafty with no baffling	0.35–0.45	0.4–0.6

SOURCE: Adapted from W. C. L. Hemeon, "Plant and Process Ventilation," 2d ed., p. 92, The Industrial Press, New York, 1963.

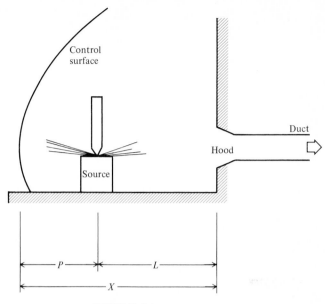

FIGURE 5-6
Control surface relative to a source with pulvation.

ventilating system. Typical values of velocity at the hood face are 7.5 to 12.5 m/s. Over any surface on which the velocity is uniform, the flow rate is given by

$$Q = AV \qquad \dot{m} = \rho AV \qquad (5\text{-}5)$$

These equations may be applied to either the control surface, in which case V is the control velocity and A the control-surface area, or the hood face, in which case A is the hood area and V the face velocity. The exhaust requirements of the hood are usually determined by applying Eq. (5-5) to the control surface, using the recommended control velocity from Table 5-1. The size of the hood opening is then determined from the flow rate and face velocity.

5-4   HOODS OF SIMPLE GEOMETRY

In this section we shall treat hoods for which the shape of the control surface can be approximated by simple geometric surfaces, such as planes, cylinders, or spheres. In such cases, the location of the control surface can be obtained by inspection or by simple calculation, and the flow rate quickly determined. Extreme accuracy usually is not obtainable by this method, but then extreme accuracy is of no practical value in ventilation work anyway. In this section, several types of hood geometries are considered, classified according to the control-surface configuration appropriate to the particular hood geometry.

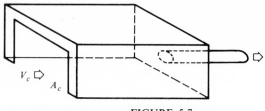

FIGURE 5-7
Hood in the form of a booth.

Booths

A booth consists of a space which is completely enclosed except for all or a part of one side. A booth is shown in Fig. 5-7, with a portion of the left side open. Air is drawn in through the open side and is exhausted through an opening or duct attached to the ceiling, floor, or one of the other walls. The process involving contamination is performed inside the booth. An example is a paint-spray booth. Equation (5-5) gives the flow rate, where A_c is the area of the open face and V_c the control velocity across that face. For a given toxicity of the contaminant, a relatively low value of control velocity can be used for a booth owing to the baffling effect of the other walls of the booth.

Semicylindrical Control Surface

A long narrow slot in a plane wall will produce a control surface having the shape of a half-cylinder. An example is a horizontal slot in the middle of a wall. Figure 5-8 shows such a control surface. We shall assume that $L \gg W$ and that

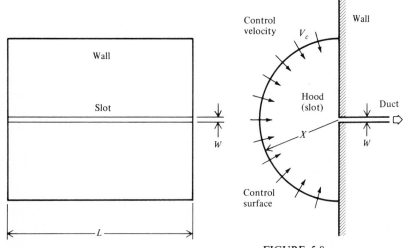

FIGURE 5-8
Semicylindrical control surface.

$X \gg W$, say, $X \geq 3W$. The area of the control surface is

$$A_c = \frac{2\pi XL}{2} = \pi XL \qquad (5\text{-}6)$$

and the expression for the volumetric flow rate becomes

$$Q = \pi X L V_c \qquad (5\text{-}7)$$

EXAMPLE 5-1   A slot is to be placed horizontally in a wall which is 10 m long. A row of barrels lines the wall below the slot. The slot is to provide dust control during filling and emptying of the barrels. If the zone of spillage extends to a point 0.7 m above the floor and 1.5 m out from the wall, and if the slot is 2.0 m above the floor, determine the X distance and the flow rate of ventilation air required. Assume there is no pulvation or projection of particles, and neglect the effect of the barrels on the flow.

SOLUTION   Figure 5-9 shows the slot and the control surface. The X distance is computed as

$$X = \sqrt{1.3^2 + 1.5^2} = 1.985 \text{ m}$$

From Eq. (5-7) the flow rate is given as

$$Q = \pi X L V_c = \pi(1.985)10 V_c = 62.4 V_c$$

Select V_c as 0.25 m/s, giving

$$Q = 62.4(0.25) = 15.6 \text{ m}^3/\text{s} \qquad Ans. \qquad ////$$

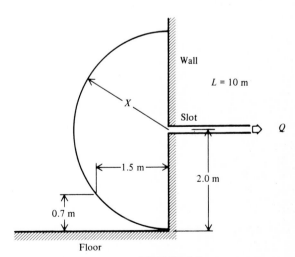

FIGURE 5-9
Control surface for Example 5-1.

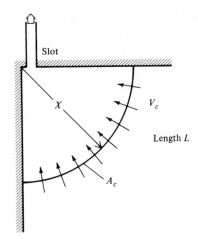

FIGURE 5-10
Quarter-cylindrical control surface.

EXAMPLE 5-2   How wide must the slot of Example 5-1 be if the face velocity is to be 10 m/s?

SOLUTION   Equation (5-5) applies to the face of the hood. It gives for the area

$$A = \frac{Q}{V} = \frac{15.6}{10} = 1.56 \text{ m}^2$$

From this the width of the slot is determined as

$$W = \frac{A}{L} = \frac{1.56}{10} = 0.156 \text{ m} = 15.6 \text{ cm} \qquad Ans. \qquad ////$$

Quarter-cylindrical Control Surface

A control surface in the shape of a quarter-cylinder is formed when a slot is cut in a wall at the point where another wall joins it at right angles, as shown in Fig. 5-10. The area of the control surface is given by

$$A_c = \frac{2\pi X L}{4} = \frac{\pi X L}{2} \qquad (5\text{-}8)$$

and the flow rate is given by

$$Q = A_c V_c = \frac{\pi X L V_c}{2} \qquad (5\text{-}9)$$

EXAMPLE 5-3   A workbench 1.3 m wide by 6.0 m long is to be ventilated by means of a slot located between the edge of the bench and the wall, running the

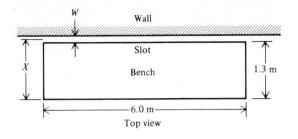
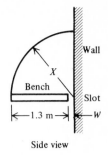

FIGURE 5-11
Figure for Example 5-3.

entire length of the bench. Find Q and W if $V_c = 0.25$ m/s and V_f (face velocity) $= 10$ m/s. Figure 5-11 shows the bench, slot, and wall configuration.

SOLUTION   The X distance is given by

$$X = 1.3 + \frac{W}{2}$$

The area of the control surface is

$$A_c = \frac{2\pi(1.3 + W/2)L}{4} + \frac{WL}{2} = 0.65\pi L + \left(\frac{1}{2} + \frac{\pi}{4}\right)WL$$

$$= (2.04 + 1.2854W)L$$

Using the control velocity as 0.25 m/s, the flow rate is

$$Q = A_c V_c = (2.04 + 1.2854W)6.0(0.25) = 3.06 + 1.928W$$

Using the velocity at the face of the slot as 10 m/s gives a second expression for flow rate in terms of slot width:

$$Q = WLV_f = 6.0(10)W = 60W$$

Solving for the slot width W from these two equations gives

$$W = 5.3 \text{ cm} \qquad Ans.$$

and solving for the flow rate Q gives

$$Q = 3.06 + 1.928(0.053) = 3.162 \text{ m}^3/\text{s} \qquad Ans. \qquad ////$$

Spherical Control Surfaces

An eighth-spherical surface results if the hood, having a cross section that is not too long relative to its width, is placed in a corner where three walls come

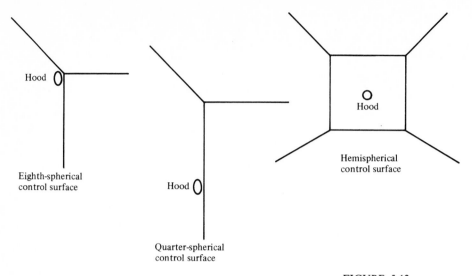

FIGURE 5-12
Spherical control surfaces.

together. Formulas for the control-surface area A_c and flow rate Q are given by

$$A_c = \frac{4\pi X^2}{8}$$

$$= \frac{\pi X^2}{2} \qquad (5\text{-}10)$$

$$Q = \frac{\pi X^2 V_c}{2} \qquad (5\text{-}11)$$

A quarter-spherical surface occurs when the hood is located in the corner between two walls. Its formulas for A_c and Q are, respectively,

$$A_c = \pi X^2 \qquad (5\text{-}12)$$
$$Q = \pi X^2 V_c \qquad (5\text{-}13)$$

A hemispherical control surface results when a hood opening is placed in a wall far removed from any corners. Its formulas are

$$A_c = 2\pi X^2 \qquad (5\text{-}14)$$
$$Q = 2\pi X^2 V_c \qquad (5\text{-}15)$$

These three control surfaces are illustrated in Fig. 5-12.

EXAMPLE 5-4   Apply the technique illustrated in the eighth-, quarter-, and hemispherical control surfaces to the case of a circular duct of diameter D projecting into a space.

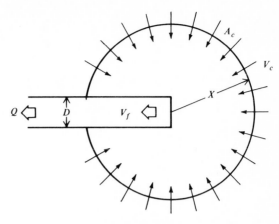

FIGURE 5-13
Control surface at the end of a duct.

SOLUTION   Figure 5-13 shows this case. The flow rate Q is given by

$$Q = A_c V_c = A_f V_f = \left(4\pi X^2 - \frac{\pi D^2}{4}\right) V_c = \frac{\pi D^2 V_f}{4}$$

Let us take $V_c = 0.25$ m/s and $V_f = 15$ m/s as representing the lowest practical value of V_c and the highest practical value of V_f, respectively. We solve for X, giving

$$4\pi X^2 = \frac{\pi D^2}{4}\left(1 + \frac{V_f}{V_c}\right) \qquad X^2 = D^2\left(1 + \frac{V_f}{V_c}\right)\frac{1}{16}$$

then

$$X = \frac{D}{4}\sqrt{1 + \frac{V_f}{V_c}} = \sqrt{1 + \frac{15}{0.25}}\frac{D}{4} = 1.95D \quad Ans. \quad ////$$

From Example 5-4, we see that the ratio of X to D is not large enough for that case to allow the assumption of a spherical control surface. It thus appears that there are cases where a more accurate representation of the control surface is necessary for a satisfactory analysis. If we take the more or less extreme values of $V_c = 0.25$ m/s and $V_f = 15$ m/s and apply them to the other cases considered so far, we obtain the values of X shown in the following table:

Control-surface shape	X
Hemisphere	2.7D
Quarter-sphere	3.9D
Eighth-sphere	5.5D
Cylinder	9.6W
Half-cylinder	19.1W
Quarter-cylinder	38.4W

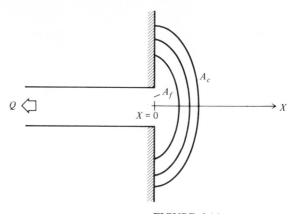

FIGURE 5-14
Experimental control surface.

The treatment of the hemisphere in this way is marginal, and for higher control velocities it would be unacceptable. The other cases can be treated as hoods of simple geometry, although the quarter-sphere case is not handled adequately for control velocities greater than about 0.38 m/s.

It must be noted that all hoods of simple geometry presuppose that the shape of the hood opening is appropriate to the shape of the assumed control surface. Although the hood opening does not require a regular geometric shape, such as a circle or slot of uniform width, it must reasonably approximate a line sink for the cylindrical cases and a point sink for the spherical cases. Thus, a long, narrow rectangle would not do for a spherical surface, and a T-shaped slot could not be used with cylindrical geometry, without further extending the method of analysis.

5-5 EXPERIMENTAL VELOCITY CONTOURS

For hoods which cannot be treated as having simple geometry, we must use other methods. For the control surface shown in Fig. 5-14, the following formula, taken from Hemeon [3], has been found to hold with acceptable accuracy for circular and rectangular openings with an aspect ratio up to about 3 : 1:

$$A_c = 10X^2 + A_f \qquad (5\text{-}16)$$

in which A_f is the area of the face opening. This formula also applies to other shapes of openings, though it should not be applied to any case in which a rectangle superimposed on the shape would have an aspect ratio greater than 3 : 1. From this equation, the flow rate is given by

$$Q = V_c(10X^2 + A_f) \qquad (5\text{-}17)$$

Equations (5-16) and (5-17) are valid only where the control surface becomes either spherical or hemispherical with increasing X. In addition, they are not valid when the value of X becomes large enough for the surface area to be adequately represented by the formula for the sphere or hemisphere; they are applicable only at small values of X.

EXAMPLE 5-5 A duct opening (grill) of 0.03 m² face area is placed in the middle of a workbench. Dust generation occurs over a region to 0.3 m above the bench and 0.3 m to the side of the opening in any direction. If $V_c = 0.25$ m/s, find Q and V_f. See Fig. 5-15.

SOLUTION Initially take $X = \sqrt{0.3^2 + 0.3^2} = 0.424$ m. Then the control-surface area is

$$A_c = 10X^2 + A_f = 10(0.18) + 0.03 = 1.83 \text{ m}^2$$

If the control surface were hemispherical, with radius equal to X above, its surface area would be $2\pi X^2 = 2\pi(0.18) = 1.13$ m². The value of X calculated above is clearly too large, as shown in Fig. 5-15, although the exact value is hard to determine.

If we note that the true value of X is between 0.3 and 0.424 m and assume an arithmetic average value, the value of X becomes 0.362 m and A_c will be

$$A_c = 10(0.362)^2 + 0.03 = 1.34 \text{ m}^2$$

For Q and V_f we have

$$Q = A_c V_c \qquad V_f = \frac{Q}{A_f} = \frac{Q}{0.03}$$

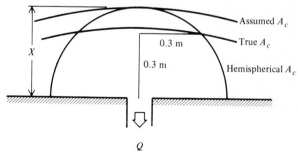

FIGURE 5-15
Control surface for Example 5-5.

For the different choices of X, the values of Q and V_f are:

X, m	Q, m³/s	V_f, m/s
0.3	0.232	7.8
0.362	0.335	11.2
0.424	0.458	15.3

////

5-6 COMPLEX HOOD DESIGN

Hood designs that cannot be handled by the simple geometric analysis using the cylindrical or spherical control surfaces and for which the experimental velocity contours resulting in Eqs. (5-16) and (5-17) are not applicable may sometimes be treated by methods that are only a little more complicated than those already considered. We shall consider several such hood designs in this section.

Symmetry Analysis

Frequently a given hood design can be taken as some integral fraction, usually one-half, of a hood for which the experimental equation (5-16) is applicable. Such a design is shown in Fig. 5-16, in which the hood opening is immediately above a solid plane surface such as a workbench or the floor. In addition, the plane of the hood is normal to the plane of the solid surface. In this case, the given hood configuration can be replaced by one twice as large, as shown in Fig. 5-16. When the larger configuration is examined, it is seen that the midplane, which coincides with the solid surface in the original configuration, is a plane of symmetry and as such no flow will occur across it. This is the same situation satisfied by the solid surface, so that the original configuration is in all respects the upper half of the larger one. We can apply Eq. (5-16) to the larger configuration and take half of it to give the control-surface area sought. Thus Eq. (5-16) is modified to

$$A_c = \frac{10X^2 + 2A_f}{2} = 5X^2 + A_f \qquad (5\text{-}18)$$

More generally, a hood for which Eq. (5-16) is applicable can be subdivided into n hoods, each having the same flow pattern by symmetry. The results will be useful if one of these subdivisions is the hood whose control-surface area is sought. The result is

$$A_c = \frac{10X^2 + nA_f}{n} = \frac{10X^2}{n} + A_f \qquad n = 1, 2, 3, \dots \qquad (5\text{-}19)$$

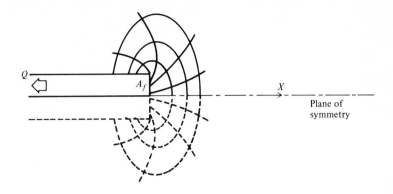

FIGURE 5-16
Symmetrical hood configuration.

EXAMPLE 5-6   A source of contamination on a workbench is to be controlled by means of a hood located on the wall just above the bench. The bench extends 1.0 m out from the wall, and the source of dust is localized at a point near the edge of the bench opposite the location of the hood. For a control velocity of 0.30 m/s, and assuming the hood is a rectangle with aspect ratio not exceeding 3 : 1, determine the flow rate and hood size.

SOLUTION   Figure 5-17 shows the workbench and hood location. The hood can be taken as half of a larger hood, whose flow pattern has the bench surface as a plane of symmetry. Equation (5-18) gives the control-surface area for this case, where X is taken as 1.0 m, the distance from the hood to the edge of the bench:

$$A_c = 5X^2 + A_f = 5(1.0)^2 + A_f = 5 + A_f$$

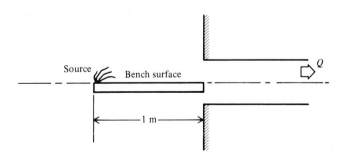

FIGURE 5-17
Hood location for Example 5-6.

The flow rate is then

$$Q = A_c V_c = (5 + A_f) \, 0.30 = 1.5 + 0.30 A_f$$

Assuming a velocity of 12.5 m/s at the hood face, the hood area is

$$A_f = \frac{Q}{12.5} = \frac{1.5 + 0.30 A_f}{12.5} = 0.12 + 0.024 A_f$$

Thus

$$A_f = \frac{0.12}{0.976} = 0.123 \text{ m}^2$$

The flow rate is then

$$Q = 1.5 + 0.30(0.123) = 1.54 \text{ m}^3/\text{s} \qquad \textit{Ans.}$$

and the dimensions of the hood opening can be 21 by 59 cm.   *Ans.*   ////

Double Parallel Slots

Consider now the case shown in Fig. 5-18 in which two long parallel slots act as the hood. Such a case can arise where a long narrow vat or open tank is ventilated by slots running down its long sides. Since the tank is open at the top, vapors can escape to the space above unless ventilation is provided. The use of slots along the sides of the tank, as discussed here, can serve this function. To be successful, the control surface must completely enclose the liquid, as shown in Fig. 5-18. A fairly accurate analysis can be performed by the technique to be developed next.

Consider the control surface of Fig. 5-18, labeled ABC. Only half of the control surface need be considered since the other half will be symmetrical. The portion AB is closely that of a quarter-cylinder, end effects being neglected. The portion BC of the control surface is roughly of the shape indicated, with the length BC being a little longer than y. The exact shape of BC is not important, as long as it remains within certain bounds (to be discussed presently). Within the limits of approximation suitable to this analysis, we shall take BC equal to y.

The plane OB, which bisects the slot, will approximately, though not exactly, bisect the flow into the slot. Thus, half the flow will cross the control surface AB and half will cross BC. Since the velocity is uniform over the control surface, this means that the area of the control surface over AB is equal to that over BC. For the cylindrical geometry, then, $AB = BC$. Figure 5-19 shows the control surface ABC and two velocity contours at slightly higher and lower values of velocity. Contour $A'B'C'$ is at a higher velocity than V_c, while contour $A''B''C''$ is at a lower velocity. The control surface is the smallest surface for which the assumption $AB = BC$ is valid and at the same time where point C lies above the surface of the tank. If $A'B'C'$ were taken as the control surface, then a

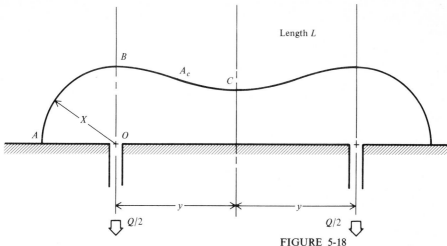

FIGURE 5-18
Control surface for two parallel slots.

portion of the tank surface would be inadequately protected; whereas if $A''B''C''$ were taken as the control surface, greater flow rates would be required than necessary and the analysis would be unnecessarily complicated. Thus, taking ABC as the control surface gives satisfactory results and can be easily treated by the approximate technique developed here.

Since we have designated in Fig. 5-18 the distance between the slots as $2y$, the area of the control surface is

$$A_c = \pi X L + 2yL \qquad (5\text{-}20)$$

Making use of the assumption that $AB = BC = y$, then

$$\frac{\pi X}{2} = y \qquad X = \frac{2y}{\pi}$$

Then the control-surface area becomes

$$A_c = 4yL \qquad (5\text{-}21)$$

EXAMPLE 5-7 A long tank is ventilated by two parallel slots 4.0 m long located 1.0 m apart. For a control velocity of 0.25 m/s, find Q.

SOLUTION Using Eq. (5-21), we obtain for A_c, where $y = 0.5$ m,

$$A_c = 4yL = 4(0.5)(4.0) = 8 \text{ m}^2$$

Then the flow rate is

$$Q = A_c V_c = 8(0.25) = 2.0 \text{ m}^3/\text{s} \qquad Ans.$$

The surface area of the tank is 4.0 m², and so the flow rate per unit area of tank surface is 0.5 m³/s/m². ////

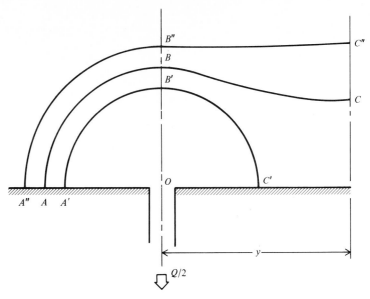

FIGURE 5-19
Control surface and neighboring velocity contours.

Circular Slot

The same method of analysis works also for the case of a circular slot around a circular tank, as shown in Fig. 5-20. Here we equate the area of the portion BC of the control surface to the area of portion BD. The area of BD is that of a 90° segment of a torus, given by

$$\pi^2 X y - 2\pi X^2$$

The area of BC is πy^2, so that solving for X gives

$$X^2 - \frac{\pi}{2} yX + \frac{y^2}{2} = 0$$

or

$$X = \frac{\pi y}{4} \pm \frac{1}{2}\sqrt{\frac{\pi^2}{4} y^2 - 2y^2} = 0.444y$$

The negative sign in the preceding equation is used since X must be less than y. Then the area of the control surface is

$$A_c = \pi y^2 + \pi^2 X y + 2\pi X^2 = 2\pi^2 X y = 8.76y^2 \qquad (5\text{-}22)$$

since the area of AB is $\pi^2 X y + 2\pi X^2$.

In the cases of the two parallel slots and of the circular slot, it was assumed that the slots were cut into a plane surface that extends beyond the control surface. More often, the slots will be at the sides of a tank which is raised above the floor so that this assumption is not met. For such a case an equivalent

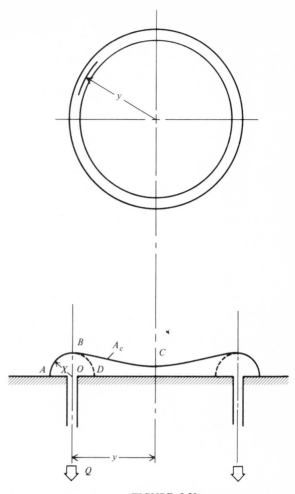

FIGURE 5-20
Control surface for a circular tank.

assumption must be made. It may well be possible to perform an analysis similar to those in this section for the particular assumption which is appropriate to the given configuration. A few such examples will be considered in the problems at the end of this chapter; otherwise, this topic will not be pursued further in this book.

5-7 DUCT DESIGN

The flow of fluids in ducts was discussed in some detail in Sec. 2-3, and sufficient information was presented in that section to handle completely the design of ducts in industrial ventilation situations. All that needs to be done is to select a suitable value of mean roughness height e, which for typical sheet-metal ductwork can be taken as 0.00012 m. With this choice, the design can be effected from the charts and formulas of Sec. 2-3.

However, the direct use of the formulas and charts of Sec. 2-3 does not afford the quickest method of design and is not the best way for repetitive computations. Since the roughness height is fairly uniform for commercial ductwork, and such variation as will be found will depend more upon the quality of the individual installation than upon choice of materials and fabrication methods likely to be used, it is worthwhile to refine the charts and formulas based upon a prior choice of roughness height. The value of 0.00012 m, mentioned previously, will be used.

Figure 5-21 shows the variation of pressure drop per unit length with flow rate and duct diameter for circular sheet-metal ducts. The velocity inside the duct is also shown. With this chart, a circular duct can be designed much more rapidly than by using Fig. 2-19. We shall extend this procedure to the case of rectangular ducts shortly. The use of Fig. 5-21 will be illustrated in several examples.

EXAMPLE 5-8   Standard air flows at the rate of 1.0 m^3/s through a 40-cm diameter duct. If the duct is 100 m long, what pressure drop will occur and what will be the velocity of the air?

SOLUTION   From Fig. 5-21, at $Q = 1.0$ m^3/s and $D = 0.40$ m, $V = 7.9$ m/s and $\Delta P/L = 1.7$ $N/m^2/m$. Then

$$\Delta P = \frac{\Delta P}{L} L = 1.7(100) = 170 \text{ N/m}^2 \qquad Ans. \qquad ////$$

EXAMPLE 5-9   Air at standard conditions flows through a circular sheet-metal duct at the rate of 3.0 m^3/s. The allowable pressure drop in 500 m of duct is 750 N/m^2. What diameter duct is required?

SOLUTION   From Fig. 5-21, for $Q = 3.0$ m^3/s and $\Delta P/L = 750/500 = 1.5$ $N/m^2/m$,

$$D = 60 \text{ cm} \qquad V = 9.6 \text{ m/s} \qquad Ans. \qquad ////$$

EXAMPLE 5-10   Air at standard conditions flows in a circular duct at the rate of 0.5 m^3/s. The duct is 30 m long, and the maximum allowable velocity is 25 m/s. What minimum duct diameter is required, and what pressure drop will occur if this diameter is used?

SOLUTION   From Fig. 5-21, using the given data, the diameter and pressure

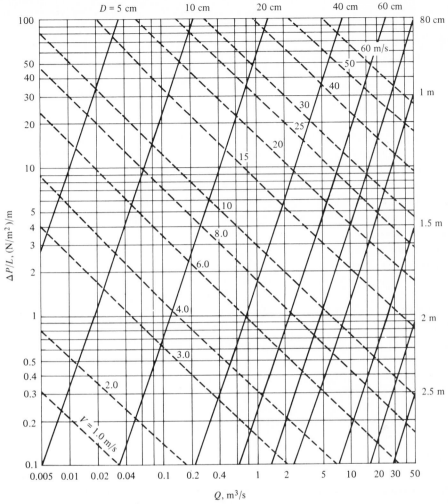

FIGURE 5-21
Pressure drop and velocity in circular sheet-metal ducts. The chart is drawn for standard air (298 K temperature and 1.0 atm pressure) and for $e = 0.00012$ m.

drop per unit length of duct are found to be 17 cm and 43 N/m²/m, respectively. The pressure drop in 30 m of duct is

$$\Delta P = 43(30) = 1290 \text{ N/m}^2 \qquad Ans. \qquad ////$$

Rectangular Ducts

The method of analysis using Fig. 5-21 is limited to circular ducts. Of equal importance in ventilation practice is the use of rectangular ducts. These cannot be handled directly from Fig. 5-21, and, in fact, to handle rectangular ducts in an analogous way would require several graphs since an additional parameter, the aspect ratio, comes into play. The aspect ratio is the ratio of width to length of the rectangular cross section. Instead of constructing a graph similar to Fig. 5-21 for each of several aspect ratios, an alternative design procedure is adopted. In this procedure the duct is designed as though it were circular and then an equivalent rectangular duct is selected. This procedure has the advantage that the particular duct shape does not influence the overall design of the ventilation system and can be selected in the final step to meet structural limitations or other requirements.

To execute this design procedure, a means for converting from a circular duct to an equivalent rectangular duct must be available. First, let us consider the type of equivalence required. To replace a section of circular duct by a rectangular duct without affecting the system requires that the new section handle the same flow rate with the same pressure drop due to friction. This differs from the equivalence implied in the use of hydraulic diameter, which is based on the use of the same formulas for the Reynolds number, pressure drop, and friction factor. Two sections having the same hydraulic diameter, in general, would have different flow rates and velocities. Let us obtain the diameter of a circular duct having the same pressure drop per unit length and the same flow rate as a rectangular duct of dimensions a by b, with $a > b$. This diameter will give us the equivalence desired.

Using Eq. (2-46), written in terms of hydraulic diameter for the rectangular duct, gives

$$\frac{\Delta P_r}{L} = \frac{f_r}{D_{hr}} \frac{\rho V_r^2}{2} \qquad (5\text{-}23)$$

Since velocity is flow rate divided by area, or

$$V_r = \frac{Q}{A_r}$$

and since for a rectangle the area is ab and the perimeter is $2(a + b)$, then

$$\frac{\Delta P_r}{L} = \frac{f_r}{4} \frac{a + b}{(ab)^3} \rho Q^2 \qquad (5\text{-}24)$$

For the circular duct the area is $\pi D^2/4$ and the perimeter is πD. The equation for the pressure drop in terms of diameter and flow rate is

$$\frac{\Delta P_c}{L} = \frac{8f_c}{\pi^2} \frac{1}{D_c^5} \rho Q^2 \qquad (5\text{-}25)$$

Setting $\Delta P_r = \Delta P_c$, equating these two equations for equal flow rate Q, and solving for D_c, we have

$$D_c = \left(\frac{32}{\pi^2}\right)^{1/5} \frac{(ab)^{0.6}}{(a + b)^{0.2}} \left(\frac{f_c}{f_r}\right)^{0.2} = 1.266 \frac{(ab)^{0.6}}{(a + b)^{0.2}} \left(\frac{f_c}{f_r}\right)^{0.2} \qquad (5\text{-}26)$$

Consider, now, the ratio of friction factors appearing in Eq. (5-26). The roughness height of the two duct shapes will be assumed equal, but the hydraulic diameters will differ somewhat. Thus, relative roughness will differ. Also the Reynolds numbers will be somewhat different in the two cases. Thus, we may expect the two friction factors to differ. However, the friction factors are only moderately sensitive to changes in relative roughness or Reynolds number, and the ratio of friction factors in Eq. (5-26) is raised to the 0.2 power. Thus, it is expected that the friction-factor ratio will have only a small influence, and we shall neglect it by assuming the ratio to equal unity. Equation (5-26) now becomes

$$D_c = 1.266 \frac{(ab)^{0.6}}{(a+b)^{0.2}} \qquad (5\text{-}27)$$

Assuming $a > b$ and solving in terms of the aspect ratio b/a, we have

$$\frac{D_c}{a} = 1.266 \frac{(b/a)^{0.6}}{(1+b/a)^{0.2}} \qquad (5\text{-}28)$$

Equation (5-28) is plotted in Fig. 5-22, along with D_c/b. Use of this graph to convert from a rectangular duct to a circular one is straightforward. However, the conversion process requires additional information about the rectangular duct sought. If its aspect ratio is known, then Fig. 5-22 can be read for D_c/a or D_c/b; and since D_c is known, a and b can be computed. In the more usual case where b is known, D_c/b can be computed and the aspect ratio b/a obtained from Fig. 5-22. Then a can be solved for. A similar procedure applies when a is known.

EXAMPLE 5-11 If the duct in Example 5-8 must have a height of 20 cm and is to be made of rectangular cross section, what must be its width?

SOLUTION From Example 5-8, $D_c = 40$ cm, and it is given that $b = 20$ cm. Then $D_c/b = 2.0$. From Fig. 5-22, the aspect ratio $b/a = 0.285$. Then

$$a = \frac{20}{0.285} = 70 \text{ cm} \qquad Ans. \qquad ////$$

EXAMPLE 5-12 Repeat Example 5-11 if the rectangular duct must have an aspect ratio of 0.5.

SOLUTION From Fig. 5-22, using $b/a = 0.5$, we read that $D_c/b = 1.6$. Then

$$b = \frac{40}{1.6} = 25 \text{ cm} \qquad Ans.$$

$$a = \frac{25}{0.5} = 50 \text{ cm} \qquad Ans. \qquad ////$$

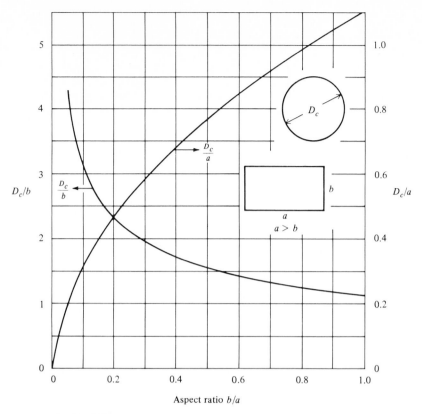

FIGURE 5-22
Equivalent rectangular and circular ducts having equal pressure drop and flow rate.

Resistance of Duct Fittings

The frictional resistance of duct fittings, such as elbows, joints, connectors, tees, and reducing and expanding sections, and abrupt changes of duct size as may occur at the entrance to the duct, at plenum chambers, and at various pieces of apparatus may add significantly to the total frictional resistance of a duct system. As pointed out in Sec. 2-3, there are two methods of treating incidental losses of this nature: In one method, the loss is expressed as a coefficient times the velocity head entering the fitting; in the other, the loss is expressed as an equivalent length of straight duct.

Table 2-3 gives some values of the loss coefficient K_L for use in the following equation, similar to Eq. (2-54):

$$\Delta P = \frac{K_L \rho V^2}{2} \qquad (5\text{-}29)$$

Table 2-3 is very brief, and the values given there may not be the most suitable for practical use. More specialized tables are available in Jennings [4] and Hemeon [5]. Although some of these tables give the loss coefficient K_L, other tables express loss in terms of equivalent length of straight duct L_e. We can develop the relation between L_e and K_L as follows: For an equivalent length of straight duct of hydraulic diameter D_h, the pressure drop is

$$\Delta P = \frac{f L_e \rho V^2}{2 D_h} \qquad (5\text{-}30)$$

For ΔP to be the same, we may equate these to obtain

$$L_e = \frac{K_L D_h}{f} \qquad (5\text{-}31)$$

The friction factor f varies with Reynolds number and relative roughness, but we can make use of the fact that at large Re, f becomes a function only of relative roughness e/D_h. An empirical relation for f as a function of e/D_h is available from Streeter [6]:

$$\sqrt{\frac{1}{f}} = 1.14 + 2 \log_{10} \frac{D_h}{e} \qquad (5\text{-}32)$$

Solving Eq. (5-32) for f and substituting into Eq. (5-31) gives, upon converting to the natural logarithm,

$$L_e = K_L D_h \left(1.14 + 0.868 \ln \frac{D_h}{e} \right)^2 \qquad (5\text{-}33)$$

Using the value of $e = 0.00012$ m, we obtain

$$L_e = K_L D_h (8.97 + 0.868 \ln D_h)^2 \qquad (5\text{-}34)$$

in which D_h for a rectangular duct is given by

$$D_h = \frac{2ab}{a+b} = 2a \frac{b/a}{1 + b/a} \qquad (5\text{-}35)$$

and $D_h = D_c$ for the equivalent circular duct.

Equation (5-34) has the drawback that equivalent length and hence total length of duct for calculation of pressure drop depend on the duct shape, which necessitates a refined calculation after duct shape has been decided upon. To avoid this, a reasonable approximation for L_e in terms of K_L can be made, which will give sufficiently accurate results if the total equivalent length of fittings is small relative to the total length of straight duct. To obtain such an approximation, let us first note that the term in Eq. (5-34) containing $\ln D_h$ will have only a small effect for normal duct sizes, which range from, say, 0.2 to 1 m. Choosing $D_h = 0.5$ m then gives

$$L_e = 70 K_L D_h \qquad (5\text{-}36)$$

The hydraulic diameter remaining in Eq. (5-36) cannot be approximated this simply, but it may be shown that at an aspect ratio of 0.2, the ratio of D_h to D_c is 0.718. The average aspect ratio is probably somewhat more than this value. A final approximation for Eq. (5-36) would be

$$L_e \approx 55 K_L D_c \qquad (5\text{-}37)$$

Equation (5-37) is not intended to be very accurate but should give satisfactory results for design purposes if the equivalent length of all fittings treated by use of that equation is of the order of 10 percent of the total equivalent length of duct.

EXAMPLE 5-13   Consider an abrupt entry to a circular duct of 25 cm diameter, carrying a flow of 1.5 m^3/s of standard air. What pressure drop will occur in the entry section due to friction? What will be the difference in static pressure between the flow inside the duct just beyond the entry and the surrounding atmosphere?

SOLUTION   From the flow rate and duct diameter the velocity is computed to be 30.6 m/s. From Table 2-3, the loss coefficient for an abrupt entry is 0.85 (Hemeon [5] gives a value of 0.5, but we shall use 0.85). The pressure drop due to friction is

$$\Delta P_1 = \frac{K_L \rho V^2}{2} = \frac{0.85(1.2)(30.6)^2}{2} = 478 \text{ N/m}^2$$

The pressure drop due to the increase in velocity head from zero to the value in the duct is

$$\Delta P_2 = \frac{\rho V^2}{2} = \frac{1.2(30.6)^2}{2} = 562 \text{ N/m}^2$$

The total pressure drop is the sum of these two, or 1040 N/m^2.   *Ans.*   ////

Design of Duct Systems

Conventional duct fabrication in the United States is based on size increments of 1 in, with $\frac{1}{2}$-in sizes inserted up to about 6 in in either linear dimension. Since these fabrication techniques are likely to continue for some time to come, it seems prudent to use dimensions, expressed in SI units, which conform closely to the fabrication practice in common use. Table 5-2 gives recommended dimensions to use for ductwork design until such time as the metric system comes into use in fabrication practice.

Recommended velocities in various parts of the ventilation system are summarized in Table 5-3, taken from Ref. [7] and converted to SI units. Most industrial ventilation values of velocity will tend toward the maximum listed, in the range of 10 to 15 m/s in ductwork.

Table 5-2 PRACTICAL EQUIVALENT DUCT SIZES IN INCHES AND IN SI UNITS

in	cm	in	cm	in	cm
3.0	7.5	16	41	48	122
3.5	9.0	17	43	50	127
4.0	10	18	46	52	132
4.5	11	19	48	54	137
5.0	13	20	51	56	142
5.5	14	21	53	58	147
6.0	15	22	56	60	152
6.5	16	23	58	62	157
7.0	18	24	61	64	163
7.5	19	25	64	66	168
8.0	20	26	66	68	173
8.5	22	27	69	70	178
9.0	23	28	71	72	183
9.5	24	29	74	74	188
10.0	25	30	76	76	193
10.5	27	32	81	78	198
11.0	28	34	86	80	203
11.5	29	36	91	82	208
12.0	30	38	97	84	213
12.5	32	40	102	86	218
13.0	33	42	107	88	224
13.5	34	44	112	90	229
14.0	36	46	117	92	234
15.0	38				

Table 5-3 RECOMMENDED AND MAXIMUM DUCT VELOCITIES*

Designation	Recommended velocities, m/s			Maximum velocities, m/s		
	Residences	Schools, theaters, public buildings	Industrial buildings	Residences	Schools, theaters, public buildings	Industrial buildings
Outside air intakes	2.5	2.5	2.5	4.1	4.6	6.1
Filters	1.3	1.5	1.8	1.5	1.8	1.8
Heating coils	2.3	2.5	3.0	2.5	3.0	3.6
Cooling coils	2.3	2.5	3.0	2.3	2.5	3.0
Air washers	2.5	2.5	2.5	2.5	2.5	2.5
Fan outlets	5.1–8.1	6.6–10.2	8.1–12.2	8.6	7.6–11.2	8.6–14.1
Main ducts	3.6–4.6	5.1–6.6	6.1–9.1	4.1–5.1	5.6–8.1	6.6–11.2
Branch ducts	3.0	3.0–4.6	4.1–5.1	3.6–5.1	4.1–6.6	5.1–9.1
Branch risers	2.5	3.0–3.6	4.1	3.3–4.1	4.1–6.1	5.1–8.1

* Converted to SI units from "ASHRAE Handbook of Fundamentals," p. 481, American Society of Heating, Refrigerating, and Air-Conditioning Engineers, New York, 1972. Used by permission.

The velocity pressure P_v is given by Eq. (2-54) as

$$P_v = \frac{\rho V^2}{2} \qquad (5\text{-}38)$$

and for standard air this is

$$P_v = 0.6 V^2 \qquad (5\text{-}39)$$

In duct systems, the duct size changes frequently, as new branches bring additional fluid into the main duct or take flow from it. The velocity in the duct is determined, of course, by the flow rate and duct area. The velocity can be adjusted higher or lower at will by changing the duct area; and a good place to change the velocity in the main duct is where a branch duct comes into the main duct and the main duct size changes anyway.

When the velocity in the main duct is decreased, the static pressure rises, as indicated by

$$\Delta P_s = K_r \left(\frac{\rho V_1{}^2}{2} - \frac{\rho V_2{}^2}{2} \right) \qquad (5\text{-}40)$$

in which ΔP_s is the rise in static pressure due to the decrease in velocity from V_1 to V_2, and K_r is the regain coefficient. For standard air, Eq. (5-40) gives

$$\Delta P_s = 0.6 K_r (V_1{}^2 - V_2{}^2) \qquad (5\text{-}41)$$

For an abrupt expansion in duct size, K_r is little greater than zero; while for a very smooth conical enlargement having an angle of less than $10°$, K_r is around 0.75. For typical commercial arrangements, the value of K_r is in the neighborhood of 0.5, and we shall use this value for design purposes. Thus Eq. (5-41) becomes

$$\Delta P_s = 0.3(V_1{}^2 - V_2{}^2) \qquad (5\text{-}42)$$

By the use of this pressure increase, referred to as *static regain*, the static pressure in the duct can be progressively raised in the direction of flow to values above that which would exist with constant velocity throughout. This pressure increase is of advantage in duct design, where branches coming in to the main duct should have roughly the same pressure drop across them, which would require that the main-duct pressure be about the same at the respective junction points. The use of static regain can approximate this condition more closely than when it is not used. To determine the amount of static regain if V_1 and V_2 are known, Eq. (5-42) may be used. If V_1 is known and ΔP_s is specified, then V_2 can be determined by solving this equation, giving

$$V_2 = \sqrt{V_1{}^2 - 3.333\,\Delta P_s} \qquad (5\text{-}43)$$

Alternatively, Fig. 5-23 may be used in place of Eqs. (5-42) or (5-43).

The value of K_r chosen in Eq. (5-40) takes into account the loss in the main duct occurring in the expansion fitting where the static regain occurs. Thus, it is not necessary to account separately for the pressure drop in the main

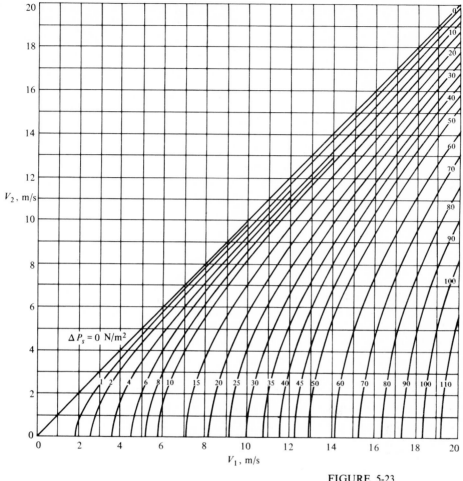

FIGURE 5-23
Static regain from Eq. (5-43).

duct at the point where the static regain is calculated. The pressure drop in the branch duct which may join the main duct at that point should be considered. And, of course, the pressure drop in the main duct through any straight run of duct and through all fittings where static regain is not applied must be accounted for.

EXAMPLE 5-14 A rectangular duct having dimensions of 51 by 173 cm carries 15 m³/s of standard air. At a point where an additional flow of 1.0 m³/s of air joins the main duct, static regain is to be applied in such a way that the pressure is increased by 15 N/m². Determine the velocity downstream of this point, and determine the duct size if the 51-cm dimension remains unchanged.

SOLUTION   The velocity upstream of the point where the branch duct joins the main duct and where static regain is to be applied is computed as 17 m/s. This is the actual velocity based on actual duct dimensions. The velocity downstream to increase the static pressure by 15 N/m^2 is read from Fig. 5-23 at $V_1 = 17$ m/s as

$$V_2 = 15.4 \text{ m/s} \qquad Ans.$$

The flow rate downstream is 16 m^3/s; and since the duct height b stays 51 cm, the duct width a becomes

$$a_2 = \frac{Q_2}{bV_2} = \frac{16}{0.51(15.4)} = 203 \text{ cm} \qquad Ans.$$

The downstream duct size is 51 by 203 cm.   ////

The static regain computations are based on the actual velocity in the duct, not on the velocity in an equivalent circular duct. Thus, if static regain methods are to be used in the duct design, it is not possible to design for circular ducts and then later convert to rectangular ducts at will.

In the design of a ventilation system, a number of hoods at various locations are served by a main duct, the hoods being connected to the main duct by branch ducts. Sometimes there will be several main ducts, joining together to form a single main duct much as limbs join the trunk of a tree. In the design, it is necessary to size all the main ducts and branches to achieve a suitable velocity distribution throughout the system. A velocity that is too low in a portion of the system indicates that a smaller duct could have been used there, resulting in a savings in initial investment. On the other hand, a velocity that is too high results in excess pressure drop and hence too much pumping power, with unnecessary expense; also, it may result in an excessive noise level. It is difficult to achieve proper velocities in a large system because some hoods are located much farther from the fan or processing unit than others. It is not uncommon for the farthest hood to be located 30 or 40 times as far away as the nearest hood. Since with equal velocity the pressure drop varies roughly as the distance from hood to processing unit, we would expect pressure drops many times as great for the farthest hood as for the nearest hood if the velocities were to be the same throughout.

To keep the same static pressure drop between all hoods and the processing unit or fan requires obviously that variations in velocity must occur. Where possible, the main ducts should be located so that the linear distance between the hood and the central processing unit be as nearly equal for all hoods as possible. Lower velocities can be used on the main ducts and branches leading to the farthest hoods, with higher velocities used in the branches to the nearer hoods. The use of static regain can help in many cases. Dampers can be used on the branches leading to the nearer hoods, except that for certain types of dust, dampers must be avoided due to their tendency to clog up. These dampers should be adjustable, and the system will need to be properly adjusted when put into operation and from time to time thereafter. The use of varying velocities, static regain, and adjustable dampers to achieve a balanced system will be illustrated by the following examples.

EXAMPLE 5-15 In Fig. 5-24 is shown a proposed duct layout for a small factory having overall dimensions of 20 by 50 m. There are seven hoods located on the side walls of the building at the points *A* to *G*; each of these hoods is to be served by a branch duct that goes up the wall to the ceiling and then across the ceiling to a main duct which runs along the centerline of the building just under the roof. A fan-and-filter unit is located in the roof structure near the center of the building. The flow rates are shown in Fig. 5-24. The branch ducts run upward a distance of 3 m along the side walls before running horizontally approximately 10 m along the ceiling to the point where they join the mains. The velocity in all the ducts must be at least 10 m/s to avoid dust buildup but must not exceed 20 m/s to prevent excessive noise and vibration.

In this example, we shall design the duct system on the left side of the building, serving hoods *A*, *B*, *E*, and *F*, relying on varying velocity in the branch ducts and on adjustable dampers to achieve a balanced system. In Example 5-16, we shall use static regain to design the right-hand duct system.

The ducts are to be rectangular with a height not exceeding 30 cm, except that the aspect ratio must not be less than 0.2, with a greater height being used if necessary to keep the aspect ratio above 0.2.

SOLUTION First consider the branch from the hood at *A* to where it joins the main duct at *I*. Although this is not stated, let us assume that the duct joins the hood horizontally and then has three elbows before joining the main duct at *I*. For this longest branch, we shall use favorable loss coefficients for the entrance to the duct at the hood and for the elbows. Let us take loss coefficients of 0.2 for these, giving a total static-pressure drop of $1.8(0.6)V^2$ due to fittings and also allowing for the necessary drop of one velocity head due to acceleration of the fluid in the duct. Let us take the velocity in the branch to be 10 m/s, the minimum acceptable amount, for this longest branch. Then, using $b = 30$ cm,

$$a = \frac{Q}{bV} = \frac{1.0}{0.30(10)} = 33 \text{ cm}$$

The aspect ratio is then $b/a = 30/33 = 0.91$. Figure 5-22 gives $D_c/a = 1.05$ for this aspect ratio, from which $D_c = 35$ cm. Then Fig. 5-21, using $Q = 1.0$ m³/s and $D_c = 35$ cm, gives

$$\frac{\Delta P}{L} = 4 \text{ N/m}^2/\text{m}$$

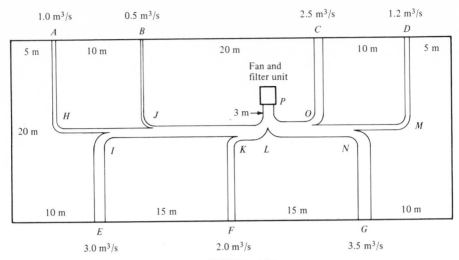

FIGURE 5-24
Ventilation system layout for Examples 5-15 and 5-16.

The total pressure drop in the branch AHI, whose length is 18 m, is then

$$\Delta P_{AI} = \frac{\Delta P}{L} L + 0.6 K_L V^2$$

$$= 4(18) + 1.8(0.6)(10^2)$$

$$= 180 \text{ N/m}^2$$

The duct size for branch AHI is 30 by 33 cm.

Next, we shall consider branch EI. The pressure drop along it must be 180 N/m². Its length is 13 m, and the total loss coefficient is assumed to be 1.8, as before. A trial-and-error solution is necessary. Try first a velocity of 11 m/s, slightly larger than for the previous branch, since this branch is a little shorter. This gives $a = 91$ cm for $b = 30$ cm, or $b/a = 0.33$, and Fig. 5-22 gives $D_e/a = 0.62$, from which $D_e = 56$ cm. Figure 5-21 then gives $\Delta P/L = 2.0 \text{ N/m}^2/\text{m}$, and the total pressure drop is

$$\Delta P_{EI} = 13(2.0) + 1.8(0.6)(11)^2 = 157 \text{ N/m}^2$$

which is a little too small. Next try $V = 12$ m/s, which gives, in turn,

$$a = 83 \text{ cm} \qquad \frac{b}{a} = 0.36 \qquad \frac{D_c}{a} = 0.65 \qquad D_c = 54 \text{ cm} \qquad \frac{\Delta P}{L} = 2.9$$

$$\Delta P_{EI} = 13(2.9) + 1.8(0.6)(12)^2 = 194 \text{ N/m}^2$$

This pressure drop is a little too large. A compromise duct size would be 30 by 88 cm.

Next consider the main duct from I to J. The flow rate is 4.0 m^3/s, and let us choose a velocity of 15 m/s. The length is 5 m. For a duct height of 30 cm, the duct width is 4.0/0.3(15) = 89 cm. Then $b/a = 0.34$, $D_e/a = 0.63$ from Fig. 5-22, $D_e = 56$ cm, $\Delta P/L = 4.2$ N/m^2/m from Fig. 5-21, and the pressure drop is 4.2(5) = 21 N/m^2. Then the total pressure drop from point A becomes

$$\Delta P_{AJ} = 180 + 21 = 201 \text{ N/m}^2$$

The dimensions of the main duct IJ are 30 by 89 cm.

Next, consider branch BJ, with length 13 m, flow rate 0.5 m^3/s, and pressure drop 201 N/m^2. Again take the total loss coefficient as 1.8. A trial-and-error solution is necessary, and the data are tabulated as follows:

V, m/s	b, cm	a, cm	b/a	D_e/a	D_e, cm	$\Delta P/L$	$\sum K_L 0.6V^2$	ΔP_{BJ}, N/m^2
12	20	21	0.95	1.07	22	9	156	273
10	20	25	0.80	0.98	25	6	108	186
11	20	23	0.87	1.02	23	7	131	222

By interpolation, use a duct having dimensions of 20 by 24 cm.

The main duct JK is considered next. The flow rate is 4.5 m^3/s, the length is 10 m, and the velocity is assumed to be 15 m/s. The width a becomes 100 cm, $b/a = 0.30$, $D_e/a = 0.58$, $D_e = 58$ cm, $\Delta P/L = 4.4$, and

$$\Delta P_{JK} = 4.4(10) = 44 \text{ N/m}^2$$
$$\Delta P_{AK} = 201 + 44 = 245 \text{ N/m}^2$$

Thus, the main duct size is 30 by 100 cm.

Next we treat the branch FK, whose flow rate is 2.0 m^3/s and whose length is 13 m. Since this branch must have a considerably greater pressure drop across it, let us use cheaper fittings having a considerably greater loss coefficient. Using an abrupt entrance from the hood gives a coefficient of 0.85 there, and we can use elbows having a smaller centerline-radius/width ratio, giving a loss coefficient of 0.4 for each elbow. This way, the total loss coefficient becomes 1.0 + 0.85 + 3(0.4) = 3.05. A trial-and-error solution is again necessary and is tabulated here:

V, m/s	b, cm	a, cm	b/a	D_e/a	D_e, cm	$\Delta P/L$	$\sum K_L 0.6V^2$	ΔP_{FK}, N/m^2
10	30	67	0.45	0.73	49	2.5	183	216
11	30	61	0.49	0.76	46	3.1	221	261

By interpolation, we have an indicated duct size of 30 by 64 cm.

Finally, for the main duct KL, using a velocity of 15 m/s and a duct height of 30 cm, as before, the same procedure results in a duct size of 30 by 144 cm, with a pressure drop of 20 N/m^2 from K to L, or a total pressure drop of 265 N/m^2 from A to L. Note that the aspect ratio of 0.21 barely exceeds the minimum allowable. ////

EXAMPLE 5-16 Using the same data and restrictions of Example 5-15, design the duct system on the right side of Fig. 5-24, making use of static regain in the design.

SOLUTION We shall start our solution in the same manner as for the left side. Take branch DMN, and assume that the total loss coefficient is 1.8, the same as for branch AHI. Assume a velocity of 12 m/s, and note that $Q = 1.2$ m³/s. Taking the duct height $b = 30$ cm; then $a = 1.2/0.3(12)$, giving 33 cm. The linear length of duct is 13 m. The aspect ratio is $b/a = 0.91$; Fig. 5-22 gives $D_c/a = 1.05$, from which $D_c = 35$ cm. Figure 5-21 gives $\Delta P/L = 5$ N/m²/m. Then the total pressure drop for this branch is

$$\Delta P_{DN} = 1.8(0.6)(12)^2 + 5(18) = 246 \text{ N/m}^2$$

and the duct size for this branch is 30 by 33 cm.

Branch GN is computed by a trial-and-error procedure similar to the ones performed in Example 5-15. We assume that $\sum K_L$ is again 1.8; and for $Q = 3.5$ m³/s and $L = 13$ m:

V, m/s	b, cm	a, cm	b/a	D_c/a	D_c, cm	$\Delta P/L$	$\sum K_L 0.6 V^2$	ΔP_{GN}, N/m²
13	30	90	0.33	0.62	56	4.0	183	235
14	30	83	0.36	0.64	53	5.5	212	284

Then the duct size will be 30 by 88 cm.

The main duct NO can be handled simply by choosing a velocity of 16 m/s. The results are

$$\Delta P_{NO} = 24 \text{ N/m}^2$$

giving

$$\Delta P_{DO} = 270 \text{ N/m}^2$$

and the required duct size is 30 by 98 cm.

Since this pressure drop is a little excessive, we can recover some of the pressure by using static regain here. Let us bring the pressure drop down to 245 N/m² for a pressure regain ΔP_s of 25 N/m². Using $V_1 = 16$ m/s, Fig. 5-23 shows that

$$V_2 = 13.2 \text{ m/s}$$

This new velocity will be used in the main duct OL, and the new pressure drop of 245 N/m² will be used in designing branch CO. A calculation similar to that for branch GN establishes that the proper duct size for branch CO is 30 by 63 cm.

The main duct OL is designed using a velocity of 13.2 m/s, flow rate of 7.2 m³/s, and length of 5 m. Take the height $b = 35$ cm and duct width $a = 156$ cm. This gives $b/a = 0.22$, $D_c/a = 0.50$, $D_c = 78$ cm, $\Delta P/L = 2.7$ N/m²/m, and $\Delta P_{OL} = 13.5$ N/m². The overall pressure drop to this point is

$$\Delta P_{DL} = 259 \text{ N/m}^2$$

and the main size is 35 by 156 cm.

At the junction point L of the two main ducts, there will be additional losses in each main duct due to the elbows and junction fitting. Take $K_L = 0.2$ on each main duct, giving

$$\Delta P_{\text{left}} = 0.2(0.6)(15)^2 = 27 \text{ N/m}^2$$

$$\Delta P_{\text{right}} = 0.2(0.6)(13.2)^2 = 21 \text{ N/m}^2$$

Then

$$\Delta P_{AL} = 265 + 27 = 292 \text{ N/m}^2$$

$$\Delta P_{DL} = 259 + 21 = 280 \text{ N/m}^2$$

We see that at the point where the main ducts join there is a small but noticeable pressure differential. As long as the pressure differential is no greater than this, its effect can be ignored in practice. However, we can correct for this effect using static regain on the left main duct, which has the greater pressure drop, to see what velocity this flow must be reduced to in order to reduce its pressure drop to that of the right main duct. Figure 5-23 gives this velocity as 13.6 m/s. The cross-sectional areas of the two main ducts at the junction point are determined from these velocities:

$$A_{\text{left}} = \frac{6.5}{13.6} = 0.48 \text{ m}^2$$

$$A_{\text{right}} = \frac{7.2}{13.2} = 0.55 \text{ m}^2$$

The total cross-sectional area of the main duct LP is then 1.03 m^2, corresponding to dimensions of 50 by 206 cm.

The pressure drop in 3 m of this main duct is then 5 N/m^2, and the total pressure drop is

$$\Delta P_{AP} = \Delta P_{DP} = 280 + 5 = 285 \text{ N/m}^2$$

which corresponds to a pressure drop of 1.14 in of water.   ////

The calculation procedure outlined in Examples 5-15 and 5-16 is somewhat tedious, especially for more complicated systems involving more hoods than those in the simplified examples. Even so, the design system is practical. Often simplifications are possible when it is realized that dampers will be used in each branch and at various points in the main duct, particularly where main ducts join together, as at point L in Fig. 5-24. Before the system is put into operation, it is necessary to balance all the main ducts and branches by adjusting the dampers so that each branch receives the proper flow. Because of this adjustment, calculations do not usually have to be as refined as the ones in the preceding examples. For instance, the trial-and-error calculations can usually be avoided for the intermediate branches. By choosing a velocity in the branch low enough so that the pressure drop in the branch is somewhat less than that required for balance, the presence of the damper will make the design satisfactory as long as the pressure drop across the damper is not too large.

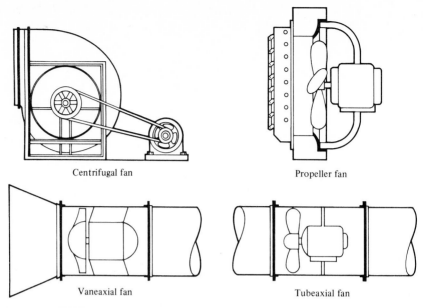

Centrifugal fan

Propeller fan

Vaneaxial fan

Tubeaxial fan

FIGURE 5-25
Classification of fan types. (*Used by permission from the Air Moving and Conditioning Association.*)

5-8 FAN SELECTION AND PERFORMANCE

Any ductwork system will require the use of a fan, blower, or other device to establish the flow through the system. There are several types of fans and blowers in common use, some of which are shown in Fig. 5-25. These fans and blowers are classified in terms of the direction of flow as axial-flow or centrifugal fans. The axial-flow fans include the propeller fan, vaneaxial fan, and tubeaxial fan. The centrifugal fan can have either forward-curved, backward-curved, or radial blades, as illustrated in Fig. 5-26. An injector is sometimes used, as shown in Fig. 5-27. The different types of fans and injectors will be discussed in more detail, especially with respect to their performance characteristics, later in this section.

Fan Characteristics

The characteristics of centrifugal fans are expressed as a series of curves having the volumetric flow rate as an abscissa. The curves are those of static and total pressure, power, and total and static efficiency. Typical curves are shown in Fig. 5-28; Fig. 5-28*a* is for forward-curved blades, Fig. 5-28*b* is for backward-curved blades, and Fig. 5-28*c* is for radial blades. The exact form and shape of the characteristic curves depends, not only on the fan itself, but also on the way in which static and total pressures are measured.

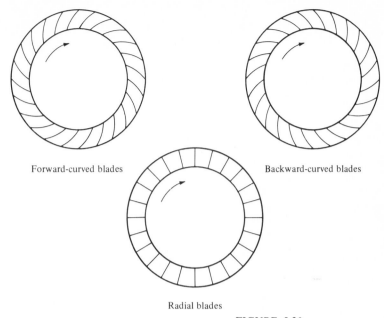

Forward-curved blades   Backward-curved blades

Radial blades

FIGURE 5-26
Blade shapes for centrifugal fans.

Figure 5-29 shows the arrangement used to develop the data in Fig. 5-28. The maximum flow rate is obtained when the fan discharges directly or through a short duct into an open space. The maximum static pressure is obtained when the flow is blocked off at the exit duct. A short duct section is shown at both the entrance and the exit of the fan to facilitate the flow pattern and allow attachment of suitable pressure gages. The flow rate can be measured by a pitot tube or by means of an anemometer placed at the end of the exit duct. For a fan which is to be used in a duct system, the inlet and exit velocities will be about the same, so that the difference between the static and total pressures will be about the same entering or leaving the fan. Thus the

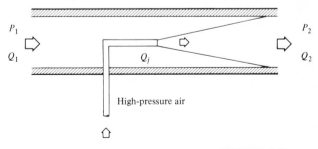

FIGURE 5-27
Cylindrical injector.

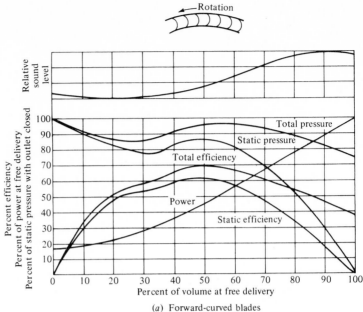

(a) Forward-curved blades

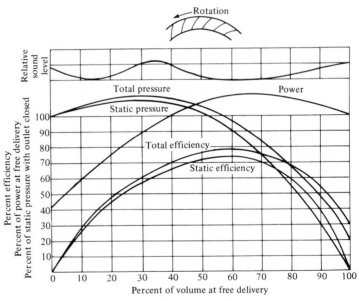

(b) Backward-curved blades

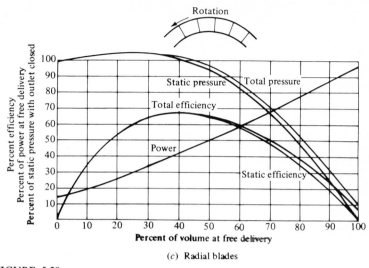

(c) Radial blades

FIGURE 5-28

Typical characteristic curves of centrifugal fans. (*From B. H. Jennings, "Environmental Engineering Analysis and Practice," pp. 464–465, Intext Educational Publishers, New York, 1970. Used by permission.*)

static and total pressure characteristic curves will almost coincide. A fan test which will more nearly duplicate this condition is shown in Fig. 5-30. The characteristic curve will then follow the total pressure curves of Fig. 5-28. The characteristic curves of an axial-flow fan are similar to those of the centrifugal fan having forward-curved blades shown in Fig. 5-28a.

A fan can be tested by using either of the arrangements shown in Figs. 5-29 and 5-30. Such a test can be performed inexpensively since only a short length of duct is required at the entrance and outlet of the fan, along with a manometer to measure the pressure rise. Also a pitot tube or anemometer is needed to measure the flow rate. For a complete test, a damper will be needed, preferably

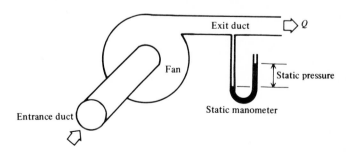

FIGURE 5-29

Test arrangement which could be used to generate data similar to that of Fig. 5-28.

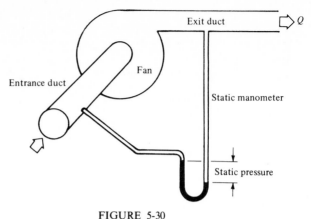

FIGURE 5-30
Test arrangement suitable for a fan in a duct.

at the outlet section, capable of regulating flow from the maximum down to no flow. However, a meaningful test can be performed if the fan outlet can be completely blocked, in addition to operating with negligible flow resistance. This produces two operating conditions: the maximum flow point, and the static no-delivery point. A rough estimate of the characteristic curve can be made from these two points alone, as shown in Fig. 5-31.

First the static no-delivery point (SND) is plotted at $Q = 0$. Then the point of maximum Q and zero pressure is plotted from the data or estimated if the maximum flow produced in the test produces a small pressure. A third point on the curve is estimated with pressure equal to the SND pressure and flow rate equal to half the maximum value. A smooth curve is then fitted through these three points. The left half of this curve is subject to considerable doubt,

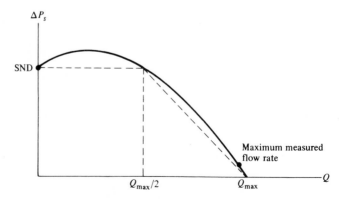

FIGURE 5-31
Approximate characteristic curve from two test points.

but the right half should be fairly close to the true curve. Since any practical operating point will lie on the right half of the curve, this procedure is suitable for cases where an existing fan is to be used in the design of a duct system and where detailed characteristics of the fan are unknown. Although greater accuracy could be obtained by performing a detailed test on the fan, the results of this approximate procedure, using the data of the rudimentary test described, will usually suffice.

EXAMPLE 5-17   An available fan is to be used in a duct. A two-point test on the fan when operating at 1000 rpm gives a flow of 12 m^3/s with a pressure rise of 50 N/m^2 approximating the maximum-flow condition. When the discharge duct is blocked off, the static pressure produced is 1000 N/m^2. Sketch the approximate characteristic curve, and estimate the pressure produced at a flow rate of 7.5 m^3/s.

   SOLUTION   The approximate curve is sketched in Fig. 5-32. The pressure rise produced at a flow rate of 7.5 m^3/s is 890 N/m^2.   *Ans.*   ////

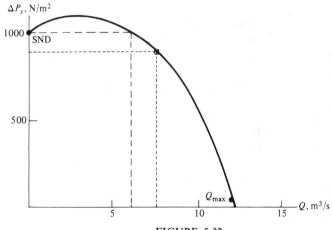

FIGURE 5-32
Characteristic curve for Example 5-17.

Characteristics of Geometrically Similar Fans

When the same fan is used under different conditions or when two geometrically similar fans are compared, certain similarities exist in the way they perform in the two cases. The characteristic curves for geometrically similar fans have the same shape, and if plotted as a function of the ratio of actual to maximum flow rate and of actual static pressure rise to static no-delivery pressure rise, the same curve holds for all such fans; this is shown in Fig. 5-33. Similar

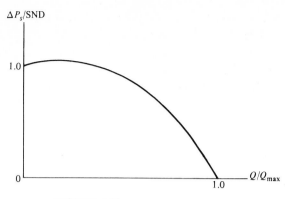

FIGURE 5-33
Characteristic curve for geometrically similar fans.

results hold for efficiency and power curves. If such a curve is available for a fan, then the same curve can be used to predict the performance of a geometrically similar fan. Two fans are geometrically similar if they have the same ratio of all dimensions and if the shape and arrangement of the blades is the same in both.

Additional relations for the same or for geometrically similar fans hold; these relations are known as the *fan laws*, which are summarized in Table 5-4. The symbol Q represents volumetric flow rate; $\dot{W}$ is power required; P is the pressure rise; N is the fan rotative speed; D is the characteristic diameter of the fan, probably the diameter of the fan rotor; ρ is the density of the fluid; and V is the characteristic velocity of the fan, probably taken as the peripheral velocity of the rotor.

The first law listed in Table 5-4, which is probably the most important one for duct design, states that if the same or an identical fan is used under different conditions but with a fluid having the same density, then Q, P, and $\dot{W}$ vary with rotative speed of the fan according to the relations

$$Q \propto N \qquad P \propto N^2 \qquad \dot{W} \propto N^3 \qquad (5\text{-}44)$$

The other fan laws can be taken in a similar way from Table 5-4. It must be understood that these laws apply to similar operation of the geometrically similar fans; that is, the operating points of both fans will coincide on the curve of Fig. 5-33. It is certainly possible to operate the two fans, even though they are similar geometrically, at different conditions so that the fan laws do not hold between those two operating states. The following examples will illustrate the use of the fan laws in the design of duct systems.

EXAMPLE 5-18   A fan has the characteristic pressure curve shown in Fig. 5-32 when turning at 1000 rpm. It is to be used in a duct system to supply 20 m³/s of standard air at a pressure rise of 2000 N/m². Determine the rotative speed required of the fan, and determine the operating point on the curve of Fig. 5-32 corresponding to the new operating point.

SOLUTION Equations (5-44) apply here. Letting subscript 1 be the corresponding operating point on Fig. 5-32 at 1000 rpm, and letting subscript 2 be the operating point in the actual design, we have

$$\frac{Q_1}{Q_2} = \frac{N_1}{N_2} \qquad \frac{\Delta P_1}{\Delta P_2} = \left(\frac{N_1}{N_2}\right)^2$$

Combining these gives

$$\Delta P_1 = \Delta P_2 \left(\frac{Q_1}{Q_2}\right)^2 = 2000\,\frac{Q_1{}^2}{(20)^2} = 5Q_1{}^2$$

The operating point on Fig. 5-32 is quickly located by trial and error as

$$Q_1 = 10.25 \text{ m}^3/\text{s} \qquad \Delta P_1 = 525 \text{ N/m}^2 \qquad Ans.$$

Table 5-4 FAN LAWS FOR GEOMETRICALLY SIMILAR FANS

Fan law	Dependent variable	Independent variable	Constant quantities
1	Q P $\dot{W}$	N N^2 N^3	D, ρ
2	Q N $\dot{W}$	D^2 $1/D$ D^2	P, ρ
3	Q P $\dot{W}$	Constant ρ ρ	V, D, N
4	Q P $\dot{W}$	D^3 D^2 D^5	N, ρ
5	Q P $\dot{W}$	D^3N D^2N^2 D^5N^3	ρ
6	Q $\dot{W}$ N	$1/\sqrt{\rho}$ $1/\sqrt{\rho}$ $1/\sqrt{\rho}$	P, D
7	P $\dot{W}$ N	ρ ρ Constant	Q, D
8	Q P $\dot{W}$ N	$1/\rho$ $1/\rho$ $1/\rho^2$ $1/\rho$	$\rho Q, D$

The new operating speed N_2 is found from

$$N_2 = N_1 \frac{Q_2}{Q_1} = 1000 \frac{20}{10.25} = 1951 \text{ rpm} \qquad Ans.$$

The power requirements in the new state are related to those in the reference condition by

$$\dot{W}_2 = \dot{W}_1 \left(\frac{N_2}{N_1}\right)^3 = 7.43\dot{W}_1 \qquad ////$$

EXAMPLE 5-19   A certain fan located in a duct pumps 30 m³/s of standard air against a static pressure rise of 1000 N/m². When the density of the air changes to 1.5 kg/m³ under conditions such that the fan speed, the volumetric flow rate, and, of course, the fan diameter remain unchanged, determine the new pressure rise.

SOLUTION   Fan law 3 in Table 5-4 applies here since N, D, and Q remain unchanged. Then

$$\Delta P_2 = \Delta P_1 \frac{\rho_2}{\rho_1} = 1000 \frac{1.5}{1.2} = 1250 \text{ N/m}^2 \qquad Ans. \qquad ////$$

EXAMPLE 5-20   A fan whose characteristics are given by Fig. 5-32 at 1000 rpm pumps 8 m³/s of standard air against a pressure rise of 830 N/m² when turning at 1000 rpm. If the density of the fluid changes to 1.6 kg/m³, and if the values of Q and ΔP are to remain unchanged, to what speed must the fan be adjusted?

SOLUTION   The given data are:

$\rho_1 = 1.2 \text{ kg/m}^3$ $\qquad \rho_2 = 1.6 \text{ kg/m}^3$ $\qquad \Delta P_1 = \Delta P_2 = 830 \text{ N/m}^2$
$N_1 = 1000 \text{ rpm}$ $\qquad Q_1 = Q_2 = 8.0 \text{ m}^3/\text{s}$ $\qquad D_2 = D_1$

Solve for N_2. We must change the speed of the fan to compensate for the change in density. We cannot expect the new operating point to correspond to the original operating point on the curve of Fig. 5-32. However, *some* operating point on that curve will correspond to the new operating condition; let us designate this point as Q_1', $\Delta P_1'$ for which

$$N_1' = N_1 \qquad \rho_1' = \rho_1 \qquad D_1' = D_1$$

Use fan law 6 from Table 5-4 to change to a state 3 for which

$$\Delta P_3 = \Delta P_1' \qquad \rho_3 = \rho_2 \qquad D_3 = D_1$$

This will take care of the change of density. Then

$$Q_3 = Q_1'\left(\frac{\rho_1}{\rho_2}\right)^{1/2}$$

$$N_3 = N_1'\left(\frac{\rho_1}{\rho_2}\right)^{1/2}$$

Now, we go to point 2 using fan law 1, giving

$$Q_2 = Q_3\frac{N_2}{N_3} \qquad \Delta P_2 = \Delta P_3\left(\frac{N_2}{N_3}\right)^2$$

These last two equations may be written as

$$Q_2 = Q_1 = Q_1'\left(\frac{\rho_1}{\rho_2}\right)^{1/2}\frac{N_2}{N_3} = Q_1'\left(\frac{\rho_1}{\rho_2}\right)^{1/2}\frac{N_2}{N_1}\frac{N_1}{N_3}$$

$$\Delta P_2 = \Delta P_1 = \Delta P_1'\left(\frac{N_2}{N_1}\right)^2\left(\frac{N_1}{N_3}\right)^2$$

From the equation relating N_3 and N_1' we have that

$$\frac{N_1}{N_3} = \left(\frac{\rho_2}{\rho_1}\right)^{1/2}$$

Then the two previous equations become

$$Q_1 = Q_1'\left(\frac{\rho_1}{\rho_2}\right)^{1/2}\left(\frac{N_2}{N_1}\right)\left(\frac{\rho_2}{\rho_1}\right)^{1/2} = Q_1'\frac{N_2}{N_1}$$

$$\Delta P_1 = \Delta P_1'\frac{N_2}{N_1}\frac{\rho_2}{\rho_1}$$

Eliminating N_2/N_1 between these equations gives

$$\Delta P_1' = \Delta P_1\frac{\rho_1}{\rho_2}\left(\frac{Q_1'}{Q_1}\right)^2$$

$$= \frac{1.2}{1.6}\frac{830}{64}Q_1'^2$$

$$= 9.727Q_1'^2$$

We can now locate the point Q_1', $\Delta P_1'$ on the curve of Fig. 5-32 by trial and error. The resulting location is at

$$Q_1' = 8.75 \text{ m}^3/\text{s} \qquad \Delta P_1' = 745 \text{ N/m}^2$$

We can then solve for the speed at the new operating condition as

$$N_2 = N_1\frac{Q_1}{Q_1'} = 1000\frac{8}{8.75} = 914 \text{ rpm} \qquad \textit{Ans.}$$

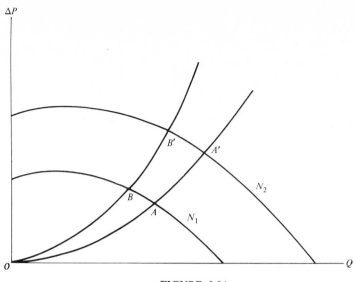

FIGURE 5-34
Characteristic curves of fan and two ducts.

This result can be checked by going from point 1' to point 2 by a different path. Go to point 4 changing density at constant Q by law 7. This gives

$$\Delta P_4 = \Delta P'_1 \frac{\rho_2}{\rho_1} = 745 \frac{1.6}{1.2} = 993 \ N/m^2$$

$$N_4 = N'_1 = 1000 \ \text{rpm}$$

$$Q_4 = Q'_4 = 8.75 \ m^3/s$$

Now we go from point 4 to point 2 at constant density, giving from law 1

$$Q_2 = Q_4 \frac{N_2}{N_4} = 8.75 \frac{914}{1000} = 8.0 \ m^3/s \qquad Check$$

$$\Delta P_2 = \Delta P_4 \left(\frac{N_2}{N_4}\right)^2 = 933(0.914)^2 = 830 \ N/m^2 \qquad Check \qquad ////$$

Interaction Between Fan and Duct System

The study of fan behavior by itself is not the whole story. It is necessary to consider what happens when the fan is installed in the ductwork. Figure 5-34 shows characteristic curves for a fan run at two different speeds, along with two characteristic curves for a ductwork system. Equations (2-46) and (2-54) indicate that pressure drop is approximately proportional to velocity squared for flow in a duct system. Thus,

$$\Delta P = K_1 Q^2 \qquad (5\text{-}45)$$

This is the equation of a parabola; curves OAA' and OBB' are two different characteristic curves for two different ductwork systems, the latter curve having a higher resistance coefficient K_1.

From fan law 1 in Table 5-4, we see that ΔP is proportional to Q^2 so that

$$\Delta P = K_2 Q^2 \qquad (5\text{-}46)$$

which forms the locus of corresponding points on the fan characteristic curves. When a fan has been matched to a duct system, then $K_1 = K_2$, and the two curves complement each other. That is, a change in fan speed changes the flow rate and pressure in such a way that the fan and duct remain matched, the fan operating at the corresponding point of the new characteristic curve. Thus, if it should be desired to increase the capacity of the system by increasing the velocities in the system, it is necessary merely to increase the rotative speed of the fan.

EXAMPLE 5-21 A duct system complete with fan carries 25 m³/s with a static pressure rise across the fan of 3000 N/m². It is desired to increase the flow rate to 35 m³/s, increasing the duct velocities accordingly, by turning the fan rotor at a higher rpm. Calculate the new fan rpm required if the original value is 1500 rpm.

SOLUTION From fan law 1, Table 5-4,

$$N_2 = N_1 \frac{Q_2}{Q_1} = 1500 \frac{35}{25} = 2100 \text{ rpm} \qquad Ans.$$

The pressure rise increases as the square of the speed ratio, and the fan power increases as the cube of the speed ratio. The new pressure rise is

$$\Delta P = 3000 \left(\frac{35}{25} \right)^2 = 5880 \text{ N/m}^2$$

The ratio of fan powers is $(35/25)^3 = 2.7$. ////

Stability of Fan-Duct Systems

A problem in matching the fan to the duct system relates to the stability of the combination. A system is stable if small changes in either flow rate, pressure, or resistance are automatically compensated for by small changes in the other quantities. If a small change in one quantity leads progressively to larger and larger changes in the other quantities, which, in turn, leads to larger and larger changes in the first quantity, then the system is unstable.

An unstable condition can result if the condition of operation of the fan is in the region where the fan characteristic curve for pressure rise has positive slope, as shown in Fig. 5-35. Operating point A, on the portion of the curve

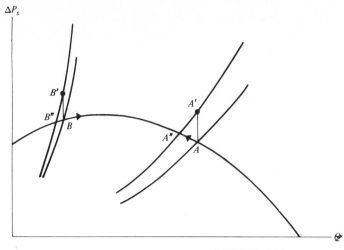

FIGURE 5-35
Stability of fan-duct system.

having negative slope, is stable. Should the duct resistance suddenly increase, for example, the necessary change in pressure will occur much more rapidly than the change in flow rate which must follow. Thus the operating point on the duct characteristic shifts to point A'. The fan cannot deliver this much pressure, and so it seeks to deliver more pressure than at point A by reducing the flow, which moves the operating point up along the fan characteristic in the direction of the arrow. Eventually both operating points coincide again at point A''. Only small changes in pressure and flow rate are involved, and so the operating point A is stable.

Consider, now, the operating point B. Should the flow resistance increase slightly, the pressure must immediately rise to that at point B'; the fan attempts to increase the pressure but can do so only by increasing the flow rate. This makes things worse, so that the fan performance must soon break down, which reduces the flow and pressure well below the fan characteristic, at which point the fan picks up again. Unless additional provision is made to render the system stable, the fan keeps hunting for its proper operating point but can never find it exactly. Since even the smallest deviation from the true operating point will cause the behavior described, the fan can never be stable operating in this region. For this reason, a fan should never be operated at a point to the left of the peak of its characteristic curve.

Fan Power Requirements

We have mentioned power requirements for the fan but have not yet discussed how to compute these requirements. Equation (2-103) may be utilized in this regard. Neglecting any changes in potential energy, let us assume constant density and take the head-loss term in that equation into account by introducing an efficiency term. In addition, let us change the sign of the work so that positive work is work done on the fan. Then Eq. (2-103) may be written

$$\dot{W} = \frac{Q}{\eta} \left[\frac{1}{2} \rho(V_2^2 - V_1^2) + P_2 - P_1 \right] \qquad (5\text{-}47)$$

If the change in kinetic energy is also negligible, this becomes

$$\dot{W} = \frac{Q \, \Delta P_s}{\eta} \qquad (5\text{-}48)$$

If the efficiency η is unity, these equations represent the amount of work which is done on the air to increase its kinetic and flow energies. The actual work done on the fan includes, in addition, losses due to fluid friction and turbulent dissipation inside the fan passages and mechanical losses in the moving parts of the fan mechanism.

Fan efficiencies in the range of 0.5 to 0.8 are common. As shown in Fig. 5-28, the efficiency curve peaks slightly to the right, usually, of the peak of the pressure characteristic curve. This is an advantage since it ensures stable operation at the point of maximum efficiency. However, care must be used since for forward-curved blades the point of maximum efficiency is probably not far enough to the right of the peak of the characteristic curve for safe operation. Therefore some sacrifice of efficiency is necessary for satisfactory performance. Even the radial-blade fan will require close design of the duct system if it is to be safely operated at the point of peak efficiency. In selection of any fan, it is important to choose the operating point carefully in order to effect a proper balance between operating stability and minimum power cost.

EXAMPLE 5-22   A radial fan with the characteristics given in Fig. 5-28c is to be used in providing a flow of 30 m³/s of standard air through a duct system, the total pressure rise being 2000 N/m². Select an operating point just to the right of the point of maximum efficiency to ensure stability. Determine the maximum flow rate and SND of the fan. If the system operates 12 h/day, 5 days/week, 50 weeks/year, how much energy is required to operate the fan for 1 year? How much will this cost at $0.008/kWh? How much would be saved if the fan were to be designed to operate at the point of maximum efficiency?

SOLUTION   From Fig. 5-28c, select the operating point at 50 percent of maximum delivery. This gives a maximum flow rate of 60 m³/s, and

$$\eta = 0.65 \qquad \frac{\Delta P_s}{\text{SND}} = 0.94$$

Then

$$\text{SND} = \frac{2000}{0.94} = 2128 \text{ N/m}^2 \qquad \textit{Ans.}$$

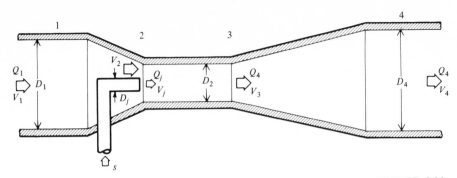

FIGURE 5-36
Venturi injector.

The power is computed as

$$\dot{W} = \frac{Q\,\Delta P_s}{\eta} = \frac{30(2000)}{0.65(1000)} = 92.3 \text{ kW}$$

The system will operate 3,000 h/year, requiring

$$92.3(3000) = 276,900 \text{ kWh/year} \qquad Ans.$$

The cost per year then is

$$S = 276,900(0.008) = \$2215/\text{year} \qquad Ans.$$

Had we operated at the point of maximum efficiency, the efficiency would have been 0.68, giving a power of 88.2 kW and an annual cost of $2118/year. The savings would then be

$$\Delta S = \$97/\text{year} \qquad \text{or 4.4 percent} \qquad Ans. \qquad ////$$

Injectors

Let us now consider the use of injectors to accomplish the same function as a fan. The injector has several inherent disadvantages. Its efficiency is usually much less than that of a fan, and it adds additional flow to the duct in which it is located, which increases the pressure drop downstream of the injector or else requires a larger duct and an increased capacity for any subsequent fans or processing equipment. Thus if an injector is to be used, it should be located as near the end of the system as possible. An advantage of the injector is its inherent simplicity and low initial and maintenance costs. For a system which operates intermittently or which needs to be put together quickly for a specific short-term objective, the injector may be best.

A cylindrical injector is shown in Fig. 5-27. A more efficient arrangement is the venturi injector shown in Fig. 5-36. We shall proceed to analyze the performance of the venturi injector, neglecting friction at the walls and assuming

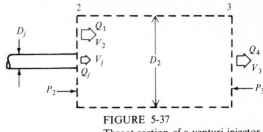

FIGURE 5-37
Throat section of a venturi injector.

that the inlet and exit velocities are small compared with those in the throat and jet. Figure 5-37 shows an enlarged view of a control surface consisting of the fluid contained within the throat at any one time. We shall assume constant density in the analysis.

The continuity equation applied to the control surface gives

$$Q_4 = Q_1 + Q_j \qquad (5\text{-}49)$$

The velocities may be related to the corresponding flow rates as

$$V_2 = \frac{4Q_1}{\pi(D_2{}^2 - D_j{}^2)} \qquad (5\text{-}50)$$

$$V_j = \frac{4Q_j}{\pi D_j{}^2} \qquad (5\text{-}51)$$

$$V_3 = \frac{4Q_4}{\pi D_2{}^2} \qquad (5\text{-}52)$$

A force-momentum balance on the control volume of Fig. 5-37, neglecting any wall friction forces, gives

$$P_3 - P_2 = \frac{4\rho}{\pi D_2{}^2}\left(Q_1 V_2 + Q_j V_j - Q_4 V_3\right) \qquad (5\text{-}53)$$

Substituting Eqs. (5-49) to (5-52) into this equation leads to

$$P_3 - P_2 = \frac{16}{\pi^2}\frac{\rho D_j{}^2}{D_2{}^4(D_2{}^2 - D_j{}^2)}\left(Q_1 - \frac{D_2{}^2 - D_j{}^2}{D_j{}^2}Q_j\right)^2 \qquad (5\text{-}54)$$

Applying Bernoulli's equation [Eq. (2-100)] to the flows between sections 1 and 2 and between sections 3 and 4, neglecting friction, work, and changes of potential energy, gives

$$P_2 - P_1 = -\frac{\rho V_2{}^2}{2} = -\frac{16\,\rho}{\pi^2\,2}\frac{Q_1{}^2}{(D_2{}^2 - D_j{}^2)^2} \qquad (5\text{-}55)$$

$$P_4 - P_3 = \frac{\rho V_3{}^2}{2} = \frac{16\,\rho}{\pi^2\,2}\frac{Q_4{}^2}{D_2{}^4} \qquad (5\text{-}56)$$

Adding Eqs. (5-54) to (5-56) leads to

$$P_4 - P_1 = \frac{16}{\pi^2} \frac{\rho}{2D_2{}^4} \left[-\frac{r^4}{(1-r^2)^2} Q_1{}^2 - 2Q_1Q_j + \frac{2-r^2}{r^2} Q_j{}^2 \right] \qquad (5\text{-}57)$$

in which r is defined as D_j/D_2.

The power input to the jet is given as the following, in which P_0 is atmospheric pressure:

$$\dot{W_j} = \frac{\rho Q_j V_j{}^2}{2} + Q_j(P_2 - P_0) \qquad (5\text{-}58)$$

Writing $P_2 - P_0$ as $P_1 - P_0 + P_2 - P_1$ and substituting Eqs. (5-51) and (5-55) reduce this to

$$\dot{W_j} = Q_j(P_1 - P_0) + \frac{16}{\pi^2} \frac{\rho}{2D_2{}^4} \left[\frac{Q_j{}^3}{r^4} - \frac{Q_j Q_1{}^2}{(1-r^2)^2} \right] \qquad (5\text{-}59)$$

If the injector operates at about atmospheric pressure, the term involving $P_1 - P_0$ can be neglected and, in addition, the second term in the brackets will usually be negligible compared with the first. This leaves

$$\dot{W_j} = \frac{16}{\pi^2} \frac{\rho}{2D_2{}^4} \frac{Q_j{}^3}{r^4} \qquad (5\text{-}60)$$

The useful power added to the mainstream is

$$\dot{W_a} = Q_1(P_4 - P_1) \qquad (5\text{-}61)$$

The efficiency of the injector is then obtained by dividing Eq. (5-61) by Eq. (5-60), subject to the approximations used in obtaining Eq. (5-60):

$$\eta = \frac{Q_1(P_4 - P_1)}{16\rho Q_j{}^3/2\pi^2 D_2{}^4 r^4} \qquad (5\text{-}62)$$

If we knew Q_j, then Eq. (5-57) would provide the characteristic curve of the injector and Eq. (5-62) would give its efficiency for different points along the characteristic curve. However, it is more productive to determine Q_j for a given operating condition in which Q_1 and $(P_4 - P_1)$ are specified. In fact, we should determine the dimensions of the injector so as to maximize, as far as possible, the efficiency for a given operating condition. If the injector is used at some other operating condition, then the efficiency will have to be what it will. To investigate whether the efficiency can be maximized, let us solve Eq. (5-57) for Q_j at the operating condition of specified Q_1 and $(P_4 - P_1)$. This gives

$$\frac{2-r^2}{r^2} Q_j{}^2 - 2Q_1Q_j - \frac{r^4}{(1-r^2)^2} Q_1{}^2 - \frac{\pi^2}{8\rho} D_2{}^4(P_4 - P_1) = 0 \qquad (5\text{-}63)$$

Solving for Q_j and using a positive sign in the quadratic formula to produce a positive value of Q_j, we have

$$Q_j = \frac{r^2 Q_1}{2 - r^2} \left\{ 1 + \left[\frac{1}{(1 - r^2)^2} + \frac{\pi^2 D_2^4}{8\rho Q_1^2} (P_4 - P_1) \frac{2 - r^2}{r^2} \right]^{1/2} \right\} \qquad (5\text{-}64)$$

Equation (5-62) for the efficiency becomes

$$\eta = \frac{(2 - r^2)^3 \mu / r^2}{[1 + \sqrt{1/(1 - r^2)^2 + (2 - r^2)\mu / r^2}]^3} \qquad (5\text{-}65)$$

in which μ has been set equal to

$$\mu = \frac{\pi^2 D_2^4 (P_4 - P_1)}{8\rho Q_1^2} = \frac{2(P_4 - P_1)}{\rho V_2'^2} \qquad (5\text{-}66)$$

In the second part of Eq. (5-66) the quantity V_2' is that velocity which would exist at section 2 if the quantity Q_1 were to flow past the section in the absence of the jet. In optimizing the design of the injector, the quantities D_2 and D_j are variable. Since the other quantities are fixed, we may instead vary r and μ. Thus, it would seem to be a simple matter of finding the maximum point in Eq. (5-65). It can easily be shown that η is zero when r is either 0 or 1, or when μ is zero, and that η approaches zero as μ approaches infinity. Thus, it would appear that a maximum exists, especially since a calculation shows that η is positive for intermediate values of r and μ. The condition of a maximum is that the partial derivatives of η ($\partial\eta/\partial r$ and $\partial\eta/\partial\mu$) are each simultaneously equal to zero. It may be shown that this condition cannot be met for $0 < r < 1$ and $0 < \mu < \infty$. The only possible location of a maximum point, it turns out, is for $r = 0$ and $\mu = 0$.

If we rewrite Eq. (5-65) as

$$\eta = \frac{(2 - r^2)^3 (\mu / r^2)}{[1 + \sqrt{1/(1 - r^2)^2 + (2 - r^2)(\mu / r^2)}]^3} \qquad (5\text{-}67)$$

we see that η does not approach zero as r and μ approach zero if μ / r^2 approaches a value other than zero. If we let

$$\mu = C r^2 \qquad (5\text{-}68)$$

then Eq. (5-67) becomes

$$\eta = \frac{(2 - r^2)^3 C}{[1 + \sqrt{1/(1 - r^2)^2 + (2 - r^2)C}]^3} \qquad (5\text{-}69)$$

and its limiting value as $r \to 0$, denoted as η_l, becomes

$$\eta_l = \frac{8C}{(1 + \sqrt{1 + 2C})^3} \qquad (5\text{-}70)$$

We may now maximize Eq. (5-70) for η_l by differentiating with respect to C and setting equal to zero. The resulting equation is

$$\sqrt{1 + 2C} = C - 1 \qquad (5\text{-}71)$$

The solution to this is $C = 4$, giving

$$\eta_l = 0.50 \qquad (5\text{-}72)$$

We see then that the maximum efficiency occurs at the physically impossible condition of $D_2 = 0$ and $D_j/D_2 = 0$. Equation (5-69) suggests that we may use positive values of r and μ without sacrificing too much efficiency provided that r and μ are properly related. The relation of Eq. (5-68) only holds in the limit as $r \to 0$; for significantly larger values of r, the relation is more complex. If we find $\partial\eta/\partial\mu$ from Eq. (5-65), we obtain

$$\frac{\partial\eta}{\partial\mu} = \frac{(2 - r^2)^3 r^{-2}[1 + \sqrt{1/(1 - r^2)^2 + (2 - r^2)\mu/r^2}]^{-4}}{\sqrt{1/(1 - r^2)^2 + (2 - r^2)\mu/r^2}}$$

$$\times \left\{ \sqrt{\frac{1}{(1 - r^2)^2} + \frac{2 - r^2}{r^2}\mu} + \frac{1}{(1 - r^2)^2} - \frac{2 - r^2}{2r^2}\mu \right\} \qquad (5\text{-}73)$$

Setting $\partial\eta/\partial\mu$ equal to zero gives

$$\sqrt{\frac{1}{(1 - r^2)^2} + \frac{2 - r^2}{r^2}\mu} + \frac{1}{(1 - r^2)^2} - \frac{2 - r^2}{2r^2}\mu = 0$$

Solving this for μ leads to

$$\mu = \frac{2r^2}{(1 - r^2)^2(2 - r^2)}[2 - 2r^2 + r^4 + \sqrt{(2 - 2r^2 + r^4)^2 - r^2(2 - r^2)}] \qquad (5\text{-}74)$$

where the positive sign has been used in the quadratic formula because a numerical calculation shows that a higher and more realistic efficiency results. Figure 5-38 shows how r and η vary with μ according to Eqs. (5-67) and (5-74); this figure can be used to determine the design operating condition and the efficiency at that point if a value of V_2' is arbitrarily selected.

Equation (5-57) for the characteristic curve can be used to obtain the SND pressure and maximum delivery for a fixed geometry and fixed value of Q_j. The results are

$$SND = \frac{16}{\pi^2} \frac{\rho}{2D_2^4} \frac{2 - r^2}{r^2} Q_j^2 \qquad (5\text{-}75)$$

$$SND = \frac{(P_4 - P_1)_o}{\mu} r^2(2 - r^2) \left[\frac{\mu}{2r^2} - \frac{r^2}{(1 - r^2)^2} \right]^2 \qquad (5\text{-}76)$$

in which $(P_4 - P_1)_o$, r, and μ are at the design operating point.

$$Q_{1_{max}} = \left(\frac{1}{r^2} - 1 \right) Q_j \qquad (5\text{-}77)$$

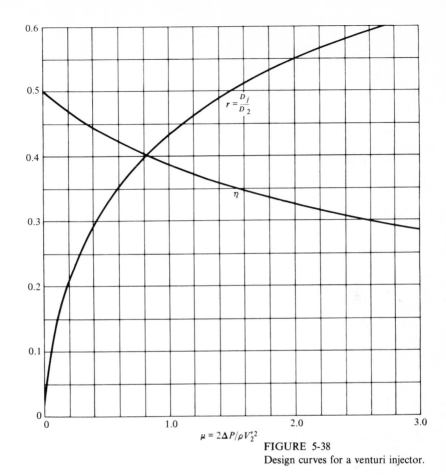

FIGURE 5-38

Design curves for a venturi injector.

EXAMPLE 5-23 An injector is designed to handle 10 m³/s of standard air with a pressure rise of 1000 N/m². Assume a superficial throat velocity V'_2 equal to 40 m/s. Determine the throat diameter, jet diameter, injection flow rate, efficiency, power necessary to supply the injection air, SND pressure, and maximum flow rate at no pressure if the design flow of injection air is used. Plot the characteristic curve, efficiency, and power curves for variable Q_1 if the design value of Q_j is supplied.

SOLUTION   From the data given, we may calculate μ as

$$\mu = \frac{2\Delta P}{\rho V_2'^2} = \frac{2(1000)}{1.2(40)^2} = 1.04$$

From Fig. 5-38, we read off values of r and η as

$$r = 0.437 \qquad r^2 = 0.191 \qquad \eta = 0.383 \qquad Ans.$$

The diameter of the throat is computed as

$$D_2 = \sqrt{\frac{4Q_1}{\pi V_2'}} = \sqrt{\frac{4(10)}{\pi(40)}} = 0.564 \text{ m} \quad Ans.$$

The jet diameter is then computed as

$$D_j = rD_2 = 0.437(0.564) = 0.246 \text{ m} \quad Ans.$$

We may compute Q_j from Eq. (5-64), after first substituting Eq. (5-66):

$$Q_j = \frac{r^2 Q_1}{2 - r^2} \left[1 + \sqrt{\frac{1}{(1 - r^2)^2} + \frac{2 - r^2}{r^2} \mu} \right]$$

$$= \frac{0.191(10)}{1.809} \left[1 + \sqrt{\frac{1}{(0.809)^2} + \frac{1.809}{0.191} 1.04} \right]$$

$$= 4.616 \text{ m}^3/\text{s} \quad Ans.$$

The jet velocity is calculated as

$$V_j = \frac{4Q_j}{\pi D_j^{\,2}} = \frac{4(4.616)}{\pi(0.246)^2} = 97.2 \text{ m/s}$$

The SND pressure is computed from Eq. (5-75) as

$$\text{SND} = \frac{16}{\pi^2} \frac{1.185}{2(0.564)^4} \frac{1.809}{0.191} (4.616)^2 = 1916 \text{ N/m}^2 \quad Ans.$$

The maximum flow rate is obtained from Eq. (5-77):

$$Q_{1_{max}} = \left(\frac{1}{r^2} - 1 \right) Q_j = \left(\frac{1}{0.191} - 1 \right) 4.616 = 19.55 \text{ m}^3/\text{s} \quad Ans.$$

Equation (5-57) gives the characteristic curve:

$$\Delta P = \frac{16}{\pi^2} \frac{\rho}{2D_2^{\,4}} \left[-\frac{r^4}{(1 - r^2)^2} Q_1^{\,2} - 2Q_j Q_1 + \frac{2 - r^2}{r^2} Q_j^{\,2} \right]$$

$$= \frac{16}{\pi^2} \frac{1.185}{2(0.564)^4} \left[-\frac{(0.191)^2}{(0.809)^2} Q_1^{\,2} - 2(4.616)Q_1 + \frac{1.809}{0.191} (4.616)^2 \right]$$

$$= 1916 - 87.64 Q_1 - 0.5291 Q_1^{\,2}$$

The work of the jet is independent of Q_1 and is given by

$$\dot{W}_j = \frac{16}{\pi^2} \frac{\rho}{2D_2^{\,4}} \frac{Q_j^{\,3}}{r^4} = \frac{16}{\pi^2} \frac{1.185}{2(0.564)^4} \frac{(4.616)^3}{(0.191)^2} = 25,593 \text{ W} \quad Ans.$$

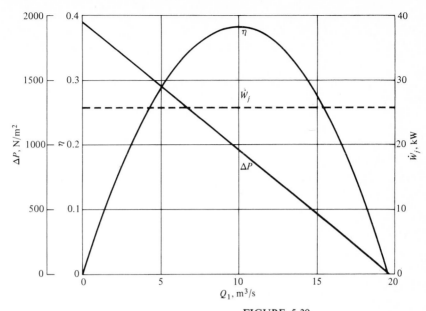

FIGURE 5-39
Characteristic curves for Example 5-22.

The efficiency then becomes

$$\eta = \frac{Q_1(P_4 - P_1)}{16\rho Q_j{}^3/2\pi^2 D_2{}^4} = \frac{Q_1 \, \Delta P}{25{,}593}$$

Figure 5-39 shows the characteristic curve, efficiency curve, and constant power curve.   ////

REFERENCES

1 Hemeon, W. C. L.: "Plant and Process Ventilation," 2d ed., pp. 46 and 50, The Industrial Press, New York, 1963.
2 Hemeon, W. C. L.: "Plant and Process Ventilation," 2d ed., p. 92, The Industrial Press, New York, 1963.
3 Hemeon, W. C. L.: "Plant and Process Ventilation," 2d ed., pp. 76–80, The Industrial Press, New York, 1963.
4 Jennings, B. H.: "Environmental Engineering Analysis and Practice," pp. 426–427, International Textbook Company, Scranton, Pa., 1970.
5 Hemeon, W. C. L.: "Plant and Process Ventilation," 2d ed., pp. 269–282, The Industrial Press, New York, 1963.
6 Streeter, V. L., ed.: "Handbook of Fluid Dynamics," pp. 3–16, McGraw-Hill Book Company, New York, 1961.

7 "ASHRAE Handbook of Fundamentals," p. 481, American Society of Heating, Refrigerating, and Air-Conditioning Engineers, New York, 1972.

8 Jennings, B. H.: "Environmental Engineering Analysis and Practice," chap. 13, International Textbook Company, Scranton, Pa., 1970.

PROBLEMS

5-1   In a workspace, methyl acetate is released into the air at the rate of 5.0 kg/h. Using the allowable concentrations of Table 1-1, what rate of ventilation air is required to keep the concentration of this particular pollutant acceptably small?

5-2   In an enclosed garage, octane is transferred from a storage tank to waiting vehicles; in the process, octane vapor is released into the air inside the garage. Using the allowable concentration shown in Table 1-1, how much ventilation air is required if 5 l/h of liquid octane are evaporated?

5-3   A conveying process injects 100 kg/h of dust into the workspace; the allowable concentration of this dust is 7.5 mg/m^3. How much ventilation air is required?

5-4   A small soldering operation, located in the middle of a ventilated room at a point where the air current has a velocity of 30 cm/s, produces fumes at a rate of 0.01 kg/h. The allowable concentration for breathing is 0.25 mg/m^3. What will be the concentration of these fumes at a distance of 0.4 m from the source? How close to the source may the worker's face be located?

5-5   A barrel of dichlorodifluoromethane is located in a room where the air current is 5 m/min. The substance evaporates at the rate of 100 g/h. Using the allowable concentration shown in Table 1-1, how far away must all individuals remain to avoid excessive exposure?

5-6   A paint-spray booth is constructed with an open face having dimensions of 1.5 by 2.0 m. If the required velocity over the face is 0.9 m/s, determine the exhaust flow rate and the exhaust diameter for a duct velocity of 13.0 m/s.

5-7   For an exhaust hood which can be analyzed as a half-cylindrical control surface, show that for $V_f = 15$ m/s and $V_c = 0.25$ m/s, then $X = 19.1$ W.

5-8   An area near a wall is to be ventilated from a small circular opening in the baseboard. If $V_c = 0.5$ m/s and $X = 3.0$ m, find Q. Assume no ceiling, corner, or obstruction is within 3 m of the opening.

5-9   A certain operation requires an air velocity of 1.0 m/s for control of dust. A hood is placed 25 cm from the source with the plane of its opening facing the dust source.
    (a) Calculate Q when the face of the hood is 5 by 8 cm; 10 by 16 cm; and 20 by 32 cm.
    (b) Calculate Q for the same three hoods when the distance from the source is reduced from 25 to 8 cm.
    (c) Tabulate the results, along with the respective face velocities, and comment on their significance.

5-10 A lateral hood has dimensions of 15 by 40 cm and is mounted on the edge of a workbench 1 m wide. The plane of the hood is normal to the plane of the bench, with the bottom edge of the hood at bench level. Taking X as 1 m and V_c as 0.38 m/s, find Q.

5-11 A ventilation system mounted in the floor consists of a large number of parallel slots 3 cm wide by 15 m long, spaced on 2-m centers. If $V_c = 0.25$ m/s, estimate the flow rate required per slot. Assume the ends of the slots extend up to the walls of the room.

5-12 A small buffing wheel is located at the end of a shaft. The null point (the point where all projected particles slow to zero velocity before being captured by the air stream) is located 8 cm from the shaft axis in the plane of the wheel. Locate an exhaust hood in each of three positions which you may choose at will. For each position calculate Q, taking V_c as 0.25 m/s.

5-13 A vacuum-cleaner attachment, which functions as a hood, is rectangular having dimensions of 5 by 20 cm. The face of the attachment is held parallel to the floor and 1 mm away from it. A face velocity of 30 m/s has been determined to be adequate for dust pickup from the floor. Estimate the flow rate through the vacuum cleaner. A sketch is shown in Fig. 5-40.

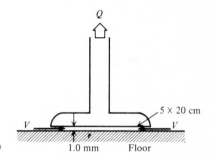

FIGURE 5-40

5-14 In order to remove fumes from a tire-vulcanizing process, a hood is to be located in the center of the bottom half of the tire mold. As shown in Fig. 5-41, the outside diameter of the mold is 1.0 m and the top half of the mold lifts straight up a distance of 0.6 m above the hood to allow removal of the tire. Estimate the flow rate through the hood to allow continuous removal of fumes if a control velocity of 0.5 m/s is required.

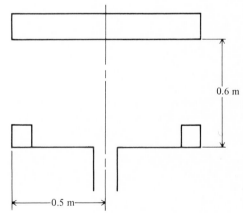

FIGURE 5-41

5-15 A soldering operation is done on electronic components on a workbench. The component is clamped in place on the bench by a foot-operated clamp for ease in doing the job. A small hood in the form of a flexible circular tube with a plain end is attached to the clamp to remove fumes generated in the soldering process; this is shown in Fig. 5-42. The electronic component is 15 by 24 by 5 cm high, with the 15 by 24 cm face resting on the bench. Estimate X and Q for this application, and determine the diameter of the tube for a velocity of 10 m/s in the tube and a control velocity of 0.25 m/s.

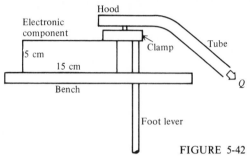

FIGURE 5-42

5-16 A small hand sander for home-workshop use (see Fig. 5-43) is arranged to be ventilated by attaching a vacuum-cleaner hose to an opening in the frame of the unit. Sandpaper is attached to an oscillating pad on the bottom of the machine, and an opening between this pad and the frame of the machine serves as the hood. Take this opening to be 1.0 cm above the plane of the surface being sanded. Assume the sander makes 10 strokes/s, that each stroke is 6 mm long, and that it is desired to collect all particles of 100 μm diameter or smaller.

(a) Estimate the pulvation distance for a particle of 100-μm diameter. Assume that the particle will travel in the plane of the sandpaper.

(b) Determine the flow rate of ventilation air required. For a velocity of 10 m/s in the vacuum-cleaner hose, what diameter should the hose be?

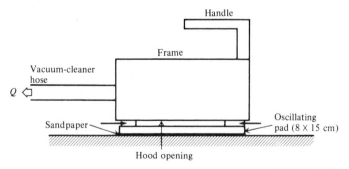

FIGURE 5-43

5-17 A long trough whose opening is at floor level is to be filled with a chemical for processing sheet stock. The trough is open at the top, and to remove vapors from it two slots are provided running parallel to the trough, one on each side. The slots are 1.2 m apart and 30 m long. For a control velocity of 0.38 m/s, what flow rate is required?

5-18 Expand the technique used for the parallel slots to the case where the tank is elevated a considerable distance above the floor, as shown in Fig. 5-44. Here *AB* will be twice the distance of *BC*. Derive an expression for the control-surface area using this assumption.

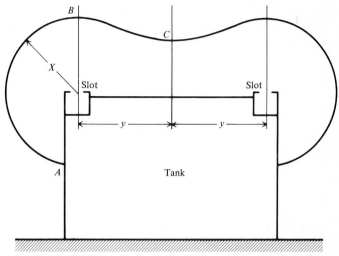

FIGURE 5-44

5-19 Repeat the analysis and derivation of Prob. 5-18 for the case of a circular tank mounted a considerable distance above the floor.

5-20 Repeat Prob. 5-18 if the tank is mounted only a short distance above the floor. Use the control surface shown in Fig. 5-45.

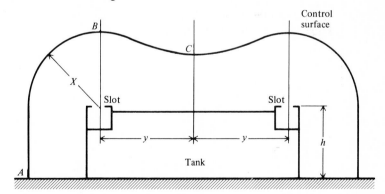

FIGURE 5-45

5-21 Repeat Prob. 5-19 if the tank is mounted only a short distance above the floor. Use the control surface shown in Fig. 5-45.

5-22 A powder is to be emptied from a barrel to a carton by use of a scoop. Since this is to be done frequently, a special place is set aside for this purpose. The powder is very dusty, and so an exhaust hood must be provided to keep down the dust from this operation. The exhaust hood is a rectangular opening, covered by a grill so that the workers do not drop the scoop into the duct, which is placed directly between the barrel and the carton, as shown in Fig. 5-46. The top of the carton is at the same elevation as the top of the barrel. The barrel top has a diameter of 0.5 m and the carton dimensions are 30 by 40 cm, with one of the long sides closest to the barrel. Dust can be formed when the powder is scooped up from the barrel and when it is dumped into the carton. Estimate X and compute from this the control-surface area and the air flow rate.

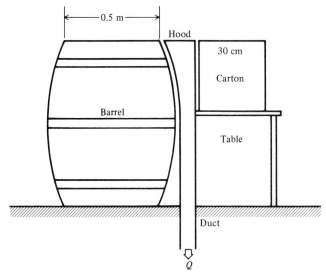

FIGURE 5-46

5-23 Show that when a rectangular duct and a circular duct are equivalent in the sense of having the same pressure drop for equal flow rate, the hydraulic diameters must be in the ratio

$$\frac{D_{hr}}{D_c} = 1.58\left(\frac{f_r}{f_c}\right)^{0.2} \frac{(b/a)^{0.4}}{(1 + b/a)^{0.8}}$$

in which D_{hr} is the hydraulic diameter of the rectangle and D_c is the diameter of the circle. Assuming the ratio of friction factors to be unity, calculate and plot this ratio as a function of aspect ratio b/a over the range from 0.1 to 1.0.

5-24 For the same conditions as in Prob. 5-23, show that the ratio of Reynolds number is given by

$$\frac{Re_r}{Re_c} = 1.99\left(\frac{f_c}{f_r}\right)^{0.2} \frac{(b/a)^{0.6}}{(1 + b/a)^{1.2}}$$

Calculate and plot this ratio as a function of b/a over the same range as in Prob. 5-23.

5-25 (a) A circular duct 50 cm in diameter has a flow rate of standard air of 2.0 m³/s. Using Fig. 5-22, determine the dimensions of a rectangular duct of aspect ratio 0.3 having the same pressure drop for equal flow.

(b) Repeat (a), taking into account the variation of friction factor (which was neglected in computing the data for Fig. 5-22) between the circular and rectangular ducts. Find the percent error.

5-26 A rectangular duct is to carry 10 m³/s of standard air for a distance of 75 m with a pressure drop not to exceed 100 N/m². The duct height is limited to 30 cm by structural considerations. Determine the duct width required. What will the velocity be in the duct?

5-27 A circular duct of 220-m length is to carry 3.5 m³/s of standard air. The velocity of flow is to be 20 m/s. What diameter duct should be used and what pressure drop will be experienced? What minimum pumping power is required to overcome this pressure drop?

5-28 A hood receives 0.1 m³/s of standard air. The branch duct from the hood to the main exhaust duct is 10 m long. If the pressure drop along this branch is to be 100 N/m², what size duct should be used and what will be the velocity in the duct?

5-29 A round duct of 50-cm diameter has a flow of 2.5 m³/s of standard air. The duct has a 90° elbow having a ratio of centerline radius to duct diameter equal to 1.5. What will be the pressure drop along the elbow?

5-30 Repeat Prob. 5-29 if the ratio of centerline radius to duct diameter is 0.75.

5-31 Repeat Prob. 5-29 for a right-angle duct.

5-32 A rectangular duct having dimensions of 60 by 80 cm has a flow of 3.55 m³/s of standard air. A 90° elbow in the duct has a ratio of centerline radius to duct width (in plane of bend) of 1.25. Assume a "hard" bend, that is, one in which the longer dimension is in the plane of the bend. Using Ref. [4] or a similar source, determine the pressure drop across the elbow.

5-33 Repeat Prob. 5-32 for a "soft" bend, in which the shorter duct dimension is in the plane of the bend.

5-34 Repeat Prob. 5-32 if the ratio of centerline radius to duct width is 0.75.

5-35 Repeat Prob. 5-32 if the duct dimensions are 30 by 180 cm. Also repeat for a soft bend (see Prob. 5-33).

5-36 A branch duct carries 2.0 m³/s and has dimensions of 25 by 40 cm. It enters the main duct at a 30° angle with the centerline of the main. What will be the pressure drop in the branch at the point of entry to the main?

5-37 Estimate the equivalent length of a 90° elbow in a rectangular duct of dimensions 30 by 45 cm.

5-38 A hood receives a flow rate of 0.4 m³/s of standard air. There is an abrupt entry to the branch duct from the hood. The branch duct is 25 m long, and at the other end enters the main in a 90° T connection. The static pressure in the main is 25 N/m² below atmospheric at the point where the branch duct joins. What minimum size of branch duct is required, if its aspect ratio is 0.3? (Be sure to allow one velocity head in addition to all friction losses.)

5-39 A circular duct 30 cm in diameter carries 0.6 m³/s of standard air. To what must its diameter be increased to increase the static pressure in the duct by 10 N/m²? Assume no change in flow rate.

5-40 Repeat Prob. 5-39 if the flow rate increases to 0.85 m³/s due to the addition of a branch duct at the point in question.

5-41 Repeat Prob. 5-39 if a branch duct at the point in question reduces the flow to 0.35 m³/s.

5-42 A main duct has dimensions of 1.0 by 3.0 m and carries standard air at a velocity of 14 m/s. At a point where branch ducts having a total flow rate of 1.5 m^3/s enter the main duct, static regain is to be used to increase the static pressure by 40 N/m^2. If the downstream duct remains 1.0 m high, how wide must it be?

5-43 A rectangular duct carrying 3.0 m^3/s of standard air has dimensions of 30 by 50 cm. At a point in the main duct, a branch duct carrying 0.5 m^3/s enters, and the dimensions of the main downstream of this point are 30 by 100 cm. What will be the increase of static pressure in the main duct between a point downstream and a point upstream of the location where the branch duct joins?

5-44 Design a duct system to handle the various flow rates given in the following table. The coordinates of the hood locations are given. The total pressure drop must be approximately 1200 N/m^2. The velocity must be at least 12.5 m/s to avoid dust buildup. The velocity should not exceed 20 m/s because of noise limitations, unless a suitable design cannot be attained otherwise. Dampers may be included at suitable locations to balance the pressure drops; show the locations of suitable dampers on the system sketch. Use circular ducts.

Location no.	Q, m^3/s	x, m	y, m	z, m
1	0.5	30	30	1
2	1.5	60	20	3
3	0.75	30	50	2
4	0.3	30	70	2
5	1.0	50	30	1
6	0.6	70	70	1
Fan		0	0	7

Design a suitable duct system showing all fittings, duct sizes, and pressure drops. Sketch a duct layout showing these pertinent values. Estimate the cost of your design based on the following data:

Straight ductwork, $S = L(10D + 1.0)$
Fittings other than elbows, $S = 10 + 20D$
Elbows, $S = 5 + 10D$

where S is the cost in dollars, D is the duct diameter in meters, and L is the duct length in meters.

5-45 Convert the design of Prob. 5-44 to the use of rectangular ducts. Keep the aspect ratio of the ducts above 0.15 in all cases. Assume that the cost data remain the same, based on equivalent diameter D_e.

5-46 A fan is to be used in a duct. A test on the fan shows that its maximum delivery when turning at 1200 rpm is 10.5 m^3/s at a static pressure of 75 N/m^2. The discharge is then blocked off, and the static pressure reading is 1500 N/m^2. Sketch approximate characteristic curves for static pressure for this fan when turning at speeds of 500, 1000, 1200, 1500, and 2000 rpm.

5-47 A fan that has a maximum flow rate of 5 m^3/s and a SND pressure rise of 500 N/m^2 when turning at 2000 rpm is to be used in a duct to deliver 7.5 m^3/s of standard air at a pressure of 750 N/m^2. At what speed should the fan turn, using an operating condition to the right of the peak on the characteristic curve?

5-48 A fan turning 500 rpm delivers 100 m^3/s at a pressure of 3000 N/m^2. At this speed its maximum flow is 250 m^3/s and its SND pressure is 2800 N/m^2. Sketch the characteristic curve if the given operating point is slightly to the right of the peak. If the fan is to operate at the corresponding point but at a speed of 750 rpm, what flow rate and pressure will it deliver?

5-49 A fan geometrically similar to that of Prob. 5-48 but with a smaller diameter and all other dimensions smaller in proportion is to produce a flow rate of 50 m^3/s at a pressure of 2000 N/m^2. What should be the ratio of diameters, and at what speed should the new fan turn?

5-50 A fan is designed to produce a flow of 7 m^3/s of standard air at a pressure of 900 N/m^2. Due to a change in operating conditions, the air flowing is now at a temperature of 250°C and still approximately atmospheric pressure. The mass flow rate and pressure must remain the same. At what speed should the fan now be operated? To estimate the characteristic curve, the maximum Q is 12 m^3/s and SND is 1000 N/m^2 at 1500 rpm, which is the original operating speed of the fan.

5-51 In Example 5-20, we considered a fan operating on air whose density changed but whose value of Q and ΔP_s remained unchanged. The speed of the fan and the operating point on the characteristic curve were changed to compensate. Derive an expression for the new speed in terms of the two densities, the original pressure and flow rate, and the slope of the characteristic curve at the original operating point. This expression should be valid if the density change is not too great, so that the characteristic curve can be replaced by its tangent.

5-52 A fan having the characteristic curve of Fig. 5-32 when operating at 1200 rpm is in use in a duct system where the flow rate is 8.0 m^3/s at a pressure of 830 N/m^2. It is desired to increase the flow rate in the duct system to 12 m^3/s. To what speed should the fan be increased, and what will be the new pressure?

5-53 A fan operating at 1200 rpm with the characteristic curve of Fig. 5-32 handles a flow rate of 8 m^3/s at a pressure of 830 N/m^2. The duct system in which this fan is located is to be extended such that the new flow rate is 10 m^3/s and the new pressure is to be 960 N/m^2. What new speed is required for the fan, and what will be the point on the old characteristic curve corresponding to the new operating point?

5-54 In Prob. 5-47 compute the power required to drive the fan, assuming a fan efficiency of 0.65.

5-55 In Prob. 5-48 compute the power required to drive the fan in each of the two operating conditions involved if the fan efficiency is 0.75 in each case.

5-56 Compute the power required in Prob. 5-49, assuming an efficiency of 0.65.

5-57 Compute the power required to operate the fan of Prob. 5-50 in both operating conditions if the efficiency is 0.60 in each case.

5-58 Compute the power required both before and after the change in the duct system in Prob. 5-53. Assume the efficiency to be 0.65 in both cases. If the system will operate 4000 h/year, and electricity costs $.009/kWh, what will be the increase in operating costs of the system per year? Suppose instead that a parallel system were installed handling 2.0 m^3/s at a pressure of 830 N/m^2, with the same fan efficiency. What would be the total yearly cost this way, and how much would be saved each year over the cost of operating the modified single system?

5-59 Obtain an expression analogous to Eq. (5-65) if none of the terms in Eq. (5-59) is considered negligible. Investigate the condition of maximum efficiency for this case, obtaining an expression similar to Eq. (5-74), if possible.

5-60 Using the equation for efficiency for the venturi jet injector derived in Prob. 5-59, compare the value calculated with that calculated from Eq. (5-65) using the same values of r and μ. Use $r = 0.2$ and μ from Eq. (5-74). Calculate for $P_1 - P_0$ equal to -1000 N/m^2, -5000 N/m^2, 10,000 N/m^2, and 0.3 MN/m^2. Assume $P_4 - P_1 = 1000$ N/m^2.

5-61 From Eq. (5-67), show that η approaches zero when $r \rightarrow 0$ or when $r \rightarrow 1$ if $\mu > 0$, and when $\mu \rightarrow 0$ or $\mu \rightarrow \infty$ if $0 < r < 1$.

5-62 From Eq. (5-65), obtain Eq. (5-73) for $\partial \eta / \partial \mu$ and a similar equation for $\partial \eta / \partial r$. Equate each of these to zero and attempt to solve for r and μ simultaneously. Show that no solution exists in the range $0 < r < 1$ and $\mu > 0$. Show in fact that the only solution possible is for $r = \mu = 0$. Sketch a three-dimensional surface of η as a function of r and μ.

5-63 Derive the equations for the venturi jet injector if wall friction is taken into account. For the inlet section, use a loss coefficient K_I, which would be almost unity except for the effect of flow around the jet duct and nozzle. For the throat use the appropriate friction factor (assume it to be constant to simplify the analysis). Also use a friction factor to allow for friction in the jet, where the jet friction factor may be different from that of the throat. (The jet friction factor may also be assumed constant.) For the divergent section 3-4, use a loss coefficient K_D. Find the point of maximum efficiency and the characteristic curve for fixed jet flow rate.

5-64 Perform an analysis similar to that done for the venturi jet injector for the case of a cylindrical injector such as the one shown in Fig. 5-27.

5-65 A venturi jet injector is to be used with a system involving the flow of 1.0 m^3/s of standard air. The velocity in the inlet duct is 15 m/s, and the pressure rise which the injector must supply is 1500 N/m^2. Design a suitable injector if r must be at least 0.2 and the superficial throat velocity V'_2 must not be greater than 70 m/s. What efficiency will be obtained, and how much power must be supplied to the jet?

5-66 In the design of Prob. 5-65, suppose the jet injector costs $300, including installation charges but not including the cost of an air compressor and motor, which are already available. Thus, the only charges to the system are the $300 plus power costs in operating the air compressor. Assume that the air compressor and motor have a combined efficiency of 0.70. Electricity costs $.009/kWh. Assuming 20 percent capital recovery each year, how much does it cost per hour to operate the system assuming it operates 2000 h/year?

It is desired to compare the cost of operating this system with that of a blower having 50 percent overall efficiency, including that of the driving motor. The cost of blower and motor installed is $750/kW of power imparted to the air. With the same capital-recovery percentage, how many hours per year must the system operate at break-even point, that is, when both systems will cost the same?

5-67 A venturi injector is available which has a throat diameter of 25 cm, an inlet diameter of 60 cm, and a jet diameter of 8 cm. It is desired to use this injector for service in a duct carrying 7.5 m^3/s of standard air, requiring a pressure rise of 2000 N/m^2. For the maximum efficiency possible with these restraints, how much should the jet flow rate be? Compute the power required under these conditions. Plot the characteristic curve of the injector operating with this jet flow rate. Compute the SND pressure rise and maximum flow rate.

6

SETTLING CHAMBERS

The settling chamber is perhaps the simplest and crudest of all pollution control devices. As such, it will serve to introduce our study of specific devices. We will give it a degree of attention probably exceeding that which its practical usefulness warrants. Not that the settling chamber is of small practical utility; in fact, it is widely used for certain purposes. Much of our motivation for studying settling chambers stems from their simplicity. The analysis of their performance is readily grasped, and yet the same principles used in their analysis apply to more sophisticated devices with more widely recognized practical value. Thus our study of settling chambers is as much as anything a prelude to the study of other devices.

The settling chamber does have wide usefulness in its own right. It can be used to separate out the larger particles in a particulate distribution. In a few cases it is sufficient to give adequate pollution control by itself. Since the settling chamber is almost the cheapest device to construct, operate, and maintain, it is likely to be used in those few cases. The main usefulness of the settling chamber, however, lies in serving as a preliminary screening device for a more efficient control device. Where the mass of larger particles is huge, the settling chamber can remove much of the mass of the particulate distribution which would otherwise choke up the other control device, impairing its operation or requiring too frequent cleaning.

FIGURE 6-1
Simple settling chamber.

The use of a settling chamber instead of another type of device will often make possible the use of a smaller model of a second device, with an overall saving in cost. In those cases where recovery of the material has economic benefit, the settling chamber can many times collect most of the mass at a location where it can be conveniently reused, with the final collection being made at some centralized location, perhaps far removed from the point of reuse but chosen for other advantages. The important thing to realize, however, is that the settling chamber by itself is seldom a solution to any air pollution problem. It can collect particles very effectively down to about 100 μm diameter. Specialized designs can collect efficiently down to 50 μm diameter. Since from an air pollution standpoint most troublesome particles have much smaller diameters than these, we see why a settling chamber by itself seldom suffices.

A cross-sectional view of the simplest form of settling chamber is shown in Fig. 6-1. The large volume of the chamber allows the air to flow at a low velocity, giving time for the particles to settle out. The force causing separation is that of gravity. The dust collects and forms a layer on the bottom of the

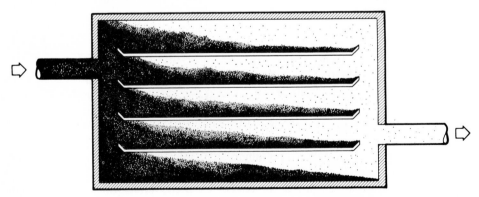

FIGURE 6-2
Settling chamber with five trays.

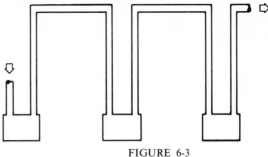

FIGURE 6-3
Combined cooler and collector.

chamber; this collected dust layer must be removed periodically. In the particular
chamber shown in Fig. 6-1, a hopper is provided to facilitate dust removal.

A more elaborate settling chamber is the one shown in Fig. 6-2 in which
trays are provided. The use of several trays improves the collection efficiency
of the device since the flow velocity remains substantially the same and yet each
particle has a much shorter distance to fall before reaching the bottom of the
passage between trays. However, the use of trays makes emptying the device
a more cumbersome process. For purposes of analysis we shall assume that n
trays are provided, with the bottom surface of the settling chamber serving as
one tray.

The device shown in Fig. 6-3 is sometimes used both as a settling chamber
and for cooling the airstream. Heat is convected and radiated from the pipes
and settling-chamber walls. The collection mechanism is a combination of gravity
settling and centrifugal force. The same principle can also be used in a
concentrator, as shown in Fig. 6-4. However, a major difficulty in using such a
device is that dust tends to collect in the bottom of the chamber, which is
undesirable.

FIGURE 6-4
Gravity concentrator.

6-1 LAMINAR FLOW IN SETTLING CHAMBERS

Consider a settling chamber containing n trays, including the bottom surface of the chamber and having the dimensions shown in Fig. 6-5. In this section we shall assume the case of laminar flow in the passageways between the trays; this condition is less common but does sometimes occur. The criterion for laminar flow is that the Reynolds number, which is based on hydraulic diameter, be less than 2300. For the passage between trays, the hydraulic diameter is given by

$$D_h = \frac{4W\,\Delta H}{2W + 2\,\Delta H} = \frac{2W\,\Delta H}{W + \Delta H} \qquad (6\text{-}1)$$

The Reynolds number $\mathrm{Re} = VD_h/v$ can be written in terms of the flow rate if it is noted that

$$V = \frac{Q}{n\,\Delta H W} \qquad (6\text{-}2)$$

The Reynolds number then becomes

$$\mathrm{Re} = \frac{Q}{vn\,\Delta H W}\frac{2W\,\Delta H}{W + \Delta H} = \frac{2Q}{nv(W + \Delta H)} \qquad (6\text{-}3)$$

If we neglect the thickness of the trays, the distance between trays is related to the height of the settling chamber as

$$\Delta H = \frac{H}{n} \qquad (6\text{-}4)$$

Then Eq. (6-3) becomes

$$\mathrm{Re} = \frac{2Q}{v(nW + H)} \qquad (6\text{-}5)$$

It is to be noted that Eq. (6-4) holds only when there is no dust layer collected on the tray. If a layer of dust of thickness H_d is present, then Eq. (6-4) is replaced by

$$\Delta H = \frac{H}{n} - H_d \qquad (6\text{-}6)$$

and Eq. (6-5) becomes

$$\mathrm{Re} = \frac{2Q}{v(nW + H - nH_d)} \qquad (6\text{-}7)$$

As a rough estimate of an upper bound on the Reynolds number, we may approximate Eq. (6-7) as

$$\mathrm{Re} \approx \frac{2Q}{vnW} \qquad (6\text{-}8)$$

A lower bound is obtained from Eq. (6-5).

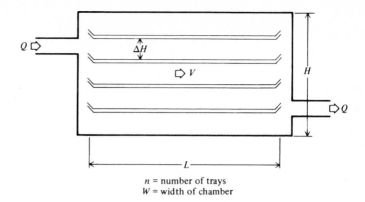

n = number of trays
W = width of chamber

FIGURE 6-5
Laminar flow in a settling chamber.

Let us now determine the efficiency of a settling chamber. Assume that a particle enters the space between two trays at a distance y above the lower tray. It will travel horizontally with a velocity V equal to that of the gas. Its vertical velocity will be the settling velocity given by an appropriate equation in Sec. 3-1. For example, in standard air the settling velocity in the laminar regime is given by Eq. (3-52). No Cunningham correction need be used since we are not considering small particles. Let us assume for simplicity that the terminal settling velocity will apply throughout the entire flow through the passage between trays. The error entailed in making this assumption is usually small, and also there is opportunity for the particle to begin settling out prior to its entry into the passage.

The time required for the particle at height y above the lower tray to settle onto the tray is given by

$$t = \frac{y}{V_t}$$

If the particular particle just settles out by the time it reaches the end of the tray, this time is also given by the time required for the particle to travel from one end of the tray to the other. This latter time is given by

$$t = \frac{L}{V}$$

Equating these two expressions gives

$$y = \frac{L V_t}{V} \qquad (6\text{-}9)$$

Now assuming that particles are uniformly distributed over the incoming stream, the efficiency of collection for particles having the diameter used in calculating V_t is given as the ratio of y to ΔH. Thus

$$\eta = \frac{y}{\Delta H} = \frac{LV_t}{V \Delta H} = \frac{nWLV_t}{Q} \qquad (6\text{-}10)$$

If the efficiency is to be equal to unity, then we may solve for L as

$$L = \frac{Q}{nWV_t} \qquad (6\text{-}11)$$

In practice, settling chambers are rarely as efficient as these equations predict even if the Reynolds number indicates laminar flow owing to irregularities in the flow pattern and other effects that are not clearly understood.

EXAMPLE 6-1 A settling chamber is to be used to collect particles of 50 μm diameter and 2000 kg/m^3 density from a stream of 10 m^3/s of standard air. If the chamber is to be 1.5 m wide, 1.5 m high, and if it will have nine trays including the bottom surface, how long must it be to give theoretically perfect collection efficiency? (a) Do the computations on the assumption that the laminar equations hold. (b) Compute the Reynolds number and decide if the laminar assumption is reasonable. If not, how many trays need be included to make the flow laminar? Recalculate the length in this case. (c) What will be the collection efficiency for particles of 25 μm diameter?

SOLUTION (a) Equation (3-52) gives the terminal velocity of the 50-μm particle as

$$V_t = 29{,}609\rho_p d^2 = 29{,}609(2000)(0.5 \times 10^{-4})^2 = 0.148 \text{ m/s}$$

Equation (6-11) then gives the length as

$$L = \frac{Q}{nWV_t} = \frac{10}{9(1.5)(0.148)} = 5.0 \text{ m} \qquad Ans.$$

(b) Equation (6-5) gives the Reynolds number as

$$Re = \frac{2Q}{v(nW + H)} = \frac{2(10)}{1.55 \times 10^{-5}[9(1.5) + 1.5]} = 86{,}022 \qquad Ans.$$

This number is well beyond the laminar range. To estimate the number of trays required to bring the Reynolds number down to 2300, other quantities remaining the same, Eq. (6-8) may be solved for n:

$$n = \frac{2Q}{vWRe} = \frac{2(10)}{1.55 \times 10^{-5}(1.5)(2300)} = 374 \qquad Ans.$$

The required length then becomes

$$L = \frac{10}{374(1.5)(0.148)} = 0.12 \text{ m} \qquad Ans.$$

Thus, it is clear that to have laminar flow in a settling chamber usually requires an awkward design. However, a very large chamber combined with a small flow may well give laminar flow in the chamber.

(c) For a particle of 25 μm diameter, the terminal settling velocity is

$$V_t = 29,609(2000)(0.25 \times 10^{-4})^2 = 0.037 \text{ m/s}$$

Then Eq. (6-10) gives the efficiency as

$$\eta = \frac{nWLV_t}{Q} = \frac{374(1.5)(0.12)(0.037)}{10} = 0.249 \qquad Ans. \qquad ////$$

6-2 TURBULENT FLOW IN SETTLING CHAMBERS

In Example 6-1 we saw that a settling-chamber design based on laminar flow requires either a very large size or an inordinately large number of trays combined with an awkward shape of chamber. Although the laminar chamber does have the advantage of giving a theoretically perfect collection efficiency for particles of the designated size, this advantage is of little practical value since the efficiency drops off rapidly with smaller particles. It is virtually impossible to collect very small particles with reasonably good efficiency using a settling chamber.

The turbulent settling chamber offers a more practical design concept, although, no matter how big it may be, theoretically the chamber will never collect all particles of a specified size. We shall see, however, that it is entirely reasonable to design a chamber in which 99 percent of the particles collected will be as large or larger than some specified diameter, for example, 50 or 100 μm. We may estimate the collection efficiency of a settling chamber having turbulent flow by a procedure to be derived next. Figure 6-6a shows a longitudinal cross section of the flow passage between trays or between the top and bottom surfaces of the chamber if no trays are used. Figure 6-6b shows the corresponding lateral cross section of the same passage. The derivation is made subject to two assumptions: First, there is a laminar layer adjacent to the bottom surface of the passage into which turbulent eddies do not penetrate, so that any particle which crosses into this layer will be captured shortly. Second, in the remainder of the flow passage the eddying motion due to turbulence will cause a uniform distribution of particles of all sizes.

Consider the element of width W, height ΔH, and length dx. If dy represents the thickness of the laminar layer, then a particle which crosses the dotted line will surely settle to the bottom while traveling a certain distance downstream in the x direction; we may as well let this distance be dx. Since the particle

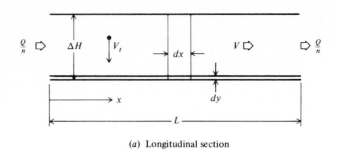

(a) Longitudinal section

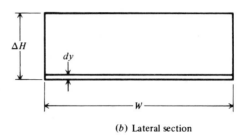

(b) Lateral section

FIGURE 6-6
Turbulent flow in the passage between trays.

settles with velocity V_t in the laminar layer, then the required distance dx for the particle to reach the bottom is given by

$$dx = V \, dt$$

where dt is the time required for the particle to settle to the bottom and is given by

$$dt = \frac{dy}{V_t}$$

Eliminating dt from the preceding two equations gives

$$dy = \frac{V_t \, dx}{V} \qquad (6\text{-}12)$$

Let us now assume a uniform distribution of all particles across the flow passage produced by the turbulent motion. At the entrance to the element dx, the fraction of the particles within the laminar layer dy will equal the ratio of the area inside the laminar layer to the total area. Since all these particles will be collected, then the fraction of the total particles of a particular diameter that are collected in the length dx is equal to the fraction of the total area that is located inside dy. Thus we may write

$$-\frac{dN}{N} = \frac{W\,dy}{W\,\Delta H} = \frac{dy}{\Delta H} = \frac{V_t}{V\,\Delta H}\,dx \qquad (6\text{-}13)$$

Now, integrating Eq. (6-13) over the length from the inlet to the passage to a particular location x along the passage gives

$$\ln N = -\frac{V_t}{V\,\Delta H}\,x + \ln C$$

or
$$N = Ce^{-V_t x/V\,\Delta H} \qquad (6\text{-}14)$$

To evaluate the constant C we note that when $x = 0$ then $N = N_0$, so that $C = N_0$. Equation (6-14) becomes

$$N = N_0 e^{-V_t x/V\,\Delta H} \qquad (6\text{-}15)$$

The collection efficiency for the total length L of the chamber may be written as

$$\eta = 1 - \frac{N_L}{N_0} \qquad (6\text{-}16)$$

Evaluating Eq. (6-15) at $x = L$ and substituting into Eq. (6-16) gives

$$\eta = 1 - e^{-V_t L/V\,\Delta H} \qquad (6\text{-}17)$$

Using Eq. (6-2), this becomes

$$\eta = 1 - e^{-nLWV_t/Q} \qquad (6\text{-}18)$$

EXAMPLE 6-2   Use the data of Example 6-1. Determine the length of settling chamber required to achieve 99 percent efficiency for 50-μm particles. What length would be required for 99.9 percent efficiency? If the first length is used, what would be the efficiency with 25-μm particles?

SOLUTION   Solving Eq. (6-18) for L and using the given data gives

$$L = \frac{-Q}{nWV_t}\ln\,(1-\eta) = -\frac{10}{9(1.5)(0.148)}\ln\,(1-0.99) = 23.0 \text{ m} \qquad Ans.$$

For an efficiency of 0.999, the value of L is

$$L = -\frac{10}{9(1.5)(0.148)}\ln\,(1-0.999) = 34.6 \text{ m} \qquad Ans.$$

For the 25-μm particle, the efficiency based on a length of 23.0 m and a settling velocity of 0.037 m/s becomes

$$\eta = 1 - \exp\left[-\frac{9(23.0)(1.5)(0.037)}{10}\right] = 0.683 \qquad Ans. \qquad /\!/\!/\!/$$

6-3   ECONOMIC SIZING OF SETTLING CHAMBERS

It can be seen from Examples 6-1 and 6-2 that much latitude is available to the designer in the selection of primary dimensions of the settling chamber. Equations (6-10) and (6-18) for the collection efficiency in laminar and turbulent flow, respectively, show that the efficiency is independent of the height of the chamber. Also, both equations have the factor nWL present as well as the flow rate and the settling velocity. Thus any choice of this product, as well as any choice of height, will produce a certain efficiency whose value will remain unchanged as long as the product nWL is the same. The question facing the designer, then, is the choice of n, W, L, and H which will produce the needed collection efficiency and satisfy any other demands of the situation. Figure 6-5 shows the pertinent dimensions of the chamber.

Since a particular value of nWL can be obtained by an infinite choice of values of n, W, and L separately, and since H can be chosen at will, other criteria are needed to guide the designer in making this choice. If no other criterion exists, the designer is free to exercise a whim in choosing the values of these quantities; however, the designer is seldom free to rely on whim if other meaningful criteria exist. We shall now discuss some possible alternative criteria which the design should meet insofar as possible.

The spacing between trays, if total height is not restricted by space limitations, will likely be determined by the desired cleaning interval of the chamber. This topic will be discussed in the next section. For the present, let us assume that ΔH has been determined and is fixed, which leaves the choice of n, W, and L. Spatial limitations may indeed fix W, L, or both, in which case the choice is considerably limited. If both are fixed, then n can be determined from the required value of nWL. Suppose, though, that no such restrictions apply. Is there any other factor which should be considered? The answer is yes. When all other constraints have been satisfied, the designer is free—in fact, obliged—to strive for minimum cost in his design. Let us consider the element of economics as it pertains to settling chambers.

It is difficult to predict closely what an item of equipment will cost, particularly a custom-made item such as a settling chamber. However, it is usually possible to predict the relative cost of alternative designs of a piece of equipment. A large piece of equipment will cost more, other factors being equal, and any unusually difficult fabricating problems will surely increase the cost. In our case, we are interested in the relative costs of different settling chambers for which nWL and ΔH are fixed, with n, W, and L separately variable. Material costs are likely to be proportional to the volume of material; and for simplicity let us assume that the thickness is the same for all competing designs, so that material costs are proportional to surface area. Labor costs can usually be predicted as equal to a constant plus an amount proportional to the size of the unit. Thus let us take the total cost as a linear function of the surface area of the unit. We shall consider only one side of each member.

The total plan area is $(n + 1)WL$, neglecting the open spaces at the ends, and the total side area is $2H(L + W)$. Thus the area is

$$A_m = (n + 1)WL + 2H(L + W) \qquad \text{(6-19)}$$

If we neglect the thickness of the trays, the total height $H = n\,\Delta H$. Then

$$A_m = (n + 1)WL + 2n\,\Delta H(L + W) \qquad \text{(6-20)}$$

Since nWL is a constant, we may rewrite Eq. (6-20) in the following form:

$$A_m = \left(1 + \frac{1}{n}\right)nWL + 2\,\Delta H\left(\frac{nWL}{W} + nW\right) \qquad \text{(6-21)}$$

We wish to minimize the area with respect to n and W, holding nWL and ΔH as constants. To do this, take the partial derivatives of A_m with respect to n and W and set these separately equal to zero. The simultaneous solution of these equations produces an optimum condition, which can be shown to be that of a minimum of A_m. The equations are

$$\frac{\partial A_m}{\partial n} = -\frac{1}{n^2}\,nWL + 2\,\Delta HW = 0 \qquad \text{(6-22)}$$

$$\frac{\partial A_m}{\partial W} = 2\,\Delta H\left(-\frac{nWL}{W^2} + n\right) = 0 \qquad \text{(6-23)}$$

These equations can be rearranged to give

$$n^2W = \frac{nWL}{2\,\Delta H} \qquad \text{(6-24)}$$

$$nW^2 = nWL \qquad \text{(6-25)}$$

Solving Eq. (6-25) for W gives

$$W = \sqrt{\frac{nWL}{n}} \qquad \text{(6-26)}$$

Substituting Eq. (6-26) back into Eq. (6-24) and solving for n gives

$$n = \frac{(nWL)^{1/3}}{(2\,\Delta H)^{2/3}} \qquad \text{(6-27)}$$

Then Eq. (6-26) gives for W

$$W = (2\,\Delta HnWL)^{1/3} \qquad \text{(6-28)}$$

and for L we obtain

$$L = (nWL)^{1/3}(2\,\Delta H)^{1/3} = W \qquad \text{(6-29)}$$

Equation (6-21) then gives the minimum area

$$A_m = nWL + 3(2\,\Delta HnWL)^{2/3} = nWL + 3W^2 \qquad \text{(6-30)}$$

EXAMPLE 6-3   Consider the problem of Examples 6-1 and 6-2. Assume that the data remain the same except that values of W and n are not given but must be determined to produce minimum surface area A_m. Determine n, W, and L for both laminar and turbulent flow.

SOLUTION   From the data of Example 6-1, $Q = 10$ m^3/s, $V_t = 0.148$ m/s, and $\Delta H = 0.167$ m. For the first attempt at a laminar-flow solution, we have

$$nWL = \frac{Q}{V_t} = \frac{10}{0.148} = 67.6$$

$$n = \frac{(nWL)^{1/3}}{(2\,\Delta H)^{2/3}} = \frac{(67.6)^{1/3}}{(0.333)^{2/3}} = 8.48 \text{ (use 8)} \qquad Ans.$$

$$W = \sqrt{\frac{nWL}{n}} = \sqrt{\frac{67.6}{8}} = 2.91 \text{ m} \qquad Ans.$$

$$L = \frac{nWL}{nW} = \frac{67.6}{8(2.91)} = 2.91 \text{ m} \qquad Ans.$$

In Example 6-1 we saw that this did not produce laminar flow. If we were to use the 386 trays needed to produce laminar flow, we would have

$$W = \sqrt{\frac{67.6}{386}} = 0.418 \text{ m} \qquad Ans.$$

$$L = \frac{67.6}{386(0.418)} = 0.418 \text{ m} \qquad Ans.$$

$$\Delta H = \frac{1.5}{386} = 3.9 \text{ mm} \qquad Ans.$$

For the turbulent solution, from Eq. (6-18),

$$nWL = -\frac{Q}{V_t} \ln(1 - \eta) = -\frac{10}{0.148} \ln(1 - 0.99) = 311$$

$$n = \frac{(311)^{1/3}}{(0.333)^{2/3}} = 14.1 \text{ (use 14)} \qquad Ans.$$

$$W = \sqrt{\frac{311}{14}} = 4.71 \text{ m} \qquad Ans.$$

$$L = W = 4.71 \text{ m} \qquad Ans.$$

$$H = n\,\Delta H = 14(0.167) = 2.34 \text{ m} \qquad Ans. \qquad ////$$

6-4   DUST REMOVAL

In this section we shall consider not so much how the collected dust is to be physically removed but rather how to predict when dust removal should be undertaken. The method of dust removal to be used, of course, depends on the particular type of settling chamber employed. In some chambers, the dust is removed by hand when it has reached a certain level in the bottom of the chamber. In others, an automatic emptying mechanism is employed which operates periodically at fixed or variable intervals. If the chamber operates at variable intervals, then most likely it will be set to operate when the accumulation of dust reaches a certain value. Still other chambers employ a continuous automatic removal system. We shall not discuss the mechanical design of automatic removal mechanisms.

The interval of dust removal can be predicted from the gas-flow rate, mass-volume concentration, and allowable accumulation just prior to removal. From Eq. (4-20), if we assume that the volume of pollutant has negligible effect on the volume of mixture, the mass-flow rate of dust is given by

$$\dot{m}_p = C_{mv} Q \qquad (6\text{-}31)$$

The dust accumulation can be given either as its volume or as its mass; the two are related by the average density of the accumulated mass of dust. It is likely that this density will be very much less than that of the dust in single-particle form, as is used in computing the settling velocity, for example. We shall designate the density of the collected dust mass as ρ_d. The mass of accumulated dust is then given as a function of time by

$$m_d = \eta \dot{m}_p t = \eta C_{mv} Q t \qquad (6\text{-}32)$$

and the volume of accumulated dust is given as

$$V_d = \frac{m_d}{\rho_d} = \frac{\eta C_{mv} Q t}{\rho_d} \qquad (6\text{-}33)$$

Solving Eqs. (6-32) and (6-33) for t gives the time of removal as

$$t = \frac{m_d}{\eta C_{mv} Q} = \frac{V_d \rho_d}{\eta C_{mv} Q} \qquad (6\text{-}34)$$

In the preceding equations, η is the overall collection efficiency based on mass. This efficiency will be considerably less than that upon which the settling chamber is designed since small particles are collected much less efficiently and also contribute to the mass of pollutant which is not collected.

EXAMPLE 6-4   In the settling chamber of Example 6-2, suppose the overall collection efficiency is 0.85. The mass-volume concentration of dust in the incoming air is 150 g/m³, and its density in the collected form is 800 kg/m³. If removal must be done when the volume of collected dust becomes 0.5 m³, what will be the time interval between cleaning operations?

SOLUTION   For a flow rate of 10 m³/s and a density of 800 kg/m³, the cleaning interval is, from Eq. (6-34),

$$t = \frac{V_d \rho_d}{\eta C_{mv} Q} = \frac{0.5(800)}{0.85(0.150)(10)(60)} = 5.23 \text{ min} \qquad Ans. \qquad ////$$

6-5 OVERALL EFFICIENCIES OF SETTLING CHAMBERS

It is of interest to estimate the overall efficiency of a settling chamber, based on the turbulent efficiency expressed by Eq. (6-18). When Eq. (3-48) for the settling velocity is substituted into this equation, the result is

$$\eta = 1 - \exp\left(-\frac{nLW\rho_p g d^2}{18\mu Q}\right) \qquad (6\text{-}35)$$

Equation (6-35) may be written in the form

$$\eta = 1 - e^{-Kd^2} \qquad (6\text{-}36)$$

in which K is defined as

$$K = \frac{nLW\rho_p g}{18\mu Q} \qquad (6\text{-}37)$$

The efficiencies based on mass, surface area, and number of particles are given in Eqs. (4-111) to (4-113), respectively, which require an approximate expression for the efficiency of the form of Eq. (4-109). With β_i equal to i, this expression may be written as

$$\eta(d) = 1 - \sum_{i=1}^{n} \frac{\alpha_i}{d^i} \qquad (6\text{-}38)$$

and also as

$$\eta(d) = 1 - \sum_{i=1}^{n} \alpha_i'\left(\frac{d_l}{d}\right)^i \qquad (6\text{-}39)$$

with

$$\alpha_i' = \frac{\alpha_i}{d_l{}^i} \qquad (6\text{-}40)$$

In Eqs. (6-39) and (6-40), d_l is the smallest diameter considered in the approximate efficiency expression, assuming that $\eta(d)$ is zero for diameters less than d_l. It is now our task to evaluate α_i so as to obtain satisfactory agreement between Eqs. (6-36) and (6-38).

To fit Eq. (6-38) to the exact equation for $\eta(d)$, let us form the integral

$$\Phi = \int_{d_l}^{\infty} \left(e^{-Kd^2} - \sum_{i=1}^{n} \frac{\alpha_i}{d^i}\right)^2 d(d) \qquad (6\text{-}41)$$

We shall achieve the best curve fit possible for a given value of n if this integral is minimized. To find the values of α_i for which the integral is a minimum, differentiate the equation with respect to α_j and set equal to zero. Thus

$$\frac{\partial}{\partial \alpha_j} \int_{d_l}^{\infty} \left(e^{-Kd^2} - \sum_{i=1}^{n} \frac{\alpha_i}{d^i}\right)^2 d(d) = 0 \qquad j = 1, 2, \dots, n-1 \qquad (6\text{-}42)$$

We shall differentiate this equation only for $n - 1$ of the values of α_j since an additional relation is available:

$$\sum_{i=1}^{n} \frac{\alpha_i}{d_i} = 1 \qquad (6\text{-}43)$$

This relation is obtained from the fact that $\eta(d)$ is assumed zero for $d \leq d_1$. Performing the differentiation inside the integral and rearranging, we obtain

$$\int_{d_1}^{\infty} \frac{e^{-Kd^2}}{d^j} \, d(d) - \sum_{i-1}^{n} \int_{d_1}^{\infty} \frac{\alpha_i}{d^{i+j}} \, d(d) = 0 \qquad (6\text{-}44)$$

We may evaluate the integral on the right, giving

$$\sum_{i=1}^{n} \frac{\alpha_i}{i+j-1} \frac{1}{d_i^i} = d_1^{j-1} \int_{d_1}^{\infty} \frac{e^{-Kd^2}}{d^j} \, d(d) \qquad j = 1, 2, \ldots, n-1 \qquad (6\text{-}45)$$

We designate the right side of Eq. (6-45) as $I_j(Kd_1^2)$; that is,

$$I_j(Kd_1^2) = d_1^{j-1} \int_{d_1}^{\infty} \frac{e^{-Kd^2}}{d^j} \, d(d) \qquad (6\text{-}46)$$

Then Eq. (6-45) becomes

$$\sum_{i=1}^{n} \frac{\alpha_i}{(i+j-1)d_1^i} = I_j(Kd_1^2) \qquad j = 1, 2, \ldots, n-1 \qquad (6\text{-}47)$$

Finally, using the shorthand notation of Eq. (6-40), we have

$$\sum_{i=1}^{n} \frac{\alpha_i'}{i+j-1} = I_j \qquad j = 1, 2, \ldots, n-1 \qquad (6\text{-}48)$$

Equation (6-43) may be written

$$\sum_{i=1}^{n} \alpha_i' = 1 \qquad (6\text{-}49)$$

Next, we must evaluate the integral of Eq. (6-46) in terms of available functions. If we integrate successively by parts, eventually we arrive at the following form:

$$d_1^{1-j} I_j = e^{-Kd_1^2} \sum_{p=1}^{m/2} \frac{(-2K)^{p-1}}{(j-1)(j-3)\cdots(j-2p+1)d_1^{j-2p+1}}$$

$$+ \frac{(-2K)^{m/2}}{(j-1)(j-3)\cdots(j-m+1)} \int_{d_1}^{\infty} \frac{e^{-Kd^2}}{d^{j-m}} \, d(d) \qquad (6\text{-}50)$$

in which m is an even integer equal to either j or $j - 1$. Equation (6-50) is valid for any integer j, though we are interested only in values up to $n - 1$.

The integral remaining in Eq. (6-50) may be evaluated for j even or for j odd. If j is even,

$$\int_{d_l}^{\infty} \frac{e^{-Kd^2}}{d^{j-m}} d(d) = \int_{d_l}^{\infty} e^{-Kd^2} d(d)$$

$$= \frac{1}{\sqrt{K}} \int_{\sqrt{K}d_l}^{\infty} e^{-z^2} dz$$

$$= \frac{\sqrt{\pi}}{2\sqrt{K}} \operatorname{erfc} \sqrt{K} \, d_l \qquad (6\text{-}51)$$

And if j is odd,

$$\int_{d_l}^{\infty} \frac{e^{-Kd^2}}{d^{j-m}} d(d) = \int_{d_l}^{\infty} \frac{e^{-Kd^2}}{d} d(d)$$

$$= \frac{1}{2} \int_{Kd^2}^{\infty} \frac{e^{-z}}{z} dz$$

$$= \frac{1}{2} E_1(Kd_l^2) \qquad (6\text{-}52)$$

In Eq. (6-52), $E_1(Kd_l^2)$ is an exponential integral of the first kind, defined as

$$E_1(x) = \int_x^{\infty} \frac{e^{-t}}{t} dt \qquad (6\text{-}53)$$

This function is available in several sources, such as Abramowitz and Stegun [1]. Equation (6-50) then reduces to

$$I_j(Kd_l^2) = e^{-Kd_l^2} \sum_{p=1}^{m/2} \frac{(-2Kd_l^2)^{p-1}}{(j-1)(j-3)\cdots(j-2p+1)}$$

$$+ \frac{1}{2} \frac{(-2Kd_l^2)^{m/2}}{(j-1)(j-3)\cdots(j-m+1)} \begin{cases} \sqrt{\dfrac{\pi}{Kd_l^2}} \operatorname{erfc} \sqrt{Kd_l^2} & \text{if } j \text{ is even} \\[2ex] E_1(Kd_l^2) & \text{if } j \text{ is odd} \end{cases}$$

$$(6\text{-}54)$$

Equation (6-54) is not clear if $j = 1$; for this case the Eq. (6-54) becomes

$$I_1 = \tfrac{1}{2} E_1(Kd_l^2) \qquad (6\text{-}55)$$

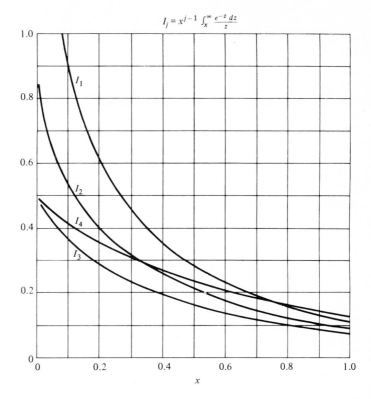

$$I_j = x^{j-1} \int_x^\infty \frac{e^{-z}\,dz}{z}$$

FIGURE 6-7
Integral function of Eq. (6-46).

The values of this integral have been worked out for the first four values of j. These results are plotted in Fig. 6-7 for $I_j(x)$ as a function of x. With the values of I_j known, the values of α'_i can be evaluated from Eqs. (6-48) and (6-49). For example, if $n = 2$, these equations reduce to

$$\alpha'_1 + \frac{\alpha'_2}{2} = I_1 \qquad (6\text{-}56)$$

$$\alpha'_1 + \alpha'_2 = 1 \qquad (6\text{-}57)$$

which can be solved to give

$$\alpha'_1 = -(1 - 2I_1) \qquad (6\text{-}58)$$
$$\alpha'_2 = 2(1 - I_1) \qquad (6\text{-}59)$$

Equations (6-58) and (6-59) are plotted as a function of $x = Kd_l^2$ in Fig. 6-8.

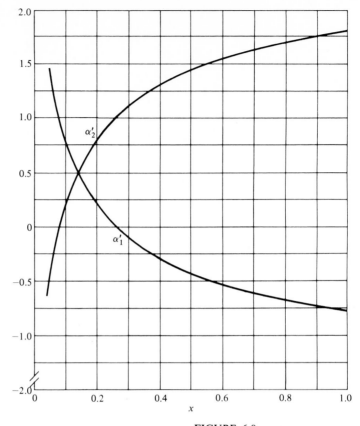

FIGURE 6-8
Coefficients in Eq. (6-39) for $n = 2$.

EXAMPLE 6-5 Using the data of Example 6-2, determine values of $\alpha'_1, \alpha'_2, \ldots, \alpha'_n$, for various values of n. Assume that $d_l = 10 \ \mu m$. Plot the curves for $\eta(d)$ for each value of n and compare with the exact curve.

SOLUTION From Eq. (6-37) and from the data of Example 6-2, we compute K as

$$K = 29{,}610 \frac{nWL\rho_p}{Q} = \frac{29{,}610(9)(1.5)(34.6)(2000)}{10} = 0.276 \times 10^{10}$$

From Fig. 6-7, or from Eqs. (6-54) and (6-55), using the appropriate table of functions, we have

$$I_1 = 0.485 \qquad I_2 = 0.327 \qquad I_3 = 0.246$$

The equation for the exact efficiency distribution is given by Eq. (6-36):

$$\eta = 1 - e^{-Kd^2} = 1 - e^{-0.276 \times 10^{10}d^2}$$

For $n = 1$, Eq. (6-49) gives $\alpha_1' = 1$ or $\alpha_1 = 10^{-5}$. Then

$$\eta(d) = 1 - \frac{10^{-5}}{d}$$

For $n = 2$, Eqs. (6-48) and (6-49) give

$$\alpha_1' + \frac{\alpha_2'}{2} = 0.485$$

$$\alpha_1' + \alpha_2' = 1$$

and the solution to these equations is

$$\alpha_1 = -0.03 \times 10^{-5}$$
$$\alpha_2 = 1.03 \times 10^{-10}$$
$$\eta(d) = 1 + \frac{0.03 \times 10^{-5}}{d} - \frac{1.03 \times 10^{-10}}{d^2}$$

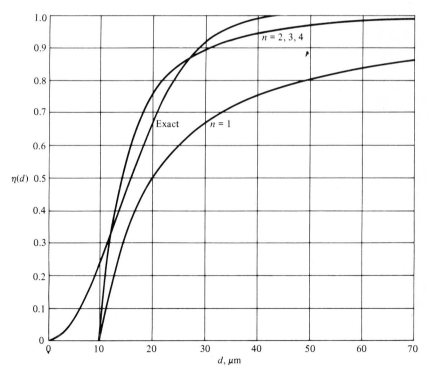

FIGURE 6-9
Efficiency approximations for Example 6-5.

For $n = 3$, the equations and their solutions are

$$\alpha_1' + \frac{\alpha_2'}{2} + \frac{\alpha_3'}{3} = 0.485$$

$$\frac{\alpha_1'}{2} + \frac{\alpha_2'}{3} + \frac{\alpha_3'}{4} = 0.327$$

$$\alpha_1' + \alpha_2' + \alpha_3' = 1$$

$$\alpha_1 = -0.012 \times 10^{-5}$$

$$\alpha_2 = 0.985 \times 10^{-10}$$

$$\alpha_3 = 0.054 \times 10^{-15}$$

$$\eta(d) = 1 + \frac{0.012 \times 10^{-5}}{d} - \frac{0.958 \times 10^{-10}}{d^2} - \frac{0.054 \times 10^{-15}}{d^3}$$

Finally, for $n = 4$, we have

$$\alpha_1' + \frac{\alpha_2'}{2} + \frac{\alpha_3'}{3} + \frac{\alpha_4'}{4} = 0.485$$

$$\frac{\alpha_1'}{2} + \frac{\alpha_2'}{3} + \frac{\alpha_3'}{4} + \frac{\alpha_4'}{5} = 0.327$$

$$\frac{\alpha_1'}{3} + \frac{\alpha_2'}{4} + \frac{\alpha_3'}{5} + \frac{\alpha_4'}{6} = 0.246$$

$$\alpha_1' + \alpha_2' + \alpha_3' + \alpha_4' = 1$$

$$\alpha_1 = -0.04 \times 10^{-5} \qquad \alpha_2 = 1.2 \times 10^{-10}$$

$$\alpha_3 = -0.42 \times 10^{-15} \qquad \alpha_4 = 0.26 \times 10^{-20}$$

$$\eta(d) = 1 + \frac{0.04 \times 10^{-5}}{d} - \frac{1.2 \times 10^{-10}}{d^2}$$

$$+ \frac{0.42 \times 10^{-15}}{d^3} - \frac{0.26 \times 10^{-20}}{d^4}$$

Figure 6-9 shows these curves along with the exact curve.   ////

EXAMPLE 6-6   Using the $n = 2$ approximation from Example 6-5, determine the overall mass efficiency for a particle distribution having $d_m = 25$ μm and $\sigma_m = 1.5$.

SOLUTION   From Eqs. (4-56) and (4-57), we have

$$\bar{u} = \ln d_m = \ln 2.5 \times 10^{-5} = -10.60$$

$$u_l = \ln d_l = \ln 1 \times 10^{-5} = -11.51$$

$$\sigma_u = \ln \sigma_m = \ln 1.5 = 0.405$$

Then Eq. (4-111) gives the overall mass efficiency as

$$\eta_m = \frac{1}{2}\,\text{erfc}\,\frac{u_l - \bar{u} - 3\sigma_u^2}{\sqrt{2}\,\sigma_u} - \frac{1}{2}\,\alpha_1 e^{-\bar{u} - 5\sigma_u^2/2}\,\text{erfc}\,\frac{u_l - \bar{u} - 2\sigma_u^2}{\sqrt{2}\,\sigma_u}$$

$$- \frac{1}{2}\,\alpha_2 e^{-\bar{u} - 8\sigma_u^2/2}\,\text{erfc}\,\frac{u_l - \bar{u} - \sigma_u^2}{\sqrt{2}\,\sigma_u}$$

$$= \frac{1}{2}\,\text{erfc}\,\frac{-11.51 + 10.60 - 3(0.164)}{0.573} - \frac{1}{2}(-0.03 \times 10^{-5})$$

$$\times \exp\left[10.60 - 2.5(0.164)\right]\text{erfc}\,\frac{-11.51 + 10.60 - 2(0.164)}{0.573}$$

$$- \frac{1}{2}(1.03 \times 10^{-10})\exp\left[2(10.60) - 4(0.164)\right]\text{erfc}\,\frac{-11.51 + 10.60 - 0.164}{0.573}$$

$$= 0.925 \qquad Ans. \hspace{3cm} ////$$

REFERENCE

1 Abramowitz, M., and I. A. Stegun, eds.: "Handbook of Mathematical Functions," Dover Publications, Inc., New York, 1965.

PROBLEMS

6-1 A settling chamber is 3 m wide, 10 m long, and has 10 trays, including the bottom surface. The tray spacing is 15 cm center to center; the tray thickness is 3 mm, and a 3-cm thick dust layer has built up on the trays. A stream of 1.5 m^3/s of standard air containing particles of 2500 kg/m^3 density flows through the chamber.
(a) Is the flow laminar or turbulent?
(b) Plot collection efficiency as a function of diameter over a range of diameters such that efficiency varies from 10 to 99 percent.

6-2 Repeat Prob. 6-1 for a flow rate of 0.4 m^3/s.

6-3 Design a settling chamber for laminar flow for a stream of 100 m^3/s of standard air containing particles having a density of 1500 kg/m^3. The chamber must not exceed 5 m in width or 6 m in height. It must collect particles of 70 μm diameter with perfect efficiency. Determine the length of the chamber required and the number of trays.

6-4 Repeat Prob. 6-3 if the flow will be turbulent and only 99 percent efficiency is required for 70-μm particles.

6-5 Design a settling chamber for use on a home central-vacuum-cleaning system. The flow rate is 1.0 dm^3/s, and the particle density is 1500 kg/m^3. It is desired to collect 90 percent of all particles of 50 μm diameter. Design for turbulent flow in the chamber.

6-6 Someone has proposed the use of an inflatable shell structure as a settling chamber to control the output of a large power plant. As a preliminary study of the feasibility of such a device, consider an application to a power plant which produces 5000 m³/s of discharge (assume standard air) having particles with a density of 500 kg/m³. Assume a chamber which is 100 m wide, 50 m high, and 1000 m long. First, determine if the flow is laminar or turbulent; then compute and plot the collection efficiency as a function of particle diameter.

6-7 Repeat Prob. 6-6 if the inflatable structure is 200 m wide, 50 m high, and 2000 m long.

6-8 A settling chamber is to be constructed in the form of a series of n trays, each tray being in the form of a disk with a hole in the center. The arrangement is shown in Fig. 6-10. Dirty air enters in the center, flows outward through the passageway between adjacent disks, and leaves from the outside. Derive an expression for the efficiency of this device in terms of the number of trays, inside and outside radii, flow rate, terminal velocity, and height between trays if needed. Assume laminar flow in the passageways, and specify the minimum value of ΔH for laminar flow.

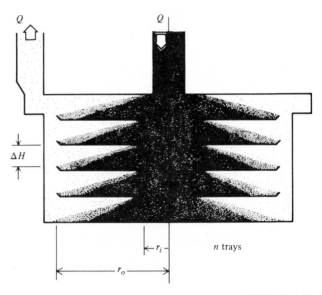

FIGURE 6-10

6-9 Repeat Prob. 6-8 if the flow is turbulent.

6-10 Repeat Prob. 6-8 if the flow is from the outside toward the center.

6-11 Repeat Prob. 6-9 if the flow is from the outside toward the center.

6-12 Repeat Prob. 6-1 if the flow rate is 6 m³/s.

6-13 A lateral cross section of a settling chamber is shown in Fig. 6-11. The sloping trays are arranged so that the collected dust will slide to the center, perhaps after rapping of the trays, and can be collected easily. Derive an expression for the collection efficiency for turbulent flow, neglecting the detrimental effect of the space in the center.

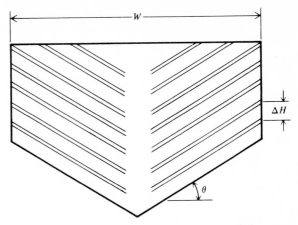

FIGURE 6-11

6-14 If the width of the space in the center is W_1, refine the derivation of Prob. 6-13 to allow for the effect of this space upon the collection efficiency.

6-15 Figure 6-12 shows a settling chamber having n sloping trays. The trays slope upward in the direction of the flow. The collected dust slides down the trays and collects in a hopper in the bottom at the front of the chamber. Assuming turbulent flow, derive an expression for the collection efficiency of the device in terms of ΔH, L, width W, and angle θ.

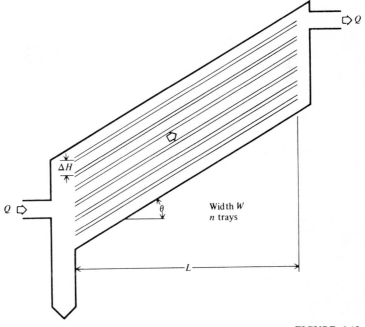

FIGURE 6-12

6-16 Design an optimum settling chamber to handle 9.5 m³/s of dusty air with dust density of 1900 kg/m³. The chamber should produce an efficiency of 0.95 for dust particles of 50 μm diameter. Assume turbulent flow in the chamber. Assume standard air.

6-17 Design an optimum settling chamber to handle 1.0 m³/s of standard air carrying dust particles of 3000 kg/m³ density. Design for laminar flow in the chamber; the efficiency should be theoretically 1.0 for all particles of 90 μm diameter or larger.

6-18 Repeat Prob. 6-5 using an optimum design.

6-19 Derive an expression for the minimum value of A_m, as given by Eq. (6-21), for the case where n and nWL are fixed and only W and L can be varied. What are the expressions for W and L?

6-20 Derive an expression for the minimum value of A_m for the case where W and nWL are fixed, with n and L being variable. Assume that ΔH is fixed but that the total height H can vary. Find expressions for n, L, and H.

6-21 Derive an expression for the minimum value of A_m if nWL is fixed and if the total height H is also fixed. The number of trays n, width W, length L, and tray spacing ΔH can vary. What are the optimum values of W, L, n, and ΔH?

6-22 In the settling chamber of Prob. 6-1, assume a mass-volume concentration of 0.35 kg/m³ and an overall collection efficiency of 0.95. At what interval will it be necessary to clean the trays of the chamber if the maximum allowable dust thickness is to be 3 cm. Assume the density of the collected dust is 1500 kg/m³.

6-23 In the configuration of Prob. 6-15, derive an expression for the cleaning interval if the dust chamber has dimensions of W by L_1 by H_1 and if the density of the collected dust is ρ_d.

6-24 Derive Eq. (6-50).

6-25 For the data of Example 6-6, compute the efficiencies based on surface area and number of particles.

6-26 Use the data of Prob. 6-4, and determine the overall efficiency based on mass. Assume that $d_m = 100$ μm and $\sigma_m = 3.0$. Use $d_l = 20$ μm.

6-27 Determine the overall efficiency for the settling chamber of Prob. 6-5 if the dust has $d_m = 70$ μm, $\sigma_m = 2.5$, and if $d_l = 10$ μm.

6-28 Repeat Fig. 6-9, if d_l is chosen as 5 μm.

6-29 The results of Fig. 6-9 suggest that choosing large values of n offers very little improvement over the use of $n = 2$, at least for that example. Thus the optimum choice of α_i in Eq. (6-38) can approximate Eq. (6-36) only so closely, regardless of how large n is. This suggests that Φ of Eq. (6-41) must have a nonzero value. Show that for the optimum choice of α_i, and for any n, the value of Φ can be written as

$$\frac{\Phi}{d_i} = \frac{1}{2}\sqrt{\frac{\pi}{2Kd_l^2}} \operatorname{erfc}\sqrt{2Kd_l^2} - \sum_{i=1}^{n-1}\sum_{j=1}^{n}\frac{\alpha_i'\alpha_j'}{i+j-1} - \alpha_n'I_n(Kd_l^2)$$

Evaluate Φ for the case of $n = 2$ and for the data of Example 6-5.

7

INERTIAL DEVICES

In Chap. 6 we saw how a device could effect the separation of particulates, making use of the force of gravity. A much more effective mechanism of separation utilizes the centrifugal force acting on the particles when the gas stream follows a curved path. Various types of cyclone collectors and baffle collectors use this mechanism. We shall study this mechanism of collection in this chapter.

7-1 CYCLONE FLOW

Before beginning our discussion of inertial devices, let us derive an equation for the velocity of flow in a circular stream, neglecting boundary-layer effects. Figure 7-1 illustrates a system consisting of a U-shaped duct in which this type of flow will occur. We shall neglect secondary flows induced by the presence of sidewalls, and we shall begin by assuming that no friction exists in the flow.

Consider an element of fluid located at the point (r, θ) or (x, y) in rectangular coordinates, as shown in Fig. 7-2. In the absence of friction, only normal pressure forces will act on the element. Consider that the flow is two-dimensional and that the element has unit depth normal to the plane of the paper. The mass of the element is given as

$$dm = \rho r \, dr \, d\theta$$

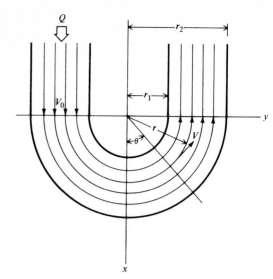

FIGURE 7-1
Cyclone flow.

The acceleration of the particle is given by

$$a = -\frac{V^2}{r}$$

The net force acting on the element due to pressure forces, neglecting the gravitational force on the gas, is

$$\sum F = \mathrm{Pr}\, d\theta - (P + dP)(r + dr)\, d\theta + 2\left(P + \frac{dP}{2}\right)\frac{d\theta}{2}\, dr$$

$$= -r\, dP\, d\theta$$

Writing that $F = a\, dm$ and substituting the preceding equation into this formula gives

$$\frac{dP}{dr} = \frac{\rho V^2}{r} \qquad (7\text{-}1)$$

Bernoulli's equation can be written from the entrance to the section at point 0 to the point at which the element is located as

$$\frac{P_0}{\rho} + \frac{V_0{}^2}{2} = \frac{P}{\rho} + \frac{V^2}{2} \qquad (7\text{-}2)$$

from which

$$\frac{dP}{dr} = -\rho V \frac{dV}{dr} \qquad (7\text{-}3)$$

Combining Eqs. (7-1) and (7-3) leads to

$$\frac{dV}{V} = -\frac{dr}{r} \qquad (7\text{-}4)$$

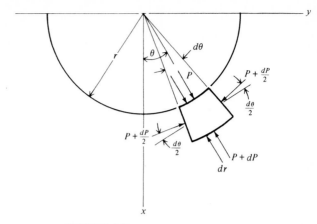

FIGURE 7-2
Forces acting on a differential element in cyclone flow.

The solution to this differential equation is

$$V = \frac{C}{r} \qquad (7\text{-}5)$$

To evaluate the constant, let us determine the flow rate over the section by integrating the expression for V

$$Q = W \int_{r_1}^{r_2} V\, dr = CW \ln \frac{r_2}{r_1}$$

Solving this equation for C, we have

$$C = \frac{Q}{W \ln r_2/r_1} \qquad (7\text{-}6)$$

Then Eq. (7-5) becomes

$$V = \frac{Q}{rW \ln r_2/r_1} \qquad (7\text{-}7)$$

Equation (7-1) gives for the pressure gradient

$$\frac{dP}{dr} = \frac{\rho Q^2}{r^3 W^2 (\ln r_2/r_1)^2} \qquad (7\text{-}8)$$

If we choose P_0 as a reference pressure at a point of radius r_0, then Eq. (7-8) leads to

$$P = P_0 + \frac{\rho Q^2}{2W^2 (\ln r_2/r_1)^2} \left(\frac{1}{r_0{}^2} - \frac{1}{r^2} \right) \qquad (7\text{-}9)$$

On the other hand, if P_0 is the pressure at the entrance section 0, where the velocity is V_0, then Eq. (7-2) gives

$$P = P_0 + \frac{\rho Q^2}{2W^2} \left[\frac{1}{(r_2 - r_1)^2} - \frac{1}{r^2 (\ln r_2/r_1)^2} \right] \qquad (7\text{-}10)$$

The velocity components may be written as follows: In cylindrical coordinates, the radial component is zero and the tangential component is equal to V; that is,

$$V_r = 0 \qquad V_\theta = V = \frac{Q}{Wr \ln r_2/r_1} \qquad (7\text{-}11)$$

Converting to rectangular coordinates gives

$$V_y = V \cos \theta \qquad V_x = -V \sin \theta$$

Since $x = r \cos \theta$ and $y = r \sin \theta$, we have

$$V_y = V \frac{x}{r} = \frac{Qx}{Wr^2 \ln r_2/r_1} = \frac{Qx}{W(x^2 + y^2) \ln r_2/r_1} \qquad (7\text{-}12)$$

$$V_x = -V \frac{y}{r} = -\frac{Qy}{Wr^2 \ln r_2/r_1} = -\frac{Qy}{W(x^2 + y^2) \ln r_2/r_1} \qquad (7\text{-}13)$$

EXAMPLE 7-1 Suppose that $r_1 = 0.25$ m and $r_2 = 0.5$ m. For a flow rate with a width of 1.0 $\text{m}^3/\text{s} \cdot \text{m}$, plot the velocity profile and determine the pressure difference between the inside and outside radii. Assume standard air.

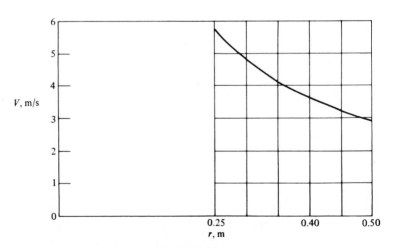

FIGURE 7-3
Velocity distribution in the flow of Example 7-1.

SOLUTION Equation (7-7) gives the velocity as a function of radius:

$$V = \frac{Q}{Wr \ln r_2/r_1} = \frac{1.0}{1.0(\ln 0.5/0.25)r} = \frac{1.44}{r} \quad Ans.$$

This equation is plotted in Fig. 7-3. Equation (7-9) gives the pressure difference between the inside and outside radii:

$$P_2 - P_1 = \frac{\rho Q^2}{2W^2(\ln r_2/r_1)^2}\left(\frac{1}{r_1{}^2} - \frac{1}{r_2{}^2}\right) = \frac{1.2(1.0)^2}{2(1)(0.693)^2}\left(\frac{1}{0.25^2} - \frac{1}{0.5^2}\right)$$

$$= 15.0 \text{ N/m}^2 \quad Ans.$$

The pressure is greater on the outside radius. $/////$

7-2 COLLECTION EFFICIENCY IN LAMINAR CYCLONE FLOW NEGLECTING GRAVITY

Figure 7-4 shows the trajectory followed by a particle of diameter d which enters the cyclonic flow section at radius r_3 and strikes the outer wall of the duct at angular position θ. Let us assume that the particle is moving at its terminal velocity at all points along its trajectory even though the terminal velocity varies as the force applied to the particle changes. The efficiency of collection for particles of diameter d is given by

$$\eta = \frac{r_2 - r_3}{r_2 - r_1} \quad (7\text{-}14)$$

The tangential velocity component is given by Eq. (7-11).

The terminal particle velocity, which is its radial component of velocity, is given by Eq. (3-19) as

$$V_t = V_r = \frac{F}{3\pi\mu d} \quad (7\text{-}15)$$

The centrifugal force acting on the particle is given as

$$F = \frac{mV_\theta{}^2}{r} = \rho_p \frac{\pi d^3}{6}\frac{V_\theta{}^2}{r} = \rho_p \frac{\pi d^3}{6}\frac{Q^2}{W^2 r^3(\ln r_2/r_1)^2} \quad (7\text{-}16)$$

Thus, Eq. (7-15) becomes

$$V_r = \frac{\rho_p Q^2 d^2}{18\mu r^3 W^2(\ln r_2/r_1)^2} \quad (7\text{-}17)$$

From Eq. (7-11) and Eq. (7-17), we can obtain a differential equation for the trajectory of the particle, noting that the ratio of tangential to radial distance traveled is equal to the ratio of tangential to radial velocities:

$$\frac{r\,d\theta}{dr} = \frac{V_\theta}{V_r} = \frac{Q/Wr(\ln r_2/r_1)}{\rho_p Q^2 d^2/18\mu r^3 W^2(\ln r_2/r_1)^2} \quad (7\text{-}18)$$

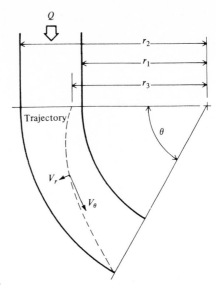

FIGURE 7-4
Trajectory of a particle in cyclone flow.

which reduces to

$$\frac{d\theta}{dr} = \frac{18\mu W \ln r_2/r_1}{\rho_p Q d^2} r \qquad (7\text{-}19)$$

Integrating Eq. (7-19) from $r = r_3$ at $\theta = 0$ to $r = r_2$ at θ, we have

$$\theta = \frac{18\mu W \ln r_2/r_1}{2\rho_p Q d^2} (r_2{}^2 - r_3{}^2) \qquad (7\text{-}20)$$

Solving this equation for r_3 gives

$$r_3 = \sqrt{r_2{}^2 - \frac{\rho_p Q d^2 \theta}{9\mu W \ln r_2/r_1}} \qquad (7\text{-}21)$$

Substituting into Eq. (7-14) gives for the efficiency

$$\eta = \frac{1 - \sqrt{1 - \rho_p Q d^2 \theta / 9\mu r_2{}^2 W (\ln r_2/r_1)}}{1 - r_1/r_2} \qquad (7\text{-}22)$$

The efficiency can be made unity by solving Eq. (7-22) for θ, designated θ_1, at which $\eta = 1$, giving

$$\theta_1 = \frac{9\mu W (r_2{}^2 - r_1{}^2) \ln r_2/r_1}{\rho_p Q d^2} \qquad (7\text{-}23)$$

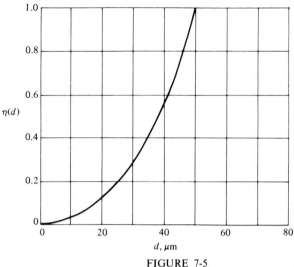

FIGURE 7-5
Efficiency curve for Example 7-2.

EXAMPLE 7-2 A stream that is 5.0 $m^3/s/m$ wide enters a cyclone flow with $r_1 = 20$ cm and $r_2 = 40$ cm. The fluid is standard air, and the particle density is 1500 kg/m^3. Through what angle θ_1 must the flow turn in the cyclone if the efficiency is to be unity for 50-μm particles? Plot efficiency as a function of diameter for this angle.

SOLUTION From the given data and Eq. (7-23), θ_1 is computed to be

$$\theta_1 = \frac{9\mu W(r_2{}^2 - r_1{}^2)\ln r_2/r_1}{\rho_p Q d^2} = \frac{9(1.84 \times 10^{-5})(0.4^2 - 0.2^2)\ln 2}{1500(5.0)(25 \times 10^{-10})}$$

$$= 0.735 \text{ rad}$$

$$= 42° \quad Ans.$$

Equation (7-22) gives the efficiency as a function of diameter:

$$\eta = \frac{1}{0.5}\left[1 - \sqrt{1 - \frac{1500(5)(0.735d^2)}{9(1.84 \times 10^{-5})(0.16)(\ln 2)}}\right]$$

$$= 2(1 - \sqrt{1 - 0.03 \times 10^{10}d^2})$$

This equation is plotted in Fig. 7-5.

EXAMPLE 7-3 Compare the force acting on a particle developed by the centrifugal force in Example 7-2 with that due to gravity.

SOLUTION Equation (7-16) gives the centrifugal force:

$$F_c = \frac{\pi \rho_p d^3 Q^2}{6 r^3 W^2 (\ln r_2/r_1)^2}$$

Equation (3-47) gives the gravitational force:

$$F_g = \frac{\pi \rho_p d^3 g}{6}$$

Combining these equations gives

$$\frac{F_c}{F_g} = \frac{Q^2}{g r^3 W^2 (\ln r_2/r_1)^2}$$

Evaluating for the data of Example 7-2, for $r = r_2$, we have

$$\frac{F_c}{F_g} = \frac{(5.0)^2}{9.81(0.4)^3(1)^2(\ln 2)^2} = 83$$

Since the data are typical of a moderate cyclone flow field, we may feel fairly comfortable in neglecting gravitational effects in most situations where centrifugal effects are predominant.   ////

7-3 COLLECTION EFFICIENCY IN TURBULENT CYCLONE FLOW

In Sec. 7-2 we considered the case where the flow is laminar. Of course, we used the formula for laminar terminal velocity of the particle in considering its motion in the radial direction due to the centrifugal force acting upon it. We shall continue to use laminar particle motion in the radial direction, but now we shall assume turbulent flow of the gas. Figure 7-6 shows the geometrical configuration used in the analysis.

For the analysis considered in the first part of this section, let us assume that the effect of the turbulent eddies is to distribute the particles uniformly over the cross section at any given angle θ. This assumption is not necessarily valid for cyclone flow, and, in fact, it may be too conservative since the centrifugal force effects may serve to damp out the turbulent eddies which naturally occur in turbulent duct flow. This area is one in which relatively little is known. Later in this section, we shall consider the possibility of a different particle distribution over the cross section.

Consider the effect of a laminar layer next to the outer edge of the cyclone, as shown in Fig. 7-6, such that all particles which enter it are captured. If we let $(r_2 \, d\theta)$ be the distance required for capture, then

$$r_2 \, d\theta = V_{\theta_2} \, dt$$

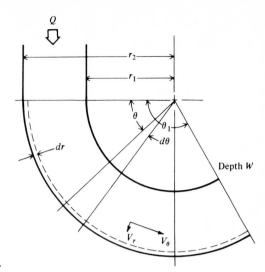

FIGURE 7-6
Turbulent cyclone flow.

The thickness of the capture zone is

$$dr = V_{r_2} \, dt = \frac{V_{r_2}}{V_{\theta_2}} r_2 \, d\theta$$

The fractional diminution of particles over the angle $d\theta$ is the fraction of the particles which lies in the capture zone, or

$$-\frac{dN}{N} = \frac{dr}{r_2 - r_1} = \frac{V_{r_2}}{V_{\theta_2}} \frac{r_2}{r_2 - r_1} d\theta$$

Integrating this equation gives

$$\ln N = -\frac{V_{r_2}}{V_{\theta_2}} \frac{r_2}{r_2 - r_1} \theta + C$$

Evaluating the constant of integration at the inlet, $\theta = 0$, where N_0 is the total number of particles there, we have

$$N = N_0 \exp\left(-\frac{V_{r_2}}{V_{\theta_2}} \frac{r_2}{r_2 - r_1} \theta\right)$$

The efficiency at the angle θ_1 is then

$$\eta = 1 - \frac{N}{N_0} = 1 - \exp\left[-\frac{V_{r_2} r_2 \theta_1}{V_{\theta_2}(r_2 - r_1)}\right] \qquad (7\text{-}24)$$

Equation (7-11) gives the tangential velocity of the particle, and Eq. (7-17) gives the radial velocity. Evaluated at the outer radius r_2, these equations become

$$V_{\theta_2} = \frac{Q}{r_2 \, W \, \ln r_2/r_1}$$

$$V_{r_2} = \frac{\rho_p Q^2 d^2}{18\mu r_2{}^3 W^2 (\ln r_2/r_1)^2}$$

Then Eq. (7-24) becomes

$$\eta = 1 - \exp\left[-\frac{\rho_p Q d^2 \theta_1}{18\mu r_2 \, W(r_2 - r_1) \ln r_2/r_1}\right] \qquad (7\text{-}25)$$

which can also be solved for θ_1:

$$\theta_1 = -\frac{18\mu r_2 \, W(r_2 - r_1)(\ln r_2/r_1) \ln (1 - \eta)}{\rho_p Q d^2} \qquad (7\text{-}26)$$

Thus, we can determine the angle of turn to achieve a given collection efficiency for a given particle size, flow rate, and cyclone geometry.

EXAMPLE 7-4   Consider the data of Example 7-2, and determine the angle of turn necessary to obtain 0.99 collection efficiency. Assume standard air.

SOLUTION   Equation (7-26) gives the angle of turn as

$$\theta = -\frac{18(1.84 \times 10^{-5})(0.4)(1.0)(0.2)(\ln 2)(\ln 0.01)}{1500(5)(25 \times 10^{-10})}$$

$$= 4.51 \text{ rad}$$

$$= 258° \qquad Ans.$$

The efficiency as a function of particle diameter for this angle of turn is

$$\eta = 1 - \exp\left[-\frac{1500(5)(4.51d^2)}{18(1.84 \times 10^{-5})(0.4)(1.0)(0.2)(\ln 2)}\right]$$

$$= 1 - e^{-0.184 \times 10^{10} d^2}$$

This curve is plotted in Fig. 7-7.

The overall efficiency can be evaluated by the method of Sec. 6-6; let us use the particle distribution of Example 6-6, with $d_l = 1.0 \times 10^{-5}$ m. Then $K d_l{}^2$ is equal to 0.184, and from Fig. 6-8

$$\alpha_1' = 0.30 \qquad \alpha_2' = 0.73$$

Then

$$\alpha_1 = \alpha_1' d_l = 0.30 \times 10^{-5} \qquad \alpha_2 = \alpha_2' d_l = 0.73 \times 10^{-10}$$

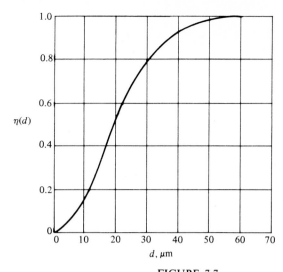

FIGURE 7-7
Efficiency for Example 7-4.

Following closely the solution of Example 6-6, the overall efficiency based on mass is determined as follows:

$$\bar{u} = -10.60 \qquad u_l = -11.51 \qquad \sigma_u = 0.405$$

$$\eta_m = \frac{1}{2} \operatorname{erfc} \frac{u_l - \bar{u} - 3\sigma_u^2}{\sqrt{2}\,\sigma_u} - \frac{1}{2} \alpha_1 e^{-\bar{u} - 5\sigma_u^2/2} \operatorname{erfc} \frac{u_l - \bar{u} - 2\sigma_u^2}{\sqrt{2}\,\sigma_u}$$

$$- \frac{1}{2} \alpha_2 e^{-\bar{u} - 4\sigma_u^2} \operatorname{erfc} \frac{u_l - \bar{u} - \sigma_u^2}{\sqrt{2}\,\sigma_u}$$

$$= \tfrac{1}{2} \operatorname{erfc} (-2.447) - 0.15 \times 10^{-5} e^{10.19} \operatorname{erfc} (-2.161)$$

$$- 0.365 \times 10^{-10} e^{20.54} \operatorname{erfc} (-1.874)$$

$$= 0.859 \qquad Ans. \qquad\qquad ////$$

Modified Cyclone Flow

The preceding analysis was based on cyclone flow as developed in Sec. 7-1. We can modify this basic flow and will do so later in order to obtain a better fit of experimental data as we proceed to examine specific devices. For now, let us consider how this modified cyclone flow will affect collection efficiency.

The particular modification to Eq. (7-5) for basic cyclone flow is

$$V_\theta = \frac{C}{r^n} \qquad (7\text{-}27)$$

in which n is a constant. We may evaluate the constant C as we did in Sec. 7-1 by integrating over the section to evaluate the flow rate and then solving for C, giving

$$Q = \int_A V_\theta \, dA = W \int_{r_1}^{r_2} V_\theta \, dr = W \int_{r_1}^{r_2} \frac{C \, dr}{r^n} = \frac{WC}{1-n}(r_2^{1-n} - r_1^{1-n})$$

$$C = \frac{(1-n)Q}{W(r_2^{1-n} - r_1^{1-n})}$$

Then the tangential velocity becomes

$$V_\theta = \frac{(1-n)Q}{W(r_2^{1-n} - r_1^{1-n})r^n} \qquad (7\text{-}28)$$

The radial velocity is reevaluated from Eq. (7-17) as

$$V_r = \frac{F}{3\pi\mu d} = \frac{mV_\theta^2}{3\pi\mu \, dr} = \frac{\rho_p d^2 V_\theta^2}{18\mu r} = \frac{\rho_p(1-n)^2 Q^2 d^2}{18\mu W^2(r_2^{1-n} - r_1^{1-n})r^{2n+1}} \qquad (7\text{-}29)$$

The collection efficiency becomes

$$\eta = 1 - \exp\left[-\frac{(1-n)\rho_p Q d^2 \theta_1}{18\mu W^2(r_2^{1-n} - r_1^{1-n})r_2^n(r_2 - r_1)}\right] \qquad (7\text{-}30)$$

Finally, the angle θ_1 required for a given collection efficiency is

$$\theta_1 = -\frac{18\mu W(r_2^{1-n} - r_1^{1-n})r_2^n(r_2 - r_1)\ln(1-\eta)}{(1-n)\rho_p Q d^2} \qquad (7\text{-}31)$$

which is valid only if $n \neq 1$.

EXAMPLE 7-5 If $n = 0.5$, find η for the data of Example 7-4, using $\theta_1 = 4.29$ rad.

SOLUTION Equation (7-30) gives

$$\eta = 1 - \exp\left\{-\frac{0.5(1500)(5)(25 \times 10^{-10})(4.29)}{18(1.84 \times 10^{-5})(1)[(0.4)^{0.5} - (0.2)^{0.5}](0.4)^{0.5}(0.2)}\right\}$$

$$= 1 - e^{-5.18}$$

$$= 0.9944 \qquad Ans.$$

We see that the efficiency is somewhat impaired for this flow relative to that for true cyclone flow. ////

Collection Efficiency Using a Special Particle-Distribution Function

In the work on turbulent flow up to this point, we have assumed that the effect of turbulent eddies is to distribute the particles uniformly over the cross section. Where strong centrifugal forces are present, however, such an assumption may

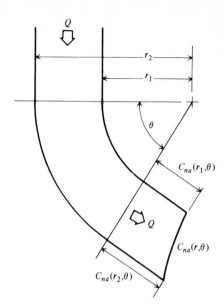

FIGURE 7-8
Variable particle density over the cross
section in cyclone flow.

be too strong. Although the exact behavior of turbulent flow is beyond the
scope of this book, it is interesting to consider what effect other particle distributions will have on the collection efficiency.

Figure 7-8 shows a general particle distribution over the cross section in
terms of a new type of concentration: the number of particles per unit area,
designated as C_{na}, which is here a function of r and θ. The total number of
particles at location θ is

$$N(\theta) = \int_{r_1}^{r_2} C_{na}(r, \theta)\, dr \qquad (7\text{-}32)$$

From the previous development

$$dr = \frac{V_{r_2}}{V_{\theta_2}} r_2\, d\theta$$

Then the fractional diminution of particles is the number in the region dr divided
by the total number, given as

$$-\frac{dN}{N} = \frac{C_{na}(r_2, \theta)\, dr}{\int_{r_1}^{r_2} C_{na}(r, \theta)\, dr} = \frac{C_{na}(r_2, \theta)}{\int_{r_1}^{r_2} C_{na}(r, \theta)\, dr} \frac{V_{r_2}}{V_{\theta_2}} r_2\, d\theta$$

Let us define a mean value of C_{na} as

$$\overline{C}_{na}(\theta) = \frac{\int_{r_1}^{r_2} C_{na}(r, \theta)\, dr}{r_2 - r_1} \qquad (7\text{-}33)$$

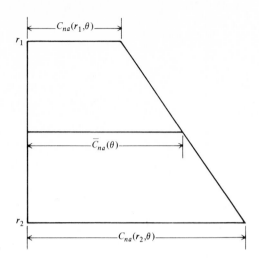

FIGURE 7-9
Linear particle-distribution function.

Then we have

$$-\frac{dN}{N} = \frac{C_{na}(r_2, \theta)}{(r_2 - r_1)\overline{C}_{na}(\theta)} \frac{V_{r_2}}{V_{\theta_2}} r_2 \, d\theta$$

Integrating this equation, and forming the expression for efficiency, as we did previously, we have

$$\eta = 1 - \exp\left[-\frac{r_2}{r_2 - r_1} \frac{V_{r_2}}{V_{\theta_2}} \int_0^\theta \frac{C_{na}(r_2, \theta)}{\overline{C}_{na}(\theta)} \, d\theta\right] \qquad (7\text{-}34)$$

Let us now choose a suitable distribution function to use in Eq. (7-34). For simplicity, we choose a function that is linear in radius. Furthermore, suppose that after a certain angle, the concentration at the inner radius r_1 will be zero; call this angle θ_2. Between $\theta = 0$ and $\theta = \theta_2$, we shall assume that the concentration at the inner radius $C_{na}(r_1, \theta)$ will vary linearly from the initial uniform concentration down to zero. This variation is illustrated in Fig. 7-9. The value of θ_2 arbitrarily will be that at which the efficiency is unity under laminar flow, as given by Eq. (7-23):

$$\theta_2 = \frac{9\mu W(r_2{}^2 - r_1{}^2) \ln r_2/r_1}{\rho_p Q d^2} \qquad (7\text{-}35)$$

For the concentration at the inner radius, we may write

$$\frac{C_{na}(r_1, \theta)}{\overline{C}_{na}(\theta)} = 1 - \frac{\theta}{\theta_2} \qquad (7\text{-}36)$$

The concentration at the outer radius is then

$$C_{na}(r_2, \theta) = 2\overline{C}_{na}(\theta) - C_{na}(r_1, \theta)$$

which may be written

$$\frac{C_{na}(r_2, \theta)}{\bar{C}_{na}(\theta)} = 1 + \frac{\theta}{\theta_2} \qquad \theta < \theta_2 \qquad (7\text{-}37)$$

For $\theta > \theta_2$, the concentration at the outer radius may be written

$$\frac{C_{na}(r_2, \theta)}{\bar{C}_{na}(\theta)} = 2 \qquad \theta > \theta_2 \qquad (7\text{-}38)$$

The integral of Eq. (7-34) may be written for the two cases as

$$\eta = 1 - \exp\left[-\frac{V_{r_2}}{V_{\theta_2}} \frac{r_2 \theta_1}{r_2 - r_1}\left(1 + \frac{\theta_1}{2\theta_2}\right)\right] \qquad 0 < \theta_1 < \theta_2 \qquad (7\text{-}39)$$

$$\eta = 1 - \exp\left[-\frac{V_{r_2}}{V_{\theta_2}} \frac{r_2 \theta_1}{r_2 - r_1}\left(2 - \frac{\theta_2}{2\theta_1}\right)\right] \qquad \theta_1 > \theta_2 \qquad (7\text{-}40)$$

Using Eqs. (7-11) and (7-17) for the tangential and radial velocities, respectively, leads to

$$\eta = 1 - \exp\left[-\frac{\rho_p Q d^2 \theta_1 (1 + \theta_1/2\theta_2)}{18\mu W r_2 (r_2 - r_1) \ln r_2/r_1}\right] \qquad 0 < \theta_1 < \theta_2 \qquad (7\text{-}41)$$

$$\eta = 1 - \exp\left[-\frac{\rho_p Q d^2 \theta_1 (2 - \theta_2/2\theta_1)}{18\mu W r_2 (r_2 - r_1) \ln r_2/r_1}\right] \qquad \theta_1 > \theta_2 \qquad (7\text{-}42)$$

Only Eq. (7-42) will be of much practical use since normally we shall expect θ_1 to exceed θ_2 for high efficiency.

Equation (7-42) may be solved for θ_1 if the efficiency is known, giving

$$\theta_1 = \frac{\theta_2}{4} - \frac{9\mu W r_2 (r_2 - r_1) \ln r_2/r_1}{\rho_p Q d^2} \ln (1 - \eta) \qquad \theta_1 > \theta_2 \qquad (7\text{-}43)$$

Using Eq. (7-35), Eqs. (7-42) and (7-43) may be cast in a simpler form:

$$\eta = 1 - \exp\left[-\left(1 + \frac{r_1}{r_2}\right)\left(\frac{\theta_1}{\theta_2} - \frac{1}{4}\right)\right] \qquad \theta_1 > \theta_2 \qquad (7\text{-}44)$$

$$\theta_1 = \theta_2 \left[\frac{1}{4} - \frac{\ln (1 - \eta)}{1 + r_1/r_2}\right] \qquad \theta_1 > \theta_2 \qquad (7\text{-}45)$$

EXAMPLE 7-6   Using the data of Example 7-4, find θ_1 to produce an efficiency of $\eta = 0.99$.

SOLUTION   Summarizing the data from Example 7-4 and previous examples:

$$\mu = 1.84 \times 10^{-5} \text{ kg/m s} \qquad r_1 = 0.2 \text{ m}$$
$$\rho_p = 1500 \text{ kg/m}^3 \qquad r_2 = 0.4 \text{ m}$$
$$Q = 5 \text{ m}^3/\text{s m} \qquad \eta = 0.99$$
$$d = 5 \times 10^{-5} \text{ m}$$

Equation (7-35) gives θ_2:

$$\theta_2 = \frac{9(1.84 \times 10^{-5})(1)(0.4^2 - 0.2^2) \ln 2}{1500(5)(25 \times 10^{-10})} = 0.735 \text{ rad} = 42°$$

Equation (7-45) gives θ_1:

$$\theta_1 = 0.735\left(0.25 - \frac{\ln 0.01}{1 + 0.5}\right) = 2.44 \text{ rad} = 140° \qquad Ans. \qquad ////$$

Collection Efficiency Using a Special Particle-Distribution Function with Modified Cyclone Flow

We may also use the modified cyclone flow in the preceding analysis. For this flow, $V_\theta = C/r^n$; for the ratio of velocities Eqs. (7-28) and (7-29) give

$$\frac{V_{r_2} r_2 \theta}{V_{\theta_2}(r_2 - r_1)} = \frac{(1 - n)\rho_p Q d^2 \theta}{18\mu W(r_2^{1-n} - r_1^{1-n})r_2^n(r_2 - r_1)} \qquad (7\text{-}46)$$

Then Eq. (7-40) gives

$$\eta = 1 - \exp\left[-\frac{(1 - n)\rho_p Q d^2 \theta_1(2 - \theta_2/2\theta_1)}{18\mu W(r_2^{1-n} - r_1^{1-n})r_2^n(r_2 - r_1)}\right] \qquad \theta_1 > \theta_2 \qquad (7\text{-}47)$$

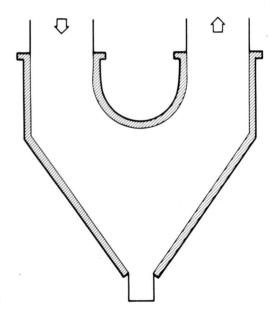

FIGURE 7-10
Centrifugal collector.

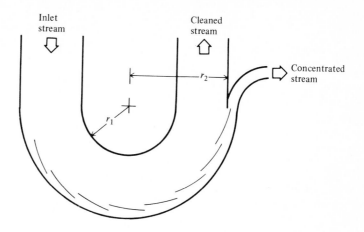

FIGURE 7-11
Centrifugal concentrator.

Solving Eq. (7-47) for θ_1, we have

$$\theta_1 = \frac{\theta_2}{4} - \frac{9\mu W (r_2^{1-n} - r_1^{1-n}) r_2^n (r_2 - r_1) \ln(1 - \eta)}{(1 - n)\rho_p Q d^2} \qquad \theta_1 > \theta_2 \qquad (7\text{-}48)$$

Notice that these equations are valid only for $\theta_1 > \theta_2$ and that θ_2 is defined the same as for regular cyclone flow.

7-4 CENTRIFUGAL COLLECTORS

The analysis of Sec. 7-3 can be simply applied to a device known as a *centrifugal collector*. A simple centrifugal collector is illustrated in Fig. 7-10. As the gas follows a more or less circular flow path, which approximates a simple cyclone flow, the particles are thrown outward by centrifugal force. When the particles enter the dead zone in the hopper beneath the flow, they will settle to the bottom. Such a device is simple to construct, has a low pressure drop, and gives substantially greater collection efficiency than a settling chamber. It can be analyzed and designed by a straightforward application of the equations and methods developed in Sec. 7-3.

A closely related device is the centrifugal concentrator, one form of which is shown in Fig. 7-11. The baffles are used to prevent particles which have been thrown to the outside from being reentrained by turbulent eddies. A suitable outer radius for use in the equations is r_2, shown in Fig. 7-11.

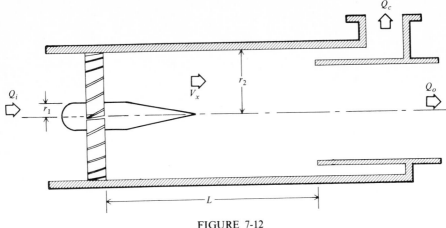

FIGURE 7-12
Straight-through cyclone concentrator with fixed blades.

7-5 STRAIGHT-THROUGH CYCLONE CONCENTRATOR

Figure 7-12 shows another practical application of cyclone flow in effecting separation of particles from the entering gas stream. This device, known as a *straight-through cyclone concentrator*, imparts a swirling motion to the entering gas stream by means of the twisted blades. The gas has both axial and tangential components of velocity in flowing through the device. The tangential velocity causes the separation, while the axial component ensures that the gas will flow through the device and exit at the other end. In the usual design, the gas will execute several complete turns in traversing the length of the separation section. At the exit, the gas near the outer wall of the tube, which contains a concentrated distribution of particles, is skimmed off and led to the dust exit. The gas near the center, which now has relatively few particles, continues on into the cleaned air duct.

To analyze the performance of the straight-through cyclone, let us first examine the vanes which impart the tangential motion to the gas. The series of vanes placed parallel to each other is called a *cascade*. Three blades of this cascade are shown in Fig. 7-13, along with the velocity diagrams of the entering and leaving gas. The gas enters the space between the vanes in the axial direction. It leaves with both an axial and a tangential component. We shall assume that the significant separating action takes place outside the core of radius r_1. The vane angle β is to be selected so that the flow leaving the vanes is cyclone flow between radii r_1 and r_2, which will require that the vane angle vary with radius. It is not imperative that the leaving flow be cyclone flow; other flow patterns could be designed for equally well. Cyclone flow is chosen for its simplicity and because the formulas already developed assume this flow distribution.

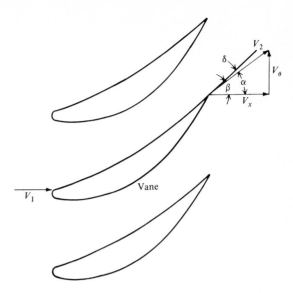

FIGURE 7-13
Vane cascade for a straight-through cyclone.

In the exit velocity diagram of Fig. 7-13, β is the blade angle, α is the exit angle, and δ is the deviation. The magnitude of the deviation is a function of the blade spacing, which is a function of radius, the turning angle, and other factors. We shall not investigate the deviation in detail but shall make a plausible assumption as to its variation with radius and exit angle. For further details of vane design, see Shepherd [1]. It can be shown that for cyclone flow downstream, the axial velocity component is uniform and essentially equal to the inlet velocity V_1. The exit angle can be made arbitrary at one value of radius, and we shall assume that the value of α is specified at the outer radius r_2. Equation (7-5) gives for cyclone flow

$$V_\theta = \frac{C}{r} = V_{\theta_2} \frac{r_2}{r} \qquad (7\text{-}49)$$

Then the exit angle is given by

$$\alpha = \tan^{-1} \frac{V_\theta}{V_x} = \tan^{-1} \frac{V_{\theta_2}}{V_1} \frac{r_2}{r} = \tan^{-1} \frac{\pi r_2{}^3 V_{\theta_2}}{rQ} \qquad (7\text{-}50)$$

which also may be written in terms of α_2 as

$$\alpha = \tan^{-1} \left(\frac{r_2}{r} \tan \alpha_2 \right) \qquad (7\text{-}51)$$

For the deviation, let us assume an empirical expression of the form

$$\delta = \delta_2 \frac{r}{r_2} + \delta_1 \frac{\alpha}{\alpha_1} \qquad (7\text{-}52)$$

in which α_1 is the exit angle at the inner radius:

$$\alpha_1 = \tan^{-1}\left(\frac{r_2}{r_1}\tan\alpha_2\right) \qquad (7\text{-}53)$$

Then the blade angle β is given by

$$\beta = \alpha + \delta = \tan^{-1}\left(\frac{r_2}{r}\tan\alpha_2\right)\left[1 + \frac{\delta_1}{\tan^{-1}(r_2\tan\alpha_2/r_1)}\right] + \delta_2\frac{r}{r_2} \qquad (7\text{-}54)$$

In order to find the total turning angle for use in the collection-efficiency equations, we shall need to estimate the effective number of times the flow executes a complete revolution in its helical motion down the cyclone. The number of revolutions will not be the same at all radii, and, in fact, at smaller radii the flow will execute more complete revolutions than at the outer radius. Thus an estimate based on the outer radius will be a conservative one. If t is the residence time of a gas molecule which traverses a streamline at the outer radius, then the particle in this time will travel an axial distance of L/V_x and a tangential distance of $r_2\theta_1/V_{\theta_2}$. Equating these values and solving for θ_1 gives

$$\theta_1 = \frac{LV_{\theta_2}}{r_2 V_x} = \frac{LV_x\tan\alpha_2}{r_2 V_x} = \frac{L}{r_2}\tan\alpha_2 \qquad (7\text{-}55)$$

This value can be used in Eqs. (7-25) and (7-26) to obtain the collection efficiency in turbulent flow. For the straight-through cyclone, the value of W is given by

$$W = \frac{2\pi L}{\theta_1} \qquad (7\text{-}55a)$$

EXAMPLE 7-7   A straight-through cyclone handles $5.0 \text{ m}^3/\text{s}$ of standard air. The cyclone diameter is 60 cm, the particle density is 3000 kg/m^3, the radius of the hub to which the blades are attached is 7.5 cm, and the cyclone effective length is 2.0 m. Using an exit angle of $45°$ from the blades at the outer radius, (a) compute and plot the blade angle as a function of radius, and (b) compute the collection efficiency of the cyclone for 30-μm particles. For the deviation, assume $\delta_1 = 3°$ and $\delta_2 = 4°$.

SOLUTION   (a) From the data

$$\delta_1 = \frac{3}{57.3} = 0.0524 \qquad \delta_2 = 0.0698 \qquad \tan\alpha_2 = 1.0$$

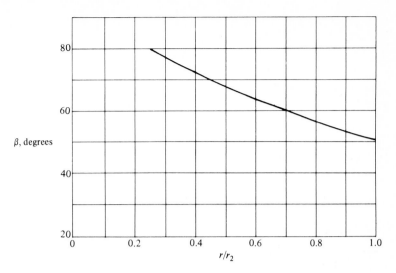

FIGURE 7-14
Vane angle for Example 7-7.

Then Eq. (7-54) gives for the vane angle

$$\beta = \left[1 + \frac{0.0524}{\tan^{-1}(30/7.5)}\right] \tan^{-1} \frac{1}{r/r_2} + 0.0698 \frac{r}{r_2}$$

$$= 1.0395 \tan^{-1} \frac{1}{r/r_2} + 0.0698 \frac{r}{r_2}$$

which is plotted in Fig. 7-14.

(b) Equation (7-25) gives for the efficiency

$$\eta = 1 - \exp\left[-\frac{3000(5.0)(30 \times 10^{-6})^2 \theta_1}{18(1.84 \times 10^{-5})(0.30)(0.225)(\ln 30/7.5)W}\right]$$

$$= 1 - \exp\left(-\frac{0.4356\theta_1}{W}\right)$$

Equation (7-55) gives for θ_1

$$\theta_1 = \frac{L}{r_2} \tan \alpha_2 = \frac{2.0(1)}{0.3} = 6.67 \text{ rad}$$

$$W = \frac{2\pi L}{\theta_1} = \frac{2\pi(2)}{6.67} = 1.88$$

Then the efficiency becomes

$$\eta = 1 - e^{-1.545} = 0.787 \qquad Ans. \qquad ////$$

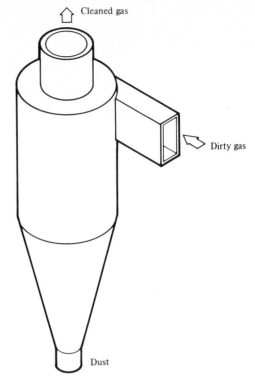

Cleaned gas

Dirty gas

Dust

FIGURE 7-15
Isometric view of a reverse-flow cyclone
collector.

7-6 REVERSE-FLOW CYCLONE COLLECTOR

Figure 7-15 shows an isometric view of a reverse-flow cyclone collector. Figure
7-16 shows a top view and a vertical cross-sectional view of the cyclone. The
dirty gas flows tangentially into the cyclone at the top. This imparts a swirl, or
cyclone, flow to the gas as it goes down near the outer radius and then back up
in the center core and out the discharge pipe at the top. The direction of swirl
(counterclockwise in Figs. 7-15 and 7-16) remains the same in both the outer annular region of flow and the inner core. The particles are slung to the outside, move
down the wall, and collect in the dust hopper at the bottom. The reverse-flow
cyclone can also be used as a concentrator, in which case some of the gas is
bled off from the main stream out the dust opening at the bottom of the cyclone,
carrying the collected dust with it. The cyclone may be mounted in virtually
any orientation since the effect of gravity on the centrifugal separation mechanism
is negligible. However, in the case of a collector, the dust opening must be at
the bottom of the cyclone in order for the dust to settle into the collection
hopper by gravity.

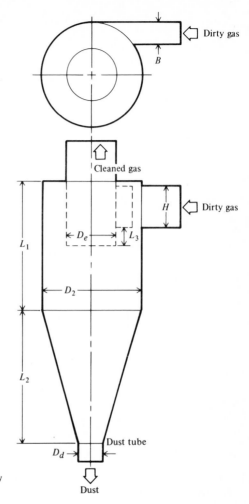

FIGURE 7-16
Front and top views of a reverse-flow
cyclone collector.

In addition to the tangential entry shown in Figs. 7-15 and 7-16, several other types of entries are used in commercial practice. Some of these entries are shown in Fig. 7-17: Figure 7-17a shows a helical entry, Fig. 7-17b shows a wrap-around entry, and Fig. 7-17c shows an axial entry. The axial entry is of particular importance, especially for small cyclones which will be used as components in a multiple-cyclone installation. Strauss [2, sec. 6-9] and Pazar [3] give further constructional details for various cyclones.

In preparation for analyzing the efficiency of a reverse-flow cyclone, let us consider the device shown in Fig. 7-18 which consists of a tube with a helical partition that induces cyclone flow in the outer annular region. The inner core can have either reverse flow or no flow, which is unimportant for the present

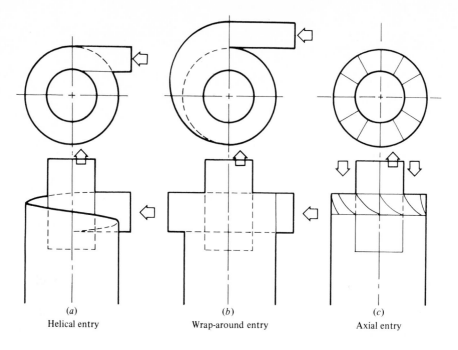

(a)
Helical entry

(b)
Wrap-around entry

(c)
Axial entry

FIGURE 7-17
Different types of entries to reverse-flow cyclone collectors.

treatment. Equations (7-25) and (7-26) may be used if true cyclone flow is present; these equations become

$$\eta = 1 - \exp\left[-\frac{\rho_p Q d^2 \theta_1}{18\mu W r_2 (r_2 - r_1)\ln r_2/r_1}\right] \quad (7\text{-}56)$$

$$\theta_1 = -\frac{18\mu W r_2 (r_2 - r_1)(\ln r_2/r_1)\ln(1-\eta)}{\rho_p Q d^2} \quad (7\text{-}57)$$

For modified cyclone flow, as given by Eqs. (7-30) and (7-31), the equations become

$$\eta = 1 - \exp\left[-\frac{(1-n)\rho_p Q d^2 \theta_1}{18\mu W (r_2^{1-n} - r_1^{1-n})r_2^{n}(r_2 - r_1)}\right] \quad (7\text{-}58)$$

$$\theta_1 = -\frac{18\mu W (r_2^{1-n} - r_1^{1-n})r_2^{n}(r_2 - r_1)\ln(1-\eta)}{(1-n)\rho_p Q d^2} \quad (7\text{-}59)$$

The angle θ_1 is given as $2\pi N$, where N is the number of turns which the gas executes in traversing the length of the cyclone.

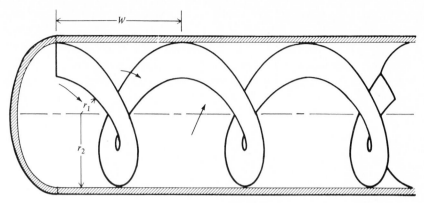

FIGURE 7-18
Cyclone flow induced by a helical partition.

EXAMPLE 7-8 How long must a cyclone be to collect particles of 10 μm diameter if the flow rate is 1.0 m^3/s of standard air and if $r_2 = 10$ cm, $r_1 = 4$ cm, and $W = 15$ cm? Use standard cyclone flow, with $\eta = 0.99$ and $\rho_p = 1000$ kg/m^3.

SOLUTION Using Eq. (7-57), we have

$$\theta_1 = -\frac{18(1.84 \times 10^{-5})(0.15)(0.1)(0.06)(\ln \tfrac{10}{4})(\ln 0.01)}{1000(1.0 \times 10^{-5})^2}$$

$$= 12.58 \text{ rad}$$

$$= 2.0 \text{ turns}$$

This gives a length of 30 cm. *Ans.* ////

In the real cyclone, the helical partition is absent. Even so, the flow will follow a helical pattern. Thus, the preceding equations may be used to predict the collection efficiency if a suitable approximate value for W can be obtained. Such a value is difficult to establish analytically; currently the best procedure available is to infer its value from the flow pattern introduced by the entry configuration. The standard entry of Figs. 7-15 and 7-16, the helical entry of Fig. 7-17a, and the wrap-around entry of Fig. 7-17b indicate a value of W equal to the height of the entry duct. For the axial entry of Fig. 7-17c, the value of W to use is obtained by the following formula:

$$W = 2\pi r_2 \cot \alpha_2 \qquad (7\text{-}60)$$

where α_2 is the exit angle with respect to the axial direction shown in Fig. 7-13, measured at the outer radius.

Although the tangential velocity distribution in a reverse-flow cyclone should theoretically be that of cyclone flow, actual measurements do not bear this out. The actual velocity distribution follows rather closely the modified cyclone-flow distribution of Eq. (7-27). The value of the exponent n has been variously determined to lie in the range from 0.5 to 0.7 for gas flow. Strauss [2, p. 175] discusses this variation and suggests that the value $n = 0.5$ be used for calculation. Following this suggestion, Eqs. (7-58) and (7-59) become

$$\eta = 1 - \exp\left[-\frac{\rho_p Q d^2 \theta_1}{36 \mu W (r_2 - \sqrt{r_1 r_2})(r_2 - r_1)}\right] \qquad (7\text{-}61)$$

$$\theta_1 = -\frac{36 \mu W (r_2 - \sqrt{r_1 r_2})(r_2 - r_1) \ln (1 - \eta)}{\rho_p Q d^2} \qquad (7\text{-}62)$$

EXAMPLE 7-9   A reverse-flow cyclone with axial entry has the following dimensions: $r_1 = 5$ cm, $r_2 = 10$ cm, $L = 1.0$ m, and the exit angle is $75°$. A flow of 1.0 m^3/s of standard air carrying particles with a density of 1000 kg/m^3 passes through the cyclone. Determine the collection efficiency as a function of particle diameter.

SOLUTION   Equation (7-60) gives the helical pitch W as

$$W = 2\pi r_2 \cot \alpha = 2\pi(10)(\cot 75°) = 16.8 \text{ cm}$$

Then the total turning angle θ_1 becomes

$$\theta_1 = \frac{2\pi L}{W} = \frac{2\pi(1.0)}{0.168} = 37.4 \text{ rad}$$

Equation (7-61) then leads to

$$\eta = 1 - \exp\left\{-\frac{1000(1.0)(37.4 d^2)}{36(1.84 \times 10^{-5})(0.168)[0.1 - \sqrt{0.05(0.1)}](0.05)}\right\}$$

$$= 1 - e^{-0.2294 \times 10^{12} d^2}$$

The following values of efficiency are then computed from the previous equation:

d, μm	η
0.5	0.056
0.75	0.121
1.0	0.205
1.5	0.403
2.0	0.601
3.0	0.873
4.0	0.975
5.0	0.997

The tangential velocity can be computed as

$$V_{\theta_{ave}} = \frac{Q}{W(r_2 - r_1)} = \frac{1.0}{0.168(0.05)} = 119 \text{ m/s} \qquad ////$$

The Cyclone of Standard Proportions

The remainder of this section will be devoted to certain practical and empirical details of cyclone usage. There is a set of standard, or conventional, cyclone proportions which tend to be approximately followed in the majority of cases. These are listed in Danielson [4], based on the dimensions shown in Fig. 7-16. In terms of the cyclone diameter D_2, the standard ratios are listed in Table 7-1.

Lapple [5] presents an empirical expression for the collection efficiency of a cyclone separator. This efficiency expression is based on the particle size for which the collection efficiency is 0.5, designated here as $d_{0.5}$ and referred to as *particle cut size* by Lapple [5]. The equation for $d_{0.5}$ is

$$d_{0.5} = \sqrt{\frac{9\mu B^2 H}{\rho_p Q \theta_1}} \qquad (7\text{-}63)$$

The efficiency curve is shown as a function of the ratio $d/d_{0.5}$ in Fig. 7-19. The value of θ in Eq. (7-63) represents the effective number of turns which the gas makes in traversing the cyclone. Although this value varies, in the absence of exact performance data, a reasonable estimate is given by

$$\theta_1 = 2\pi \frac{L_1 + L_2/2}{H} = \frac{\pi}{H}(2L_1 + L_2) \qquad (7\text{-}64)$$

Equation (7-63) applies to tangential, helical, or wrap-around entries; for the axial entry, the following equation may be used:

$$d_{0.5} = \sqrt{\frac{27\pi\mu B^3}{\rho_p Q \theta_1 \tan \alpha}} \qquad (7\text{-}65)$$

Equation (7-64) gives $\theta_1 = 12\pi$ for a cyclone of standard proportions.

**Table 7-1   STANDARD CYCLONE PRO-
PORTIONS**

Length of cylinder	$L_1 = 2D_2$
Length of cone	$L_2 = 2D_2$
Diameter of exit	$D_e = \frac{1}{2}D_2$
Height of entrance	$H = \frac{1}{2}D_2$
Width of entrance	$B = \frac{1}{4}D_2$
Diameter of dust exit	$D_d = \frac{1}{4}D_2$
Length of exit duct	$L_3 = \frac{1}{8}D_2$

SOURCE: Danielson [4].

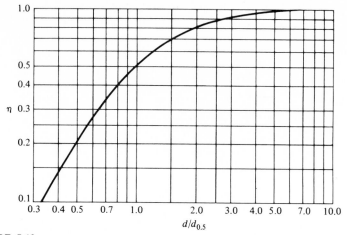

FIGURE 7-19
Empirical collection efficiency as a function particle diameter for reverse-flow collectors. (*From C. E. Lapple, Processes Use Many Collection Types,* Chem. Eng., *vol. 58, no. 5, p. 147, May 1951. Used by permission.*)

EXAMPLE 7-10 A cyclone collector having standard proportions with an outside diameter $D_2 = 1.0$ m handles 4.0 m^3/s of standard air carrying particles with a density of 2365 kg/m^3. Using $\theta = 12\pi$, determine the collection efficiency as a function of particle diameter using Eqs. (7-56), (7-61), and (7-63) in conjunction with Fig. 7-20.

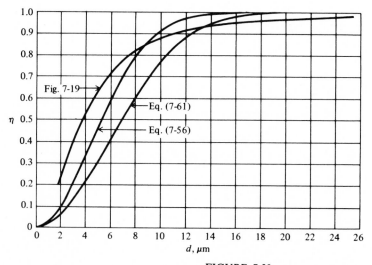

FIGURE 7-20
Efficiency curves for Example 7-10.

SOLUTION   Using $W = H = D_2/2 = 0.5\,\text{m}$, $B = D_2/4 = 0.25\,\text{m}$, and the other data given, Eqs. (7-56) and (7-61) give, respectively,

$$\eta = 1 - \exp\left[-\frac{2365(4.0)(37.7d^2)}{18(1.84 \times 10^{-5})(0.5)(0.5)(0.25)(0.693)}\right]$$

$$= 1 - e^{-2.486 \times 10^{10}d^2}$$

$$\eta = 1 - \exp\left[-\frac{2365(4.0)(37.7d^2)}{36(1.84 \times 10^{-5})(0.5)(0.5 - \sqrt{0.125})(0.5)}\right]$$

$$= 1 - e^{-1.471 \times 10^{10}d^2}$$

For the cut diameter $d_{0.5}$ Eq. (7-63) gives

$$d_{0.5} = \sqrt{\frac{9(1.84 \times 10^{-5})(0.0625)(0.5)}{2365(4.0)(37.7)}} = 3.80\ \mu m$$

The empirical efficiency is then obtained from Fig. 7-20. The numerical values of efficiency obtained by these three methods are plotted in Fig. 7-20.   ////

7-7 MULTIPLE CYCLONES

Equation (7-56) and the examples have shown that the efficiency of a cyclone increases as the diameter of the cyclone is reduced even if the tangential velocity remains constant. To achieve higher efficiencies dictates the use of smaller cyclones. However, the pressure drop increases as the tangential velocity increases; keeping the velocity constant then serves to keep the pumping power reasonable. But if the cyclone is made smaller while the tangential velocity remains about the same, the flow which the cyclone can handle is reduced by the square of the cyclone diameter, suggesting that multiple cyclone collectors or concentrators be used. These consist of a large number of small cyclones placed in parallel; Fig. 7-21 shows such an arrangement of reverse-flow cyclones.

In the case of reverse-flow cyclones care must be taken to ensure that the pressure drop across each cyclone is just about the same. If not, back flow may occur through one or more of the cyclones, undoing all that the remaining cyclones have accomplished. This problem does not occur with multiple straight-through cyclones, though these have the disadvantage that they can act only as concentrators. Figure 7-22 shows a multiple concentrator using straight-through cyclones. The reverse-flow collector of Fig. 7-21 can be easily modified to act as a concentrator also.

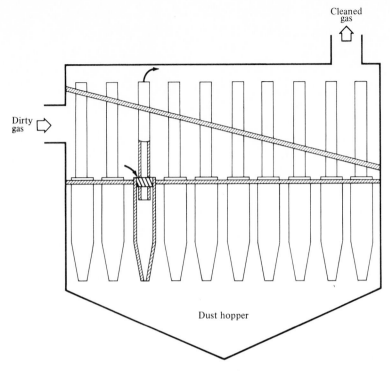

FIGURE 7-21
Multiple reverse-flow cyclone collector.

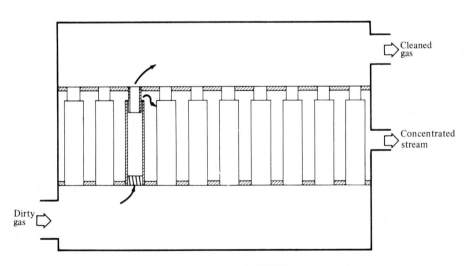

FIGURE 7-22
Straight-through multiple cyclone concentrator.

7-8 PRESSURE DROP AND POWER REQUIREMENTS

A major factor in the evaluation of cyclone design and performance is the pressure drop undergone by the gas in traversing the cyclone. The power which must be expended somewhere in the duct system to overcome this pressure drop is given by

$$\dot{W} = Q\,\Delta P \qquad (7\text{-}66)$$

The prediction of pressure drop in a cyclone not yet tested is at best only an approximation to the actual value. Thus, a performance test of the actual cyclone or of a model is required to determine the pressure drop accurately. Even so, it may still be useful to give a numerical prediction, as inaccurate as it may be for a given cyclone configuration.

An empirical method given by Caplan [6], based on the work of Alexander [7], is chosen for presentation here. The pressure drop is given as

$$\Delta P = C\,\frac{BH}{D_e^2}\frac{\rho V_1^2}{2} \qquad (7\text{-}67)$$

in which V_1 is the inlet velocity to the cyclone. This equation may be written in terms of the flow rate as

$$\Delta P = C\,\frac{\rho Q^2}{2D_e^2 BH} \qquad (7\text{-}68)$$

The constant C is given by

$$C = 4.62\,\frac{D_e}{D_2}\left\{\left[\left(\frac{D_2}{D_e}\right)^{2n}-1\right]\frac{1-n}{n}+f\left(\frac{D_2}{D_e}\right)^{2n}\right\} \qquad (7\text{-}69)$$

where f is given in terms of n as

n	f
0	1.90
0.2	1.94
0.4	2.04
0.6	2.21
0.8	2.40

Figure 7-23 shows the value of n for various cyclone diameters and gas temperatures. Figure 7-24 is a plot of Eq. (7-69) from which the value of C can be read immediately.

Substituting Eq. (7-68) into Eq. (7-66) gives for the power

$$\dot{W} = C\,\frac{\rho Q^3}{2D_e^2 BH} \qquad (7\text{-}70)$$

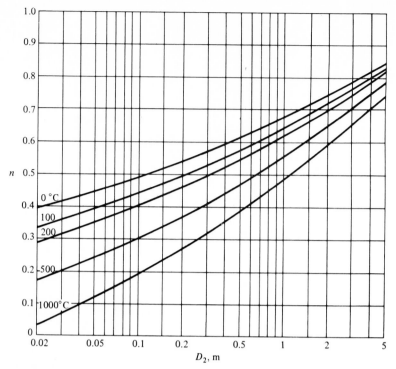

FIGURE 7-23
Constant n for Eq. (7-69). (*Plotted from data and equations given by R. M. Alexander, Fundamentals of Cyclone Design and Operation,* Proc. Australas. Inst. Min. Metall., *vol. 152–153, pp. 202–228, 1949.*)

Equations (7-68) and (7-70) can be rewritten for a cyclone of standard proportions:

$$\Delta P = \frac{16 C \rho Q^2}{D_2{}^4} \qquad (7\text{-}71)$$

$$\dot{W} = \frac{16 C \rho Q^3}{D_2{}^4} \qquad (7\text{-}72)$$

EXAMPLE 7-11 A cyclone treats 100 m³/s of air at 300°C. The cyclone is of standard proportions and has an outer diameter D_2 of 4.0 m. Compute the pressure drop and power required.

SOLUTION   Figure 2-8 gives the density of the air as 0.62 kg/m³. Figure 7-23 gives the value of n as 0.79; Fig. 7-24 then gives $C = 17.5$. Equation (7-71) gives the pressure loss as

$$\Delta P = \frac{16(17.5)(0.6)(100)^2}{(4.0)^4} = 6563 \text{ N/m}^2 \qquad Ans.$$

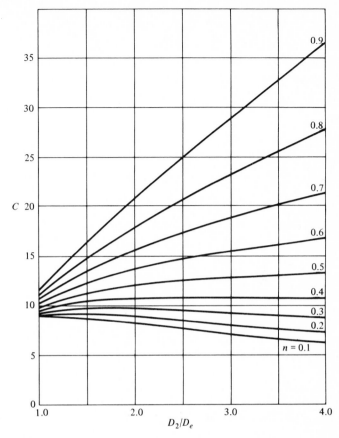

FIGURE 7-24
Constant C for Eqs. (7-68) and (7-69). (*Plotted from data and equations given by R. M. Alexander, Fundamentals of Cyclone Design and Operation,* Proc. Australas. Inst. Min. Metall., *vol. 152–153, pp. 202–28, 1949.*)

Then Eq. (7-66) gives for the power

$$\dot{W} = Q\,\Delta P = 100(6563) = 656.3 \text{ kW} \qquad Ans. \qquad ////$$

Equations (7-67) to (7-69), as presented in Ref. [6], are not particularly intended for use with cyclones having an axial entry. Despite this fact, it is suggested here that Eqs. (7-71) and (7-72) be used for such cyclones in the absence of performance data. Of course, it is recommended that true performance data be obtained wherever possible. Since multiple cyclones are usually of the axial-entry type, it is important to have a means of predicting the pressure drop through such a system.

In a multiple cyclone, let us use Q to represent the total flow rate through the unit. Then Q/N will be the flow rate through each individual cyclone if N is the total number of cyclones included. Equations (7-71) and (7-72) may then be written as

$$\Delta P = \frac{16C\rho Q^2}{N^2 D_2{}^4} \qquad (7\text{-}73)$$

$$\dot{W} = \frac{16C\rho Q^3}{N^2 D_2{}^4} \qquad (7\text{-}74)$$

Equation (7-74) suggests that the power will be minimized, and, in fact, it becomes zero if the number of cyclones is increased without limit. However, increasing the number of cyclones is not practical since cyclones that are too small tend to lose collection efficiency and the total initial cost will increase when the cyclone becomes too small. Nevertheless, the use of multiple cyclones offers real advantages. Typically, in such installations the cyclone diameter ranges from 5 to 30 cm (2 to 12 in).

Cost Analysis for Multiple Cyclones

It is possible to estimate the most economical operating condition for a multiple cyclone. To do this, let us assume that the cyclone has fixed proportions, except for its length, and that the size of the cyclone can be specified by its length and outer radius r_2. The number of cyclones can be varied independently. Let us assume further that the cost of each cyclone is proportional to its length and radius, with an additional component of cost which is proportional to the total number present. We may then write for the total annual cost

$$S = K_1 N r_2 L + K_2 \dot{W} t + K_3 N + K_4 \qquad (7\text{-}75)$$

Here, K_1 is determined as the annual capitalization rate times the cost of a representative cyclone divided by its length and radius; K_2 is the power cost per hour and t is the number of hours per year the unit will be in operation; K_3 represents that portion of the cost of assembly, installation, and maintenance which can be assigned as proportional to the number of individual cyclones present; and K_4 provides for the remainder of the cost, converted to an annual basis.

In Eq. (7-57), let $\theta_1 = 2\pi L/W$ and note that r_1/r_2 and W/r_2 are fixed by virtue of the fixed proportions of the cyclone. Then this equation may be written as

$$\frac{L}{N r_2{}^4} = -\frac{18\mu(W/r_2)^2(1 - r_1/r_2)(\ln r_2/r_1)\ln(1 - \eta)}{2\pi \rho_p Q d^2} \qquad (7\text{-}76)$$

Equation (7-74) may be written as

$$\dot{W} = \frac{C\rho Q^3}{N^2 r_2^4} \qquad (7\text{-}77)$$

And Eq. (7-75) may be written as

$$S = K_1' N^2 r_2^5 + \frac{K_2'}{N^2 r_2^4} + K_3 N + K_4 \qquad (7\text{-}78)$$

in which

$$K_1' = K_1 \frac{L}{N r_2^4} \qquad (7\text{-}79)$$

$$K_2' = K_2 t C\rho Q^3 \qquad (7\text{-}80)$$

To minimize the cost, take the partial derivatives of Eq. (7-78) with respect to N and r; these give, respectively,

$$\frac{\partial S}{\partial N} = 2K_1' N r_2^5 - \frac{2K_2'}{N^3 r_2^4} + K_3 \qquad (7\text{-}81)$$

$$\frac{\partial S}{\partial r_2} = 5K_1' N^2 r_2^4 - \frac{4K_2'}{N^2 r_2^5} \qquad (7\text{-}82)$$

When these derivatives are set equal to zero and the resulting equations solved for r_2 and N, we have

$$r_2 = \frac{1.313 K_3^{4/11}}{K_1'^{3/11} K_2'^{1/11}} \qquad (7\text{-}83)$$

$$\dot{N} = \frac{0.5125 K_1'^{4/11} K_2'^{5/11}}{K_3^{9/11}} \qquad (7\text{-}84)$$

Equation (7-76) then gives L; substituting for r_2 and N into Eq. (7-78) gives the annual cost.

EXAMPLE 7-12 A multiple cyclone is to handle 100 m³/s of standard air carrying particles of 1400 kg/m³ density. The cyclone must produce a collection efficiency of 0.99 for 8-μm particles. The proportions of the cyclone are $r_1/r_2 = 0.5$ and $W/r_2 = 1.0$. The cost of a cyclone having $r_2 = 5.0$ cm and $L = 30$ cm is \$5 when mass-produced. The power cost is \$.009/kWh, t is 3000 h/year, and the capitalization rate is 0.12/year. Assume $K_3 = \$5$/unit and K_4 is \$2000. For optimum cost operation, compute N, r_2, L, and the total annual cost.

SOLUTION Equation (7-76) gives for L/Nr_2^4

$$\frac{L}{Nr_2^4} = -\frac{18(1.84 \times 10^{-5})(1)^2(0.5)(\ln 2)(\ln 0.01)}{2\pi(1400)(100)(64 \times 10^{-12})} = 9.39$$

The constant K_1 is computed as

$$K_1 = \frac{0.12(5.00)}{0.30(0.05)} = 40$$

Equation (7-79) gives $K_1' = 40(9.39) = 376$. Using a diameter of 29 cm (found from a previous trial to be almost right), Fig. 7-23 gives $n = 0.56$ and Fig. 7-24 then gives $C = 13$. K_2' is found from Eq. (7-80) as

$$K_2' = 9 \times 10^{-6}(3000)(13)(1.185)(10)^6 = 4.16 \times 10^5$$

Equation (7-83) gives for r_2

$$r_2 = \frac{1.313(5)^{4/11}}{376^{3/11}(4.16 \times 10^5)^{1/11}} = 0.144 \text{ m} \qquad Ans.$$

which gives $D_2 = 29$ cm, as assumed.

Equation (7-84) gives for N

$$N = \frac{0.5125(376)^{4/11}(4.16 \times 10^5)^{5/11}}{5^{9/11}} = 425 \qquad Ans.$$

L is found from the preceding result as

$$L = 9.39Nr_2{}^4 = 9.39(425)(0.144)^4 = 1.72 \text{ m} \qquad Ans.$$

The power is given by Eq. (7-77) as

$$\dot{W} = \frac{13(1.185)(10)^6}{(425)^2(0.144)^4} = 198 \text{ kW}$$

Equation (7-75) gives the annual cost as

$$S = 40(425)(0.144)(1.72) + 9 \times 10^{-6}(198,000)(3000) + 5(425) + 2000$$
$$= \$13,682 \qquad Ans. \qquad\qquad ////$$

REFERENCES

1 Shepherd, D. G.: "Principles of Turbomachinery," The Macmillan Company, New York, 1956.
2 Strauss, W.: "Industrial Gas Cleaning," Pergamon Press, New York, 1966.
3 Pazar, C.: "Air and Gas Cleanup Equipment," chap. 2, Noyes Data Corporation, Park Ridge, N.J., 1970.
4 Danielson, J. A., ed.: Air Pollution Engineering Manual, p. 92, *Public Health Service Publ. No.* 999-AP-40, U.S. Government Printing Office, 1967.
5 Lapple, C. E.: Processes Use Many Collection Types, *Chem. Eng.*, vol. 58, no. 5, pp. 144–51, May 1951.
6 Caplan, K. J.: Source Control by Centrifugal Force and Gravity, in A. C. Stern, ed., "Air Pollution," vol. 3, chap. 43, pp. 366–377, Academic Press, Inc., New York, 1968.
7 Alexander, R. McK.: Fundamentals of Cyclone Design and Operation, *Proc. Australas. Inst. Min. Metall.*, vol. 152/3, pp. 202–228, 1949.

PROBLEMS

7-1 Standard air is flowing in a U bend at the rate of 2.25 m^3/s · m width. The inner and outer radii are 10 and 30 cm, respectively. Determine the x and y components of velocity at a point of radius 20 cm located 30° downstream of the midsection of the U.

7-2 A stream of 15 m^3/s of standard air flows around a bend in laminar flow. The radii are 0.5 and 1.0 m, and the duct width is 2.0 m. If the maximum angle is 180°, calculate and plot efficiency as a function of particle size. Assume the particle density is 2000 kg/m^3.

7-3 In Prob. 7-2, what angle of turn is necessary to collect particles 10 μm in diameter with perfect efficiency? If $r_2 - r_1$ remains the same but r_2 is varied, plot the angle θ_1 as a function of r_2 in order to collect 10-μm particles perfectly.

7-4 Repeat Prob. 7-2 for turbulent flow.

7-5 Repeat Prob. 7-3 but for turbulent flow and for an efficiency of 0.99.

7-6 In Example 7-4, determine the overall efficiencies based on surface area and on number of particles.

7-7 A stream of standard air enters a cyclone having a modified cyclone flow with $n = 0.75$. The inlet velocity is 25 m/s, $r_1 = 1.0$ m, and $r_2 = 1.7$ m. What angle of turn is required to achieve a collection efficiency of 0.99 for 20-μm particles having a density of 1800 kg/m^3? Assume $W = 1.0$ m.

7-8 Repeat Prob. 7-7 using the analysis of Eq. (7-47).

7-9 Repeat Prob. 7-2 for turbulent flow using the analysis of Eq. (7-44).

7-10 Repeat Prob. 7-3 for turbulent flow and for an efficiency of 0.99 using the analysis of Eq. (7-45).

7-11 Consider a cyclone flow having an inlet velocity of 50 m/s, an angle of turn of 20π, and inner and outer radii of 2.5 and 5.0 cm, respectively. Assume standard air, turbulent flow following the analysis of Eq. (7-44), a particle density of 1500 kg/m^3, and a particle distribution given by

$$d_m = 20 \ \mu m \qquad \bar{u} = \ln d_m = -10.81$$

$$d_l = 1 \ \mu m \qquad u_l = \ln d_l = -13.81$$

$$\sigma_m = 2.0 \qquad \sigma_u = \ln \sigma_m = 0.693$$

Compute and plot the collection efficiency as a function of particle diameter. Compute the overall collection efficiencies based on mass, surface area, and number of particles.

7-12 Design a centrifugal collector similar to that shown in Fig. 7-10 to remove 99 percent of particles of 30 μm diameter and 2000 kg/m^3 density from a stream of 25 m^3/s of standard air. Determine suitable values for the major dimensions of the collector.

7-13 Design a concentrator similar to that of Fig. 7-11. It is to have an efficiency of 0.99 for particles of 25 μm diameter and 1200 kg/m^3 density. Assume standard air, and assume that the concentrated stream has 0.1 of the flow rate of the inlet stream. The flow rate at the inlet is 50 m^3/s. Determine r_1 and r_2, and design the baffles, paying particular attention to the radial spacing between adjacent baffles.

7-14 Use Eqs. (7-51) and (7-53) in Eq. (7-52) for the deviation. Obtain an expression for the radius at which the deviation is a maximum, and obtain an expression for this maximum deviation.

7-15 Derive an equation similar to Eq. (7-54) if the tangential component of velocity is assumed to vary linearly with radius, being maximum at the outer radius. The axial component will no longer be uniform in this case.

7-16 Repeat Example 7-7 if the exit angle at the outer radius is to be 75°.

7-17 In Example 7-7 what must be the exit angle if the collection efficiency is to be 0.99 for 30-μm particles, other data remaining the same?

7-18 A straight-through cyclone is to be designed to handle 100 m^3/s of standard air carrying a particulate substance of 1750 kg/m^3 density. The collection efficiency must be 0.99 for 20-μm particles, and the axial velocity component is to be 20 m/s. The gas exit angle at the outer radius is to be 75°. Determine the diameter and length of the cyclone.

7-19 A straight-through cyclone concentrator has a flow rate of 0.5 m^3/s of standard air with an axial velocity of 50 m/s. The particle density is 2000 kg/m^3. If the angle of flow leaving the vane at the outer radius is 70° and the length of the cyclone is 1.0 m, plot collection efficiency as a function of particle diameter.

7-20 Derive an expression for $d_{0.5}$ from Eq. (7-56) for a cyclone of standard proportions. Compare with Eq. (7-63).

7-21 Derive an expression for $d_{0.5}$ from Eq. (7-61) for a cyclone of standard proportions.

7-22 A cyclone collector with a tangential entry handles 10 m^3/s of standard air containing dust particles of density equal to 900 kg/m^3. The cyclone outer diameter is 0.5 m, its inner diameter is 0.28 m, and its length is 3.0 m. Plot collection efficiency as a function of particle diameter.

7-23 A reverse-flow cyclone collector has standard proportions with an outer diameter of 2.0 m. It handles 100 m^3/s of standard air with particles whose density is 1600 kg/m^3. Determine its efficiency for particles of 10 μm diameter using Eqs. (7-56) and (7-61), and Fig. 7-19. Repeat for particle diameters of 20, 30, and 50 μm.

7-24 Design a reverse-flow cyclone having standard proportions to handle 1000 m^3/s of standard air containing particles whose density is 1050 kg/m^3 and to provide a collection efficiency of 0.99 for a particle size of 25 μm.

7-25 An axial-entry, reverse-flow cyclone collector has standard proportions and a gas-entry angle of 75°. The outer diameter is 35 cm, and it handles 3.0 m^3/s of standard air containing particles whose density is 2800 kg/m^3. What collection efficiency is expected for 10-μm particles?

7-26 Design an axial-entry, reverse-flow cyclone having standard proportions to provide 0.99 collection efficiency for 5.0-μm particles of a dust whose density is 1200 kg/m^3 in a stream of 0.75 m^3/s of standard air.

7-27 A stream of air at 300°C and atmospheric pressure contains particles whose density is 1500 kg/m^3; it must pass through a collector whose efficiency is 0.99 for 2.0-μm particles. Design an axial-entry, reverse-flow cyclone to achieve this. What is the flow rate through the cyclone which you have designed?

7-28 Repeat Prob. 7-23 for the case of an axial entry with a gas angle of 70°.

7-29 Design a multiple cyclone collector to treat 100 m^3/s of standard air carrying particles of 1500 kg/m^3 density. It must provide 0.99 collection efficiency for 10-μm particles. Determine the dimensions and the entry angle for each individual cyclone and the number of cyclones required. What will be the approximate size of the collector?

7-30 Repeat Prob. 7-29 for a concentrator using straight-through cyclones.

7-31 Derive Eq. (7-76).

7-32 Derive Eqs. (7-81) and (7-82).

7-33 Derive Eqs. (7-83) and (7-84).

7-34 Determine the pressure drop and power required for the straight-through cyclone of Prob. 7-18. Assume the loss will be the same as for a reverse-flow cyclone having the same outer diameter and flow rate.

7-35 Compute the pressure loss and power requirement for the cyclone of Prob. 7-22.

7-36 Compute the pressure loss and power requirement for the cyclone of Prob. 7-23.

7-37 Compute the pressure loss and power requirement for the cyclone of Prob. 7-24.

7-38 Compute the pressure loss and power requirement for the cyclone of Prob. 7-25.

7-39 Compute the pressure loss and power requirement for the cyclone of Prob. 7-26.

7-40 Compute the pressure loss and power requirement for the cyclone of Prob. 7-27.

7-41 Compute the pressure loss and power requirement for the cyclone of Prob. 7-29.

7-42 A multiple cyclone is to be used to provide 0.99 efficiency for a stream of standard air flowing at the rate of 750 m^3/s carrying particles having a density of 800 kg/m^3 and a diameter of 5.0 μm. Determine the optimum cyclone dimensions and the total annual cost for a capitalization rate of 0.15, power charge of \$.01/kWh, $r_1/r_2 = 0.5$, $W/r_2 = 0.75$, 2500 h/year of operation, $K_3 = \$15$/unit, and $K_4 = \$30,000$. A single unit having $D_2 = 30$ cm and $L = 1.0$ m costs \$35.

7-43 Design a reverse-flow multiple cyclone collector to provide 0.99 efficiency for collection of 1.0-μm particles of dust having a density of 900 kg/m^3 in a flow of 2000 m^3/s of air at 300°C. Design for minimum cost and determine the dimensions of each cyclone, number of cyclones, power cost per year, and total cost per year. Use the proportions and cost data given in Prob. 7-42.

7-44 Consider the cyclone of Prob. 7-25, and assume that the dust has a size distribution specified by $d_m = 15$ μm, $\sigma_m = 1.5$, and $d_l = 2$ μm. Determine the overall collection efficiency based on mass.

7-45 Repeat Prob. 7-44, determining the overall collection efficiency based on number of particles.

7-46 Using the result of Prob. 7-43, determine the overall collection efficiency based on number of particles if the dust has a size distribution specified by $d_m = 10$ μm, $\sigma_m = 1.0$, and $d_l = 0.2$ μm.

8

ELECTROSTATIC PRECIPITATORS

The electrostatic precipitator is one of the most widely used collection devices for particulates. It has many advantages: Its range of size is enormous; it is used on the largest fossil-fuel-fired electric generating plants and in small household air-conditioning systems as well. It is versatile enough to provide virtually complete collection of particles of many substances, both solids and liquids. It can operate at high temperatures and pressures, and its power requirements are low. For these reasons, the electrostatic precipitator is often the preferred method of collection where high efficiency is required with small particles.

However, the precipitator has some serious drawbacks: For one thing, there are certain dusts which can be collected by a precipitator only with great difficulty. Also, the precipitator cannot collect gaseous pollutants. In fact, its use with particulates may require that certain gaseous pollutants be present or that other difficulties be imposed in order to achieve good collection efficiency. For instance, a power-plant precipitator will not satisfactorily collect fly ash from low-temperature flue gases unless a certain amount of sulfur trioxide is present in the gas; this difficulty can be overcome by operating the precipitator on hot flue gas. Despite its drawbacks, the electrostatic precipitator continues to be widely used.

8-1 BASIC PRINCIPLES OF OPERATION

In the electrostatic precipitator, the basic force which acts to separate the particles from the gas is electrostatic attraction. In the first part of the process, an electrostatic charge is imparted to the particles. It is true that many particles are charged already as a result of prior processes, but this charge is generally too weak for practical use. When the particles have been sufficiently charged, an electric field is applied to the flow region, exerting an attractive force to the particles and causing them to migrate toward the oppositely charged electrode at right angles to the gas-flow direction. The particles collect on the electrode. If the particles are of a liquid, after collecting on the electrode the liquid then flows down the electrode by action of gravity and collects at the bottom. If the particles are of a solid, the collected particles are removed from the electrode by shaking it in a process known as *rapping*. The collected material then accumulates in a hopper located beneath the collector section. In some applications, the collected particles, either liquid or dust, are removed from the collecting plates by flushing them with a liquid, usually water.

The particle-charging process is done by means of a corona surrounding a highly charged electrode, such as a wire. (The corona will be described in detail in Sec. 8-3.) The electric field used in collecting the particles is set up between two electrodes. Two electrodes are used in the charging process, and two electrodes are used in the collecting process, with an electric field existing between the electrodes in both cases. The same pair of electrodes may serve both the particle-charging and collecting functions; if this is the case, the precipitator is referred to as a *single-stage precipitator*. If two pairs of electrodes are used, one for particle charging and one for collection, then the precipitator is a two-stage precipitator.

Figure 8-1 shows a cross-sectional view of a cylindrical single-stage electrostatic precipitator. The dirty gas enters near the bottom and flows upward through the cylindrical collector portion of the precipitator. As the gas flows upward, electrostatic forces cause the dust particles to migrate to the collector electrode where they stick. The cleaned gas then emerges at the top. The collected dust is removed periodically from the collector electrode by rapping it; this dust then falls to the dust hopper and is accumulated there for periodic removal. The discharge electrode consists of a wire suspended from an insulator at the top and held in position by a weight at the bottom. It is important that the wire be maintained in proper alignment along the axis of the cylinder; the weight performs this function. A power supply furnishes a large dc voltage on the order of 50 kV, which is either steady or pulsed; modern usage favors the pulsed dc voltage. Although Fig. 8-1 shows only a single collector, obviously a large number of these collectors can be placed in parallel in a single housing. Although it is not as widely used as the parallel-plate precipitator, which will be discussed next, the multiple cylindrical precipitator is nevertheless frequently employed; it is of particular advantage in collecting liquid particles.

The parallel-plate precipitator is shown in Fig. 8-2. Here the gas flows

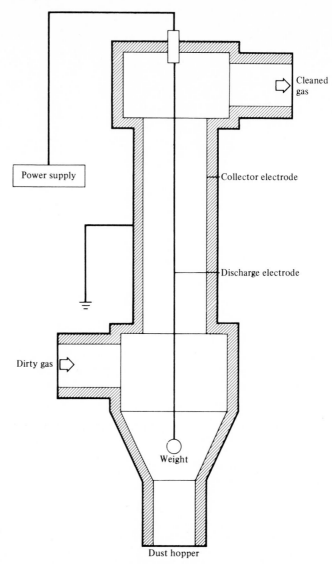

FIGURE 8-1
Cylindrical single-stage electrostatic precipitator.

between two vertical parallel plates between which are suspended a number of vertical wires held in place by weights attached at the bottom. These wires form the discharge electrode, while the vertical plates form the collection electrode. The electric field that is maintained between the wires and the plates charges the particles and provides the collecting force acting on the particles. As in the case of the cylindrical precipitator, the collected dust is removed by rapping the

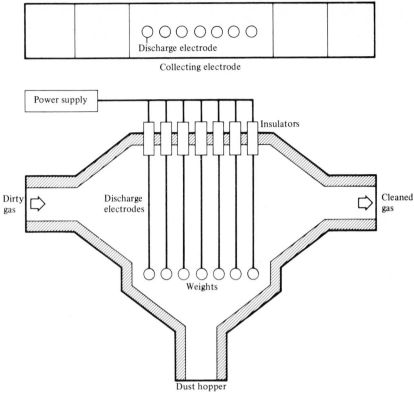

FIGURE 8-2
Parallel-plate single-stage electrostatic precipitator.

collecting electrode; this dust accumulates in the dust hopper at the bottom of the unit. A number of these parallel-plate units can be located side by side in a single housing to form a multiple unit.

In the two-stage precipitator, separate electrode pairs are provided to perform the particle-charging and the collecting functions. Upon entering the active zone of the precipitator, the gas first encounters the charging section. Here a corona is maintained, which imparts a charge to the particles. In most applications of two-stage precipitators, a positive corona is used; that is, the wire is positively charged relative to the cylinder or plate. The charging section is short, providing a gas-residence time of only a fraction of a second, so that very little collection takes place in this section. Next, the gas enters the collection region, in which an electric field is maintained between the collecting-field electrode and the collector electrode. If a positive corona is used, the collecting-field electrode will be positively charged also. The length of the collecting section will be 10 or more times that of the charging section to allow sufficient time for the collection to take place.

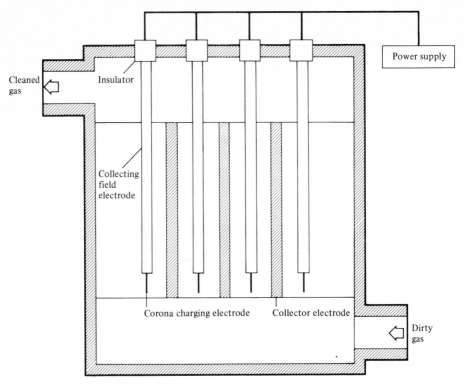

FIGURE 8-3
Cylindrical two-stage electrostatic precipitator.

Figure 8-3 shows a cylindrical two-stage precipitator arranged vertically. The corona electrode is located at the bottom of the collector cylinder; it consists of a small-diameter wire, in this case attached directly to the collecting-field electrode. The collecting-field electrode has a considerably larger diameter. In the case shown in Fig. 8-3, both the corona electrode and the collecting-field electrode are charged to the same voltage; the smaller diameter of the corona electrode induces the formation of a corona, which is absent from the collecting-field electrode due to its larger diameter. In other cases, the corona electrode and the collecting-field electrode are charged to different voltages by separate power supplies, the collecting-field electrode having the lower voltage. The collector electrode, which is normally grounded, extends over both the charging and collecting sections and is the same electrode for both sections.

Figure 8-4 shows a parallel-plate two-stage precipitator. The corona electrodes consist of vertical wires; the collecting-field electrodes consist of plates; and the collector electrodes also consist of plates that are parallel to the collecting-field electrodes, extending over both the charging and collecting sections. Here the corona electrode is maintained at a higher voltage than the collecting-field electrode by its separate power supply.

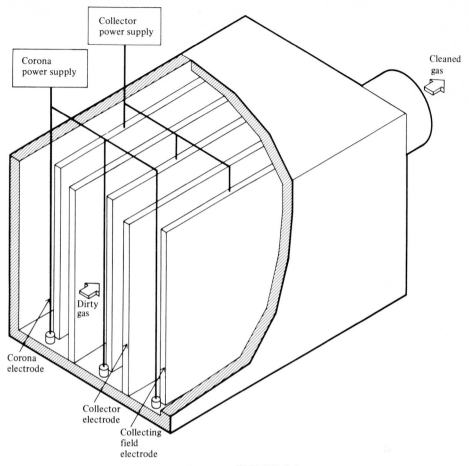

FIGURE 8-4
Parallel-plate two-stage electrostatic precipitator.

Two-stage precipitators are used for most air-conditioning applications, whereas single-stage precipitators are used for most industrial applications, especially the large ones. The single-stage units are generally operated with negative corona, while the two-stage units usually employ positive corona. There is a reason for this practice: Due to the negative corona, the single-stage unit produces a significant amount of ozone during operation. The amount of ozone produced is not objectionable when the cleaned gas is discharged into the atmosphere, as in industrial applications; however, the amount of ozone is objectionable when the cleaned gas is discharged into a living space, as in air-conditioning applications. Positive corona produces less ozone than negative corona and is preferable for air-conditioning applications. Two-stage units are used in small industrial applications, and there is no reason why these units cannot be operated with negative corona.

In air-conditioning applications, the collector electrodes are sprayed with an adhesive coating to which the dust particles adhere. Cleaning is accomplished by washing the accumulated dust off the electrode surfaces. The surfaces are then sprayed with a new coat of the adhesive material before the unit is placed back in service.

8-2 COLLECTION EFFICIENCY FOR ELECTRICALLY CHARGED PARTICLES

In this section, we shall be concerned with the collection efficiency of a precipitator. Charged particles migrate toward the collector electrode under the action of the electric field set up between the electrodes. The magnitude of the charge attached to the particle, as well as the strength of the electric field between the electrodes, will be studied in detail in later sections. For the present, these quantities are assumed to be known. Figure 8-5 shows the collection section of a precipitator of fairly arbitrary geometry in which the section consists of a cylinder of arbitrary cross section. Both the cylindrical and parallel-plate types fit into this category. The analysis will also hold for either a single- or two-stage precipitator.

When a charged particle possessing charge q_p is located in a region where an electric field of strength E_c is also present, a force F will act on the particle. The magnitude of this force is given by

$$F = q_p E_c \qquad (8\text{-}1)$$

which can be verified from any elementary physics book. For a small particle, its terminal velocity will soon be reached under the action of this force. The terminal velocity is given by Eq. (3-61), which becomes

$$V_t = \frac{FC}{3\pi\mu d} = \frac{q_p E_c C}{3\pi\mu d} \qquad (8\text{-}2)$$

in which the Cunningham correction factor C is included. For standard air, Eq. (8-2) becomes

$$V_t = 5766 \frac{FC}{d} = 5766 \frac{q_p E_c C}{d} \qquad \text{standard air} \qquad (8\text{-}3)$$

The Cunningham correction factor is given by Eq. (3-55), and for standard air it is tabulated in Table 3-1. For standard air, Eq. (8-3) becomes

$$V_t = \frac{5766 q_p E_c}{d^2} (d + 0.168 \times 10^{-6} + 5.336 \times 10^{-8} e^{-8.246 \times 10^6 d}) \qquad (8\text{-}4)$$

As shown in Fig. 8-5, the terminal velocity of the particle is perpendicular to the velocity of the gas. The electric force acting on the particle will act in the direction of the electric field, which is also perpendicular to the gas-flow direction. The particle will migrate toward the oppositely charged electrode.

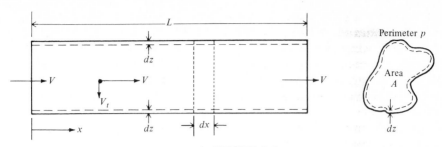

FIGURE 8-5
Collection section of an electrostatic precipitator.

The collection efficiency is evaluated by a method similar to that of Sec. 6-3 for the settling chamber. Referring to Fig. 8-5, a particle which enters the capture zone adjacent to the wall will migrate a distance dz toward the wall while it moves a distance dx in the axial direction. Since the time required to move these two distances is the same, we may write

$$dz = V_t \, dt = V_t \frac{dx}{V} \qquad (8\text{-}5)$$

We shall assume that the effect of turbulent flow in the duct will be to distribute the particles uniformly across any section. Also, we shall allow the field strength at the wall E_c to vary along the length of the collection section. The fraction of particles which will be captured in distance dx is the ratio of the area within the capture zone to the total cross-sectional area:

$$-\frac{dN}{N} = \frac{p \, dz}{A} = \frac{p}{A} \frac{V_t}{V} dx \qquad (8\text{-}6)$$

which integrates to the following equation if we note that $N = N_0$ when $x = 0$ and allow V_t to vary with x:

$$N = N_0 \exp\left(-\frac{p}{AV} \int_0^x V_t \, dx\right) \qquad (8\text{-}7)$$

The efficiency then becomes

$$\eta = 1 - \exp\left(-\frac{p}{AV} \int_0^L V_t \, dx\right) \qquad (8\text{-}8)$$

Using Eq. (8-2) gives

$$\eta = 1 - \exp\left(-\frac{p}{AV} \frac{q_p C}{3\pi\mu d} \int_0^L E_c \, dx\right) \qquad (8\text{-}9)$$

Noting that $AV = Q$ and that the product pL is equal to the collector-surface area A_c, Eq. (8-9) may be written as

$$\eta = 1 - \exp\left(-\frac{q_p C A_c}{3\pi\mu dQ}\frac{1}{L}\int_0^L E_c\, dx\right) \qquad (8\text{-}10)$$

Next, we define the mean electric field strength at the wall E_{cm} as

$$E_{cm} = \frac{1}{L}\int_0^L E_c\, dx \qquad (8\text{-}11)$$

Then Eq. (8-10) becomes

$$\eta = 1 - \exp\left(-\frac{q_p C A_c E_{cm}}{3\pi\mu dQ}\right) \qquad (8\text{-}12)$$

in which C is given by Eq. (3-55). For standard air, Eq. (8-12) becomes

$$\eta = 1 - \exp\left(-\frac{5766 q_p C A_c E_{cm}}{Qd}\right) \qquad (8\text{-}13)$$

EXAMPLE 8-1   An electrostatic precipitator for use with standard air containing dust particles of 1.0 μm diameter is in the form of a cylinder 0.3 m in diameter and 2.0 m long. The flow rate is 0.075 m^3/s. If $E_{cm} = 100{,}000$ V/m at the outer diameter, and if $q_p = 0.3$ fC, compute the collection efficiency.

SOLUTION   From Table 3-1, the Cunningham correction factor C for standard air is 1.168. The collector surface area A_c is $\pi DL = \pi(0.3)(2.0) = 1.88$ m^2. Equation (8-13) gives the efficiency as

$$\eta = 1 - \exp\left[-\frac{5766(0.3 \times 10^{-15})(10^5)(1.168)(1.88)}{0.075(10^{-6})}\right] = 0.9937 \qquad Ans. \qquad ////$$

8-3   THE ELECTRIC FIELD

In this section, let us consider the electric field which is set up between the charging and collector electrodes or between the collecting-field electrode and the collector electrode in the case of the second stage of a two-stage precipitator.

Parallel-Plate Electrodes

First, let us consider the electric field between two parallel plates in the case where the gas contains charged particles along with ions carrying the same charge polarity as the particles and where no current flows through the gas. Strictly speaking, if a voltage difference exists between the plates, charged particles will

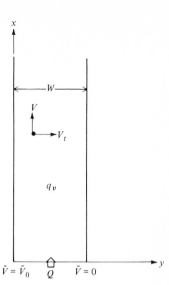

FIGURE 8-6
Electric field between two parallel plates.

migrate toward one of the plates, thus inducing at least a localized current. If the particles retain their charge when they reach the collecting plate, and if all ions present in the gas attach themselves to particles before reaching the collecting plate, then no current will flow into the collecting plate. We refer to this situation as involving no current flow.

Figure 8-6 shows the geometry of the two plate electrodes between which the gas flows. We shall let q_v be the charge density, or total charge which exists in the gas per unit volume. To analyze the electric field within the gas, we return to Gauss' law expressed by Halliday and Resnick [1, pp. 598 and 633]:

$$\oint \mathbf{E} \cdot d\mathbf{A} = \frac{q}{\varepsilon_0} \qquad \varepsilon_0 = 8.85 \times 10^{-12} \text{ A}^2\text{s}^4/\text{kg} \cdot \text{m}^3 \qquad (8\text{-}14)$$

This equation states that the integral of the product of field strength normal to a surface times the surface area is proportional to the charge contained in the volume enclosed by that surface.

Take the volume as that of a parallelepiped whose dimension in the y direction is dy, as shown in Fig. 8-7. The electric field E is distributed uniformly over the faces A; the integral of the dot product $\mathbf{E} \cdot d\mathbf{A}$ on the face to the left is $-E(y)A$ and on the face to the right is $E(y + dy)A$. Over the other faces the dot product is zero since the area and the field vector are at right angles. Equation (8-14) may then be written as

$$E(y + dy)A - E(y)A = \frac{q}{\varepsilon_0} = \frac{q_v A \, dy}{\varepsilon_0} \qquad (8\text{-}15)$$

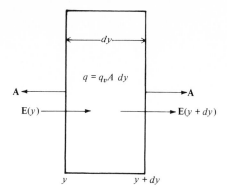

FIGURE 8-7
Volume element in an electric field.

Expanding $E(y + dy)$ in a Taylor series, neglecting the higher-order terms, and simplifying gives

$$\frac{dE}{dy} = \frac{q_v}{\varepsilon_0} \qquad (8\text{-}16)$$

Next, we relate the field strength to the voltage (see Ref. [1]):

$$E = -\frac{d\tilde{V}}{dy} \qquad (8\text{-}17)$$

Then the resulting differential equation for the voltage is

$$\frac{d^2\tilde{V}}{dy^2} = -\frac{q_v}{\varepsilon_0} \qquad (8\text{-}18)$$

Using the boundary conditions shown in Fig. 8-6, that $\tilde{V}$ is equal to $\tilde{V}_0$ when y is zero and $\tilde{V}$ is equal to zero when y is W, leads to the following solution to Eq. (8-18):

$$\tilde{V} = \tilde{V}_0 + \left(\frac{q_v}{\varepsilon_0}\frac{W}{2} - \frac{\tilde{V}_0}{W}\right)y - \frac{q_v}{\varepsilon_0}\frac{y^2}{2} \qquad (8\text{-}19)$$

Equation (8-17) gives the field strength as a function of y:

$$E = \left(\frac{\tilde{V}_0}{W} - \frac{q_v}{2\varepsilon_0}\frac{W}{}\right) + \frac{q_v}{\varepsilon_0}y \qquad (8\text{-}20)$$

At the collecting surface, where $y = W$, the field strength E_c is

$$E_c = \frac{\tilde{V}_0}{W} + \frac{q_v W}{2\varepsilon_0} \qquad (8\text{-}21)$$

EXAMPLE 8-2   If the voltage $\tilde{V}_0$ is 2.0 kV, plot curves for E and $\tilde{V}$ for the space between two parallel plates 2.0 cm apart for the two cases $q_v = 0.2$ mC/m^3 and $q_v = 0$.

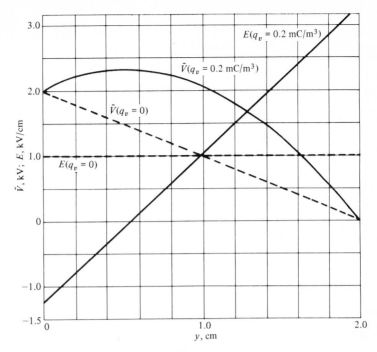

FIGURE 8-8
Voltage and field strength between parallel plates.

SOLUTION   For $q_v = 0.2$ mC/m^3, Eqs. (8-19) and (8-20) give

$$\tilde{V} = 2000 + 0.126 \times 10^6 y - 11.3 \times 10^6 y^2$$
$$E = -0.126 \times 10^6 + 22.6 \times 10^6 y$$

The value of E_c is 0.326 kV/mm. For $q_v = 0$, $\tilde{V} = 2000 - 10^5 y$ and $E = 100{,}000$ V/m $= 0.1$ kV/cm. The curves for $\tilde{V}$ and E are plotted in Fig. 8-8 for the two cases.   ////

Cylindrical Electrodes

A similar derivation can be made for the electric field between two concentric cylinders. This geometry, which is shown in Fig. 8-9, can arise in the collecting stage of a two-stage precipitator, as in Fig. 8-3, or it can represent the space between the corona surrounding the discharge electrode and the collector electrode in a single-stage unit. An analysis similar to that used to develop Eq. (8-16), when applied to the cylindrical element shown in Fig. 8-9, leads to the following differential equation:

$$\frac{d}{dr}(rE) = r\frac{q_v}{\varepsilon_0} \qquad (8\text{-}22)$$

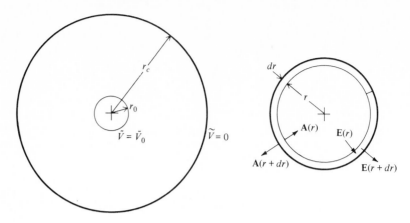

FIGURE 8-9
Electric field between two concentric cylinders.

The integral of Eq. (8-22) is

$$E = \frac{q_v r}{2\varepsilon_0} + \frac{C_1}{r} \qquad (8\text{-}23)$$

Using the relation of Eq. (8-18) for field strength in terms of the gradient of voltage, $E = -d\tilde{V}/dr$, gives

$$\frac{d\tilde{V}}{dr} = -\frac{q_v}{2\varepsilon_0} r - \frac{C_1}{r} \qquad (8\text{-}24)$$

whose solution is

$$\tilde{V} = -\frac{q_v r^2}{4\varepsilon_0} - C_1 \ln r + C_2 \qquad (8\text{-}25)$$

Using the boundary conditions that $\tilde{V} = \tilde{V}_0$ when $r = r_0$ and that $\tilde{V} = 0$ when $r = r_c$ enables us to solve for the constants C_1 and C_2. When these are substituted back into Eq. (8-25), the result is

$$\tilde{V} = \frac{q_v}{4\varepsilon_0}(r_c{}^2 - r^2) + \left[\tilde{V}_0 - \frac{q_v}{4\varepsilon_0}(r_c{}^2 - r_0{}^2)\right]\frac{\ln r_c/r}{\ln r_c/r_0} \qquad (8\text{-}26)$$

and Eq. (8-23) gives for the field strength

$$E = \frac{q_v r}{2\varepsilon_0} + \frac{\tilde{V}_0 - q_v(r_c{}^2 - r_0{}^2)/4\varepsilon_0}{\ln r_c/r_0}\frac{1}{r} \qquad (8\text{-}27)$$

The field strength at the collector surface is then

$$E_c = \frac{q_v r_c}{2\varepsilon_0} + \frac{\tilde{V}_0 - q_v(r_c{}^2 - r_0{}^2)/4\varepsilon_0}{r_c \ln r_c/r_0} \qquad (8\text{-}28)$$

And the field strength at the surface of the inner cylinder is

$$E_0 = \frac{q_v r_0}{2\varepsilon_0} + \frac{\tilde{V}_0 - q_v(r_c^2 - r_0^2)/4\varepsilon_0}{r_0 \ln r_c/r_0} \qquad (8\text{-}29)$$

EXAMPLE 8-3 Using the following data, evaluate the voltage and electric field strength for two concentric cylinders when $q_v = 0.25$ mC/m^3 and when $q_v = 0$:

$$r_0 = 1.0 \text{ cm} \qquad r_c = 10 \text{ cm} \qquad \tilde{V}_0 = 15 \text{ kV}$$

SOLUTION For $q_v = 0.25$ mC/m^3, Eq. (8-26) gives the following (in kilovolts):

$$\tilde{V} = 70.62 - 0.7062r^2 - 23.874 \ln \frac{r_c}{r}$$

Equation (8-27) becomes (in kilovolts per centimeter)

$$E = 1.412r - \frac{23.874}{r}$$

In both the preceding equations, r is given in centimeters.

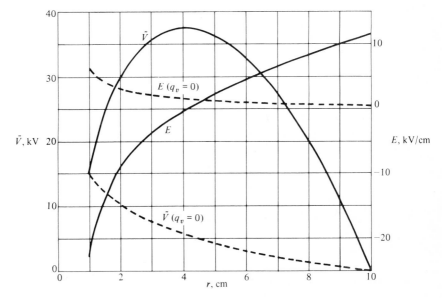

FIGURE 8-10
Voltage and field strength between two concentric cylinders.

For $q_v = 0$, the corresponding two equations are (r is again expressed in centimeters)

$$\tilde{V} = 6.52 \ln \frac{r_c}{r}$$

$$E = \frac{6.52}{r}$$

These equations are plotted in Fig. 8-10.   ////

Cylindrical Electrodes with Current Flow

So far, we have not considered any electric currents in the field, except for localized currents brought about by turbulent motion of the charged particles. The effect of these currents is absorbed in the assumption that the particle-charge distribution is uniform over any cross section. In single-stage precipitator practice, particularly, an ionic current will flow in the direction of the electric field. This current consists of ions that are charged with the same polarity as the charging electrode and moving toward the collecting electrode, which has the opposite polarity. As the ions traverse the gas-flow region, their charge adds to the space charge induced by the particles; this space charge changes the voltage and field-strength distributions across the region and requires that the charging electrode be maintained at a much higher potential than would otherwise be necessary. In the charging region, the ions are adsorbed onto the particles; after the particles become charged, the ions can no longer attach to particles, and hence they traverse the entire space and enter the collecting electrode. It is also possible to have a flow of free electrons in certain cases. Because the electrons move at such high velocities, unless they attach to atoms to form ions, their residence time is so short that there is an insufficient number present to contribute substantially to the space charge.

The velocity of the ions is great enough so that they will not be distributed uniformly over the space by turbulent action of the gas; we can ignore the effect of turbulent diffusion on ion distribution. We continue to assume that the particles are uniformly distributed over the space by turbulent action. The presence of an ion current assumes that a source of ions is present. This will not automatically be the case but requires the presence of a corona, as will be discussed in the next section. For the time being, let us assume that ions will be available in sufficient numbers to provide the ionic current called for by the voltage imposed on the charging electrode.

To begin our consideration of the effect of the ionic current, let us derive an expression for the space charge contributed by the ions present in the space. Assume that a current i/L per unit length enters the cylindrical element of Fig. 8-9 at radius r, and that this same current flows out at radius $r + dr$. Current is defined as charge flow per unit time. To determine the charge contained in the volume of thickness dr at any one time, imagine that the flow of current is shut off for a time; it is then allowed to flow, and the time required for the first particle to reach the other side of the element is noted. The amount of charge which flows in this time is the total charge contained in the volume; after this time has elapsed, charge will flow out at the same rate it flows in. Thus the charge contained in the element whose volume is $2\pi r L\, dr$ is

$$q = i\,\frac{dr}{|V_i|} \qquad (8\text{-}30)$$

in which i is the current and V_i is the average ion velocity.

The ionic velocity is given by means of a quantity called the *ion mobility K*, defined as the ion velocity per unit of field strength:

$$K = \frac{V_i}{E} \qquad (8\text{-}31)$$

so that $q = i\,dr/K|E|$. The charge per unit volume due to the ion flow is then given by

$$q_{vi} = \frac{q}{2\pi r L\, dr} = \frac{i/L}{2\pi r K |E|} \qquad (8\text{-}32)$$

in which we have designated this charge as q_{vi}. The charge due to the particles per unit volume can be denoted as q_{vp}, so that the total space charge per unit volume becomes

$$q_v = q_{vp} + \frac{i/L}{2\pi r K |E|} \qquad (8\text{-}33)$$

Using Eq. (8-33), Eq. (8-22) for the field strength becomes

$$\frac{d}{dr}(rE) = \frac{r q_{vp}}{\varepsilon_0} + \frac{i/L}{2\pi\varepsilon_0 K |E|} \qquad (8\text{-}34)$$

Equation (8-34) must be integrated twice in conjunction with the relation $E = -d\tilde{V}/dr$, and the solution must be evaluated to fit the boundary conditions: When $r = r_0$, $E = E_0$; and when $r = r_c$, $\tilde{V} = 0$. As it stands, Eq. (8-34) can be integrated the first time readily enough, but a second integral then appears impossible. To integrate Eq. (8-34) twice to obtain an expression for voltage, an approximation will be made. We are interested primarily in a solution valid for small values of the current i/L, so that the second term on the right of the equation will be small compared with the first term. If we simply ignored the second term, we should have the solution for no current, that is, Eq. (8-26).

The main difficulty in integrating Eq. (8-34) is the term E in the denominator of the second term on the right. Since the solution for a small current will vary only a moderate amount from the solution for no current, and since the second term is fairly small anyway, let us replace the E in the second term on the right by the value obtained from the solution for no current, Eq. (8-27), designated as E_1 for this solution:

$$E_1 = \frac{q_{vp}}{2\varepsilon_0} r + \frac{\tilde{V}_0 - q_v(r_c^2 - r_0^2)/4\varepsilon_0}{\ln r_c/r_0} \frac{1}{r} \qquad (8\text{-}35)$$

Equation (8-34) may then be written

$$\frac{d}{dr}(rE) = \frac{q_{vp}}{\varepsilon_0} r + \frac{(i/L)/2\pi\varepsilon_0 K}{q_{vp} r/2\varepsilon_0 + [\tilde{V}_0 - q_{vp}(r_c^2 - r_0^2)/4\varepsilon_0]/r(\ln r_c/r_0)} \qquad (8\text{-}36)$$

To make the mathematics easier to write, let us define

$$a_1 = \frac{q_{vp}}{\varepsilon_0} \qquad (8\text{-}37)$$

$$a_2 = \frac{i/L}{2\pi\varepsilon_0 K} \qquad (8\text{-}38)$$

$$a_3 = 2\frac{\tilde{V}_0 - q_{vp}(r_c^2 - r_0^2)/4\varepsilon_0}{\ln r_c/r_0} \qquad (8\text{-}39)$$

so that Eq. (8-36) then becomes

$$\frac{d}{dr}(rE) = a_1 r + \frac{2a_2 r}{a_1 r^2 + a_3} \qquad (8\text{-}40)$$

A first integral to Eq. (8-40) is

$$rE = \frac{a_1 r^2}{2} + \frac{a_2}{a_1} \ln(a_1 r^2 + a_3) + C_1 \qquad (8\text{-}41)$$

The constant of integration is evaluated from the condition that when $r = r_0$, $E = E_0$, and this constant is then substituted back into Eq. (8-41). This particular boundary condition is used because at the edge of the corona the field strength, not the voltage, is known. The voltage, in fact, will increase as the current increases, while the field strength at that point will remain essentially constant. The resulting equation is

$$rE = \frac{a_1}{2}(r^2 - r_0^2) + \frac{a_2}{a_1} \ln\frac{a_1 r^2 + a_3}{a_1 r_0^2 + a_3} + r_0 E_0 \qquad (8\text{-}42)$$

Using the relation $E = -d\tilde{V}/dr$, then

$$\frac{d\tilde{V}}{dr} = -\frac{a_1}{2}\left(r - \frac{r_0^2}{r}\right) - \frac{r_0 E_0}{r} - \frac{a_2}{a_1}\frac{1}{r}\ln\frac{a_1 r^2 + a_3}{a_1 r_0^2 + a_3} \qquad (8\text{-}43)$$

Integrating this equation from r_0 to r_c, with the respective limits on $\tilde{V}$ of $\tilde{V}_0$ and $\tilde{V}_c = 0$,

$$\int_{\tilde{V}_0}^{0} d\tilde{V} = -\frac{a_1}{2}\int_{r_0}^{r_c}\left(r - \frac{r_0^2}{r}\right)dr - r_0 E_0 \int_{r_0}^{r_c}\frac{dr}{r}$$
$$- \frac{a_2}{a_1}\int_{r_0}^{r_c}\frac{1}{r}\ln\frac{a_1 r^2 + a_3}{a_1 r_0^2 + a_3}dr \qquad (8\text{-}44)$$

which may be partially integrated to give

$$\tilde{V}_0 = \frac{a_1}{4}\left(r_c^2 - r_0^2\right) - \frac{a_1 r_0^2}{2}\ln\frac{r_c}{r_0} + r_0 E_0 \ln\frac{r_c}{r_0} + \frac{a_2}{a_1}\int_{r_0}^{r_c}\frac{1}{r}\ln\frac{a_1 r^2 + a_3}{a_1 r_0^2 + a_3}\,dr \quad (8\text{-}45)$$

We may neglect r_0^2 in comparison with r_c^2 in the first term on the right, and we may also neglect $a_1 r_0^2$ in comparison with a_3 in the argument of the logarithm. Furthermore, let $u = a_1 r^2/a_3$ in the integral; Eq. (8-45) may now be written

$$\tilde{V}_0 = \frac{a_1 r_c^2}{4} + r_0 E_0 \ln\frac{r_c}{r_0} + \frac{a_2}{2a_1}\int_{a_1 r_0^2/a_3}^{a_1 r_c^2/a_3}\frac{\ln(u+1)}{u}\,du \quad (8\text{-}46)$$

We may also neglect $a_1 r_0^2/2$ in comparison with $r_0 E_0$. All these assumptions can be justified by calculation when the final results are obtained, based on usual precipitator geometries and operating conditions and with the further restriction to small values of current.

The integral in Eq. (8-46) cannot be expressed in closed form; it may be found in a standard table of integrals as

$$\int\frac{\ln(x+1)}{x}\,dx = x - \frac{x^2}{2^2} + \frac{x^3}{3^2} - \cdots = \frac{(\ln x)^2}{2} - \frac{1}{x} + \frac{1}{2^2 x^2} - \frac{1}{3^2 x^3} + \cdots \quad (8\text{-}47)$$

The first series is convergent for $x < 1$, the second for $x > 1$. Let us evaluate this integral between the limits of x_0 and x_c, where we assume that $x_0 < 1$ and $x_c > 1$:

$$\int_{x_0}^{x_c}\frac{\ln(x+1)}{x}\,dx = \int_{x_0}^{1}\frac{\ln(x+1)}{x}\,dx + \int_{1}^{x_c}\frac{\ln(x+1)}{x}\,dx$$

$$= \left[x - \frac{x^2}{2^2} + \frac{x^3}{3^2} - \cdots\right]_{x_0}^{1} + \left[\frac{(\ln x)^2}{2} - \frac{1}{x} + \frac{1}{2^2 x^2} - \frac{1}{3^2 x^3} + \cdots\right]_{1}^{x_c}$$

$$= \frac{(\ln x_c)^2}{2} - \frac{1}{x_c} + \frac{1}{2^2 x_c^2} - \frac{1}{3^2 x_c^3} + \cdots - x_0 + \frac{x_0^2}{2^2} - \frac{x_0^3}{3^2} + \cdots$$

$$+ 2\left(1 - \frac{1}{2^2} + \frac{1}{3^2} - \cdots\right)$$

The series in parentheses in the last equation has the value of $\pi^2/12 = 0.822$. Using this numerical value, retaining only the terms through x_c^2 and x_0^2, letting $x_c = a_1 r_c^2/a_3$ and $x_0 = a_1 r_0^2/a_3$, and simplifying, gives for $\tilde{V}_0$

$$\tilde{V}_0 = \frac{a_1 r_c^2}{4} + r_0 E_0 \ln\frac{r_c}{r_0} + 0.822\frac{a_2}{a_1} + \frac{a_2}{4a_1}\left(\ln\frac{a_1 r_c^2}{a_3}\right)^2 - \frac{a_2 a_3}{2a_1^2 r_c^2} - \frac{a_2 r_0^2}{2a_3} \quad (8\text{-}48)$$

Solving for a_2, and from this for i/L, leads to

$$\frac{i}{L} = \frac{4\pi\varepsilon_0 K a_1(\tilde{V}_0 - a_1 r_c^2/4 - r_0 E_0 \ln r_c/r_0)}{1.64 + (\ln a_1 r_c^2/a_3)^2/2 - a_3/a_1 r_c^2 - a_1 r_0^2/a_3} \quad (8\text{-}49)$$

Equation (8-42) gives for E approximately

$$E = \frac{a_1}{2}\left(r - \frac{r_0^2}{r}\right) + \frac{a_2}{a_1}\frac{1}{r}\ln\left(\frac{a_1}{a_3}r^2 + 1\right) + \frac{r_0}{r}E_0 \qquad (8\text{-}50)$$

For the collecting field strength E_c at $r = r_c$

$$E_c = \frac{a_1 r_c}{2} + \frac{r_0 E_0}{r_c} + \frac{a_2}{a_1 r_c}\ln\left(\frac{a_1 r_c^2}{a_3} + 1\right) \qquad (8\text{-}51)$$

Equation (8-35) gives for E_1

$$E_1 = \frac{a_1 r}{2} + \frac{a_3}{2r} \qquad (8\text{-}52)$$

Equation (8-49) gives the basic solution for the voltage applied to the charging electrode $\tilde{V}_0$ as a function of the ionic current per unit length i/L. This relation is implicit in $\tilde{V}_0$ in that Eq. (8-49) cannot be solved directly for $\tilde{V}_0$ since $\tilde{V}_0$ is tied up in the term a_3. It is an explicit equation in i/L, however. In solving for the current as a function of voltage, it is well to note that Eq. (8-49) gives meaningful results only for a narrow range of values of $\tilde{V}_0$. The lowest value of $\tilde{V}_0$, that is, its smallest magnitude without regard to sign, must be the value for no current flow obtained by setting the numerator equal to zero in Eq. (8-49):

$$\tilde{V}_0(i = 0) = \frac{a_1 r_c^2}{4} + r_0 E_0 \ln\frac{r_c}{r_0} \qquad (8\text{-}53)$$

The voltage can then range upward a few thousand volts from this lowest magnitude. If the voltage becomes too much larger than the no-current value, the current will become too large for the small-current assumption in the derivation to hold and the equations break down. The solution obtained here is a reasonably good approximation for values of current up to 1 or 2 mA/m.

For operation with a positive corona within the range of validity of the assumptions made, the quantities $\tilde{V}_0$, i, a_1, and E_0 are all positive. If the corona is negative, then all these quantities become negative.

EXAMPLE 8-4   Consider a positive corona with standard air containing charged particles flowing in the space between the corona and the cylindrical collector electrode. The following data apply:

$$q_{vp} = 0.25 \times 10^{-3}\ \text{C/m}^3 \qquad E_0 = 4.6 \times 10^6\ \text{V/m}$$
$$r_0 = 0.003\ \text{m} \qquad\qquad r_c = 0.10\ \text{m}$$
$$K = 2.2 \times 10^{-4}\ \text{m}^2/\text{V}\cdot\text{s} \qquad \varepsilon_0 = 8.85 \times 10^{-12}\ \text{C/V}\cdot\text{m}$$

Plot i/L, E_c, E_{1_c}, and E_{1_0} as functions of $\tilde{V}_0$.

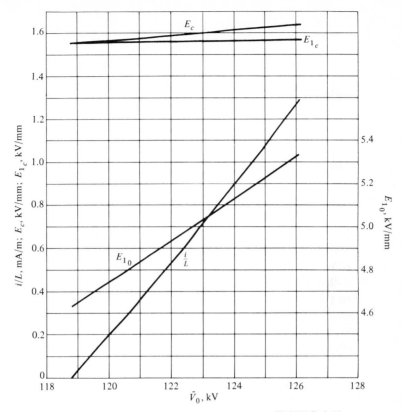

FIGURE 8-11
Results from Example 8-4.

SOLUTION   Equations (8-37) and (8-39) give for a_1 and a_3

$$a_1 = \frac{0.25 \times 10^{-3}}{8.85 \times 10^{-12}} = 2.82 \times 10^7 \text{ V/m}^2$$

$$a_3 = \frac{2}{\ln 0.1/0.003} \left[\tilde{V}_0 - \frac{0.25 \times 10^{-3}(0.01)}{4(8.85 \times 10^{-12})} \right] = 0.571\tilde{V}_0 = 40,355$$

Equation (8-53) gives the voltage for the no-current case:

$$\tilde{V}_0(i = 0) = \frac{2.82 \times 10^7(0.01)}{4} + 0.003(4.6 \times 10^6) \ln \frac{0.1}{0.003}$$

$$= 118,800 \text{ V}$$

When simplified, Eq. (8-49) gives,

$$\frac{i}{L} = \frac{6.88 \times 10^{-7}(\tilde{V}_0 - 118,800)}{1.64 + (\ln 282,000/a_3)^2/2 - a_3/282,000 - 254/a_3}$$

FIGURE 8-12
Wire-and-plate geometry.

The last two terms in the denominator amount to small corrections; they can be approximated using

$$a_3 \approx 0.571(120{,}000) - 40{,}355 = 28{,}165 \text{ V}$$

Then, the equation for i/L becomes

$$\frac{i}{L} = \frac{6.88 \times 10^{-7}(V_0 - 118{,}800)}{1.53 + (\ln 282{,}000/a_3)^2/2}$$

Equation (8-51) gives for the collecting field strength

$$E_c = 1.55 \times 10^6 + \left[\frac{a_2}{2.82 \times 10^6} \ln \frac{282{,}000}{a_3} + 1 \right]$$

Equation (8-52) gives for E_{1_c} and E_{1_0}

$$E_{1_c} = 1.41 \times 10^6 + 5a_3$$
$$E_{1_0} = 42{,}300 + 167a_3$$

Figure 8-11 shows i/L, E_c, E_{1_c}, and E_{1_0} plotted as functions of $\tilde{V}_0$. The last two quantities indicate the degree of error committed by replacing Eq. (8-34) by Eq. (8-36).   ////

In the preceding example, if the corona were negative, then q_{vp} and E_0 would be negative. In the calculations, a_1 and a_3 would be negative and also i/L and $\tilde{V}_0$ would be negative. If the magnitudes of these quantities were unchanged, then the curves of Fig. 8-11 would apply, provided that the signs of both ordinate and abscissa were reversed.

The geometrical configuration shown in Fig. 8-12, consisting of a series of parallel wires stretched between two parallel plates, is of great practical importance since the majority of industrial precipitators are constructed this way. The mathematical analysis of this situation is beyond what can be treated reasonably here. For elementary design purposes, it is suggested that the current and voltage be computed by the formulas for the wire-and-cylinder geometry. Of course, in designing the power supply, an ample factor of safety should be provided to allow for greater voltage and current than the values calculated. Example 8-4 indicates that the collecting-field strength, which will be of major concern in this treatment, is only slightly affected by the current flow. Assuming this condition holds true for the wire-and-plate geometry also, for the collecting-field strength we may use the value calculated for no current flow between two parallel plates, given by Eq. (8-21).

8-4 GENERATION OF THE CORONA

Successful operation of an electrostatic precipitator depends on a copious supply of ions, which will then attach themselves to the particles and, in conjunction with the applied electric field, apply forces to the particles causing them to migrate to the collecting electrode. Although there are several mechanisms of formation of ions in a gas, the only mechanism found practicable for use in an electrostatic precipitator is the corona. The corona is a phenomenon which occurs under certain conditions where electricity flows in a gas. First we shall discuss some features of the corona in general, and then we shall discuss positive and negative coronas in greater detail.

Under ordinary conditions, when subjected to moderate voltage gradients, a gas will conduct electricity in minute amounts only. The small current which flows is due to the presence of a few ionized gas atoms; these ions are formed from the gas atoms by incident background radiation which knocks an electron off from the atom. Thus, a few free electrons and an equal number of positive ions will be present in the gas at all times. As the electrons migrate toward the positive electrode and the positive ions toward the negative electrode, a small current flows. This current is completely negligible as far as the corona is concerned, but the presence of the small number of free electrons is of crucial importance. These few free electrons plus the presence of a strong electric field create the necessary conditions for formation of a corona.

When an electron is present in an electric field, it will be accelerated in the direction of the field toward the region of more positive voltage. This acceleration will continue until the electron collides with an atom, another electron, or a particle; collisions with atoms will occur much more frequently than those with other electrons or particles. When an electron collides with an atom or a molecule of the gas, it may bounce off with no change in energy and continue to be accelerated by the electric field. But if the electron possesses sufficient energy, upon collision with the molecule it may knock an electron off from the molecule, thus producing a second free electron, which will also be accelerated by the electric field, as well as leaving behind a positive ion. There are now two free electrons which will be accelerated until each one, upon collision with gas molecules, produces an extra free electron, giving four at this point. This process continues very rapidly until each original free electron produces many thousands of free electrons. During this process, for each free electron formed a positive ion is also created. This process occurs within the corona region, which is a narrow region surrounding the wire electrode, as shown in Fig. 8-13.

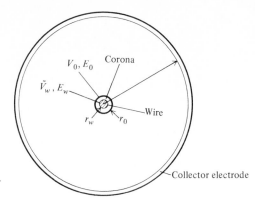

FIGURE 8-13
Corona in a wire-and-cylinder precipitator.

Under suitable conditions, such as those for which a spark discharge will occur, the large number of free electrons formed will move to the positive electrode. But there is another possibility: If the electrons move into a region of much lower electric field strength, they will be slowed down by collisions and will be unable to accelerate to their former high velocities. Molecules of a number of gases have the property that they absorb electrons which collide with them especially if the electrons have relatively low energies upon impact. Thus negative ions are formed, and the number of free electrons is reduced. Under typical conditions of negative corona operation, the free electrons are almost entirely eliminated and a large number of negative ions is formed.

The process of free-electron production is referred to as *electron avalanche.* This process is accompanied by emission of light and sometimes by generation of sound. It is this process which produces the typical corona, which can be sensed visibly and often audibly as well. The absorption of electrons and consequent negative ion formation, if it occurs at all, take place mostly in the region outside the corona.

Let us now look at the positive corona formation in the light of the picture of general corona formation just presented. In this case, the wire of Fig. 8-13 is positively charged relative to the collector electrode; the electrons then migrate toward the wire. The region of large field strength is confined to the corona region, and the field strength is greatest at the wire surface. The avalanche process commences at the edge of the corona region and proceeds all the way to the wire surface. There is then no opportunity for the electrons to be absorbed, and hence no negative ions are produced. However, there is a large number of positive ions produced in the avalanche process, and these migrate out of the corona region toward the collector electrode. These positive ions then attach to the particles present, charging them positively and causing them to migrate toward the collector electrode. The free electrons which rush into the wire are carried away as a current in the wire. The formation of the required number of free electrons to initiate the avalanche process occurs at the edge of the corona and is due in part to background radiation. It appears that the number of free electrons formed by background radiation, although sufficient to start a corona, is not enough to sustain it. Additional free electrons are formed near the edge of the corona by ultraviolet emission from the corona itself.

The negative corona operates somewhat differently. The electron avalanche begins at the surface of the wire and proceeds radially outward to about the edge of the corona region. The free electrons begin to attach to gas molecules near the edge of the corona, and this attachment process continues for some distance beyond the corona. In the attachment process, negative ions are formed; these ions attach to the particles and cause the particles, now negatively charged, to migrate toward the collecting electrode, which is at a positive voltage relative to the wire. The positive ions formed in the electron avalanche process move toward the wire and are neutralized there by electrons conducted in by the wire. Free electrons are formed at the wire surface to initiate the avalanche process by ultraviolet radiation and possibly by dislodgement from the wire upon bombardment by the positive ions striking the wire surface.

A negative corona cannot occur unless the free electrons generated in the corona are able to attach themselves to gas molecules to form negative ions. If they are unable to do this readily, most of the free electrons will move to the positive electrode and a spark occurs. A spark represents a breakdown of the corona, much as a capacitor breaks down if charged to too high a voltage. Some gases, particularly nitrogen, do not readily absorb electrons; such gases are unsuitable for negative corona operation unless electron-absorbing gases are also present. Of those gases commonly appearing in industrial exhaust gases, sulfur dioxide is one of the best electron-absorbing gases. Oxygen, water vapor, and carbon dioxide are also good electron-absorbing gases. Since one or another of these gases frequently appears in exhaust gases in a sizable concentration, the negative corona is usually suitable, although there are exceptions.

The mechanism of corona formation discussed suggests that the edge of the corona is defined by the electric field strength there, with a higher strength inside the corona and a lower strength outside. Empirical results confirm this conclusion. An empirical expression for the field strength at the edge of the corona has been given by Peek and is presented in White [2, pp. 91 and 92]. This equation is

$$E_0 = \pm 3 \times 10^6 f \left(\frac{T_0 P}{T P_0} + 0.03 \sqrt{\frac{T_0 P}{T P_0 r_0}} \right) \qquad (8\text{-}54)$$

In Eq. (8-54), $P_0 = 1.0$ atm and $T_0 = 293$ K are room pressure and temperature. The quantity r_0 is the radius at the edge of the corona, and f is a roughness factor, which takes into account any rough spaces on the wire surface or any dust specks there. Since the effect of any rough spot is to reduce the field strength required to form the corona in the vicinity of the spot, the factor f is less than or equal to unity. For clean smooth wires, $f = 1$; for wires encountered in practice, $f = 0.6$ is a reasonable value to use in the absence of exact data. Equation (8-54) may be used for either positive or negative corona; the positive sign is used for positive corona and the negative sign for negative corona.

In Sec. 8-3, equations were developed for the electric field strength and voltage across the space outside the corona. These equations are important in determining the charge imparted to the particles in this region. It is possible to develop similar equations for the region inside the corona, although this would require further insight into the electron avalanche and ion transport mechanisms than has yet been presented. Such equations are of some use in precipitator analysis and design; the principal benefit is to give values for the voltage and current supplied to the wire to maintain the corona. Although the insight and mathematical complexity required to develop equations for field strength and voltage inside the corona are beyond the scope of this book, we can make a rough estimate of the voltage and current, which can be used in the design of a power supply to the precipitator.

The current required to neutralize the positive ions flowing to the wire, in the case of a negative corona, or that due to the incoming electrons, in the case of a positive corona, can be found as the product of the charge density produced in the gas times the volumetric flow rate of gas:

$$i = q_v Q \qquad (8\text{-}55)$$

The voltage at the wire surface can be estimated very roughly as the product of the mean field strength across the corona times the width of the corona, this product being added to the voltage at the edge of the corona. A rough estimate of the field strength at the wire surface is given by

$$E_w \approx \frac{r_0 E_0}{r_w} \qquad (8\text{-}56)$$

so that the mean field strength across the thickness of the corona is

$$\bar{E} \approx \frac{E_0 + r_0 E_0/r_w}{2} = \frac{E_0}{r_w} \frac{r_0 + r_w}{2} \qquad (8\text{-}57)$$

Then the voltage at the wire surface is

$$\tilde{V}_w \approx \tilde{V}_0 + \bar{E}(r_0 - r_w) \approx \tilde{V}_0 + E_0 \frac{r_0^2 - r_w^2}{2r_w} \qquad (8\text{-}58)$$

Since these equations are only very rough approximations, an adequate factor of safety must be allowed in designing the power supply.

EXAMPLE 8-5   Determine E_0 if $P = 1.0$ atm, $T = 300°C$, $r_0 = 2$ mm, and f is assumed equal to 0.7.

SOLUTION   Equation (8-54) gives

$$E_0 = 3 \times 10^6 (0.7) \left[\frac{293(1.0)}{573(1.0)} + 0.03 \sqrt{\frac{293(1.0)}{573(1.0)(0.002)}} \right]$$

$$= 2.08 \times 10^6 \text{ V/m}$$

$$= 20.8 \text{ kV/cm} \qquad Ans.$$

////

EXAMPLE 8-6   In Example 8-4, the voltage $\tilde{V}_0$ was computed as 118,800 V. The other data included

$$E_0 = 4.6 \times 10^6 \text{ V/m} \qquad r_0 = 0.003 \text{ m}$$

Assume $r_w = 0.001$ m and compute $\tilde{V}_w$.

   SOLUTION   Equation (8-58) gives for $\tilde{V}_w$

$$\tilde{V}_w = 118,800 + 4.6 \times 10^6 \frac{9 \times 10^{-6} - 1 \times 10^{-6}}{2(0.001)}$$

$$= 137,200 \text{ V}$$

$$= 137.2 \text{ kV} \qquad Ans. \qquad\qquad ////$$

   Because of the different operation of the positive and negative coronas, certain differences in their characteristics may be expected. For one thing, for successful operation the negative corona requires the presence of an electron-absorbing gas in the region surrounding it. This is not true of the positive corona, which can therefore be used anytime. Experience has shown the negative corona to be more stable and to give better collection performance if an electron-absorbing gas is present; for this reason, it is preferred in most industrial applications. However, the negative corona generates a significant amount of ozone during operation, which makes it objectionable for air-conditioning applications (but it is not a serious polluter of the atmosphere in this respect). Thus, for air-conditioning applications, the positive corona is normally used. In order to reduce ozone generation still further, two-stage precipitators are normally used, with the collecting stage designed to operate without a corona.

   The radius at the outer edge of the corona has not yet been treated here and will not be treated rigorously since its determination requires knowledge of the field-strength variation across the corona. Cobine [3, p. 258] indicates that the following empirical equation may be used to estimate r_0 in terms of r_w, where both quantities are expressed in meters:

$$r_0 = r_w + 0.03\sqrt{r_w} \qquad (8-59)$$

The radius r_0 is not necessarily at the edge of the glow region of the corona, which may extend out farther, but is the edge of the region in which the field strength exceeds the value given by Eq. (8-54), conforming to our previous analysis.

8-5   PARTICLE CHARGING

The first step in the collection process consists of applying an electric charge to the particles coming in with the gas. This charge is imparted by ions which move in the electric field toward the collecting electrode and encounter particles to which they become attached. In this section, we shall be concerned with the mechanism by which the ions come in contact with the particles; it is a reasonable assumption that if an ion actually touches a particle, it will remain attached since an opposite charge is induced in the particle at the vicinity of the ion, which provides a force that tends to hold the ion to the particle.

In moving between the electrodes, the ions follow the electric flux lines, which are curves everywhere tangent to the electric field vector. Unless the particle is perfectly insulating, these flux lines are deflected in the vicinity of a particle. When the particle is uncharged, these flux lines deflect toward the particle, resulting in the capture of a larger number of ions than would be captured if the ions followed straight lines. As the particle becomes charged, the flux lines deflect the other way, so that fewer ions are captured, until eventually the ions are deflected completely away from the particle. At this point, the particle carries a saturation charge, and no further charging is possible.

Field Strength in the Vicinity of a Particle

To study the charging process in detail, let us first obtain an equation for the electric field strength in the space immediately surrounding a particle if the field strength is uniform at some distance away from the particle. Since the influence of the particle extends out only a few particle diameters, then for small particles we may neglect the variation of field strength other than that due to the influence of the particle itself. Let us consider a spherical, dielectric, uncharged particle as shown in Fig. 8-14. That the particle is uncharged means that there are no ions attached to it; induced charges will be present in the particle, which are allowed for in the equations and which are balanced so that the particle is electrically neutral.

Equation (8-14), as modified, is applicable to this derivation. For any region not containing charges, such as region 1 inside the particle or region 2 outside the particle, Eq. (8-14) can be written in the form of the laplacian of potential $\nabla^2 \tilde{V}$:

$$\nabla^2 \tilde{V} = 0 \qquad (8\text{-}60)$$

In spherical coordinates, with axial symmetry, the laplacian can be written as

$$\nabla^2 \tilde{V} = \frac{1}{r^2} \frac{\partial}{\partial r} \left(r^2 \frac{\partial \tilde{V}}{\partial r} \right) + \frac{1}{r^2 \sin \theta} \frac{\partial}{\partial \theta} \left(\sin \theta \frac{\partial \tilde{V}}{\partial \theta} \right) = 0 \qquad (8\text{-}61)$$

For regions 1 and 2, Eq. (8-61) may be written

$$\frac{1}{r^2} \frac{\partial}{\partial r} \left(r^2 \frac{\partial \tilde{V}_1}{\partial r} \right) + \frac{1}{r^2 \sin \theta} \frac{\partial}{\partial \theta} \left(\sin \theta \frac{\partial \tilde{V}_1}{\partial \theta} \right) = 0 \qquad (8\text{-}62)$$

$$\frac{1}{r^2} \frac{\partial}{\partial r} \left(r^2 \frac{\partial \tilde{V}_2}{\partial r} \right) + \frac{1}{r^2 \sin \theta} \frac{\partial}{\partial \theta} \left(\sin \theta \frac{\partial \tilde{V}_2}{\partial \theta} \right) = 0 \qquad (8\text{-}63)$$

We must also be concerned with the boundary conditions necessary for a solution. At the origin $r = 0$, the voltage must remain finite; this is the first

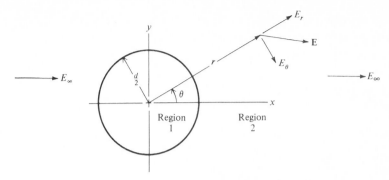

FIGURE 8-14
Electric field in the vicinity of a particle.

boundary condition we shall consider. At a large distance from the particle, the x component of field strength approaches E_∞, or

$$E_r \to E_\infty \cos \theta \qquad (8\text{-}64)$$

Since $E_r = -\partial \tilde{V}/\partial r$, Eq. (8-64) may be written as

$$\frac{\partial \tilde{V}_2}{\partial r} \to -E_\infty \cos \theta \qquad \text{as } r \to \infty \qquad (8\text{-}65)$$

Let us now consider the boundary conditions at an arbitrary point $(d/2, \theta)$ on the surface of the particle. The potential is continuous at this surface, so that

$$\tilde{V}_1 = \tilde{V}_2 \qquad \text{when } r = \frac{d}{2} \qquad (8\text{-}66)$$

Figure 8-15 shows a portion of the surface of the particle, along with a control volume surrounding this surface. Equation (8-14) may be modified if q is the free charge at the surface (zero in this case), which does not include any induced charges in the particle:

$$\oint \kappa \mathbf{E} \cdot d\mathbf{A} = \frac{q}{\varepsilon_0} \qquad \kappa = \frac{\varepsilon}{\varepsilon_0} \qquad (8\text{-}67)$$

This modification is derived by Halliday and Resnick [1, pp. 661–664].

In Eq. (8-67), κ is called the *dielectric constant* and is a function of the material. A perfectly insulating substance, which is closely approximated by air, has $\kappa = 1$; whereas a perfectly conducting substance has $\kappa = \infty$. Applying Eq. (8-67) to the control volume of Fig. 8-15, with $q = 0$, gives

$$-\kappa_1 E_{r_1}\, dA + \kappa_2 E_{r_2}\, dA = \frac{q}{\varepsilon_0} = 0$$

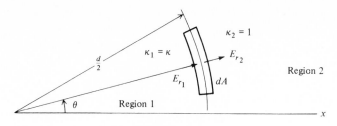

FIGURE 8-15
Control volume at the surface of a sphere.

Since $\kappa_2 = 1$ and κ_1 can be denoted simply as κ, this equation becomes

$$-\kappa E_{r_1} = -E_{r_2}$$

which, in terms of voltage, becomes

$$\kappa \frac{\partial V_1}{\partial r} = \frac{\partial V_2}{\partial r} \quad \text{when } r = \frac{d}{2} \quad (8\text{-}68)$$

To solve Eqs. (8-62) and (8-63), try solutions of the form

$$\tilde{V}_1 = f_1(r) \cos \theta \qquad \tilde{V}_2 = f_2(r) \cos \theta$$

Substituting these back into the differential equations leads to the following ordinary differential equations:

$$r^2 f''_1(r) + 2rf'_1(r) - 2f_1(r) = 0$$
$$r^2 f''_2(r) + 2rf'_2(r) - 2f_2(r) = 0$$

These give the solutions

$$\tilde{V}_1 = \left(C_1 r + \frac{C_2}{r^2} \right) \cos \theta \qquad (8\text{-}69)$$

$$\tilde{V}_2 = \left(C_3 r + \frac{C_4}{r^2} \right) \cos \theta \qquad (8\text{-}70)$$

The first boundary condition requires that $C_2 = 0$. The second condition, Eq. (8-65), requires that $C_3 = -E_\infty$. Equation (8-66) leads to

$$C_1 = -E_\infty + \frac{8C_4}{d^3} \qquad (8\text{-}71)$$

while Eq. (8-68) leads to

$$\kappa C_1 = -E_\infty - \frac{16C_4}{d^3} \qquad (8\text{-}72)$$

The simultaneous solution of Eqs. (8-71) and (8-72) produces

$$C_1 = -\frac{3E_\infty}{\kappa + 2} \qquad (8\text{-}73)$$

$$C_4 = \frac{\kappa - 1}{\kappa + 2}\frac{E_\infty d^3}{8} \qquad (8\text{-}74)$$

which then gives for the voltages inside and outside the sphere

$$\tilde{V}_1 = -\frac{3E_\infty}{\kappa + 2} r \cos \theta \qquad (8\text{-}75)$$

$$\tilde{V}_2 = -E_\infty r \cos \theta + \frac{\kappa - 1}{\kappa + 2}\frac{E_\infty d^3}{8r^2} \cos \theta \qquad (8\text{-}76)$$

The radial component of field strength is obtained from Eq. (8-76) for the region outside the sphere:

$$E_r = -\frac{\partial \tilde{V}_2}{\partial r} = E_\infty \cos \theta + \frac{1}{4}\frac{\kappa - 1}{\kappa + 2} E_\infty \frac{d^3}{r^3} \cos \theta \qquad (8\text{-}77)$$

At the surface of the sphere, Eq. (8-77) becomes

$$E_{r_s} = \frac{3\kappa}{\kappa + 2} E_\infty \cos \theta \qquad (8\text{-}78)$$

Figure 8-16, taken from White [2, p. 130], shows the potential curves and flux lines obtained from Eq. (8-76) applicable to the region outside the sphere.

As the particle remains in the electric field with ions traversing along the flux lines, many of the ions will travel to the particle surface and be collected by the particle. This charges the particle, so that the assumption of no free charge on the particle surface is no longer valid, and the equations must be modified to include the effect of free charge on the particle surface. Since the particle is in turbulent motion in the gas stream, we shall assume that during the collection process the particle rotates so that the charges are uniformly distributed around the surface of the particle. This assumption will make the analysis much simpler than a more rigorous treatment. Figure 8-17 shows a control surface surrounding the particle; let E_2 denote the field strength surrounding the particle due to the particle charge only. Equation (8-14) applies to this control surface and leads to the following result:

$$E_2 = \frac{q_p}{4\pi\varepsilon_0 r^2} \qquad (8\text{-}79)$$

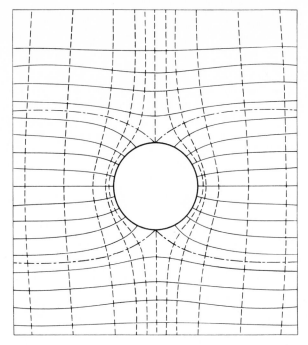

FIGURE 8-16
Electric potential and flux curves near a spherical particle. (*From H. J. White,
"Industrial Electrostatic Precipitation," p. 130, Addison-Wesley Publishing Company,
Inc., Reading, Mass., 1963. Used by permission.*)

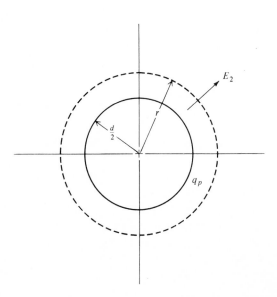

FIGURE 8-17
Field strength due to the charge on the
particle.

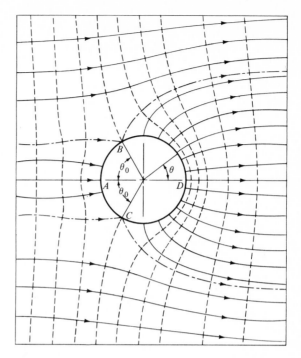

FIGURE 8-18
Electric potential and flux curves near a partially charged spherical particle.
(*From H. J. White, "Industrial Electrostatic Precipitation," p. 132, Addison-Wesley Publishing Company, Inc., Reading Mass., 1963. Used by permission.*)

Since Eq. (8-63) is linear, the sum of two or more solutions is itself a solution. Thus, Eqs. (8-77) and (8-79) can be added to produce a solution which takes into account the uniform field at infinity as well as the surface charge q_p:

$$E_r = E_\infty \cos \theta + \frac{1}{4} \frac{\kappa - 1}{\kappa + 2} E_\infty \cos \theta \frac{d^3}{r^3} + \frac{q_p}{4\pi\varepsilon_0 r^2} \qquad (8\text{-}80)$$

At the surface of the sphere the result is

$$E_{rs} = \frac{3\kappa}{\kappa + 2} E_\infty \cos \theta + \frac{q_p}{\pi\varepsilon_0 d^2} \qquad (8\text{-}81)$$

Figure 8-18, also taken from White [2, p. 132] shows the potential curves and the electric flux lines for the partially charged particle.

As the charge on the particle builds up, the area from which ions can move in to the surface is diminished, as can be seen by comparing Figs. 8-16 and 8-18. Eventually, no ions will strike the surface, and the particle will then have received its saturation charge, which is the maximum charge it can have at the particular field strength E_∞. The magnitude of this saturation charge can be obtained from Eq. (8-81) by letting $E_{rs} = 0$ at the point $\theta = \pi$, giving

$$q_{ps} = \frac{3\kappa}{\kappa + 2} \pi \varepsilon_0 d^2 E_\infty \qquad (8\text{-}82)$$

A quantity of interest, which can also be obtained from Eq. (8-81), is the electric flux which intersects the particle. Flux lines enter the surface of the sphere over the region CAB of Fig. 8-18 and leave over the region BDC; we are interested in those lines which enter over the region CAB for which θ ranges from π to $\pi - \theta_0$. First, to evaluate θ_0, set the right side of Eq. (8-81) equal to zero and solve for θ_0, giving

$$\cos \theta_0 = \frac{q_p(\kappa + 2)}{\pi \varepsilon_0 d^2 3\kappa E_\infty} \qquad (8\text{-}83)$$

Next, to evaluate the electric flux over this region, use the basic definition of electric flux:

$$\Phi = \int_A \mathbf{E} \cdot d\mathbf{A} \qquad (8\text{-}84)$$

Applying this definition to the incoming flux at the surface of the sphere gives

$$\Phi = \int_\pi^{\pi - \theta_0} E_{rs} 2\pi \frac{d}{2} \sin \theta \frac{d}{2} \, d\theta \qquad (8\text{-}85)$$

Using Eq. (8-81) and performing the integration and substituting the limits, we have

$$\Phi = \frac{\pi d^2}{4} \frac{3\kappa}{\kappa + 2} E_\infty \left(1 - \frac{\kappa + 2}{3\kappa} \frac{q_p}{\pi \varepsilon_0 d^2 E_\infty} \right)^2 \qquad (8\text{-}86)$$

Using Eq. (8-82), Eq. (8-86) may be written in the form

$$\Phi = \frac{q_{ps}}{4\varepsilon_0} \left(1 - \frac{q_p}{q_{ps}} \right)^2 \qquad (8\text{-}87)$$

Equation (8-86) shows that if the particle is completely uncharged, the area at infinity over which particles will strike the surface of the sphere is $3\kappa/(\kappa + 2)$ times the projected area of the sphere since Φ at infinity is equal to $A_\infty E_\infty$. When the particle is saturated $q_p = q_{ps}$, and Eq. (8-87) shows that no more ions will come to the particle.

Our primary interest in this section is in determining the magnitude of the saturation charge and the rate at which this charge is imparted to the particle. The first goal is achieved in Eq. (8-82); we now must look to the rate at which charge is added to the sphere. If N is the number of ions per unit volume in the space adjacent to the particle and $A_\infty(t)$ is the area at a considerable distance

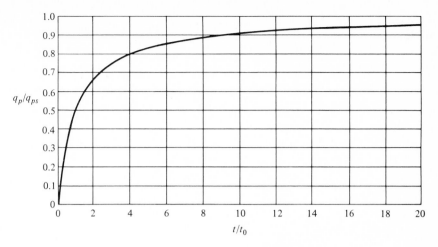

FIGURE 8-19
Charging rate of a spherical particle.

from the particle over which ions will be captured, as used in the preceding paragraph, then the current flowing into the particle surface is given by

$$i = N|q_e|KE_\infty A_\infty(t) \qquad (8\text{-}88)$$

Using Eq. (8-87), $A_\infty(t)$ becomes

$$A_\infty(t) = \frac{\Phi}{E_\infty} = \frac{q_{ps}}{4\varepsilon_0 E_\infty}\left(1 - \frac{q_p}{q_{ps}}\right)^2$$

Then Eq. (8-88) gives

$$i = \frac{dq_p}{dt} = Nq_e K \frac{q_{ps}}{4\varepsilon_0}\left(1 - \frac{q_p}{q_{ps}}\right)^2 \qquad (8\text{-}89)$$

This differential equation can be integrated, subject to the initial condition that $q_p = 0$ when $t = 0$, to give

$$t = \frac{4\varepsilon_0}{Nq_e K}\frac{q_p/q_{ps}}{1 - q_p/q_{ps}} \qquad (8\text{-}90)$$

If we let t_0 be the time constant of this process, such that $q_p = 0.5q_{ps}$ when $t = t_0$, Eq. (8-90) becomes

$$t = t_0 \frac{q_p/q_{ps}}{1 - q_p/q_{ps}} \qquad t_0 = \frac{4\varepsilon_0}{Nq_e K} \qquad (8\text{-}91)$$

In Eqs. (8-88 to 8-91), q_e is the electron charge, 1.6×10^{-19} C, and K is the ion mobility. Equation (8-91) is presented in Fig. 8-19, which shows that charging is rapid at first but becomes very slow as q_{ps} is approached.

EXAMPLE 8-7 Estimate the time constant t_0 for the following data:

$$\frac{i}{L} = 1.0 \text{ mA/m} \qquad r = 3.0 \text{ cm} \qquad E = 2 \times 10^6 \text{ V/m}$$

SOLUTION From Eq. (8-91) for t_0 and using the fact that $Nq_e = q_{vi}$ expressed in Eq. (8-32), we have

$$t_0 = \frac{4\varepsilon_0}{q_{vi} K} = \frac{4\varepsilon_0}{K} \frac{2\pi r K E}{i/L} = \frac{8\pi\varepsilon_0 r E}{i/L}$$

Evaluating this equation gives

$$t_0 = \frac{8\pi(8.85 \times 10^{-12})(0.03)(2 \times 10^6)}{0.001} = 0.0133 \text{ s} \qquad Ans.$$

The procedure given in this example does not produce the best value for N, the number of ions per unit volume, since i/L is a strong function of r in the charging region as the ions are adsorbed onto the large number of initially unchanged particles present in this region. ////

Diffusion Charging

The field-charging process just described is only one method by which particles are charged in the electric field, although it is the dominant process for particles larger than 0.5 μm. The second process of importance is the diffusion-charging process, which will now be discussed. The ions present in the field continually collide with other ions, with neutral molecules, and with the particles present. When an ion collides with a particle, it tends to remain attached to the particle, as we have already mentioned, due to electric image forces set up within the particle. Diffusion charging occurs when the ions which collide with the particles do so because of their random motion rather than their motion along the lines of electric flux. The analysis of this process is founded in kinetic theory, which best describes the random motion of the ions.

To derive the equations for diffusion charging, let us begin with some results from kinetic theory, taken from Cobine [3, pp. 6–20]. Let N be the number of ions per unit volume located at a fair distance from the particle, and let N_1 be the number of ions per unit volume in the region immediately surrounding the particle; then

$$N_1 = Ne^{-q_e \tilde{V}/kT} \qquad (8\text{-}92)$$

in which q_e is the charge of the ion, $\tilde{V}$ is the voltage at the point in question due to the charge attached to the particle, k is Boltzmann's constant, and T is absolute temperature. Let N_2 be the rate at which ions cross the unit area of a surface; then

$$N_2 = N_1 \sqrt{\frac{kT}{2\pi m}} \qquad (8\text{-}93)$$

in which m is the mass of the ion.

Equation (8-79) gives the field strength in the region immediately surrounding the particle; when this equation is integrated with respect to r, the voltage is obtained as

$$\tilde{V} = \frac{q_p}{4\pi\varepsilon_0 r} \qquad (8\text{-}94)$$

Then Eq. (8-92) becomes

$$N_1 = N \exp\left(-\frac{q_e q_p}{4\pi\varepsilon_0 kTr}\right) \qquad (8\text{-}95)$$

At the surface of the particle this equation is

$$N_{1s} = N \exp\left(-\frac{q_e q_p}{2\pi\varepsilon_0 kTd}\right) \qquad (8\text{-}96)$$

The rate at which charge is imparted to the particle is given by Eq. (8-93), evaluated at the surface of the particle; using Eq. (8-96) also, there results

$$\frac{dq_p}{dt} = q_e N_{2s} \pi d^2$$

$$= \pi q_e d^2 N_{1s} \sqrt{\frac{kT}{2\pi m}}$$

$$= q_e \sqrt{\frac{\pi kT}{2m}} d^2 N \exp\left(-\frac{q_e q_p}{2\pi\varepsilon_0 kTd}\right) \qquad (8\text{-}97)$$

The integration of Eq. (8-97) is elementary. Using the condition that $q_p = 0$ when $t = 0$, the result is

$$q_p = \frac{2\pi\varepsilon_0 kTd}{q_e} \ln\left(1 + \frac{q_e^2 N dt}{2\varepsilon_0 \sqrt{2m\pi kT}}\right) \qquad (8\text{-}98)$$

This equation ignores the effect of field charging. A more accurate treatment would have to consider the effect of field charging and would allow for the effect of the distortion of the electric field, such as that shown in Fig. 8-18, upon the diffusion-charging mechanism. Such a treatment requires numerical integration of the differential equations for the field strength and will not be attempted here. However, according to Robinson [4, p. 262], reasonable agreement with experimental results can be obtained by simply adding together the saturation field charge and the diffusion charge, that is, by summing Eqs. (8-82) and (8-98) to give

$$q_p = \frac{3\kappa}{\kappa + 2} \pi \varepsilon_0 d^2 E_\infty + \frac{2\pi \varepsilon_0 kTd}{q_e} \ln \left(1 + \frac{q_e^2 N dt}{2\varepsilon_0 \sqrt{2m\pi kT}}\right) \qquad (8\text{-}99)$$

It is understood that Eq. (8-99) does not hold until the saturation charge from field charging has been closely approached, that is, for values of t less than about ten to twenty times the value of t_0 given in Eq. (8-91).

EXAMPLE 8-8 Using the following data, determine q_p as a function of time for combined field and diffusion charging. Use particle diameters of 0.1, 0.5, and 1.0 μm.

$$\kappa = 5 \qquad E = 3 \times 10^6 \text{ V/m} \qquad T = 300 \text{ K}$$
$$N = 2 \times 10^{15} \text{ ions/m}^3 \qquad m = 5.3 \times 10^{-26} \text{ kg}$$

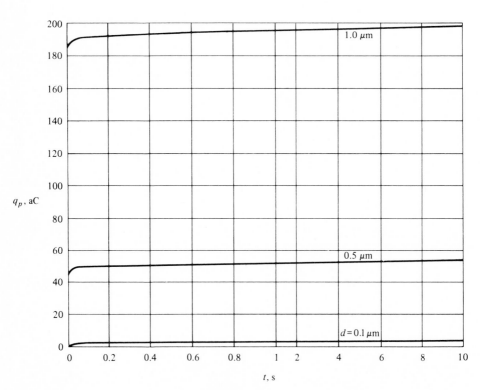

FIGURE 8-20
Combined field and diffusion charging from Example 8-8.

SOLUTION   Equation (8-91) gives the time constant t_0 as

$$t_0 = \frac{4(8.85 \times 10^{-12})}{2 \times 10^{15}(1.6 \times 10^{-19})(2.2 \times 10^{-4})} = 0.000503 \text{ s}$$

Equation (8-99) then becomes

$$q_p = \frac{3(5)}{7}\pi(8.85 \times 10^{-12})(3 \times 10^6 d^2) + \frac{2\pi(8.85 \times 10^{-12})(1.38 \times 10^{-23})(300d)}{1.6 \times 10^{-19}}$$

$$\times \ln\left[1 + \frac{(1.6 \times 10^{-19})^2(2 \times 10^{15})td}{2(8.85 \times 10^{-12})\sqrt{2(5.3 \times 10^{-26})(\pi)(1.38 \times 10^{-23})(300)}}\right]$$

$$= 178.7 \times 10^{-6} d^2 + 1.438 \times 10^{-12} d \ln(1 + 7.79 \times 10^{10} td)$$

Figure 8-20 shows how q_p increases with time for the three cases. The saturation charges for the three respective diameters are 1.787, 44.7, and 178.7 aC.   ////

8-6   SINGLE-STAGE PRECIPITATORS

At this point, it is desirable to put together the equations derived, along with any additional equations needed, so that design of the different types of precipitators may be made.

Cylindrical Precipitator with Relatively Low Dust Concentration

Consider first the cylindrical single-stage precipitator. The equations already derived, with some modification, will suffice to design this precipitator if the particle concentration is relatively small for the entering gas. It is assumed that the particles are charged essentially to saturation very quickly after entering the precipitator. Thereafter, the volumetric charge density in the cylinder decreases as particles are collected but the charge on each particle remains the same. Our equations need to take this fact into account; for this reason some additional derivation is in order.

The field strength at the edge of the corona is fixed and is given by Eq. (8-54). Equations (8-27) and (8-29) can be combined, giving for E and $\tilde{V}_0$.

$$E = \frac{q_{vp}}{2\varepsilon_0}\left(r - \frac{r_0^2}{r}\right) + \frac{r_0 E_0}{r} \tag{8-100}$$

$$\tilde{V}_0 = \frac{q_{vp}}{4\varepsilon_0}\left(r_c^2 - r_0^2 - 2r_0^2 \ln\frac{r_c}{r_0}\right) + r_0 E_0 \ln\frac{r_c}{r_0} \tag{8-101}$$

Since normally r_0 is much less than r_c, Eq. (8-101) may be written as

$$\tilde{V}_0 = \frac{q_{vp}r_c^2}{4\varepsilon_0} + r_0 E_0 \ln\frac{r_c}{r_0} \tag{8-102}$$

The saturation charge acquired by the particles depends on the field strength to which they are exposed. The magnitude of this field strength will differ depending on the location of the particle. For this analysis, an average field strength will be used, and it is assumed that each particle receives a saturation charge corresponding to this average field strength. Actually, the charge should be much greater than the value calculated this way since the turbulent action of the gas flowing in the cylinder will cause a large fraction of the particles to reside in the high-strength portion of the tube for a time, thus imparting a greater-than-average charge to the bulk of the particles. This effect will be ignored here.

The mean charge is determined from the following equation:

$$E_m = \frac{\int_{r_0}^{r_c} E 2\pi r \, dr}{\int_{r_0}^{r_c} 2\pi r \, dr} \qquad (8\text{-}103)$$

Substituting Eq. (8-100) into Eq. (8-103) leads to

$$E_m = \frac{q_{vp} r_c}{3\varepsilon_0} + \frac{2E_0 r_0 (r_c - r_0)}{r_c^2} \qquad (8\text{-}104)$$

At the entrance to the precipitator, the mean field strength is

$$E_{m_0} = \frac{q_{vp_0} r_c}{3\varepsilon_0} + \frac{2E_0 r_0 (r_c - r_0)}{r_c^2} \qquad (8\text{-}105)$$

in which q_{vp_0} is the charge density the particles acquire when they reach saturation at the mean field strength existing at the entrance before many particles have been collected.

The saturation charge is given by Eq. (8-82), which becomes

$$q_{ps} = \frac{3\kappa}{\kappa + 2} \pi \varepsilon_0 d^2 E_m \qquad (8\text{-}106)$$

Combining Eq. (8-106) with Eq. (8-105) and noting that charge density is related to the saturation charge and the particle density C_{nv} as

$$q_{vp} = q_{ps} C_{nv} \qquad q_{vp_0} = q_{ps} C_{nv_0} \qquad (8\text{-}107)$$

the following equation for initial charge density results:

$$q_{vp_0} = \frac{2E_0 r_0 (r_c - r_0)}{r_c^2 [(\kappa + 2)/3\kappa\pi\varepsilon_0 \, d^2 C_{nv_0} - r_c/3\varepsilon_0]} \qquad (8\text{-}108)$$

The saturation charge becomes

$$q_{ps} = \frac{q_{vp_0}}{C_{nv_0}} = \frac{2E_0 r_0 (r_c - r_0)\varepsilon_0}{r_c^2 [(\kappa + 2)/3\kappa\pi d^2 - r_c C_{nv_0}/3]} \qquad (8\text{-}109)$$

Equation (8-107) then gives the charge density at any point along the precipitator:

$$q_{vp} = \frac{2E_0 r_0(r_c - r_0)\varepsilon_0 C_{nv}}{r_c^2[(\kappa + 2)/3\kappa\pi d^2 - r_c C_{nv_0}/3]} \qquad (8\text{-}110)$$

Equation (8-100) gives the field strength:

$$E = \frac{E_0 r_0(r_c - r_0)C_{nv}(r - r_0^2/r)}{r_c^2[(\kappa + 2)/3\kappa\pi d^2 - r_c C_{nv_0}/3]} + \frac{r_0 E_0}{r} \qquad (8\text{-}111)$$

Equation (8-102) gives the voltage at the edge of the corona in terms of the particle density along the cylinder:

$$\tilde{V}_0 = \frac{E_0 r_0(r_c - r_0)C_{nv}}{2[(\kappa + 2)/3\kappa\pi d^2 - r_c C_{nv_0}/3]} + r_0 E_0 \ln \frac{r_c}{r_0} \qquad (8\text{-}112)$$

Equations (8-104) and (8-105) then give the mean field strength:

$$E_m = \frac{2E_0 r_0(r_c - r_0)C_{nv}}{r_c[(\kappa + 2)/\kappa\pi d^2 - r_c C_{nv_0}]} + \frac{2E_0 r_0(r_c - r_0)}{r_c^2} \qquad (8\text{-}113)$$

$$E_{m_0} = \frac{2E_0 r_0(r_c - r_0)C_{nv_0}}{r_c[(\kappa + 2)/\kappa\pi d^2 - r_c C_{nv_0}]} + \frac{2E_0 r_0(r_c - r_0)}{r_c} \qquad (8\text{-}114)$$

When Eqs. (8-110) and (8-112) are substituted into Eq. (8-29), the collecting field strength becomes

$$E_c = \frac{E_0 r_0(r_c - r_0)C_{nv}}{r_c[(\kappa + 2)/3\kappa\pi d^2 - r_c C_{nv_0}/3]} + \frac{r_0 E_0}{r_c} \qquad (8\text{-}115)$$

Figure 8-21 shows the arrangement of corona and collecting cylinder. The particle density C_{nv} varies with distance along the cylinder as particles are collected, which causes the field strength E, collecting field strength E_c, and voltage at the edge of the corona V_0 to vary in the x direction. Since E_c depends on the particle density C_{nv}, as shown in Eq. (8-115), it will be best to derive the equation for collection efficiency directly for this situation instead of using Eq. (8-12). From the derivation which led to Eq. (8-12), the rate of diminution of particles crossing a given section at position x is given by

$$-\frac{dN}{N} = \frac{2q_{ps} E_c r_c C}{3\mu dQ} dx$$

The rate at which particles cross a given section N is related to the particle density by

$$N = C_{nv} Q \qquad (8\text{-}116)$$

Using Eqs. (8-116) and (8-115) in the equation preceding (8-116) leads to

$$\frac{dC_{nv}}{C_{nv}} = -\frac{2q_{ps} C r_0 E_0}{3\mu dQ} \left[\frac{3(r_c - r_0)C_{nv}}{(\kappa + 2)/\kappa\pi d^2 - r_c C_{nv_0}} + 1 \right] dx \qquad (8\text{-}117)$$

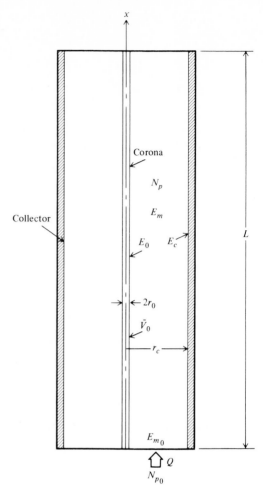

FIGURE 8-21
Analysis of a cylindrical single-stage
precipitator.

This equation may be readily integrated from C_{nv_0} at $x = 0$ to C_{nv_L} at $x = L$, where $C_{nv_L} = C_{nv_0}(1 - \eta)$, leading to

$$L = \frac{1}{\beta} \ln \frac{\alpha C_{nv_0} + 1/(1 - \eta)}{\alpha C_{nv_0} + 1} \qquad (8\text{-}118)$$

in which α and β are defined as

$$\alpha = \frac{3(r_c - r_0)}{(\kappa + 2)/\kappa \pi d^2 - r_c C_{nv_0}} \qquad \text{m}^3 \qquad (8\text{-}119)$$

$$\beta = \frac{2q_{ps} C r_0 E_0}{3\mu d Q} \qquad \text{m}^{-1} \qquad (8\text{-}120)$$

For standard air, Eq. (8-120) may be written

$$\beta = \frac{36{,}231 q_{ps} C r_0 E_0}{dQ} \qquad (8\text{-}121)$$

The use of these equations is illustrated in the following example.

EXAMPLE 8-9   Design a single-stage cylindrical precipitator to handle an air-stream containing solid particles if the following data are given:

$$Q_{total} = 1000 \text{ m}^3/\text{s} \qquad T = 300 \text{ K}$$
$$C_{mv} = 0.01 \text{ kg/m}^3 \qquad \kappa = 4$$
$$d = 10 \ \mu\text{m} \qquad P = 1.0 \text{ atm}$$
$$\rho_p = 1500 \text{ kg/m}^3 \qquad \eta = 0.99$$

In addition, the following dimensions and other data are chosen:

$$r_c = 0.25 \text{ m} \qquad C = 1.0$$
$$r_0 = 0.002 \text{ m} \qquad \mu = 1.8 \times 10^{-5} \text{ kg/m s}$$
$$f = 0.6 \qquad V = 2.0 \text{ m/s}$$

SOLUTION   The flow rate through each tube is computed as

$$Q = AV = \frac{\pi}{4} (0.5)^2 (2.0) = 0.393 \text{ m}^3/\text{s}$$

From this equation, the number of tubes required is $1000/0.393 = 2545$. The Reynolds number is

$$Re = \frac{V D_c \rho}{\mu} = \frac{2.0(0.5)(1.185)}{1.8 \times 10^{-5}} = 65{,}833$$

The incoming particle density is computed from C_{mv} and the mass of each particle as

$$C_{nv_0} = \frac{6 C_{mv}}{\pi \rho_p d^3} = \frac{6(0.01)}{\pi (1500)(10^{-15})} = 1.27 \times 10^{10} \text{ particles/m}^3$$

The field strength at the edge of the corona is computed from Eq. (8-54):

$$E_0 = -3 \times 10^6 (0.6)\left(1 + \frac{0.03}{\sqrt{0.002}}\right) = -3.01 \times 10^6 \text{ V/m}$$

Equation (8-108) gives q_{vp_0}:

$$q_{vp_0} = \frac{2(-3.01 \times 10^6)(0.002)(0.248)}{(0.25)^2 [6/12(8.85 \times 10^{-12})(10^{-10})(1.27 \times 10^{10})\pi - 0.25/3(8.85 \times 10^{-12})]}$$
$$= -1.01 \times 10^{-5} \text{ C/m}^3$$

and Eq. (8-109) gives for q_{ps}

$$q_{ps} = \frac{q_{vp_0}}{C_{nv_0}} = \frac{-1.01 \times 10^{-5}}{1.27 \times 10^{10}} = -0.795 \times 10^{-15} \text{ C}$$

Equation (8-111) gives for the field strength

$$E = \frac{-3.01 \times 10^6 (0.002)(0.248 C_{nv})(r - 4 \times 10^{-6}/r)}{0.0625[6/12(\pi \times 10^{-10}) - 0.25(1.27 \times 10^{10})/3]} + \frac{0.002(-3.01 \times 10^6)}{r}$$

$$= -0.451 \times 10^{-4} C_{nv} \left(r - \frac{4 \times 10^{-6}}{r} \right) - \frac{6022}{r} \quad \text{V/m}$$

At the entrance $(x = 0)$, this becomes

$$E_{x=0} = -0.573 \times 10^6 r - \frac{6022}{r} \quad \text{V/m}$$

Equation (8-112) gives the voltage at the edge of the corona:

$$\tilde{V}_0 = \frac{-3.01 \times 10^6 (0.002)(0.248 C_{nv})}{2[6/12(\pi \times 10^{-10}) - 0.25(1.27 \times 10^{10})/3]} + 0.002(-3.01 \times 10^6) \ln \frac{0.25}{0.002}$$

$$= -0.141 \times 10^{-5} C_{nv} - 29{,}077 \quad \text{V}$$

At the entrance this is $\tilde{V}_{00} = -46{,}984$ V; at the exit, the value becomes $\tilde{V}_{0L} = -29{,}256$ V. Equation (8-113) gives for E_m

$$E_m = \frac{2(-3.01 \times 10^6)(0.002)(0.248) C_{nv}}{0.25[6/4(\pi \times 10^{-10}) - 0.25(1.27 \times 10^{10})]} + \frac{2(-3.01 \times 10^6)(0.002)(0.248)}{0.0625}$$

$$= -0.75 \times 10^{-5} C_{nv} - 47{,}774 \quad \text{V/m}$$

At the entrance, the value of E_m becomes $E_{m_0} = -143{,}024$ V/m. Equation (8-115) gives for the collecting field strength

$$E_c = \frac{-3.01 \times 10^6 (0.002)(0.248 C_{nv})}{0.25(0.053 \times 10^{10})} + \frac{0.002(-3.01 \times 10^6)}{0.25}$$

$$= -0.113 \times 10^{-4} C_{nv} - 24{,}080 \quad \text{V/m}$$

At the entrance, this becomes $E_{c_0} = -167{,}590$ V/m. Most of the preceding calculations are included here for instructional interest; it is unnecessary to perform all these calculations in the actual design of a precipitator.

We must compute the required length of the precipitator; Eqs. (8-119), (8-121), and (8-118) give for the length

$$\alpha = \frac{3(0.248)}{6/4(\pi \times 10^{-10}) - 0.25(1.27 \times 10^{10})} = 4.68 \times 10^{-10} \text{ m}^3$$

$$\beta = \frac{36{,}231(-0.795 \times 10^{-15})(1.0)(0.002)(-3.01 \times 10^6)}{10^{-5}(0.393)} = 0.0441 \text{ m}^{-1}$$

$$L = \frac{1}{0.0441} \ln \frac{4.68 \times 10^{-10}(1.27 \times 10^{10}) + 1/0.01}{4.68 \times 10^{-10}(1.27 \times 10^{10}) + 1} = 61.8 \text{ m} \quad \textit{Ans.} \qquad ////$$

In Example 8-9, the choice of r_0, the radius at the edge of the corona, was arbitrary. In Example 8-10, we shall see that the choice of this radius makes a tremendous difference in the required length of the precipitator. Although undoubtedly there is a practical limit to r_0, Eq. (8-59) suggests that the only limit is the wire diameter. This equation can be solved for r_w, yielding

$$r_w = r_0 + 0.00045 - 0.03\sqrt{r_0 + 0.000225} \qquad (8\text{-}122)$$

in which r_0 and r_w are expressed in meters. For Example 8-9, the value of r_w is computed to be 0.001034 m. Consider now the result of doubling r_0, as shown in the next example.

EXAMPLE 8-10   Repeat Example 8-9 if r_0 is chosen to be 0.004 m, the other data remaining the same.

SOLUTION   From Example 8-9, we have $Q = 0.393$ m³/s and $C_{nv_0} = 1.27 \times 10^{10}$ particles/m³. E_0 is obtained as before from Eq. (8-54):

$$E_0 = -3 \times 10^6(0.6)\left(1 + \frac{0.03}{\sqrt{0.004}}\right) = -2.65 \times 10^6 \text{ V/m}$$

From Eq. (8-109), the particle charge q_{ps} is,

$$q_{ps} = \frac{2(-2.65 \times 10^6)(0.004)(0.246)(8.85 \times 10^{-12})}{0.0625[6/12(\pi \times 10^{-10}) - 0.25(1.27 \times 10^{10})/3]} = -1.383 \times 10^{-15} \text{ C}$$

The voltage $\tilde{V}_{00}$ at the inlet at the edge of the corona is obtained from Eq. (8-112):

$$\tilde{V}_{00} = \frac{-2.65 \times 10^6(0.004)(0.246)(1.27 \times 10^{10})}{2[6/12(\pi \times 10^{-10}) - 0.25(1.27 \times 10^{10})/3]} + 0.004(-2.65 \times 10^6) \ln \frac{0.25}{0.004}$$

$$= -0.749 \times 10^5 \text{ V}$$

Equations (8-119) and (8-121) give for α and β

$$\alpha = \frac{3(0.246)}{6/4(\pi \times 10^{-10}) - 0.25(1.27 \times 10^{10})} = 4.612 \times 10^{-10} \text{ m}^3$$

$$\beta = \frac{36{,}231(-1.383 \times 10^{-15})(1.0)(0.004)(-2.65 \times 10^6)}{10^{-5}(0.393)} = 0.135 \text{ m}^{-1}$$

Then, from Eq. (8-118), the length of the tube is

$$L = \frac{1}{0.135} \ln \frac{4.612 \times 10^{-10}(1.27 \times 10^{10}) + 1/0.01}{4.612 \times 10^{-10}(1.27 \times 10^{10}) + 1} = 20.3 \text{ m} \qquad Ans.$$

Equation (8-122) gives for the radius of the wire

$$r_w = 0.004 + 0.00045 - 0.03\sqrt{0.004 + 0.000225} = 0.0025 \text{ m} \qquad Ans. \qquad ////$$

Cylindrical Precipitator with Relatively High Dust Concentration

Examples 8-9 and 8-10 are based on a rather small inlet particle concentration. If C_{nv_0} were larger than that of Example 8-9, the other data remaining the same, rather peculiar results would be obtained. For instance, if C_{nv_0} were taken to be 1.8×10^{10} particles/m^3, Eq. (8-115) gives for E_c at $x = 0$ the value of -6.5×10^6 V/m; such a high value of E_c would require the existence of a corona at the collection electrode. If C_{nv_0} were 10^{11} particles/m^3, then E_c is computed to be 64,300 V/m, giving a reversal of sign from E_0 and thus producing an electric field that is unsuitable for collection of particles. With these relatively high particle concentrations, if all the particles are charged to saturation, the space charge distorts the electric field to the point where the device can no longer function as a precipitator. The secret is not to charge the particles to saturation in the inlet section; indeed, this is probably impossible. Instead, saturation will be achieved only after most of the particles have been collected at a considerable distance down the precipitator. To treat this situation, it will be necessary to analyze the precipitator performance in a somewhat different way.

Equation (8-54) gives the value of E_0 as before. The value of E_c is now assumed to be a specified function of position x, although the choice of this function will be made later. Equations (8-27) to (8-29) can be combined to express E, q_v, and $\tilde{V}_0$ as functions of E_0 and E_c, with E also a function of r. The following equations result from this process:

$$E = \frac{E_c r_c - E_0 r_0}{r_c^2 - r_0^2} r + \frac{E_0 r_c^2 r_0 - E_c r_c r_0^2}{r_c^2 - r_0^2} \frac{1}{r} \qquad (8\text{-}123)$$

$$q_{vp} = 2\varepsilon_0 \frac{E_c r_c - E_0 r_0}{r_c^2 - r_0^2} \qquad (8\text{-}124)$$

$$\tilde{V}_0 = \frac{1}{2}(E_c r_c - E_0 r_0) + \frac{(E_0 r_c - E_c r_0) r_c r_0}{r_c^2 - r_0^2} \ln \frac{r_c}{r_0} \qquad (8\text{-}125)$$

Using the fact that $r_0 \ll r_c$, these equations may be written as

$$E = \frac{E_c}{r_c} r - \frac{E_0 r_0}{r_c^2} r + \frac{E_0 r_0}{r} \qquad (8\text{-}126)$$

$$q_{vp} = \frac{2\varepsilon_0}{r_c^2}(E_c r_c - E_0 r_0) \qquad (8\text{-}127)$$

$$\tilde{V}_0 = \frac{1}{2}(E_c r_c - E_0 r_0) + r_0 E_0 \ln \frac{r_c}{r_0} \qquad (8\text{-}128)$$

The mean field strength E_m, as given by Eq. (8-103) and using Eq. (8-126), gives

$$E_m = \frac{2}{r_c^2} \int_{r_0}^{r_c} Er\, dr = \frac{2}{3}E_c + 2E_0 \frac{r_0}{r_c}\left(\frac{2}{3} - \frac{r_0}{r_c}\right) \qquad (8\text{-}129)$$

The particle concentration will continuously decrease as particles are collected. At a certain distance x_s downstream from the entrance, the particles which remain in the stream will reach saturation charge. The fractional diminution of particle concentration is given by

$$-\frac{dC_{nv}}{C_{nv}} = \frac{2r_c C q_p E_c}{3\mu dQ} dx$$

Combining this equation with Eqs. (8-107) and (8-127) gives

$$dC_{nv} = -\frac{4\varepsilon_0 C}{3\mu dQr_c} (E_c^2 r_c - E_0 E_c r_0) dx$$

This equation can now be integrated between the limits of C_{nv_0} at $x = 0$ and C_{nv_s} at $x = x_s$, where C_{nv_s} is the particle concentration at the point where the particles first have saturation charge. When solved for x_s, the result is

$$x_s = \frac{3\mu dQr_c(C_{nv_0} - C_{nv_s})}{4\varepsilon_0 C(E_c^2 r_c - E_0 E_c r_0)} \qquad (8\text{-}130)$$

Combining Eqs. (8-82), (8-107), and (8-127), and solving for C_{nv_s} gives

$$C_{nv_s} = \frac{2(\kappa + 2)}{3\kappa\pi r_c^2 d^2} \frac{E_c r_c - E_0 r_0}{2E_c/3 + 2E_0(\frac{2}{3} - r_0/r_c)r_0/r_c} \qquad (8\text{-}131)$$

When Eq. (8-131) is substituted into Eq. (8-130), x_s becomes

$$x_s = \frac{3\mu dQr_c}{4\varepsilon_0 CE_c}\left[\frac{C_{nv_0}}{E_c r_c - E_0 r_0} - \frac{2(\kappa + 2)/3\kappa\pi r_c^2 d^2}{2E_c/3 + 2E_0(\frac{2}{3} - r_0/r_c)r_0/r_c}\right] \qquad (8\text{-}132)$$

Particles will be collected beyond the point x_s; in fact, the behavior will be very similar to that of the precipitator with low particle concentration, analyzed previously. Using Eq. (8-115), with C_{nv_s} substituted for C_{nv_0}, an equation for L similar to Eq. (8-118) is obtained:

$$L = x_s + \frac{1}{\beta} \ln \frac{C_{nv_s}[\alpha' C_{nv_0} + 1/(1 - \eta)]}{C_{nv_0}(\alpha' C_{nv_s} + 1)} \qquad (8\text{-}133)$$

in which α' is given by

$$\alpha' = \frac{3(r_c - r_0)}{(\kappa + 2)/\kappa\pi d^2 - r_c C_{nv_s}} \quad m^3 \qquad (8\text{-}134)$$

and β is given by Eq. (8-120). The quantity q_{ps} in Eq. (8-120) is obtained from Eqs. (8-127) and (8-131):

$$q_{ps} = \frac{q_{vp}}{C_{nv_s}} = \frac{3\kappa\pi\varepsilon_0 d^2}{\kappa + 2}\left[\frac{2}{3}E_c + 2E_0\frac{r_0}{r_c}\left(\frac{2}{3} - \frac{r_0}{r_c}\right)\right] \qquad (8\text{-}135)$$

The value of E_c chosen for the first section of the precipitator is rather arbitrary. The value chosen should not be too close to E_0 since a back corona is highly undesirable. A layer of dust attached to the collector electrode will reduce the voltage drop across the gas and tend to form a back corona, particularly if the dust resistivity is high. However, a large value of E_c will improve the collection efficiency, provided that detrimental effects are not introduced, such as a back corona. A suitable value of E_c will probably lie in the range of 0.1 to 0.3 times E_0.

EXAMPLE 8-11 Use the data of Example 8-9, except that $C_{mv} = 0.1$ kg/m³, and design the precipitator for this case.

SOLUTION From Example 8-9, $E_0 = -3.01 \times 10^6$ V/m. The initial particle concentration is $C_{nv_0} = 1.27 \times 10^{11}$ particles/m³. Equation (8-131) gives for the particle concentration at saturation

$$C_{nv_s} = \frac{2(6)}{3(4)(\pi)(0.0625)(10^{-10})}$$

$$\times \frac{0.25E_c - 0.002(-3.01 \times 10^6)}{0.667E_c + 2(-3.01 \times 10^6)(0.667 - 0.008)(0.002/0.25)}$$

$$= 1.909 \times 10^{10} \frac{E_c + 24{,}080}{E_c - 47{,}582} \text{ particles/m}^3$$

Note that the collection field strength E_c has not yet been specified.
For the location at which saturation occurs, Eq. (8-130) applies:

$$x_s = \frac{3(1.8 \times 10^{-5})(10^{-5})(0.393)(0.25)(12.7 \times 10^{10} - C_{nv_s})}{4(8.85 \times 10^{-12})(1.0)[0.25E_c + 3.01 \times 10^6(0.002)]E_c}$$

$$= 6 \frac{12.7 \times 10^{10} - C_{nv_s}}{(E_c + 24{,}080)E_c} \quad \text{m}$$

Equation (8-135) gives the particle charge at saturation:

$$q_{ps} = \frac{12\pi}{6}(8.85 \times 10^{-12})(10^{-10})[0.667E_c + 2(-3.01 \times 10^6)(0.008)(0.659)]$$

$$= 3.71 \times 10^{-21}(E_c - 47{,}582) \quad \text{C}$$

α' and β are obtained from Eqs. (8-120) and (8-134) as

$$\alpha' = \frac{3(0.248)}{6/4(\pi \times 10^{-10}) - 0.25C_{nv_s}} = \frac{1.56 \times 10^{-10}}{1 - 0.524 \times 10^{-10}C_{nv_s}} \quad \text{m}^3$$

$$\beta = \frac{2(1.0)(0.002)(-3.01 \times 10^6)q_{ps}}{3(1.8 \times 10^{-5})(10^{-5})(0.393)} = -0.0567 \times 10^{15}q_{ps} \quad \text{m}^{-1}$$

Equation (8-133) then gives the length of precipitator required:

$$L = x_s + \frac{1}{\beta} \ln \frac{C_{nv_s}(12.7 \times 10^{10}\alpha' + 100)}{12.7 \times 10^{10}(\alpha'C_{nv_s} + 1)}$$

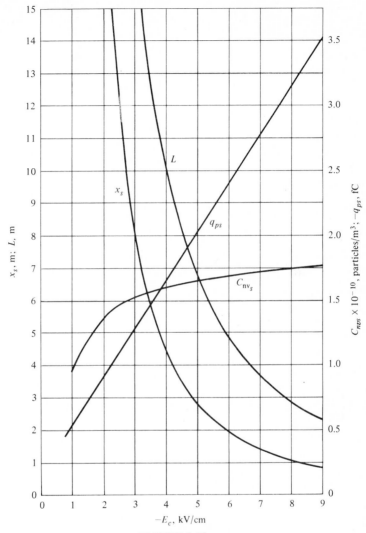

FIGURE 8-22
Parameters from Example 8-9 as a function of E_c.

The preceding quantities are plotted as functions of E_c in Fig. 8-22. We see from this figure that L drops off rapidly as E_c increases in magnitude. Let us choose E_c as $-600,000$ V/m; then the length of the precipitator section becomes 4.85 m. Equation (8-128) gives for the voltage at the edge of the corona

$$\tilde{V}_0 = 0.5[-0.6 \times 10^6(0.25) + 3.01 \times 10^6(0.002)] + 0.002(-3.01 \times 10^6) \ln \frac{0.25}{0.002}$$

$$= -101,006 \text{ V} \qquad Ans.$$

Equation (8-58) gives an estimate of the wire voltage. Using $r_w = 0.001034$ m, as calculated previously, the result is

$$\tilde{V}_w = \tilde{V}_0 + E_0 \frac{r_0{}^2 - r_w{}^2}{2r_w} = -105{,}272 \text{ V} \qquad Ans. \qquad ////$$

Wire-and-Plate Precipitator with Relatively Low Dust Concentration

An exact analysis of the electric field formed in the wire-and-plate precipitator involves partial differential equations and for this reason will not be tackled here. Instead, let us seek a means for using the formulas already developed for the case of the cylindrical precipitator to give reasonably satisfactory results in this case also. The collecting field strength E_c can be taken as constant along the length of the precipitator. With low dust concentration, the particles can be imparted their saturation charge in a short distance from the entrance, and thus charge may be assumed to be constant also. Equations (8-12) and (8-13) then apply, with $E_{cm} = E_c$.

To determine the charge on the particles, the equation developed for use with a cylindrical precipitator, Eq. (8-109), will also be used here after r_c has been suitably modified. The modification which will be used is to let the plan area associated with one wire $2Wl$ be equal to the cross-sectional area of the cylindrical precipitator $\pi r_c{}^2$, giving

$$r_c = \sqrt{\frac{2Wl}{\pi}} = 0.8\sqrt{Wl} \qquad (8\text{-}136)$$

Equation (8-109) then becomes

$$q_{ps} = \frac{6E_0 \, \varepsilon_0 \, r_0 (0.8\sqrt{Wl} - r_0)}{0.64 Wl[(\kappa + 2)/\kappa \pi d^2 - 0.8\sqrt{Wl} C_{nv_0}]} \qquad (8\text{-}137)$$

Figure 8-23 shows the geometry of the wire-and-plate configuration.

EXAMPLE 8-12 Using the data of Example 8-9, design a wire-and-plate precipitator to produce the same collection efficiency. Take $W = 0.25$ m, H (height of collection plate) $= 10$ m, and l (spacing between wires) $= 0.4$ m. Also use a velocity of 4 m/s through the precipitator.

SOLUTION From Example 8-9, $E_0 = -3.01 \times 10^6$ V/m and $C_{nv_0} = 1.27 \times 10^{10}$ particles/m³. Let us take E_c as 0.2 times E_0, or -6×10^5 V/m. Equation (8-137) gives for the particle charge

$$q_{ps} = \frac{6(-3.01 \times 10^6)(8.85 \times 10^{-12})(0.002)[0.8\sqrt{0.25(0.4)} - 0.002]}{0.64(0.25)(0.4)[6/4(\pi \times 10^{-10}) - 0.8\sqrt{0.25(0.4)}1.27 \times 10^{10}]}$$

$$= -0.803 \times 10^{-15} \text{ C}$$

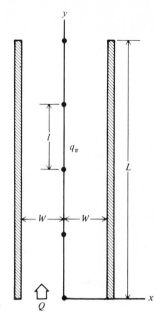

FIGURE 8-23
Wire-and-plate precipitator.

The flow rate through the half section is given by

$$Q = WHV = 0.25(10)(4) = 10 \text{ m}^3/\text{s}$$

Equation (8-12) gives the efficiency in terms of $A_c = HL$; solving for L gives

$$L = -\frac{3\pi\mu dQ \ln(1-\eta)}{q_{ps} CHE_c}$$

$$= -\frac{3\pi(1.84 \times 10^{-5})(10^{-5})(10) \ln 0.01}{-0.803 \times 10^{-15}(1.0)(10)(-6 \times 10^5)} = 16.6 \text{ m} \qquad Ans.$$

The number of sections required for a total flow rate of 1000 m³/s is $\frac{1000}{20} = 50$, which gives a total width for the precipitator of 25 m, not counting the extra space required for the housing. ////

Wire-and-Plate Precipitator with Relatively High Dust Concentration

The case of high dust concentration is handled here somewhat similarly to that of low dust concentration. The charge on the particles is obtained from the equations developed for the cylindrical precipitator by replacing the radius r_c by its equivalent value given by Eq. (8-136). The collection-efficiency equation will be derived using the plate geometry, however.

The equation for fractional diminution of particle concentration is given by

$$\frac{dC_{nv}}{C_{nv}} = -\frac{q_p E_c C}{3\pi\mu dWV} dx$$

Using Eqs. (8-107), (8-127), and (8-136), the particle charge is given by

$$q_p = \frac{2\varepsilon_0}{0.64 C_{nv} Wl}(0.8 E_c \sqrt{Wl} - E_0 r_0) \qquad (8\text{-}138)$$

Then the rate of change of particle concentration is given by

$$dC_{nv} = -0.332 \frac{\varepsilon_0 E_c C}{\mu dW^2 lV}(0.8 E_c \sqrt{Wl} - E_0 r_0) \, dx$$

Integrating this equation from the inlet to the point at the location x_s where the charge becomes saturated produces

$$x_s = \frac{\mu dW^2 lV(C_{nv_0} - C_{nv_s})}{0.332\varepsilon_0 E_c C(0.8 E_c \sqrt{Wl} - E_0 r_0)} \qquad (8\text{-}139)$$

After using Eq. (8-136) for r_c, Eq. (8-131) yields

$$C_{nv_s} = \frac{0.212(\kappa + 2)}{\kappa d^2} \frac{0.8 E_c \sqrt{Wl} - E_0 r_0}{0.427 Wl E_c + 2 E_0 r_0 (0.533\sqrt{Wl} - r_0)} \qquad (8\text{-}140)$$

For the remaining length of the precipitator, assuming that the collecting field strength E_c remains constant throughout, the fractional diminution of particle concentration is given by

$$\frac{dC_{nv}}{C_{nv}} = -\frac{q_{ps} E_c C}{3\pi\mu dWV} dx$$

Upon integrating from x_s and C_{nv_s} to L and $C_{nv_L} = (1 - \eta)C_{nv_0}$ and solving for L, the result is

$$L = x_s + \frac{3\pi\mu dWV}{q_{ps} E_c C} \ln \frac{C_{nv_s}}{(1 - \eta)C_{nv_0}} \qquad (8\text{-}141)$$

Equation (8-135) gives an expression for q_{ps}, again using the equivalent value for r_c:

$$q_{ps} = \frac{3\kappa\pi\varepsilon_0 d^2}{\kappa + 2}\left[\frac{2}{3} E_c + \frac{2.5 E_0 r_0}{\sqrt{Wl}}\left(\frac{2}{3} - \frac{1.25 r_0}{\sqrt{Wl}}\right)\right] \qquad (8\text{-}142)$$

The value of E_0 can be obtained from Eq. (8-54), as has been done in all cases, and E_c must be assumed, for instance, as 0.2 to 0.3 times the value of E_0. Although this procedure will produce only a rough approximation, it gives the best easily workable solution currently available.

EXAMPLE 8-13   Repeat Example 8-12 using a dust concentration of $C_{mv} = 0.1$ kg/m^3.

SOLUTION   From Example 8-11, $C_{nv_0} = 1.27 \times 10^{11}$ particles/m^3. The value of r_c from Eq. (8-136) is 0.253 m. Here we take $W = 0.25$ m and $l = 0.4$ m. The velocity through the precipitator is 2.0 m/s. Equation (8-135) gives q_{ps} as

$$q_{ps} = \frac{3(4)(\pi)(8.85 \times 10^{-12})(10^{-10})}{6}$$

$$\times \left[\frac{2}{3}(-6 \times 10^5) + 2(-3.01 \times 10^6)\frac{0.002}{0.253}\left(\frac{2}{3} - \frac{0.002}{0.253}\right) \right]$$

$$= -2.4 \times 10^{-15} \text{ C}$$

Equation (8-140) gives C_{nv_s}:

$$C_{nv_s} = \frac{0.212(6)}{4 \times 10^{-10}}[0.8(-6 \times 10^5)(0.3162) - (-3.01 \times 10^6)(0.002)]$$

$$\times \{0.427(0.1)(-6 \times 10^5) + 2(-3.01 \times 10^6)(0.002)$$

$$\times [0.533(0.3162) - 0.002]\}^{-1}$$

$$= 1.675 \times 10^{10} \text{ particles/m}^3$$

Equation (8-139) gives x_s:

$$x_s = [1.8 \times 10^{-5}(10^{-5})(0.25)^2(0.4)(2)(1.27 \times 10^{11} - 1.675 \times 10^{10})]$$

$$\times \{0.332(8.85 \times 10^{-12})(-6 \times 10^5)(1.0)[0.8(-6 \times 10^5)(0.3162)$$

$$+ 3.01 \times 10^6(0.002)]\}^{-1}$$

$$= 0.371 \text{ m}$$

Then Eq. (8-141) gives the length L of the precipitator:

$$L = 0.371 + \frac{3\pi(1.8 \times 10^{-5})(10^{-5})(0.25)(2)}{-2.4 \times 10^{-15}(-6 \times 10^5)(1.0)} \ln \frac{1.675 \times 10^{10}}{0.01(1.27 \times 10^{11})}$$

$$= 1.89 \text{ m} \qquad Ans. \qquad\qquad ////$$

8-7 TWO-STAGE PRECIPITATORS

At this time, let us turn our attention to two-stage precipitators. Since these devices are normally used for relatively low dust concentrations, as is typical in ventilation practice, we shall not consider higher concentrations. Thus, we shall assume that the charge supplied to each particle is not limited by overall space-charge effects; instead, the particle can achieve a substantial fraction of its saturation charge, limited only by the electric field strength in the region occupied by the particle and the time during which the particle travels through the charging stage. We shall assume that a substantial current flows between the electrodes and that the space charge due to ions is much greater than that due to the charged particles in the region.

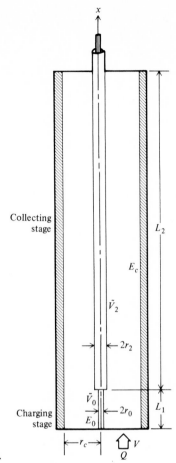

FIGURE 8-24
Cylindrical two-stage precipitator.

Cylindrical Precipitators

For the cylindrical two-stage precipitator, Fig. 8-24 shows the geometry and nomenclature involved in the analysis. Equation (8-33) may be written as follows:

$$q_v = q_{vp} + \frac{i}{2\pi K r L_1 |E|} \qquad (8\text{-}143)$$

Neglecting the space charge due to the particles,

$$q_v = \frac{i}{2\pi K L_1 r |E|} \qquad (8\text{-}144)$$

When the particle space charge is neglected, Eq. (8-34) gives

$$\frac{\partial}{\partial r}(rE) = \frac{i}{2\pi\varepsilon_0 KL_1|E|} \qquad (8\text{-}145)$$

If we assume that the current i is constant and that the field strength is a function of radius only, Eq. (8-145) becomes

$$\frac{d}{dr}(rE) = \frac{i}{2\pi\varepsilon_0 KL_1|E|} \qquad (8\text{-}146)$$

which integrates to

$$(rE)^2 = \frac{|i|}{2\pi\varepsilon_0 KL_1}r^2 + C \qquad (8\text{-}147)$$

When we note that at radius r_0, the field strength is E_0, given by Eq. (8-54), we obtain

$$C = r_0{}^2E_0{}^2 - \frac{|i|}{2\pi\varepsilon_0 KL_1}r_0{}^2$$

Then Eq. (8-147) becomes

$$E = \pm\frac{1}{r}\left[\frac{|i|}{2\pi\varepsilon_0 KL_1}(r^2 - r_0{}^2) + r_0{}^2E_0{}^2\right]^{1/2} \qquad (8\text{-}148)$$

The positive sign applies to a positive corona, while the negative sign signifies the more unusual case of negative corona.

Equation (8-90) gives the time required to impart a given charge on a particle and is rewritten as

$$t = \frac{4\varepsilon_0}{N_i q_e K}\frac{q_p/q_{ps}}{1 - q_p/q_{ps}} \qquad (8\text{-}149)$$

The time available for charging is the residence time of the gas in the charging stage, given by

$$t = \frac{L_1}{V} \qquad (8\text{-}150)$$

The charge residing in an elemental volume, such as the one shown in Fig. 8-9, considering only the ions present, is given as the current flow times the time required for an ion to traverse the width dr of the element:

$$q_i = \frac{|i|\,dr}{KE} \qquad (8\text{-}151)$$

This charge is also equal to the number of free ions in the element times the ionic charge:

$$q_i = \pm N_i q_e 2\pi r L_1\,dr \qquad (8\text{-}152)$$

Equating Eqs. (8-151) and (8-152) and considering q_e to be a positive quantity, the product $N_i q_e$ is given as

$$N_i q_e = \frac{|i|}{2\pi r K L_1 |E|} \qquad (8\text{-}153)$$

Substituting Eqs. (8-150) and (8-153) into Eq. (8-149) leads to

$$\frac{q_p}{q_{ps}} = \frac{1}{1 + 8\pi\varepsilon_0 V |E| r/|i|} \qquad (8\text{-}154)$$

It can be seen that the degree of charging of the particles is independent of the length of the charging section since L_1 cancels out of the equation. This length can be arbitrarily selected; a value at least as large as the radius of the collecting section is suggested. The current i is independent of the value of L_1 selected.

When Eq. (8-82) for q_{ps} is substituted into Eq. (8-154), the result is

$$q_{p1} = \frac{3\kappa}{\kappa + 2} \pi\varepsilon_0 d^2 \frac{E}{1 + 8\pi\varepsilon_0 V |E| r/|i|} \qquad (8\text{-}155)$$

The quantity q_{p1} represents the charge acquired by the particle at the exit from the charging stage. We can reasonably assume that a particle stays at the same radius through the charging section, although in the collecting stage we shall assume random distribution due to turbulent action. The charge on the particle will then depend on its radial location in the charging stage.

We must determine the mean charge on the particles. This is done by the following averaging process:

$$\bar{q}_{p1} = \frac{\int_{r_0}^{r_c} q_{p1} C_{nv_0} 2\pi r L_1 \, dr}{\int_{r_0}^{r_c} C_{nv_0} 2\pi r L_1 \, dr} = \frac{2}{r_c^2} \int_{r_0}^{r_c} r q_{p1} \, dr \qquad (8\text{-}156)$$

Using Eq. (8-155) in Eq. (8-156), we obtain

$$\bar{q}_{p1} = \frac{2}{r_c^2} \frac{3\kappa}{\kappa + 2} \pi\varepsilon_0 d^2 \int_{r_0}^{r_c} \frac{Er \, dr}{1 + 8\pi\varepsilon_0 V |E| r/|i|} \qquad (8\text{-}157)$$

At this point, the mean value $\overline{Er}$ of the product of the field strength E and radius r is defined as

$$\overline{Er} = \frac{1}{r_c - r_0} \int_{r_0}^{r_c} Er \, dr = \pm \frac{\sqrt{|i|/2\pi\varepsilon_0 K L_1}}{r_c - r_0} \int_{r_0}^{r_c} \sqrt{r^2 + r_0^2 \left(\frac{2\pi\varepsilon_0 K L_1 E_0^2}{|i|} - 1 \right)} \, dr$$

$$(8\text{-}158)$$

Then Eq. (8-157) may be written

$$\bar{q}_{p1} = \frac{2}{r_c^2} \frac{3\kappa}{\kappa + 2} \pi\varepsilon_0 d^2 \frac{(r_c - r_0)\overline{Er}}{1 + 8\pi\varepsilon_0 V \overline{Er}/|i|} \qquad (8\text{-}159)$$

The integral of Eq. (8-158) can be evaluated, and using the fact that $r_0^2 \ll r_c^2$, the final result is

$$\overline{Er} = \pm \frac{1}{2(r_c - r_0)} \left[r_c^2 \sqrt{\frac{|i|}{2\pi\varepsilon_0 \, KL_1} + \left(E_0 \frac{r_0}{r_c} \right)^2} - r_0^2 |E_0| \right.$$

$$+ r_0^2 \left(\sqrt{\frac{2\pi\varepsilon_0 \, KL_1}{|i|}} \, E_0^2 - \sqrt{\frac{|i|}{2\pi\varepsilon_0 \, KL_1}} \right)$$

$$\left. \times \ln \frac{r_c}{r_0} \frac{1 + \sqrt{1 + 2\pi\varepsilon_0 \, KL_1 (E_0 r_0/r_c)^2/|i|}}{1 + |E_0| \sqrt{2\pi\varepsilon_0 \, KL_1/|i|}} \right] \tag{8-160}$$

The voltage at the edge of the corona can be obtained from Eqs. (8-17) and (8-148). The differential equation for the voltage is

$$\frac{d\tilde{V}}{dr} = -E = \mp \frac{1}{r} \left[\frac{|i|}{2\pi\varepsilon_0 \, KL_1} (r^2 - r_0^2) + r_0^2 E_0^2 \right]^{1/2} \tag{8-161}$$

Upon integrating between the limits of V_0 at r_0 and 0 at r_c,

$$\tilde{V}_0 = \pm \int_{r_0}^{r_c} \sqrt{\frac{|i|}{2\pi\varepsilon_0 \, KL_1} r^2 + \left(E_0^2 - \frac{|i|}{2\pi\varepsilon_0 \, KL_1} \right) r_0^2} \, \frac{dr}{r}$$

Upon performing the integration and substituting in the limits,

$$\pm \tilde{V}_0 = r_c \sqrt{\frac{|i|}{2\pi\varepsilon_0 \, KL_1} + \left(E_0 \frac{r_0}{r_c} \right)^2} - r_0 |E_0| - r_0 \sqrt{E_0^2 - \frac{|i|}{2\pi\varepsilon_0 \, KL_1}}$$

$$\times \ln \frac{\sqrt{1 + 2\pi\varepsilon_0 \, KL_1 (E_0 r_0/r_c)^2/|i|} + (r_0/r_c)\sqrt{2\pi\varepsilon_0 \, KL_1 E_0^2/|i| - 1}}{|E_0| \sqrt{2\pi\varepsilon_0 \, KL_1/|i|} + \sqrt{2\pi\varepsilon_0 \, KL_1 E_0^2/|i| - 1}}$$

$$\tag{8-162}$$

Equation (8-122) will give the radius of the wire in the charging stage, and Eq. (8-58) will give the voltage of the wire $\tilde{V}_w$.

We turn now to the collecting stage of the precipitator. The collection field strength is given by Eq. (8-29) as

$$E_c = \bar{q}_{p1} \left(\frac{r_c}{2\varepsilon_0} - \frac{r_c^2 - r_2^2}{4\varepsilon_0 r_c \ln r_c/r_2} \right) C_{nv} + \frac{V_2}{r_c \ln r_c/r_2} \tag{8-163}$$

The fractional diminution of particles along the precipitator is given by

$$-\frac{dC_{nv}}{C_{nv}} = \frac{p}{A} \frac{V_t}{V} dx = \frac{2\pi r_c}{\pi r_c^2} \frac{V_t}{V} dx$$

Using Eqs. (8-2), (8-107), and (8-163), this differential equation becomes

$$\frac{dC_{nv}}{C_{nv}} = -\frac{2\bar{q}_{p1} C}{3\pi\mu d V r_c} \left[\bar{q}_{p1} \left(\frac{r_c}{2\varepsilon_0} - \frac{r_c^2 - r_2^2}{4\varepsilon_0 r_c \ln r_c/r_2} \right) C_{nv} + \frac{V_2}{r_c \ln r_c/r_2} \right] dx \tag{8-164}$$

At the entrance where $x = L_1$, the particle concentration is assumed to be C_{nv_0} since relatively few particles will be collected in the charging section. At the exit where $x = L_2$, the concentration is $(1 - \eta)C_{nv_0}$. Integrating Eq. (8-164) between these respective limits yields

$$L_2 = \frac{3\pi\mu dVr_c^2 \ln r_c/r_2}{2\bar{q}_{p1}C\tilde{V}_2}\left[\ln \frac{1}{1-\eta} - \alpha \ln \frac{\alpha C_{nv_0} + V_2/(r_c \ln r_c/r_2)}{\alpha(1-\eta)C_{nv_0} + V_2/(r_c \ln r_c/r_2)}\right] \qquad (8\text{-}165)$$

in which α has been defined as

$$\alpha = \frac{\bar{q}_{p1}}{2\varepsilon_0}\left(r_c - \frac{r_c^2 - r_2^2}{2r_c \ln r_c/r_2}\right) \qquad V \cdot m \qquad (8\text{-}166)$$

EXAMPLE 8-14   A cylindrical two-stage precipitator is to be designed with the following data. Determine the length L_2 required.

$$r_c = 0.10 \text{ m} \qquad\qquad V = 1.0 \text{ m/s}$$
$$r_0 = 0.002 \text{ m} \qquad\qquad \kappa = 6$$
$$r_2 = 0.02 \text{ m} \qquad\qquad K = 2.2 \times 10^{-4} \text{ m}^2/\text{V} \cdot \text{s}$$
$$L_1 = 0.1 \text{ m} \qquad\qquad f = 0.75$$
$$d = 5 \text{ } \mu m \qquad\qquad \eta = 0.99$$
$$C_{nv_0} = 10^9 \text{ particles/m}^3 \qquad i = 10^{-4} \text{ A}$$

SOLUTION   Equation (8-54) gives for E_0

$$E_0 = 3 \times 10^6 (0.75)\left(1 + \frac{0.03}{\sqrt{0.002}}\right) = 3.76 \times 10^6 \text{ V/m}$$

For later use, note that

$$\frac{i}{2\pi\varepsilon_0 KL_1} = \frac{10^{-4}}{2\pi(8.85 \times 10^{-12})(2.2 \times 10^{-4})(0.1)} = 8.18 \times 10^{10} \text{ V}^2/\text{m}^2$$

$$\frac{2\pi\varepsilon_0 KL_1}{i} = 1.22 \times 10^{-11} \text{ m}^2/\text{V}^2$$

$$E_0^2 \frac{r_0^2}{r_c^2} = \left(3.76 \times 10^6 \frac{0.002}{0.10}\right)^2 = 0.5655 \times 10^{10} \text{ V}^2/\text{m}^2$$

Equation (8-160) gives $\bar{E}r$:

$$\bar{E}r = \frac{1}{2(0.098)}\bigg|0.01\sqrt{8.18 \times 10^{10} + 0.566 \times 10^{10}} - 4 \times 10^{-6}(3.76 \times 10^6)$$

$$+ 4 \times 10^{-6}[\sqrt{12.2 \times 10^{-12}} (3.76 \times 10^6)^2 - \sqrt{8.18 \times 10^{10}}]$$

$$\times \ln \frac{0.01}{0.002} \frac{1 + \sqrt{1 + 1.22 \times 10^{-11}(0.566 \times 10^{10})}}{1 + 3.76 \times 10^6\sqrt{12.2 \times 10^{-12}}}\bigg|$$

$$= 17,000 \text{ V}$$

Equation (8-159) gives $\bar{q}_{p1}$:

$$\bar{q}_{p1} = \frac{2}{0.01}\frac{18}{8}\pi(8.85 \times 10^{-12})25 \times 10^{-12}\frac{0.098(17,000)}{1 + 8\pi(8.85 \times 10^{-12})(1.0)(17,000)/10^{-4}}$$

$$= 5.02 \times 10^{-16} \text{ C}$$

Equation (8-162) gives $\tilde{V}_0$, the voltage at the edge of the corona in the charging stage:

$$\tilde{V}_0 = 0.1\sqrt{8.18 \times 10^{10} + 0.566 \times 10^{10}} - 0.002(3.76 \times 10^6)$$

$$- 0.002\sqrt{(3.76)^2 10^{12}} - 0.0818 \times 10^{12} \times$$

$$\ln \frac{\sqrt{1 + 1.22 \times 10^{-11}(0.566 \times 10^{10})} + 0.002\sqrt{1.22 \times 10^{-11}(3.76)^2(10^{12})} - 1/0.1}{3.76 \times 10^6\sqrt{12.2 \times 10^{-12}} + \sqrt{1.22 \times 10^{-11}(3.76)^2(10^{12})} - 1}$$

$$= 44,580 \text{ V}$$

The radius of the wire in the charging stage r_w is given by Eq. (8-122):

$$r_w = 0.002 + 0.00045 - 0.03\sqrt{0.002 + 0.000225} = 0.001034 \text{ m}$$

Equation (8-58) then gives a fair approximation for the required voltage of the wire:

$$\tilde{V}_w = 44,580 + 3.76 \times 10^6\frac{4 \times 10^{-6} - (1.034)^2(10^{-6})}{2(0.001034)} = 49,909 \text{ V}$$

We assume that the voltage of the wire in the collecting section $\tilde{V}_2$ is equal to the wire voltage in the charging section, say, 50,000 V. Equation (8-166) gives

$$\alpha = \frac{5.02 \times 10^{-16}}{2(8.85 \times 10^{-12})}\left[0.1 - \frac{0.01 - 0.0004}{2(0.1)\ln 0.1/0.02}\right] = 1.99 \times 10^{-6} \text{ V} \cdot \text{m}$$

Finally, Eq. (8-165) gives L_2, the length of the collecting section:

$$L_2 = \frac{3\pi(1.8 \times 10^{-5})(5 \times 10^{-6})(1.0)(0.01)(1.61)}{2(5.02 \times 10^{-16})(1.0)(50,000)}$$

$$\times \left[\ln 100 - 1.99 \times 10^{-6}\ln\frac{10^9(1.99 \times 10^{-6}) + 50,000/0.1(1.61)}{10^7(1.99 \times 10^{-6}) + 50,000/0.1(1.61)}\right]$$

$$= 1.25 \text{ m} \quad Ans. \qquad\qquad ////$$

Parallel-Plate Precipitator

The parallel-plate two-stage precipitator treated here is the one shown in Fig. 8-4. The charging section consists of a vertical wire suspended midway between the two grounded collector plates; the collecting section has a charged plate midway between and parallel to the two collecting plates. The dimensions of a typical charging and collecting section are shown in Fig. 8-25.

In the charging section, a corona surrounds the charged wire; the voltage at the edge of the corona is $\tilde{V}_0$, and that at the wire is $\tilde{V}_w$. We shall assume

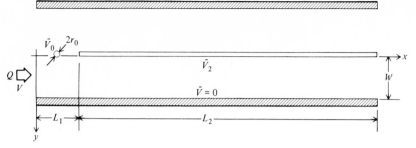

FIGURE 8-25
Plate-type two-stage precipitator.

that the equations developed for the charging section of the cylindrical two-stage precipitator apply here also. Thus, Eq. (8-160) gives $\overline{Er}$, Eq. (8-159) gives $\bar{q}_{p1}$, Eq. (8-162) gives $\tilde{V}_0$, Eq. (8-122) gives r_w, and Eq. (8-58) gives $\tilde{V}_w$ approximately. These quantities are all that are needed for the charging section if the charging current i is given or can be assumed.

We turn now to the collecting section. Normally the voltage of the plate $\tilde{V}_2$ will not equal that of the charging wire $\tilde{V}_w$; the value of $\tilde{V}_2$ will be determined by the following analysis. Equation (8-20) gives the field strength in the space between the plates as

$$E = \frac{\tilde{V}_2}{W} - \frac{q_{vp}}{\varepsilon_0}\left(\frac{W}{2} - y\right) \qquad (8\text{-}167)$$

Using Eq. (8-107), this becomes

$$E = \frac{\tilde{V}_2}{W} - \frac{\bar{q}_{p1}C_{nv}}{\varepsilon_0}\left(\frac{W}{2} - y\right) \qquad (8\text{-}168)$$

At the center plate, where E is given by E_2 and $\tilde{V}$ by $\tilde{V}_2$, Eq. (8-168) becomes

$$E_2 = \frac{\tilde{V}_2}{W} - \frac{\bar{q}_{p1}C_{nv}\,W}{2\varepsilon_0} \qquad (8\text{-}169)$$

At the collecting plate, the collecting-field strength E_c is given by

$$E_c = \frac{\tilde{V}_2}{W} + \frac{\bar{q}_{p1}C_{nv}\,W}{2\varepsilon_0} \qquad (8\text{-}170)$$

The fractional diminution of particle concentration is given by

$$-\frac{dC_{nv}}{C_{nv}} = \frac{p}{A}\frac{V_t}{V}\,dx = \frac{1}{W}\frac{V_t}{V}\,dx$$

Combining this equation with Eqs. (8-2) and (8-170) gives

$$\frac{dC_{nv}}{C_{nv}} = -\frac{\bar{q}_{p1}{}^2 C}{6\pi\varepsilon_0\,\mu dV}\left(C_{nv} + \frac{2\varepsilon_0\,\tilde{V}_2}{\bar{q}_{p1}W^2}\right)dx \qquad (8\text{-}171)$$

Integrating Eq. (8-171) between the limits of C_{nv_0} at L_1 and $(1 - \eta)C_{nv_0}$ at $L_1 + L_2$ leads to the following equation for L_2 :

$$L_2 = \frac{3\pi\mu dV W^2}{\bar{q}_{p1}CV_2}\ln\frac{1 - \eta + 2\varepsilon_0\,\tilde{V}_2/\bar{q}_{p1}W^2C_{nv_0}}{(1 - \eta)(1 + 2\varepsilon_0\,\tilde{V}_2/\bar{q}_{p1}W^2C_{nv_0})} \qquad (8\text{-}172)$$

In this equation, $\tilde{V}_2$ must be selected more or less arbitrarily. If $\tilde{V}_2$ is selected too small, the length L_2 will be excessive; if it is too great, the field strength will be too high for satisfactory operation. Thus, a wide range of values of $\tilde{V}_2$ may be chosen. An initial choice of $\tilde{V}_2$ in the range of 0.1 to 0.5 times the cylindrical corona voltage is suggested; this value can be modified later if needed.

Returning to the charging section for a moment, the current required to impart the charge $\bar{q}_{p1}$ to the particles is given by

$$i' = \bar{q}_{p1}C_{nv_0}Q = \bar{q}_{p1}C_{nv_0}WHV \qquad (8\text{-}173)$$

in which H is the height of the collector section. In deriving the equations which apply to the charging section, it was assumed that the actual current supplied to the charging section greatly exceeds the value given by Eq. (8-173). For instance, in Example 8-14 the value of i' is computed to be 0.158×10^{-7} A, which is 0.016 percent of the current assumed in that example. An a posteriori evaluation of i' should be made in all cases to ensure that $i \gg i'$.

EXAMPLE 8-15   Use the data of Example 8-14 to compute L_2 for a parallel-plate, two-stage precipitator. Use $W = 0.10$ m and $H = 1.0$ m, and assume realistic values of i and V_2.

SOLUTION   As a first solution, let us choose $i = 10^{-3}$ A and $\tilde{V}_2 = 30,000$ V. The method used in Example 8-14 gives

$$E_0 = 3.76 \times 10^6 \text{ V/m} \qquad \overline{Er} = 47,209 \text{ V}$$
$$\bar{q}_{p1} = 14.3 \times 10^{-16} \text{ C} \qquad \tilde{V}_0 = 98,069 \text{ V}$$

Equation (8-172) then gives

$$\frac{2\varepsilon_0\,\tilde{V}_2}{\bar{q}_{p1}W^2C_{nv_0}} = \frac{2(8.85 \times 10^{-12})(30,000)}{14.3 \times 10^{-16}(0.01)(10^9)} = 37.2$$

$$L_2 = \frac{3\pi(1.8 \times 10^{-5})(5 \times 10^6)(1.0)(0.01)}{14.3 \times 10^{-6}(1.0)(30,000)}\ln\frac{0.01 + 37.2}{0.01(1 + 37.2)}$$

$$= 0.91 \text{ m} \qquad Ans.$$

As a second solution, take $i = 10^{-2}$ A and $\tilde{V}_2 = 30,000$ V. Then

$$\overline{Er} = 146,024 \text{ V} \qquad \bar{q}_{p1} = 44.6 \times 10^{-16} \text{ C} \qquad \tilde{V}_0 = 282,169 \text{ V}$$

$$\frac{2\varepsilon_0 V_2}{\bar{q}_{p1} W^2 C_{nv_0}} = \frac{2(8.85 \times 10^{-12})(30,000)}{44.6 \times 10^{-16}(0.01)(10^9)} = 11.9$$

$$L_2 = \frac{3\pi(1.8 \times 10^{-5})(5 \times 10^{-6})(1.0)(0.01)}{44.6 \times 10^{-16}(1.0)(30,000)} \ln \frac{0.01 + 11.9}{0.01(12.9)}$$

$$= 0.288 \text{ m} \qquad Ans.$$

For the first solution

$$i' = \bar{q}_{p1} C_{nv_0} WHV = 14.3 \times 10^{-16}(10^9)(0.1)(1.0)(1.0) = 1.43 \times 10^{-7} \text{ A}$$

Therefore $i \gg i'$. ////

REFERENCES

1 Halliday, D., and R. Resnick: "Physics for Students of Science and Engineering," part II, 2d ed., John Wiley & Sons, Inc., New York, 1962.
2 White, H. J.: "Industrial Electrostatic Precipitation," Addison-Wesley Publishing Company, Inc., Reading, Mass., 1963.
3 Cobine, J. D.: "Gaseous Conductors, Theory and Engineering Applications," Dover Publications, Inc., New York, 1958.
4 Robinson, M.: Electrostatic Precipitation, in W. Strauss, ed., "Air Pollution Control," part I, Interscience Publishers, a division of John Wiley & Sons, Inc., New York, 1971.

PROBLEMS

8-1 A parallel-plate precipitator has plates 2.0 m high by 4.0 m long. Each section of the precipitator handles 0.7 m^3/s of standard air containing 0.5-μm-diameter dust particles. For $E_{cm} = 300,000$ V/m and $q_p = 100$ aC, determine the collection efficiency.

8-2 A cylindrical precipitator having a diameter of 1.0 m, $E_{cm} = 200,000$ V/m at the surface, is to produce 0.99 efficiency when handling 0.25 m^3/s of air at 500°C and 1.0 atm containing 2.5-μm particles for which $q_p = 1.0$ fC. Determine the required length of the cylinder.

8-3 A precipitator is to be constructed with each collector section having the shape of an equilateral triangle, each side of which is 40 cm wide. The length of the collector section is 3.0 m. The total flow rate of fluid, which is standard air, through the precipitator is 10 m^3/s; there are 100 sections; and $E_{cm} = 150,000$ V/m at the collector surface. If $q_p = 0.25$ fC, what will be the collection efficiency for 1.0-μm particles?

8-4 Plot the curves for $\tilde{V}$ and E for the space between two parallel plates 10 cm apart if $\tilde{V}_0 = 100$ kV and $q_v = 0.1$ mC/m^3. Assume no current flow.

8-5 Plot curves for $\tilde{V}$ and E as functions of radius for the space between two concentric cylinders if the two radii are 0.2 and 20 cm. Assume that $\tilde{V}_0 = 135$ kV, and $q_v = 0.25$ mC/m^3, and that there is no current flow.

8-6 A negative corona surrounds the smaller cylinder of a cylindrical precipitator. The radius of the corona is 2 mm, and that of the outer cylinder is 15 cm. If $q_v = -0.15$ mC/m^3, $E_0 = -5$ kV/mm, and $K = 2.2 \times 10^{-4}$ m^2/V $\cdot$ s, determine the voltage for no current and compute and plot current as functions of voltage.

8-7 Using the results of Example 8-4, verify the assumptions made in writing Eq. (8-46) from Eq. (8-45).

8-8 Using the results of Prob. 8-6, verify the assumptions made in writing Eq. (8-46) from Eq. (8-45).

8-9 A negative corona forms about a wire of 2 mm diameter in standard air. Estimate the outer diameter of the corona and the field strength E_0 at that point. Assume $f = 0.6$.

8-10 From Eq. (8-76), determine an equation for E_θ, the field strength in the tangential direction.

8-11 Rewrite Eqs. (8-75) and (8-76) in x, y coordinates. From these, determine E_x and E_y.

8-12 Determine, in x, y coordinates, the equations of the equipotential curves and of the flux curves in Fig. 8-16.

8-13 Determine the equation for voltage in the vicinity of a charged particle. From this equation, write equations for field strength in r, θ and in x, y coordinates.

8-14 Repeat Prob. 8-12 for the case of the charged particle, as shown in Fig. 8-18.

8-15 Determine the time constant and saturation charge for field charging of particles having a dielectric constant of 4 if the ion concentration is 10^{16} ions/m^3 and the field strength is 0.5 kV/mm. Take $d = 5.0$ μm.

8-16 Using the data of Prob. 8-5, plot a curve for q_{ps} as a function of radius for 15-μm particles having $\kappa = 7.5$ if the ion concentration is 10^{15} ions/m^3 across the space between the two cylinders. Assume that the particle concentration is small.

8-17 Repeat Prob. 8-16 if the ion concentration varies linearly with radius from 10^{16} ions/m^3 at the inner radius to 10^{14} ions/m^3 at the outer radius.

8-18 Plot saturation charge and time constant as functions of particle diameter if $E = 1.5$ kV/mm, $N = 10^{17}$ ions/m^3, and $\kappa = 2.5$. Assume K is 2.2×10^{-4} m^2/V $\cdot$ s.

8-19 A particle collection in air consists of 10^{10} particles/m^3 having an average diameter of 10 μm and $\kappa = 3$. The air flows at a velocity of 2.0 m/s parallel to the corona. On the average, the particles are to reach 0.9 of their saturation charge while traveling 10 cm. What average value of N is required? Estimate the current flow from the corona wire to achieve this result.

8-20 Derive an equation for the ion concentration across the space between a corona wire and the concentric collecting cylinder if N_0 is the ion concentration at the edge of the corona, C_{nv} is the particle concentration, d is the particle diameter, and κ is the dielectric constant of the particles. Assume that flow turbulence causes the particles to have equal charge across the radius at any axial position. The ion concentration N is to be a function of both r and x. Assume that the field strength distribution $E(r)$ is specified.

8-21 Derive a more accurate expression for diffusion charging combined with field charging using the result of Prob. 8-13 in treating the diffusion-charging mechanisms and taking into account the effect of diffusion charging on the electric field as it influences the field-charging process.

8-22 Compute q_p for diffusion charging only for a particle having $d = 0.05$ μm if $m = 6.0 \times 10^{-26}$ kg, $T = 300$ K, and $N = 1.5 \times 10^{15}$ ions/m^3. Plot q_p as a function of time.

8-23 Rework Prob. 8-22 if field charging is also considered, using $\kappa = 4$ and $E = 1.0$ kV/mm.

8-24 Repeat Example 8-9 using a velocity through the cylinder of 1.0 m/s, the other data remaining the same.

8-25 Repeat Example 8-9 varying r_c over a range of values from 0.1 to 1.0 m, the other data remaining the same.

8-26 In Prob. 8-25, assuming constant pipe-wall thickness, what value of r_c should result in minimum material cost of the tubes?

8-27 In Example 8-9, what current flow is required to charge the particles?

8-28 Derive an equation for the current flow per unit length along the wire for the case where the particle concentration is high. Also, determine the total current flow. The current will vary with x between $x = 0$ and $x = x_s$ and will be zero for $x > x_s$.

8-29 A cylindrical electrostatic precipitator is to be designed to treat 40 m³/s of air at 300°C and 1.0 atm to remove 99.9 percent of the particles originally contained in the stream. Tests on the dust stream indicate that the initial particle concentration is $C_{mv} = 0.003$ kg/m³, that the particle density is 1750 kg/m³, that the mean particle diameter is 5 μm, and that the dielectric constant $\kappa = 3.0$. In designing the precipitator, assume $r_0 = 0.005$ m, $r_c = 0.1$ m, $f = 0.7$, $C = 1.0$, and the velocity through the cylinder is 1.5 m/s. Determine the length of the cylinder, the number of cylinders required, the wire and corona voltages required, and the current which must be supplied to the wire.

8-30 Repeat Example 8-11, using $r_0 = 0.004$ m, the other data remaining the same.

8-31 Using the data of Example 8-11, design the precipitator with a range of values of r_c from 0.1 to 1.0 m. For a cylinder wall thickness of 0.5 cm, what value of r_c should result in the least cost for cylinder material?

8-32 Determine the total current flow to the wire required to charge the particles in Example 8-11.

8-33 A cylindrical single-stage precipitator is to treat 100 m³/s of air with 0.995 efficiency. The air is at standard conditions, with $C_{mv} = 0.25$ kg/m³ and with particles having an average diameter of 25 μm. Take $\kappa = 2$, $f = 0.6$, $C = 1$, and $\rho_p = 3000$ kg/m³. Design a precipitator having $r_0 = 0.005$ m, $r_c = 0.2$ m, and a gas velocity equal to 3.0 m/s.

8-34 A cylindrical single-stage precipitator is to be used to clean the hot combustion gases leaving a coal-fired combustion chamber prior to entering a gas turbine. At this point, $T = 800$ K, $P = 10^6$ N/m², $C_{mv} = 0.01$ kg/m³, the mean particle diameter is 40 μm, and the particle density is 1500 kg/m³. To limit erosion in the turbine blades, 99.9 percent of the particles having diameter above 1.0 μm must be removed by the precipitator. Take $r_0 = 0.005$ m, $r_c = 0.15$ m, $V = 1.5$ m/s, $f = 0.6$, $\kappa = 4.5$, and $C = 1.0$. Determine the required length of the precipitator. Also compute the required wire voltage.

8-35 A cylindrical single-stage precipitator is to be used for atmospheric sampling. It is to provide 0.95 efficiency for a flow rate of 0.001 m³/s for particles of 5 μm diameter, 2000 kg/m³ density, $\kappa = 4$, and $C_{mv} = 0.00005$ kg/m³. Use a velocity of 10 cm/s in a single tube, with $r_0 = 0.1r_c$ and $f = 0.65$. Determine the required length of tube and the necessary wire voltage.

8-36 Solve Prob. 8-29 using a wire-and-plate precipitator.

8-37 Design a wire-and-plate precipitator for the data of Prob. 8-34.

8-38 A wire-and-plate precipitator is to be designed to provide 0.996 collection efficiency for a stream of 75 m³/min of standard air, having $C_{mv} = 0.002$ kg/m³, with $\rho_p = 1450$ kg/m³ and $\kappa = 3.5$. Assume $r_0 = 0.004$ m, $f = 0.7$, $C = 1.0$, and the velocity of the gas is 3.5 m/s. The particle size is 2.0 μm. Determine suitable dimensions for the collector plates, and determine the number of plates required.

8-39 Repeat Prob. 8-38 if the particle diameter is 0.3 μm.

8-40 Repeat Prob. 8-33 for a wire-and-plate precipitator. Use $W = 0.2$ m, $H = 2.0$ m, and $l = 0.4$ m.

8-41 Design a cylindrical two-stage precipitator for the atmospheric sampler of Prob. 8-35. Choose $r_2 = 0.2r_c$.

8-42 Solve Prob. 8-29 using a two-stage electrostatic precipitator. Let $r_2 = 0.02$ m and $L_1 = 0.2$ m.

8-43 Solve Prob. 8-34 using a cylindrical two-stage precipitator. Let $r_2 = 0.02$ m and $L_1 = 0.2$ m.

8-44 A cylindrical two-stage precipitator is to be used in residental service to clean 4.0 m^3/s of standard air containing 3-μm particles of 900 kg/m^3 density, with $C_{mv} = 0.001$. Use $V = 1.5$ m/s, $r_0 = 0.05r_c$, $r_2 = 0.1r_c$, $r_c = 0.1$ m, and $L_1 = r_c$. Compute L_2 for an efficiency of 0.96 for the 3-μm particle.

8-45 Repeat Prob. 8-44 if the collection efficiency must be 0.96 for 0.2-μm particles, the other data remaining the same.

8-46 Derive Eq. (8-173).

8-47 Compute the value of i' for Example 8-14.

8-48 A parallel-plate, two-stage precipitator has $L_2 = 30$ cm, $H = 40$ cm, $r_0 = 2$ mm, and $W = 1.0$ cm. For $d = 5$ μm, $C_{nv_0} = 10^9$ particles/m^3, $V = 3.0$ m/s, $\kappa = 6$, $K = 2.2 \times 10^{-4}$ m^2/V · s, $f = 0.75$, $i = 10^{-3}$ A, and $\tilde{V}_2 = 20,000$ V, estimate the collection efficiency of the precipitator.

8-49 Design a precipitator to handle 4 m^3/s of air, with the restriction that the precipitator must fit within a duct having dimensions of 0.5 by 1.0 m and with the further stipulation that the length of the precipitator must not exceed 0.7 m. Use the data in Prob. 8-48 as to dust properties. Determine the size and number of the precipitator plates and the voltage to be maintained on the plates and wires. Specify the wire current also.

8-50 Work Prob. 8-34 using a parallel-plate, two-stage precipitator. Take $W = 0.15$ m and $H = 1.0$ m. Assume an appropriate value for $\tilde{V}_2$.

9

PARTICULATE SCRUBBERS

This chapter deals with the subject of scrubbers as applied to the collection of particulates. Scrubbers can also be used to collect gaseous pollutants, as we shall see in Chap. 11. There are many different types of scrubbers available; we shall discuss some of these types here. As usual, our discussion will center on the processes involved, in this case the collection of particles by liquid drops as they move through the airstream, which is the process by which all wet scrubbers operate.

9-1 INTRODUCTION

In all wet scrubbers particles, or drops, of the scrubbing medium are formed, generally much larger than the particles to be collected, and these particles sweep through the gas to be cleaned, collecting particles of the substance to be removed. In most cases the scrubbing medium is water; occasionally a different substance is used. Different types of scrubbing devices employ different means of forming the water droplets and different means of ensuring a relative velocity between the droplets and the gas to be cleaned. In all cases the cleaning mechanism involves attachment of the particulates to the droplets. The droplets are then collected and drained to a sump. Further processing is then required to remove the collected substance from the water before the water is discharged or reused.

The types of scrubbers in common use include: the spray chamber, or spray tower; the centrifugal, or cyclone, scrubber; the orifice, or self-induced spray, scrubber; impingement-plate and packed-bed scrubbers; disintegrator scrubbers; and the venturi scrubber. Several of these types will be discussed in some detail in this chapter.

Our study will focus largely on the mechanism by which water droplets (or droplets of some other liquid or even of solid particles) collect and retain smaller particles, either solid or liquid, and in this way remove them from the airstream. Three basic mechanisms are involved: interception, inertial impaction, and diffusion. Each mechanism will first be treated separately, and then the three mechanisms will be combined for a single droplet. Finally, the effect of a large number of droplets sweeping across the gas will be considered. Afterward, the results of our study of the collection mechanisms will be applied to the various scrubber types.

One unfortunate by-product of scrubber systems when collecting solid particulates is a sludge which is usually expensive to dispose of. Much water pollution has been caused in the past by discharging this sludge directly into natural streams. Current practice strongly disfavors such a disposal method, so that an expensive water treatment must be devised to handle the sludge. In some cases a salable product is obtained, but more often additional expense is entailed in disposing of the solid products obtained from the water treatment.

Dry scrubbing has been proposed occasionally but has been seldom used to date. No widely satisfactory process has yet been developed, although future possibilities of the emergence of some dry-scrubbing process do exist. Such a process is very attractive since it would eliminate the water-treatment process now required. If the scrubbing particles could be of the same substance as that being collected, the benefit would be quite obvious.

9-2 INTERCEPTION AND INERTIAL IMPACTION ON SPHERICAL DROPS

Wet scrubbers typically operate by forming droplets of the scrubbing liquid and causing these to move through the particulate-air mixture. The droplets then sweep up the particles as the air moves past the droplet. Most of the particles flow past the droplet along with the air, but some of the particles touch the surface of the droplet and attach to it. Thus, the droplet will collect some of the particles in its path but by no means all of them.

We shall now consider a collection efficiency for a single droplet. This efficiency, designated η_d, is defined as the ratio of the number of particles collected to the number of particles initially contained in the volume swept through by the droplet. To the extent that we can rationally predict the behavior

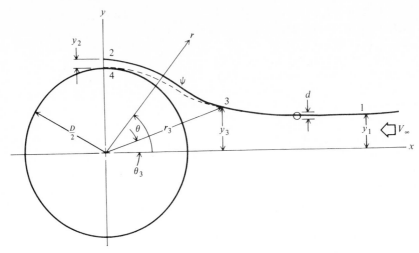

FIGURE 9-1
Flow around a sphere.

of the particles as they flow around and into the droplet, we can say that a particle of given diameter and density will strike the droplet if it lies initially within a certain distance y_1 of the axis of motion of the droplet. This is shown in Fig. 9-1. If the particle lies farther away from the axis than this, it will pass by the droplet and not be collected. This reasoning applies only to the interception and inertial impaction mechanisms treated in this section, not to the diffusion mechanism treated in Sec. 9-3; the determination of the value of y_1 will follow shortly.

The collection efficiency of the individual droplet due to interception and inertial impaction combined η_{di} can be defined as the ratio of the area of the circle having radius y_1 to the projected area of the droplet. This ratio is then modified by an attachment coefficient σ, which allows for the fact that some particles may strike the droplet and bounce off and also for the fact that some particles which were predicted to strike the droplet may not do so. The value of σ must also reflect the fact that some of the particles which bounce off the droplet will strike the droplet again and attach the second or a subsequent time and also the fact that some of the particles that were not predicted to strike the droplet will do so and attach to it. From the previous considerations, the efficiency is defined as

$$\eta_{di} = \frac{\pi y_1{}^2 \sigma}{\pi D^2/4} = \frac{4\sigma y_1{}^2}{D^2} \qquad (9\text{-}1)$$

No attempt will be made to evaluate σ in this book; instead, some reasonable values will be assumed in working problems.

Consider the particle of diameter d moving along the streamline, as shown in Fig. 9-1. As the gas streamline bends away from the droplet, the particle cannot accelerate sufficiently to follow the streamline. As a result it follows a curved path which, in this case, brings it barely into contact with the droplet. Particles which lie closer to the axis will strike the droplet nearer to its front. When the particle strikes the droplet due to this lateral motion as the particle attempts to follow the gas streamline, the collection mechanism is referred to as *inertial impaction*. This would be the sole mechanism if the particle had its own mass but zero diameter. However, a particle having diameter but no mass would still strike the droplet if its initial position were close enough to the axis since the particle extends a distance $d/2$ inside the streamline. When a particle is collected by this means, the mechanism is called *interception*. Actually, the two mechanisms work together and are treated here as a single mechanism.

Potential Flow around a Drop

Let us now begin an approximate analysis to estimate the value of y_1 for which a particle having diameter d and density ρ_p, which initially lies this distance away from the axis, will just barely touch the droplet at the top. In Fig. 9-1, let ψ represent the streamline which the particle initially follows. At point 1, the particle moves essentially parallel to the axis. At point 3, the path of the particle begins to diverge from the gas streamline. The path of the particle then crosses the y axis a distance $d/2$ above the droplet surface. Point 2 is where the gas streamline crosses the y axis and is a distance y_2 above the surface of the droplet.

We shall make a number of approximations in this analysis. To begin, let us assume that the gas streamline can be adequately described using the results of inviscid flow theory, also called *potential flow*. We shall ignore, for the moment, the effect of the boundary layer adjacent to the surface of the droplet and also assume that the wake formed downstream of the droplet will not affect the shape of the streamlines on the forward half of the sphere. The equation of the streamline, from Milne-Thomson [1], is given by

$$\psi = -\frac{1}{2} V_\infty \sin^2 \theta \left(r^2 - \frac{D^3}{8r} \right) \qquad (9\text{-}2)$$

in which ψ is the stream function. The velocity components are given in terms of ψ by

$$u_r = \frac{1}{r^2 \sin \theta} \frac{\partial \psi}{\partial \theta} \qquad u_\theta = -\frac{1}{r \sin \theta} \frac{\partial \psi}{\partial r} \qquad (9\text{-}3)$$

These velocity components then become

$$u_r = -V_\infty \cos \theta \left(1 - \frac{D^3}{8r^3} \right) \qquad (9\text{-}4)$$

$$u_\theta = \frac{1}{2} V_\infty \sin \theta \left(2 + \frac{D^3}{8r^3} \right) \qquad (9\text{-}5)$$

In these equations, r and θ are the axisymmetric spherical coordinates. At the surface of the sphere, the velocities are

$$u_r = 0 \qquad u_\theta = \tfrac{3}{2}V_\infty \sin\theta \qquad (9\text{-}6)$$

Also, at $\theta = \pi/2$, $u_\theta = 1.5\,V_\infty$. At a considerable distance upstream the velocity components are

$$u_r \to -V_\infty \cos\theta \qquad u_\theta \to V_\infty \sin\theta \qquad (9\text{-}7)$$

The equation of the streamline which passes through point 2 is

$$\psi = -\frac{V_\infty}{8}\left[(D + 2y_2)^2 - \frac{D^3}{D + 2y_2}\right] \qquad (9\text{-}8)$$

Eq. (9-8) may be written as follows, with the sum in the denominator of the last term expanded in a binomial series:

$$\psi = -\frac{V_\infty}{8}\left[D^2 + 4Dy_2 + 4y_2{}^2 - D^3\left(\frac{1}{D} - \frac{2y_2}{D^2} + \frac{4y_2{}^2}{D^3} - \frac{8y_2{}^2}{D^4} + \cdots\right)\right]$$

Now, if y_2 is small compared with $D/2$, the preceding equation may be approximated as

$$\psi = -\tfrac{3}{4}\,Dy_2\,V_\infty \qquad (9\text{-}9)$$

Combining Eqs. (9-2) and (9-9) gives

$$\sin^2\theta\left(r^2 - \frac{D^3}{8r}\right) = \tfrac{3}{2}Dy^2 \qquad (9\text{-}10)$$

Approximate Particle Motion

Equation (9-10) is the equation of the streamline which passes through points 1 to 3 in Fig. 9-1. Starting at point 3, the particle begins to diverge significantly from the streamline path owing to the acceleration force applied to the particle as it attempts to follow the streamline. The divergence of the particle path from the streamline is the same as the motion of a particle acted upon by a force which is equal to the mass of the particle times its acceleration. To determine the exact acceleration of the particle requires the solution of a cumbersome set of differential equations; we shall seek a reasonable approximation to this acceleration.

The time required for the gas at point 3 to travel to point 2 can be estimated since the velocity at point 3 is just over V_∞ and that at point 2 is a little less than $1.5V_\infty$. The average velocity of the gas can be taken as $1.25V_\infty$, and the velocity of the particle is assumed to be the same. The distance traveled by the particle in moving from point 3 to point 4 is approximately $(r_3 \cos\theta_3)$. In traveling from point 3 to point 4, the particle moves a distance of $D/2 + d/2 - y_3$. The preceding information is sufficient to estimate the acceleration of the particle in the y direction.

The average velocity in the x direction is

$$\bar{u}_x = \tfrac{5}{4} V_\infty \qquad (9\text{-}11)$$

and the time required for the particle to travel from point 3 to point 4 is

$$t = \frac{r_3 \cos \theta_3}{\bar{u}_x} = \frac{4}{5} \frac{r_3 \cos \theta_3}{V_\infty} \qquad (9\text{-}12)$$

Then, assuming constant acceleration a_y and neglecting $d/2$ with respect to $D/2$,

$$a_y = \frac{D/2 - y_3}{t^2/2} = \frac{D - 2y_3}{t^2} \qquad (9\text{-}13)$$

The force acting on the particle due to this acceleration is

$$F = ma_y = \frac{\pi d^3 \rho_p a_y}{6} \qquad (9\text{-}14)$$

Assuming that the lateral motion of the particle is laminar, Eq. (3-61) then gives for the terminal velocity

$$V_t = \frac{FC}{3\pi d\mu} = \frac{\rho_p d^2 a_y C}{18\mu} \qquad (9\text{-}15)$$

We have assumed that the acceleration a_y is constant, which means that the terminal velocity with which the particle diverges from the streamline is also constant. Then $V_t t$ represents the vertical distance between the streamline and the path of the particle at time t. From Fig. 9-1, this distance should equal $y_2 - d/2$ when the particle is at point 4; Eq. (9-15) then gives

$$y_2 = \frac{d}{2} + V_t t = \frac{d}{2} + \frac{\rho_p d^2 a_y t C}{18\mu} \qquad (9\text{-}16)$$

Applying Eq. (9-13) to this gives, upon simplification,

$$y_2 = \frac{d}{2} + \frac{5}{72} \frac{\rho_p d^2 V_\infty C}{\mu r_3 \cos \theta_3} (D - 2y_3) \qquad (9\text{-}17)$$

The distance y_3 can be expressed as a fraction of the diameter of the droplet:

$$y_3 = \alpha D \qquad (9\text{-}18)$$

From Fig. 9-1,

$$\sin \theta_3 = \frac{y_3}{r_3} = \frac{\alpha D}{r_3} \qquad (9\text{-}19)$$

Since Eq. (9-10) applies to the streamline passing through point 3, we may write it as

$$\sin^2 \theta_3 \left(r_3{}^2 - \frac{D^3}{8r_3} \right) = \tfrac{3}{2} D y_2 \qquad (9\text{-}20)$$

Substituting Eqs. (9-18) and (9-19) into Eq. (9-20) and solving for r_3 gives

$$r_3 = \frac{D}{(8 - 12y_2/\alpha^2 D)^{1/3}} \qquad (9\text{-}21)$$

Then Eq. (9-19) gives for $\sin\theta_3$

$$\sin\theta_3 = \alpha\left(8 - 12\frac{y_2}{\alpha^2 D}\right)^{1/3} \qquad (9\text{-}22)$$

and from this we have

$$\cos\theta_3 = \sqrt{1 - \sin^2\theta_3} = \sqrt{1 - \alpha^2\left(8 - 12\frac{y_2}{\alpha^2 D}\right)^{2/3}} \qquad (9\text{-}23)$$

and, using Eq. (9-21),

$$r_3\cos\theta_3 = \frac{D}{(8 - 12y_2/\alpha^2 D)^{1/3}}\sqrt{1 - \alpha^2\left(8 - 12\frac{y_2}{\alpha^2 D}\right)^{2/3}} \qquad (9\text{-}24)$$

Then, substituting Eqs. (9-18) and (9-24) into Eq. (9-17), we obtain

$$\frac{y_2}{D} - \frac{1}{2}\frac{d}{D} = \frac{5}{72}\frac{\rho_p d^2 V_\infty C}{\mu D}\frac{(8 - 12y_2/\alpha^2 D)^{1/3}}{\sqrt{1 - \alpha^2(8 - 12y_2/\alpha^2 D)^{2/3}}}(1 - 2\alpha) \qquad (9\text{-}25)$$

We define the quantity β as

$$\beta = \frac{5}{72}\frac{\rho_p d^2 V_\infty C}{\mu D} \qquad (9\text{-}26)$$

Then, Eq. (9-25) may be written

$$\frac{y_2}{D} - \frac{1}{2}\frac{d}{D} = \frac{\beta(8 - 12y_2/\alpha^2 D)^{1/3}(1 - 2\alpha)}{\sqrt{1 - \alpha^2(8 - 12y_2/\alpha^2 D)^{2/3}}} \qquad (9\text{-}27)$$

The choice of α in Eq. (9-27) is essentially arbitrary as long as it remains between 0 and $\frac{1}{2}$. In fact, it makes sense to choose α so as to maximize y_2. If α is small, the point 3 will lie far out on the incoming streamline and the time for the particle to move from y_3 to $D/2$ will be large; if α is near 0.5, the time during which the particle deviates from the streamline will be small. Either way, the value of y_2 will be much smaller than for some intermediate value of α. This is an unusual situation in which optimization is applied, not to meet some physical requirement, but instead to obtain the best approximation which the method will give.

To optimize Eq. (9-27) is somewhat more difficult than necessary; an approximate value of the optimum α can be obtained if Eq. (9-27) is modified to

$$\frac{y_2}{D} = \beta\left(8 - 12\frac{y_2}{\alpha^2 D}\right)^{1/3}(1 - 2\alpha) \qquad (9\text{-}28)$$

Upon differentiating Eq. (9-28) with respect to α and setting $dy_2/d\alpha = 0$, the following equation is obtained:

$$4\,\frac{y_2}{\alpha^3 D}(1 - 2\alpha)\left(8 - 12\,\frac{y_2}{\alpha^2 D}\right)^{-2/3} - \left(8 - 12\,\frac{y_2}{\alpha^2 D}\right)^{1/3} = 0 \qquad (9\text{-}29)$$

When Eq. (9-29) is solved for y_2 in terms of α, we have

$$y_2 = \frac{2\alpha^3 D}{1 + \alpha} \qquad (9\text{-}30)$$

$$8 - 12\,\frac{y_2}{\alpha^2 D} = \frac{8(1 - 2\alpha)}{1 + \alpha} \qquad (9\text{-}31)$$

Substituting back into Eq. (9-28) gives

$$\beta = \frac{\alpha^3}{(1 + \alpha)^{2/3}(1 - 2\alpha)^{4/3}} \qquad (9\text{-}32)$$

In a typical problem, β will be known and α desired. The value of α giving the approximate maximum value of y_2/D can be obtained from Fig. 9-2, which is based on Eq. (9-32). For small values of β, it can be shown that

$$\alpha \to \beta^{1/3} \qquad \text{for } \beta \to 0 \qquad (9\text{-}33)$$

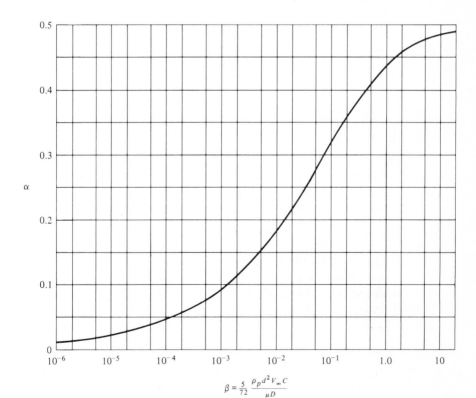

$$\beta = \frac{5}{72}\,\frac{\rho_p d^2 V_\infty C}{\mu D}$$

FIGURE 9-2
Optimum value of α as a function of β.

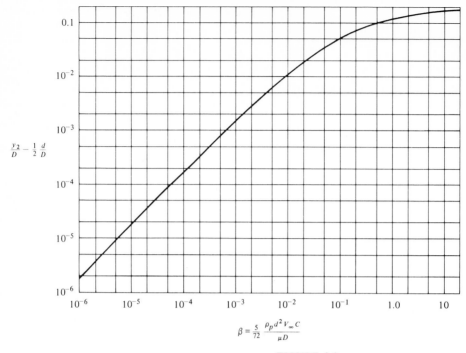

$$\beta = \frac{5}{72} \frac{\rho_p d^2 V_\infty C}{\mu D}$$

FIGURE 9-3
Maximum value of y_2 as a function of β.

while for values of α approaching 0.5,

$$\alpha \to 0.5 - \frac{0.086}{\beta^{3/4}} \qquad \text{for } \alpha \to 0.5 \qquad (9\text{-}34)$$

Figure 9-3 gives the values of $(y_2/D - 0.5d/D)$ obtained from Eq. (9-27) after the optimum values of α have been computed from Eq. (9-32). These values are tabulated in Table 9-1, along with empirical equations accurate to about 1 percent over the total range of β.

Boundary-Layer Effects

So far, we have ignored the effect of the boundary layer which grows on the surface of the sphere. Owing to the relatively high velocity between the sphere and the air, and to the normally small diameter of the sphere, the boundary layer will be quite thin on the front half of the sphere. Even so, its presence can have a large effect on the collection efficiency of the drop, especially at small values of efficiency. Thus, we must include the effect of the boundary layer in relating y_1 to y_2, as shown in Fig. 9-1, which influences the collection efficiency from Eq. (9-1). The development of the boundary layer on an axisymmetric body is treated very lucidly by Kays [2].

Table 9-1   SOLUTION TO EQ. (9-27) BASED ON OPTIMUM VALUES OF α

$\beta \times 10^6$	$(y_2/D - \frac{1}{2}d/D) \times 10^6$	$\beta \times 10^5$	$(y_2/D - \frac{1}{2}d/D) \times 10^5$
1	1.9411	1	1.8760
2	3.8524	2	3.6907
3	5.7475	3	5.4728
4	7.6305	4	7.2308
5	9.5036	5	8.9691
6	11.368	6	10.691
7	13.226	7	12.398
8	15.076	8	14.092
9	16.921	9	15.775
$\beta \times 10^4$	$(y_2/D - \frac{1}{2}d/D) \times 10^4$	$\beta \times 10^3$	$(y_2/D - \frac{1}{2}d/D) \times 10^3$
1	1.7446	1	1.4966
2	3.3699	2	2.7810
3	4.9335	3	3.9621
4	6.4523	4	5.0707
5	7.9354	5	6.1230
6	9.3884	6	7.1286
7	10.815	7	8.0943
8	12.219	8	9.0258
9	13.602	9	9.9263
β	$y_2/D - \frac{1}{2}d/D$	β	$y_2/D - \frac{1}{2}d/D$
0.01	0.01080	0.1	0.05276
0.02	0.01842	0.2	0.07458
0.03	0.02468	0.3	0.08817
0.04	0.03007	0.4	0.09775
0.05	0.03483	0.5	0.10496
0.06	0.03909	0.6	0.11063
0.07	0.04295	0.7	0.11523
0.08	0.04649	0.8	0.11906
0.09	0.04974	0.9	0.12230
1	0.12508	10	0.15968
2	0.14062	11	0.16038
3	0.14740	12	0.16079
4	0.15132	13	0.16121
5	0.15389	14	0.16168
6	0.15569	15	0.16192
7	0.15710	16	0.16214
8	0.15814	17	0.16259
9	0.15902	18	0.16275
19	0.16299	35	0.16472
20	0.16319	40	0.16511
25	0.16385	45	0.16521
30	0.16420	50	0.16544
		∞	0.16650

Table 9-1 SOLUTION TO EQ. (9-27) BASED ON OPTIMUM VALUES OF α *(Continued)*

Empirical equations

Range	Equation
$0 < \beta \leq 0.01$	$\dfrac{y_2}{D} = \dfrac{2\beta(1 - 2\beta^{1/3})^{4/3}}{(1 + \beta^{1/3})^{1/3}} + 2.543 \times 10^{37} \exp\left(-\dfrac{85.705}{\beta^{0.01625}}\right)$
$0.01 \leq \beta \leq 0.6$	$\dfrac{y_2}{D} = \exp\left(-\dfrac{2.352}{\beta^{0.1422}}\right) + 0.03128 \sin 0.15112(\ln \beta + 4.6052)^{1.661}$ $- 0.00331 \sin 0.6538(\ln \beta + 4.6052)^{1.1136}$
$0.6 \leq \beta \leq 5$	$\dfrac{y_2}{D} = \exp\left(-\dfrac{2.352}{\beta^{0.1422}}\right) + 0.03128 \sin 0.15112(\ln \beta + 4.6052)^{1.661}$
$\beta \geq 5$	$\dfrac{y_2}{D} = 0.1665 - 0.235 \exp(-1.8964\beta^{0.2694})$

Figure 9-4 shows the boundary layer which has finite thickness at the stagnation point and increasing thickness along the surface. Although separation and wake effects may occur on the back half of the sphere, these should not be a serious problem in determining the boundary-layer thickness at the midpoint $\theta = \pi/2$. The free-stream velocity immediately outside the boundary layer can be taken from the potential-flow theory and is given by Eq. (9-6).

Kays [2] gives the following equation for the boundary-layer thickness:

$$\delta_m = \frac{0.667v^{1/2}}{Ru_\infty{}^3}\left(\int_0^x u_\infty{}^5 R^2 \, dx\right)^{1/2} \qquad (9\text{-}35)$$

in which δ_m is the momentum thickness; for a laminar boundary layer, the boundary-layer thickness is related to the momentum thickness as

$$\frac{\delta}{\delta_m} = \frac{5}{0.664} = 7.53 \qquad (9\text{-}36)$$

Then the equation for δ becomes

$$\delta = \frac{5.02v^{1/2}}{Ru_\infty{}^3}\left(\int_0^x u_\infty{}^5 R^2 \, dx\right)^{1/2} \qquad (9\text{-}37)$$

Inside the integral, the following substitutions are made:

$$u_\infty = \frac{3}{2} V_\infty \sin \theta \qquad R = \frac{D}{2} \sin \theta \qquad x = \frac{D}{2} \theta$$

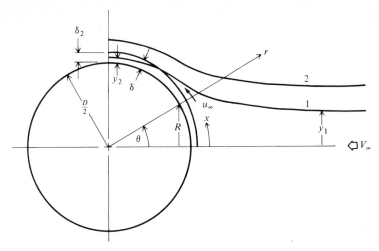

FIGURE 9-4
Boundary layer on the sphere.

In the denominator of Eq. (9-37), $R = D/2$ and $u_\infty = 3V_\infty/2$. Substituting these values into Eq. (9-37) and evaluating for $\delta = \delta_2$, which occurs when $\theta = \pi/2$, we obtain

$$\delta_2 = 2.90 \sqrt{\frac{vD}{V_\infty}} \left(\int_0^{\pi/2} \sin^7 \theta \, d\theta \right)^{1/2} \qquad (9\text{-}38)$$

The value of the integral in Eq. (9-38) is $\frac{16}{35}$; the value of δ_2 is as follows:

$$\delta_2 = 1.958 \sqrt{\frac{vD}{V_\infty}} \qquad (9\text{-}39)$$

The velocity profile in the boundary layer can be approximated as

$$u = u_\infty \left[\frac{3}{2}\frac{y}{\delta} - \frac{1}{2}\left(\frac{y}{\delta}\right)^3 \right] = \frac{3}{2} V_\infty \left[\frac{3}{2}\frac{y}{\delta} - \frac{1}{2}\left(\frac{y}{\delta}\right)^3 \right] \qquad (9\text{-}40)$$

By continuity of mass flow rate, we may relate y_1 to y_2:

$$\pi y_1^2 = \frac{2\pi}{V_\infty} \int_0^{y_2} \left(y + \frac{D}{2} \right) u \, dy$$

Substituting this equation into Eq. (9-1) gives for the collection efficiency

$$\eta_{di} = \frac{8\sigma}{D^2 V_\infty} \int_0^{y_2} \left(y + \frac{D}{2} \right) u \, dy \qquad (9\text{-}41)$$

By substituting Eq. (9-40) into Eq. (9-41) and performing the integration, the result for the case in which $y_2 < \delta_2$ is

$$\eta_{di} = 8.811\sigma \sqrt{\frac{v}{V_\infty D}} \left\{ \left(\frac{y_2}{\delta_2}\right)^2 - \frac{1}{6}\left(\frac{y_2}{\delta_2}\right)^4 + \frac{4}{3}\frac{\delta_2}{D}\left[\left(\frac{y_2}{\delta_2}\right)^3 - \frac{1}{5}\left(\frac{y_2}{\delta_2}\right)^5\right]\right\} \qquad (9\text{-}42)$$

In all cases of interest, δ_2/D is quite small, so that Eq. (9-42) may be approximated as

$$\eta_{di} = 8.811\sigma \sqrt{\frac{v}{V_\infty D}} \left[\left(\frac{y_2}{\delta_2}\right)^2 - \frac{1}{6}\left(\frac{y_2}{\delta_2}\right)^4\right] \qquad y_2 < \delta_2 \qquad (9\text{-}43)$$

When $y_2 = \delta_2$, Eq. (9-43) becomes

$$\eta_{di} = 7.342\sigma \sqrt{\frac{v}{V_\infty D}} \qquad y_2 = \delta_2 \qquad (9\text{-}44)$$

For the case in which $y_2 > \delta_2$, Eq. (9-41) becomes, using the approximation made in Eq. (9-44),

$$\eta_{di} = 7.342\sigma \sqrt{\frac{v}{V_\infty D}} + \frac{4\sigma}{D^2}\int_{\delta_2}^{y_2}\left[2\left(y + \frac{D}{2}\right) + \frac{D^3}{8(y + D/2)^2}\right] dy \qquad (9\text{-}45)$$

Performing this integration and neglecting δ_2 as compared with D in certain places leads to

$$\eta_{di} = 7.342\sigma \sqrt{\frac{v}{V_\infty D}} + 2\sigma \frac{(y_2/D - \delta_2/D)[3 + 6y_2/D + 4(y_2/D)^2]}{1 + 2y_2/D} \qquad y_2 > \delta_2$$

$$(9\text{-}46)$$

Equations (9-43) and (9-46) then give approximations to the collection efficiency for the single drop. In Sec. 9-5, we shall see how to combine the individual collection efficiencies to obtain a collection efficiency for a series of drops.

EXAMPLE 9-1   A spherical drop has a diameter of 0.75 mm and moves through standard air at a velocity of 15 m/s. The air contains particles having a density of 1800 kg/m^3 over a range of diameter from 0.5 to 10 μm. Compute and plot the collection efficiency due to interception and inertial impaction as a function of particle diameter. Assume $\sigma = 0.7$ and $C = 1.0$.

SOLUTION   Equation (9-26) gives the value of β as

$$\beta = \frac{5}{72}\frac{1800(15)(1.0)(d^2)}{1.8 \times 10^{-5}(7.5 \times 10^{-4})} = 0.1389 \times 10^{12}d^2$$

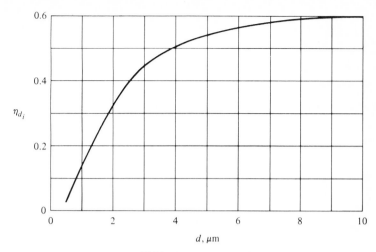

FIGURE 9-5
Collection efficiency for a droplet in Example 9-1.

The value of δ_2 is obtained from Eq. (9-39)

$$\delta_2 = 1.958 \sqrt{\frac{1.5 \times 10^{-5}(7.5 \times 10^{-4})}{15}} = 5.365 \times 10^{-5} \text{ m}$$

and the collection efficiency for $y_2 = \delta_2$ is found from Eq. (9-44):

$$\eta_{di} = 7.342(0.7)\sqrt{\frac{1.5 \times 10^{-5}}{15(7.5 \times 10^{-4})}} = 0.1877 \qquad y_2 = \delta_2$$

Equation (9-43) gives the efficiency if $y_2 < \delta_2$:

$$\eta_{di} = 0.227\left(\frac{y_2}{\delta_2}\right)^2\left[1 - \frac{1}{6}\left(\frac{y_2}{\delta_2}\right)^2\right] \qquad y_2 < \delta_2$$

And Eq. (9-46) gives the efficiency if $y_2 > \delta_2$:

$$\eta_{di} = 0.1877 + 1.4\frac{y_2/D - 0.0715}{1 + 2y_2/D}\left[3 + 6\frac{y_2}{D} + 4\left(\frac{y_2}{D}\right)^2\right] \qquad y_2 > \delta_2$$

Values of $y_2/D - d/2D$ are read from Fig. 9-3 after β has been computed for each value of d; these values are then used to compute the collection efficiency. Figure 9-5 shows the collection efficiency as a function of particle diameter.

////

It should be noted that the analysis used in this section is based on the assumption that the droplets have a large velocity. There is also a regime in which the droplets have a small velocity relative to the air, such that Stokes' law applies to the motion of the droplets as well as to the particles. We shall not treat the latter condition here since it has only limited application to scrubbers.

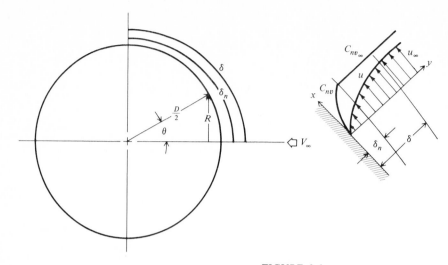

FIGURE 9-6
Velocity and diffusion boundary layers.

9-3 DIFFUSION TO SPHERICAL DROPS

In addition to the interception and inertial impaction mechanisms of collection, the mechanism of diffusion becomes important as the particle size becomes smaller and is the dominant collection mechanism for particles under 0.1 μm in diameter. In this section the diffusion mechanism is treated as though it were the sole collection mechanism present; in Sec. 9-4 the collection efficiency due to combined mechanisms will be presented. We shall derive an approximate equation for the collection efficiency due to diffusion onto a spherical drop. The approach used here is to apply boundary-layer theory to the front half of the sphere. Figure 9-6 shows the velocity boundary layer of thickness δ and a thinner concentration boundary layer of thickness δ_n, based on the number of particles per unit volume C_{nv}.

Let us first derive an equation for the thickness of the velocity boundary layer.

Starting with Eq. (9-37) and making the same substitutions used to obtain Eq. (9-38) we have

$$\delta = \frac{2.90}{\sin^4 \theta} \sqrt{\frac{vD}{V_\infty}} \left(\int_0^\theta \sin^7 \theta \, d\theta \right)^{1/2} \qquad (9\text{-}47)$$

Performing the integration and simplifying gives

$$\delta \sqrt{\frac{V_\infty}{vD}} = \frac{2.90}{\sin^4 \theta} [0.457 - (0.457 + 0.229 \sin^2 \theta + 0.171 \sin^4 \theta$$

$$+ 0.143 \sin^6 \theta) \cos \theta]^{1/2} \qquad (9\text{-}48)$$

This function is plotted in Fig. 9-7.

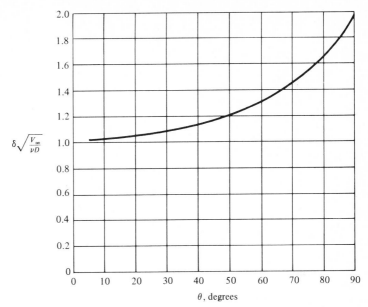

FIGURE 9-7
Velocity boundary-layer thickness around a sphere.

The deposition of particles by diffusion follows the laws of brownian motion; the basic principles of brownian motion are discussed by Lee, Sears, and Turcotte [3] and by Fuchs [4]. The rate of deposition follows Fick's law and may be expressed as

$$\frac{\dot{N}}{A} = \mathscr{D}\left(\frac{\partial C_{nv}}{\partial y}\right)_{y=0} \qquad (9\text{-}49)$$

in which $\dot{N}$ represents the rate at which particles are deposited on a surface of area A for a given value of the number concentration gradient $(\partial C_{nv}/\partial y)_{y=0}$. The diffusion coefficient is given by the following formula, taken from Davies [5]:

$$\mathscr{D} = \frac{kTC}{3\pi d\mu} \qquad (9\text{-}50)$$

in which k is Boltzmann's constant and C is the Cunningham correction factor.

The velocity inside the boundary layer is given by Eq. (9-40). A similar equation is used for the number concentration inside the concentration boundary layer:

$$C_{nv} = C_{nv_\infty}\left[\frac{3}{2}\frac{y}{\delta_n} - \frac{1}{2}\left(\frac{y}{\delta_n}\right)^3\right] \qquad (9\text{-}51)$$

Let us assume for the purpose of this analysis that the concentration boundary-layer thickness is a constant fraction of the velocity boundary-layer thickness, which can be expressed as

$$\delta_n = \beta\delta \qquad (9\text{-}52)$$

Then Eq. (9-51) may be written as

$$\dot{C}_{nv} = C_{nv\infty}\left[\frac{3}{2\beta}\frac{y}{\delta} - \frac{1}{2\beta^3}\left(\frac{y}{\delta}\right)^3\right] \qquad (9\text{-}53)$$

and the concentration gradient at the surface of the sphere is

$$\left(\frac{\partial C_{nv}}{\partial y}\right)_{y=0} = \frac{3}{2\beta\delta} C_{nv\infty} \qquad (9\text{-}54)$$

Using the element of area on the surface of the sphere given by $2\pi(D/2)\sin\theta(D/2)\,d\theta$, as indicated in Fig. 9-8, and combining this value with Eqs. (9-49) and (9-54), we have

$$d\dot{N} = \frac{3\pi}{4\beta}D^2\mathscr{D}C_{nv\infty}\frac{\sin\theta}{\delta}\,d\theta \qquad (9\text{-}55)$$

Eq. (9-55) can be integrated over the front half of the sphere:

$$\dot{N} = \frac{3\pi}{4\beta}D^2\mathscr{D}C_{nv\infty}\int_0^{\pi/2}\frac{\sin\theta}{\delta}\,d\theta \qquad (9\text{-}56)$$

which may be rewritten as

$$\dot{N} = \frac{3\pi}{4\beta}D^2\mathscr{D}C_{nv\infty}\sqrt{\frac{V_\infty}{\nu D}}\int_0^{\pi/2}\frac{\sin\theta\,d\theta}{\delta\sqrt{V_\infty/\nu D}} \qquad (9\text{-}57)$$

The integral in Eq. (9-57) can be evaluated numerically by Simpson's rule, using Eq. (9-48) in the integrand; this process leads to the value of 0.7566 for the integral. Equation (9-57) then becomes

$$\dot{N} = 1.783\frac{D^2\mathscr{D}C_{nv\infty}}{\beta}\sqrt{\frac{V_\infty}{\nu D}} \qquad (9\text{-}58)$$

where $\dot{N}$ represents the rate at which particles are attached to the surface of the front half of the sphere.

Next, consider the flow past the section of the boundary layer at $\theta = \pi/2$. The rate at which particles flow past the section in the absence of diffusion onto the front surface of the sphere less the rate at which they actually flow past the section is equal to the rate at which they are collected, that is, $\dot{N}$. The number of particles flowing past a section is given by the concentration times the volumetric flow rate, which leads to the following equation:

$$\dot{N} = C_{nv\infty}\int_0^{\delta_n}u2\pi\frac{D}{2}\,dy - \int_0^{\delta_n}C_{nv}2\pi\frac{D}{2}u\,dy \qquad (9\text{-}59)$$

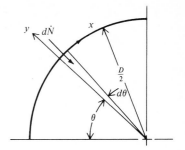

FIGURE 9-8
Diffusion onto an element of area on the
surface of a sphere.

Using Eqs. (9-40) and (9-53) and simplifying gives

$$\dot{N} = \frac{3}{2} \pi D V_\infty C_{nv_\infty} \int_0^{\beta\delta} \left[\frac{3}{2} \frac{y}{\delta} - \frac{1}{2} \left(\frac{y}{\delta} \right)^3 \right] \left[1 - \frac{3}{2\beta} \frac{y}{\delta} + \frac{1}{2\beta^3} \left(\frac{y}{\delta} \right)^3 \right] dy$$

Writing $\varepsilon = y/\delta$ and expanding the integrand gives

$$\dot{N} = \frac{3}{2} \pi D V_\infty C_{nv_\infty} \delta_2 \int_0^\beta \left(\frac{3}{2} \varepsilon - \frac{9}{4\beta} \varepsilon^2 + \frac{3}{4\beta^3} \varepsilon^4 - \frac{1}{2} \varepsilon^3 + \frac{3}{4\beta} \varepsilon^4 - \frac{1}{4\beta^3} \varepsilon^6 \right) d\varepsilon \quad (9\text{-}60)$$

Upon performing the integration, we have

$$\dot{N} = D C_{nv_\infty} V_\infty \delta_2 (0.7069\beta^2 - 0.05049\beta^4) \quad (9\text{-}61)$$

Using Eq. (9-39),

$$\dot{N} = D C_{nv_\infty} V_\infty \sqrt{\frac{vD}{V_\infty}} (1.384\beta^2 - 0.09886\beta^4) \quad (9\text{-}62)$$

There are now two equations for $\dot{N}$; equating (9-58) and (9-62) and solving for $v/\mathscr{D}$, which by definition is the Schmidt number Sc, gives

$$Sc = \frac{1.288}{\beta^3(1 - 0.07142\beta^2)} \quad (9\text{-}63)$$

Since β is expected to be much smaller than unity, Eq. (9-63) may be solved approximately for β:

$$\beta = \frac{1.088}{Sc^{1/3}} \quad (9\text{-}64)$$

Substituting Eq. (9-64) into Eq. (9-58) gives

$$\dot{N} = 1.639 D^2 \mathscr{D} C_{nv_\infty} Sc^{1/3} \sqrt{\frac{V_\infty}{vD}} \quad (9\text{-}65)$$

Because of expected wake effects, the rear half of the sphere will be extremely difficult to analyze accurately. Let us not attempt to do so here; instead we shall make the arbitrary assumption that the same number of particles will be collected on the rear half as on the front half of the sphere. Then when the coefficient is doubled, Eq. (9-65) becomes

$$\dot{N} = 3.28 \frac{D^2 V_\infty C_{nv_\infty}}{\sqrt{vD V_\infty}} \mathscr{D} Sc^{1/3} \qquad (9\text{-}66)$$

The collection efficiency is obtained by dividing the number of particles collected per unit time by the rate at which particles flow past a circular section whose diameter is that of the sphere. The denominator in the efficiency equation is

$$\dot{N}_0 = \frac{\pi}{4} D^2 V_\infty C_{nv_\infty} \qquad (9\text{-}67)$$

Dividing Eq. (9-66) by Eq. (9-67) gives for the collection efficiency

$$\eta_{dd} = \frac{4.18\sigma}{Sc^{2/3}} \sqrt{\frac{v}{V_\infty D}} = \frac{4.18\sigma}{Sc^{2/3} Re^{1/2}} \qquad (9\text{-}68)$$

in which the attachment coefficient σ is also included. Equation (9-50) gives the Schmidt number:

$$Sc = \frac{v}{\mathscr{D}} = \frac{3\pi d\mu^2}{kTC\rho} \qquad (9\text{-}69)$$

For standard air, this may be written

$$Sc = 6.55 \times 10^{11} \frac{d}{C} \qquad (9\text{-}69a)$$

EXAMPLE 9-2 Determine the collection efficiency for a spherical drop 0.5 mm in diameter moving at 10 m/s through standard air. Evaluate for a range of particle diameters. Assume $\sigma = 1$.

SOLUTION The mean free path λ in standard air is 6.57×10^{-8} m. Equation (3-55) gives the Cunningham correction factor C:

$$C = 1 + \{0.1652 + 0.05256 \exp\left[-8.371(d \times 10^6)\right]\} \frac{1}{(d \times 10^6)}$$

The Reynolds number is computed as

$$Re = \frac{V_\infty D}{v} = \frac{10(0.5 \times 10^{-3})}{1.55 \times 10^{-5}} = 323$$

Equation (9-69) gives the Schmidt number as

$$Sc = 6.55 \times 10^5 \frac{(d \times 10^6)}{C}$$

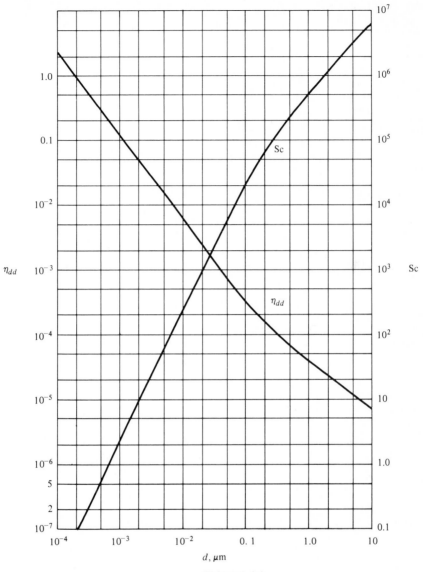

FIGURE 9-9
Schmidt number and efficiency for Example 9-2.

Then Eq. (9-68) gives an expression for the collection efficiency:

$$\eta_{dd} = \frac{0.219}{Sc^{2/3}}$$

The variation of efficiency with particle diameter is plotted in Fig. 9-9.   ////

9-4 COMBINED MODES OF COLLECTION

In Sec. 9-2 we derived an equation for the collection efficiency on a single drop due to inertial impaction and interception combined. In Sec. 9-3, we obtained a similar equation for the collection efficiency due to diffusion, or brownian motion, effects. The inertial impaction and interception effects dominate at large particle diameters, while the diffusion effects dominate at small particle diameters. There is, however, a range of particle diameters for which both effects occur simultaneously. In this range, the efficiencies calculated for each effect need to be combined into a single efficiency. The actual combination of effects is quite complex; each effect reduces the number of particles upon which the other effect can act. If the collection efficiencies due to both effects are each small, we may combine them approximately by assuming that each effect acts on the average number of particles not collected by the other effect. For this we use the arithmetic average of the number of particles approaching the drop and the number uncollected by the other effect as the gas leaves the drop.

Writing N_c as the number of particles collected and N_0 as the original number, the number of particles collected due to inertial impaction is given by

$$N_{c_i} = \eta_{di}\tfrac{1}{2}[N_0 + N_0(1 - \eta_{dd})]$$

and the number collected due to diffusion is given by

$$N_{c_d} = \eta_{dd}\tfrac{1}{2}[N_0 + N_0(1 - \eta_{di})]$$

The sum of these, divided by N_0, gives the collection efficiency for the drop:

$$\eta_d = \frac{N_{c_i} + N_{c_d}}{N_0}$$

$$= \frac{1}{2}\eta_{di}(2 - \eta_{dd}) + \frac{1}{2}\eta_{dd}(2 - \eta_{di})$$

$$= \eta_{di} + \eta_{dd} - \eta_{di}\eta_{dd} \tag{9-70}$$

EXAMPLE 9-3   Standard air containing particles having densities of 1000 kg/m^3 flows at 12 m/s past a spherical drop with a diameter of 1.0 mm. Assume $\sigma = 0.75$. Calculate and plot the separate and combined efficiencies over the range of particle diameters from 0.01 to 10 μm.

SOLUTION   The Cunningham correction factor is computed as in Example 9-2. The parameter β becomes

$$\beta = 0.0463C(d \times 10^6)^2$$

δ_2 is computed to be 69.23 μm. For $y_2 < \delta_2$,

$$\eta_{di} = 0.2336\left[\left(\frac{y_2}{\delta_2}\right)^2 - \frac{1}{6}\left(\frac{y_2}{\delta_2}\right)^4\right]$$

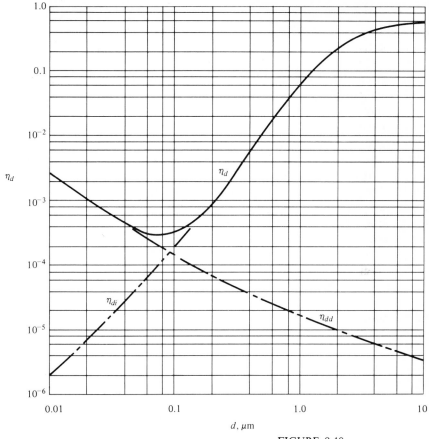

FIGURE 9-10
Efficiency curves for Example 9-3.

and for $y_2 > \delta_2$,

$$\eta_{di} = 0.1947 + 1.5 \frac{y_2/D - 0.06923}{1 + 2y_2/D} \left[3 + 6y_2/D + 4\left(\frac{y_2}{D}\right)^2 \right]$$

The Schmidt number is computed also as in Example 9-2. The diffusion efficiency is given by

$$\eta_{dd} = \frac{0.1108}{Sc^{2/3}}$$

and the combined efficiency is computed from Eq. (9-70):

$$\eta_d = \eta_{di}(1 - \eta_{dd}) + \eta_{dd}$$

Figure 9-10 shows plots of the individual and combined efficiencies.   ////

9-5 COLLECTION EFFICIENCY FOR A SERIES OF DROPS

No scrubber can achieve a satisfactory overall collection efficiency unless each small group of particles is exposed to the collecting influence of a large number of drops. Thus, any successful scrubber requires a large number of drops in its operation. We must account for the repeated exposure of the particles to the action of a series of drops. Consider a long cylinder of diameter D, where D is the diameter of the drops, and examine what happens to the particles which initially enter this cylinder with the airstream. Assume that n drops also pass along this cylinder in the opposite direction to the air and particle flow, with a relative velocity such as to produce a combined collection efficiency η_d for each drop.

Let N_0 be the number of particles originally entering this cylinder in some time interval, let N_1 be the number of particles which pass the first drop, and so on, until N_n particles remain in the cylinder after passing n drops. Then, after the first drop,

$$N_1 = N_0(1 - \eta_d)$$

After the second drop,

$$N_2 = N_0(1 - \eta_d)^2$$

After n drops,

$$N_n = N_0(1 - \eta_d)^n$$

The overall collection efficiency after n drops is

$$\eta = 1 - \frac{N_n}{N_0} = 1 - (1 - \eta_d)^n \qquad (9\text{-}71)$$

Before Eq. (9-71) can be applied, the value of n must be found; this value will be studied in connection with specific types of scrubbers.

EXAMPLE 9-4 In a scrubber, the relative velocity and drop size is such that the collection efficiency for a single drop due to inertial impaction, interception, and diffusion combined is 3×10^{-4}. The number of drops which a given group of particles must pass in order not to be collected is 1000. What will be the overall collection efficiency?

SOLUTION Equation (9-71) applies, giving

$$\eta = 1 - (1 - 3 \times 10^{-4})^{1000} = 0.2592 \qquad Ans. \qquad ////$$

A somewhat more sophisticated approach to predicting the efficiency of a collection device is to allow for a certain fraction b of the incoming air to escape the scrubbing process. This approach is physically reasonable since certain portions of the flow field may receive less than the average intensity of droplets; this situation may occur in corners and near the walls of the device or in other dead zones. Assume that a fraction b of the incoming air is not cleaned while the remaining fraction $(1 - b)$ is cleaned to the extent considered previously; then the number of particles leaving the device is

$$N_n = (1 - b)N_0(1 - \eta_d)^n + bN_0$$

From this, the collection efficiency becomes

$$\eta = (1 - b)[1 - (1 - \eta_d)^n] \qquad (9\text{-}72)$$

9-6 SPRAY CHAMBERS

The spray chamber is perhaps the simplest illustration of the principle of wet scrubbing as applied to removal of particulates from an airstream. Figure 9-11 shows a cross-sectional view of a simple spray chamber. The spray fluid is customarily either clean water from the normal water supply or recirculated slurry formed from clean water and the pollutant substance being collected, diluted by a fraction of clean water. This fluid is sprayed into the chamber from a series of nozzles located at the top of the chamber while the air-particle mixture enters the bottom of the chamber and flows upward, encountering the drops formed from the sprays which fall to the bottom by gravity. The drops remove the particles by the scrubbing action already discussed. They then collect at the bottom, with their load of foreign substance, to form the slurry which flows out from the bottom of the chamber. The slurry may then be sent directly to a processing plant for treatment, or, as indicated previously, a small fraction of the slurry may be sent for treatment, with the bulk of the slurry mixed with water and recirculated through the sprays.

To analyze the performance of a spray chamber, values of V_∞ and n are required. The value of V_∞, which may be used in Eqs. (9-43), (9-46), and (9-68) to compute the drop collection efficiencies, is simply the terminal velocity for free fall under the influence of gravity, as computed from the equations in Chap. 3. Alternatively, the terminal velocity of water drops falling through standard air under the action of gravity is given in Table 9-2, along with the terminal Reynolds number. The upward velocity of the air in the chamber, given by

$$V_a = \frac{Q}{A_{sc}} \qquad (9\text{-}73)$$

must not exceed the drop velocity V_d to prevent the air from carrying the drops out of the top of the chamber. Equation (3-47) gives the force acting on the drop:

$$F = \frac{\pi \rho_d D^3 g}{6} = 5.135 \rho_d D^3 \qquad (9\text{-}74)$$

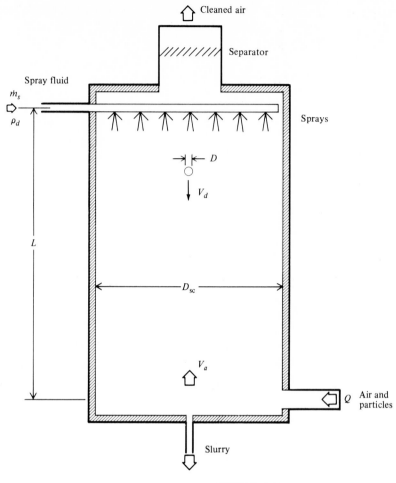

FIGURE 9-11
Schematic diagram of a spray chamber.

If water is the spray fluid, Eq. (9-74) becomes

$$F = 5135D^3 \qquad (9\text{-}75)$$

Two of the three cases dealt with in Chap. 3 will be considered here. If the Reynolds number for the drop motion is between 10 and 700, Eq. (3-29) applies, giving

$$V_\infty = \frac{4.8}{\rho D}\sqrt{447\mu^2 + \frac{\rho\rho_d D^3 g}{6}} - 20.4\mu \qquad 10 < \mathrm{Re}_d < 700 \qquad (9\text{-}76)$$

For standard air as the gas and water as the spray fluid, Eq. (9-76) becomes

$$V_\infty = \frac{178.3}{D}\sqrt{0.7814 \times 10^{-10} + D^3} - \frac{1.520 \times 10^{-3}}{D} \qquad 10 < \mathrm{Re}_d < 700 \qquad (9\text{-}77)$$

The Reynolds number becomes

$$\mathrm{Re}_d = \frac{V_\infty D}{\nu} = 11.50 \times 10^6 \sqrt{D^3 + 0.7814 \times 10^{-10}} - 98.06 \qquad (9\text{-}78)$$

If the Reynolds number is greater than 700, then Eq. (3-30) gives

$$V_\infty = \frac{2.4}{D}\sqrt{\frac{\pi \rho_d D^3 g}{6\rho}} = 5.440\sqrt{\frac{\rho_d D}{\rho}} \qquad \mathrm{Re}_d > 700 \qquad (9\text{-}79)$$

and for standard air and water

$$V_\infty = 158\sqrt{D} \qquad \mathrm{Re}_d > 700 \qquad (9\text{-}80)$$

The Reynolds number is given by

$$\mathrm{Re}_d = 1.020 \times 10^7 D^{3/2} \qquad (9\text{-}81)$$

Table 9-2 TERMINAL VELOCITIES OF WATER DROPS FALLING IN STANDARD AIR

D, mm	V_t, m/s	Re
0.05	0.0740	0.239
0.10	0.259	1.67
0.20	0.682	8.80
0.30	1.02	19.8
0.40	1.51	39.0
0.50	2.04	65.7
0.60	2.56	99.0
0.70	3.05	138
0.80	3.51	181
0.90	3.93	228
1.00	4.33	279
1.25	5.20	420
1.50	5.96	577
1.75	6.61	746
2.00	7.06	912
2.5	7.90	1274
3.0	8.65	1675
4.0	9.99	2579
5.0	11.2	3604
6.0	12.2	4738
7.0	13.2	5970
8.0	14.1	7294
9.0	15.0	8703

Next, consider the evaluation of n, the number of drops encountered by a group of particles. Let $\dot{N}_d$ be the rate of drop formation, in number per second, in the entire chamber. The number encountered by a group of particles will be the number of drops which fall through a cylinder of diameter D during a certain period of time. This number is the total number formed during that time period times the ratio of areas $\pi D^2/4A_{sc}$. To determine the time period involved, consider a particle which just enters the bottom of a cylinder of diameter D and length L. Drops have been entering the top of the cylinder and continue to do so, falling down the cylinder and emerging at the bottom. The particle will encounter all those drops which are in the cylinder at the time when the particle enters plus all those drops which enter the cylinder during the time required for the particle (if it is not collected) to travel to the top of the cylinder.

Since the drops fall with velocity V_d, where $V_d = V_\infty - V_a$, and the particles travel upward with velocity V_a, the time period becomes $L/V_a + L/V_d$. Then n becomes

$$n = \dot{N}_d \frac{\pi D^2/4}{A_{sc}} \left(\frac{L}{V_a} + \frac{L}{V_d} \right) \qquad (9\text{-}82)$$

The total drop formation rate is related to the mass flow rate of the spray fluid as

$$\dot{N}_d = \frac{6\dot{m}_s}{\pi D^3 \rho_d} \qquad (9\text{-}83)$$

Substituting Eq. (9-83) into Eq. (9-82) and simplifying gives

$$n = \frac{1.5\dot{m}_s L}{\rho_d D} \left[\frac{1}{Q} + \frac{1}{A_{sc}(V_\infty - V_a)} \right] \qquad (9\text{-}84)$$

For water as the spray fluid Eq. (9-84) becomes

$$n = \frac{0.0015\dot{m}_s L}{D} \left[\frac{1}{Q} + \frac{1}{A_{sc}(V_\infty - V_a)} \right] \qquad (9\text{-}85)$$

If the spray chamber is circular, then $A_{sc} = \pi D_{sc}^2/4$. The value of n computed from this equation, along with the value of η_d computed from Eq. (9-70), may be used in Eq. (9-71) to compute the overall collection efficiency for the spray chamber.

EXAMPLE 9-5   A spray chamber uses water as the spray fluid, treating standard air containing particles with a density of 1500 kg/m³. The flow rate of the air is 10 m³/s; that of the water is 0.01 m³/s. The average drop diameter is 1.0 mm. Determine the collection efficiency for a range of particle sizes. Neglect the diffusion mechanism, and assume $C = 1$ and $\sigma = 1$. Take $D_{sc} = 2.0$ m and $L = 5.0$ m.

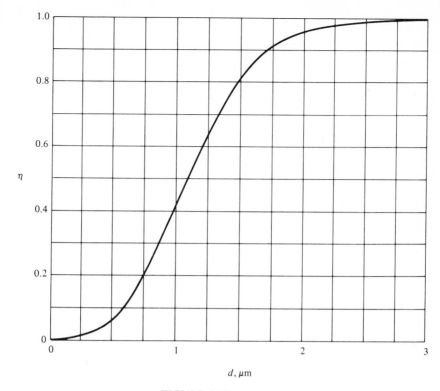

FIGURE 9-12
Collection efficiency for the spray chamber of Example 9-5.

SOLUTION Equation (9-78) gives the value of Re_d as 279.5, which is in the range of validity of that equation. Then the relative velocity between the drops and the air V_∞ is

$$V_\infty = Re_d \frac{v}{D} = 4.33 \text{ m/s}$$

The cross-sectional area of the spray chamber and the air velocity are given by

$$A_{sc} = \frac{\pi D_{sc}^2}{4} = 3.14 \text{ m}^2$$

$$V_a = \frac{Q}{A_{sc}} = \frac{10}{\pi} = 3.183 \text{ m/s}$$

The drop velocity $V_d = 4.33 - 3.18 = 1.15$ m/s.
Equation (9-84) gives the value of n as

$$n = \frac{1.5(0.01)(5)}{0.001} \left[\frac{1}{10} + \frac{1}{\pi(4.33 - 1.15)} \right] = 28.3$$

Equation (9-39) gives

$$\delta_2 = 1.958 \sqrt{\frac{1.55 \times 10^{-5}(0.001)}{4.33}} = 117 \; \mu m$$

Equation (9-43) gives the collection efficiency for a single drop due solely to interception and inertial impaction for $y_2 < \delta_2$:

$$\eta_{di} = 0.527 \left[\left(\frac{y_2}{\delta_2} \right)^2 - \frac{1}{6} \left(\frac{y_2}{\delta_2} \right)^4 \right]$$

It turns out that all of our calculations are in this range. Then Eq. (9-71) gives the overall collection efficiency:

$$\eta = 1 - (1 - \eta_{di})^{28.3}$$

The results are plotted in Fig. 9-12. Very good collection efficiencies are indicated for particles of 2.5 μm and larger in diameter.   ////

   The question of the best drop size to use in a spray chamber is one that should be investigated. Although several considerations are pertinent, one of the most important relates to the individual drop collection efficiency η_{di} for which only the mechanisms of inertial impaction and interception are considered important in spray chambers. Since the drops fall by gravity, very small drops fall very slowly, requiring small air velocities and large chambers and producing a small collection efficiency due to the small velocity of air around the drop. However, a large drop diameter produces a reduced collection efficiency due to the decreased acceleration of particles flowing around the drop. Thus it is expected that some optimum drop size will produce a maximum value of η_{di}. Calculations bear this out, as shown in Fig. 9-13 for collection of small water droplets onto drops of water in a spray chamber. The value of η_{di} reaches a maximum for drop diameters of 2.0 to 4.0 mm for all particle sizes, although larger drop sizes produce little deterioration in collection efficiency. As expected, the collection efficiency increases with increasing particle diameter. However, the graph in Fig. 9-13 does not tell the whole story since the number of drops available for a given water flow rate depends very strongly on the drop diameter.

Power Requirements

Next, consider the power required to pump the air through the spray chamber. Assume that with no drops present the pressure drop for a given flow rate Q is ΔP_{ns}. This pressure drop necessitates a power expenditure of $Q \Delta P_{ns}$. The presence of the drops will greatly increase the pressure drop across the unit and hence the power requirement. To evaluate the power expended as the air flows around the drop, it is noted that power is given by the product of the force acting on the drop times the relative velocity between the air and the drop times the number of drops in the chamber at any one time.

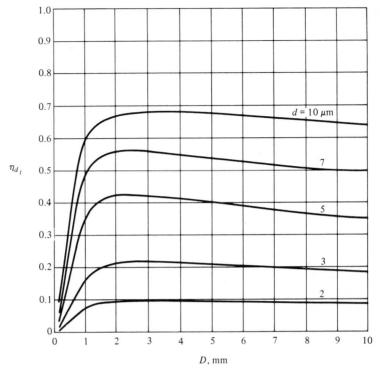

FIGURE 9-13
Collection efficiency for a single drop as a function of the drop diameter for various particle diameters and for $\rho_d = \rho_p = 1000$ kg/m³, $C = \sigma = 1$.

The force acting on the drop is given by Eq. (9-74). The relative velocity can be computed, along with the Reynolds number, from Eqs. (9-76) to (9-81). The power required becomes

$$\dot{W} = 5.135\rho_d D^3 V_\infty N_d + Q\,\Delta P_{ns} \qquad (9\text{-}86)$$

in which N_d is the number of drops in the chamber. This quantity is equal to the drop formation rate $\dot{N}_d$, given by Eq. (9-83) rewritten as

$$\dot{N}_d = \frac{6Q_s}{\pi D^3} \qquad (9\text{-}87)$$

multiplied by the time required for a drop to fall the length L of the chamber. This time is given by L/V_d in which V_d is given by $V_\infty - V_a$ or by $V_\infty - Q/A_{sc}$. Thus

$$N_d = \dot{N}_d \frac{L}{V_\infty - Q/A_{sc}} = \frac{6Q_s A_{sc} L}{\pi D^3 (A_{sc} V_\infty - Q)} \qquad (9\text{-}88)$$

Using Eq. (9-88), Eq. (9-86) gives for the air pumping power

$$\dot{W} = 9.807 \, \frac{\rho_d Q_s A_{sc} L}{A_{sc} - Q/V_\infty} + Q \, \Delta P_{ns} \qquad (9\text{-}89)$$

EXAMPLE 9-6   Determine the air pumping power required in Example 9-5 if $\Delta P_{ns} = 200 \text{ N/m}^2$.

SOLUTION   From Example 9-5:

$$D_{sc} = 2 \text{ m} \qquad V_\infty = 4.33 \text{ m/s}$$
$$A_{sc} = \pi \text{ m}^2 \qquad Q = 10 \text{ m}^3/\text{s}$$
$$L = 5 \text{ m} \qquad Q_s = 0.01 \text{ m}^3/\text{s}$$

Equation (9-89) gives

$$\dot{W} = 9.807 \, \frac{1000(0.01)(\pi)(5)}{\pi - 10/4.33} + 10(200) = 3.85 \text{ kW} \qquad \textit{Ans.} \qquad ////$$

EXAMPLE 9-7   Determine the air pumping power for the spray chamber of Example 9-5 for a range of drop diameters, the other quantities remaining the same. Use $\Delta P_{ns} = 200 \text{ N/m}^2$.

SOLUTION   With all the data except the drop diameter being the same, only the value of V_∞ will vary. Equation (9-89) leads to

$$\dot{W} = 2000 + \frac{1540.5}{\pi - 10/V_\infty}$$

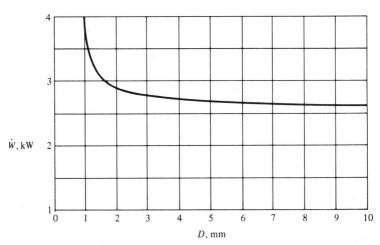

FIGURE 9-14
Variation of pumping power with drop diameter for the data of Example 9-7.

Equations (9-78) and (9-81) give Re_d, from which V_∞ is easily calculated. It is noted that the power becomes infinite when $V_\infty = 10/\pi$; this result occurs because with the given flow rates and spray-chamber diameter the drops are suspended in the stream. With continual introduction of new drops and no removal of them, the number of drops present becomes infinite. This condition must be avoided, of course, since it leads to all kinds of problems. For larger drop diameters, the power required is shown in Fig. 9-14.   ////

The liquid pumping power is an additional consideration in the design of a scrubber, which, unfortunately, cannot be evaluated from theoretical considerations. The power required vastly exceeds the sum of all identifiable energies residing in the spray that is formed; most of the excess energy is lost due to fluid friction in the spray-nozzle passages [6]. Typical water pressures at the entrance to the spray nozzle lie in the range from 0.5 to 1.0 MN/m². The pumping power for a given pressure and flow rate is given by

$$\dot{W} = Q_s \, \Delta P_s \qquad (9\text{-}90)$$

The water pumping power bears little relation to the design of the spray chamber, except that more power is required to produce smaller drops. Better performance must come from improved design of spray nozzles, a subject which must remain beyond the scope of this book.

It is not feasible here to develop a detailed theory for the optimization of the spray chamber. The complexity of the process for determining η_{di} makes the mathematical process of optimization too cumbersome to consider. It will be necessary to evaluate several designs to see which is the most economical. In assessing the economy of competing designs, the initial cost of the equipment, air pumping power, water pumping power, and cost of water treatment are the main factors to be considered. In most plants, the water-treatment cost and the water pumping power are the major costs; neither of these costs will be treated in detail.

9-7 COLLECTION OF PARTICLES BY AN ACCELERATING DROP

Heretofore, our analysis has assumed that the drop travels with uniform velocity V_∞ relative to the air flowing around it. For the simple spray chamber, this assumption is adequate; for more elaborate scrubbers the relative velocity between the drop and the air changes too much during the process to allow this assumption. Typically, the drop is injected into the airstream with a small absolute velocity but with maximum velocity relative to the airstream. The drag force caused by the air flowing around the drop then accelerates the drop to air velocity. With respect to a coordinate system attached to the flowing airstream, the drop first appears with a high velocity and is decelerated to zero velocity. If the coordinate system is attached to the drop, the air velocity is initially large and then attenuates to zero. Such a coordinate system is not inertial, but we shall use it anyway, assuming that any changes brought about by the acceleration of the coordinate system are small compared with those due to the flow around the drop. Figure 9-1 then applies to this situation if V_∞ is taken as a function of time t.

Particle Collection Rate and Efficiency

The collection efficiency for the drop at any instant is given by Eq. (9-43) or Eq. (9-46), depending on whether y_2 is less than or greater than δ_2. The quantity δ_2 is given by Eq. (9-39), while y_2 is obtained from Eq. (9-26) and Table 9-1. Thus the instantaneous collection efficiency η_{di} is readily obtained, neglecting the diffusion mechanism of collection. The instantaneous collection efficiency varies with V_∞, and so we must seek some overall collection efficiency. This can best be done by computing the number of particles collected by the drop as it decelerates for a given number density of particles in the air. For this analysis, assume that the number density is constant throughout the path of the drop.

The number of particles collected depends on how far the drop travels or rather on how its distance of travel varies with its velocity. The distance of travel is given by Eqs. (3-117) to (3-122). Those equations give the distance traveled by the particle starting from its point of injection; let us define a distance s' starting from the end point of its motion as

$$s' = s_{\max} - s \qquad (9\text{-}91)$$

Thus s' becomes zero when the particle accelerates to the gas velocity.

Combining Eqs. (3-119) and (3-120), we obtain

$$\frac{s'\rho}{\rho_d D} = 0.4041 - \frac{2}{3}\ln\frac{22}{\text{Re} + 12} \qquad \text{Re} \le 10 \qquad (9\text{-}92)$$

Combining Eqs. (3-118) and (3-120) gives

$$\frac{s'\rho}{\rho_d D} = 6.313 + 1.938\ln\frac{\text{Re}^2 + 195.6\,\text{Re} - 835.5}{626{,}084}$$

$$- 1.858\ln\frac{1.293(\text{Re} - 4.18)}{\text{Re} + 200} \qquad 10 \le \text{Re} \le 700 \qquad (9\text{-}93)$$

Combining Eqs. (3-117), (3-120), and (3-121), we have

$$\frac{s'\rho}{\rho_d D} = 3.03\ln\frac{\text{Re}}{700} + 6.313 \qquad \text{Re} \ge 700 \qquad (9\text{-}94)$$

Equations (9-92) to (9-94) apply to the Reynolds number ranges indicated. The motion of the drop is traced starting from its initial Reynolds number through the sequence of equations until the Reynolds number becomes zero.

Let C_{nv} be the number density of particles in the region through which the drop travels, and let ΔN be the number of particles collected by a single drop. Then

$$\Delta N = \frac{\pi D^2 \sigma}{4} C_{nv} \int_0^{s_{max}} \frac{\eta_{di}}{\sigma} \, ds' \qquad (9\text{-}95)$$

Now let $\dot{N}_d$ be the rate at which drops are injected into the airstream, and let $\Delta \dot{N}$ be the rate at which particles are collected by the $\dot{N}_d$ drops. Then

$$\Delta \dot{N} = \frac{\pi D^2}{4} C_{nv} \dot{N}_d \sigma \int_0^{s_{max}} \frac{\eta_{di}}{\sigma} \, ds' \qquad (9\text{-}96)$$

This integral can be broken up into small intervals and written as a sum:

$$\frac{\Delta \dot{N} \rho}{C_{nv} \sigma \dot{N}_d \rho_d D^3} = \frac{\pi}{4} \sum \frac{\eta_{di_{ave}}}{\sigma} \left(\frac{\rho}{\rho_d D} \Delta s' \right) \qquad (9\text{-}97)$$

Letting ΔRe be the interval of Reynolds number in the sum, Eq. (9-97) may be written

$$\frac{\Delta \dot{N} \rho}{C_{nv} \sigma \dot{N}_d \rho_d D^3} (\text{Re} + \Delta \text{Re}) = \frac{\Delta \dot{N} \rho}{C_{nv} \sigma \dot{N}_d \rho_d D^3} \text{Re} + \frac{\pi}{4\sigma} \eta_{di_{ave}} \Delta \frac{\rho s'}{\rho_d D} \qquad (9\text{-}98)$$

Letting the subscript n indicate the number of terms which have been summed, and letting $n + 1$ represent the addition of one more term, Eq. (9-98) may also be written as

$$\left(\frac{\Delta \dot{N} \rho}{C_{nv} \sigma \dot{N}_d \rho_d D^3} \right)_{n+1} = \left(\frac{\Delta \dot{N} \rho}{C_{nv} \sigma \dot{N}_d \rho_d D^3} \right)_n$$

$$+ \frac{\pi}{4} \frac{\eta_{di_{n+1}} + \eta_{di_n}}{2\sigma} \left[\left(\frac{\rho s'}{\rho_d D} \right)_{n+1} - \left(\frac{\rho s'}{\rho_d D} \right)_n \right] \qquad (9\text{-}99)$$

Note that the summation begins at the point at which the particle has reached zero velocity relative to the gas and proceeds in the direction opposite to the particle motion.

The results of Eq. (9-99) are tabulated in Table 9-3 for various values of $\sqrt{\rho_p/\rho} \, d/D$ and for various values of the Reynolds number. The tabulated quantity

$$\frac{\Delta \dot{N} \rho}{C_{nv} \sigma \dot{N}_d \rho_d D^3}$$

can be determined for any value of the initial Reynolds number by simply entering the table at that value of Reynolds number. Furthermore, the value of the tabulated quantity can be obtained for the case where the drop starts at one value of Re and slows to a second value of Re, not zero; simply enter the table at each of these values of Re and take the difference. A few examples will illustrate the procedure.

If $\dot{N}$ is the rate at which particles enter the scrubber or a stage of a scrubber, then $C_{nv} = \dot{N}/Q$. The quantity tabulated in Table 9-3 then becomes

$\Delta \dot{N} \rho Q / \sigma \dot{N} \dot{N}_d \rho_d D^3$. Writing the collection efficiency for the scrubber or for the stage of the scrubber as $\Delta \dot{N} / \dot{N}$ and making use of Eq. (9-87), the preceding quantity becomes

$$\frac{\eta \rho Q}{\sigma \dot{N}_d \rho_d D^3} = \frac{\pi}{6} \frac{\eta \rho Q}{\sigma \rho_d Q_s} \qquad (9\text{-}99a)$$

Table 9-3   PARTICLE-COLLECTION RATE FOR A DECELERATING DROP*

Re $\sqrt{\rho_p/\rho}\, d/D$	0.005	0.01	0.02	0.05	0.1
0	0	0	0	0	0
10	1.191(−9)†	1.747(−8)	2.421(−7)	6.594(−6)	0.0000609
20	6.055(−9)	8.736(−8)	1.175(−6)	2.957(−5)	0.0002865
40	4.648(−8)	6.504(−7)	8.290(−6)	1.776(−4)	0.001745
60	1.534(−7)	2.106(−6)	2.593(−5)	5.692(−4)	0.004949
80	3.636(−7)	4.912(−6)	5.870(−5)	0.001275	0.01031
100	7.114(−7)	9.478(−6)	1.103(−4)	0.002323	0.01798
200	6.226(−6)	7.800(−5)	7.958(−4)	0.01544	0.09242
300	1.900(−5)	2.284(−4)	0.002158	0.04052	0.2083
400	4.106(−5)	4.780(−4)	0.004627	0.07714	0.3504
500	7.384(−5)	8.352(−4)	0.008246	0.1236	0.5055
600	1.184(−4)	0.001305	0.01284	0.1782	0.6647
700	1.755(−4)	0.001887	0.01847	0.2390	0.8225
800	2.451(−4)	0.002575	0.02512	0.3038	0.9751
900	3.260(−4)	0.003352	0.03261	0.3699	1.119
1000	4.192(−4)	0.004222	0.04100	0.4368	1.255
1100	5.241(−4)	0.005176	0.05020	0.5032	1.384
1200	6.407(−4)	0.006209	0.06015	0.5688	1.506
1300	7.701(−4)	0.007326	0.07088	0.6335	1.622
1400	9.111(−4)	0.008449	0.08226	0.6968	1.732
1500	0.001065	0.009927	0.09432	0.7590	1.837
1600	0.001231	0.01169	0.1070	0.8200	1.938
1700	0.001410	0.01353	0.1203	0.8797	2.034
1800	0.001600	0.01545	0.1340	0.9379	2.126
1900	0.001803	0.01747	0.1482	0.9949	2.215
2000	0.002018	0.01958	0.1627	1.050	2.301
2250	0.002612	0.02535	0.2007	1.184	2.501
2500	0.003282	0.03175	0.2400	1.311	2.686
2750	0.004026	0.03882	0.2805	1.532	2.857
3000	0.004845	0.04639	0.3205	1.545	3.014
3250	0.005733	0.05494	0.3624	1.658	3.167
3500	0.006695	0.06365	0.4021	1.760	3.304
3750	0.007724	0.07313	0.4424	1.860	3.436
4000	0.008817	0.08280	0.4811	1.953	3.557
4250	0.009980	0.09359	0.5218	2.047	3.680
4500	0.01120	0.1043	0.5599	2.134	3.790
4750	0.01245	0.1152	0.5973	2.216	3.895
5000	0.01384	0.1272	0.6361	2.300	4.001
6000	0.02089	0.1758	0.7810	2.599	4.372
7000	0.02982	0.2270	0.9187	2.865	4.696
8000	0.03956	0.2780	1.046	3.099	4.977
9000	0.05053	0.3298	1.168	3.315	5.232
10,000	0.06240	0.3808	1.282	3.512	5.462

(*Continued*)

Table 9-3 PARTICLE COLLECTION RATE FOR A DECELERATING DROP
(*Continued*)*

Re $\sqrt{\rho_p/\rho}\, d/D$	0.2	0.5	1	2	5	10
0	0	0	0	0	0	0
10	0.000600	0.00564	0.01526	0.02419	0.03026	0.03149
20	0.002402	0.01813	0.04281	0.06390	0.07696	0.07945
40	0.01161	0.06113	0.1163	0.1570	0.1786	0.1822
60	0.02860	0.1221	0.2066	0.2627	0.2901	0.2942
80	0.05323	0.1951	0.3070	0.3751	0.4066	0.4111
100	0.08471	0.2760	0.4126	0.4905	0.5251	0.5298
200	0.3136	0.7172	0.9393	1.048	1.090	1.095
300	0.5936	1.148	1.421	1.544	1.590	1.595
400	0.8782	1.541	1.846	1.977	2.024	2.029
500	1.150	1.895	2.220	2.356	2.404	2.409
600	1.405	2.214	2.553	2.693	2.741	2.746
700	1.642	2.502	2.853	2.995	3.044	3.049
800	1.861	2.763	3.122	3.266	3.315	3.320
900	2.061	2.996	3.363	3.508	3.557	3.562
1000	2.245	3.209	3.581	3.727	3.776	3.781
1100	2.415	3.404	3.780	3.927	3.976	3.981
1200	2.574	3.582	3.963	4.110	4.160	4.164
1300	2.722	3.749	4.132	4.281	4.330	4.335
1400	2.862	3.904	4.290	4.439	4.488	4.493
1500	2.993	4.049	4.438	4.587	4.636	4.641
1600	3.119	4.187	4.577	4.726	4.776	4.780
1700	3.238	4.316	4.708	4.858	4.907	4.912
1800	3.351	4.438	4.832	4.982	5.031	5.036
1900	3.459	4.555	4.950	5.100	5.149	5.154
2000	3.561	4.666	5.062	5.212	5.261	5.266
2250	3.801	4.922	5.321	5.471	5.520	5.525
2500	4.017	5.153	5.553	5.704	5.753	5.758
2750	4.217	5.364	5.766	5.917	5.966	5.971
3000	4.397	5.554	5.958	6.108	6.158	6.162
3250	4.573	5.738	6.142	6.293	6.343	6.347
3500	4.728	5.900	6.306	6.457	6.506	6.511
3750	4.877	6.056	6.462	6.613	6.662	6.667
4000	5.013	6.197	6.604	6.755	6.804	6.809
4250	5.149	6.339	6.746	6.897	6.946	6.951
4500	5.272	6.466	6.874	7.025	7.074	7.079
4750	5.388	6.586	6.994	7.145	7.194	7.199
5000	5.504	6.706	7.114	7.265	7.315	7.319
6000	5.907	7.121	7.530	7.681	7.730	7.735
7000	6.256	7.477	7.887	8.038	8.088	8.093
8000	6.554	7.782	8.193	8.349	8.393	8.398
9000	6.825	8.058	8.469	8.620	8.669	8.674
10,000	7.067	8.303	8.714	8.865	8.915	8.920

* The tabulated quantity is

$$\frac{\Delta \dot{N} \rho}{C_{nt}\, \sigma \dot{N}_d\, \rho_d\, D^3} = \frac{\eta \rho Q}{\sigma \dot{N}_d\, \rho_d\, D^3} = \frac{\pi}{6}\, \frac{\eta \rho Q}{\sigma \rho_d\, Q_s}$$

† Numbers in parentheses represent powers of 10. Thus $1.191(-9) = 1.191 \times 10^{-9}$.

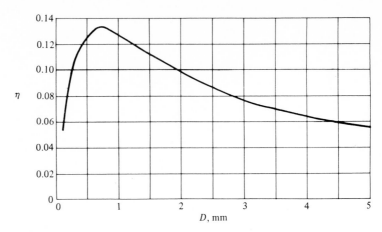

FIGURE 9-15
Collection efficiency of an injection scrubber as a function of drop size for the data of Example 9-8.

EXAMPLE 9-8   A stream of air contains particles with diameters of 2.0 μm and densities of 1500 kg/m^3. Water is injected into the airstream with a volumetric flow rate of 0.0001 times that of the air, forming droplets having a uniform diameter D. If the airstream velocity is 15 m/s, if the drops are injected with essentially zero velocity, and if it is assumed that no parcel of air is exposed to more than one drop, determine the collection efficiency for a range of drop diameters. Assume $\sigma = 1$.

SOLUTION   The initial Reynolds number is based on the relative velocity between the air and the drop at the time of injection and is

$$\text{Re}_0 = \frac{V_{\infty_0} D}{\nu} = 10^6 D$$

The parameter $\sqrt{\rho_p/\rho}\, d/D$ is given as $0.07115/D \times 10^3$; Table 9-3 gives

$$\frac{\pi}{6} \frac{\eta \rho Q}{\rho_d Q_s} = \frac{\pi(1.185)(\eta)}{6(1000)(0.0001)} = 6.205\eta$$

Figure 9-15 shows the collection efficiency of the scrubber as a function of the drop diameter.   ////

Distance Traveled by the Drop

For the proper design of equipment, knowledge of the distance traveled by the drop is needed. Since the drop requires infinite time to reach exactly the air velocity, during most of which it will travel at almost the air velocity, the drop will have traveled an infinite distance by the time it completely reaches the air velocity. As a practical matter, the drop will almost reach the air velocity in a reasonably short distance. To estimate this distance, in general, let us write Eqs. (3-112) and (3-113), respectively, as

$$\frac{t\mu}{\rho_d D^2} = 3.03\left(\frac{1}{\text{Re}} - \frac{1}{\text{Re}_0}\right) \qquad \text{Re} \geq 700 \qquad (9\text{-}100)$$

$$\frac{t\mu}{\rho_d D^2} = 3.03\left(\frac{1}{700} - \frac{1}{\text{Re}_0}\right) - \frac{1}{52.63}\ln\frac{1.293(\text{Re} - 4.18)}{\text{Re} + 200} \qquad 10 \leq \text{Re} \leq 700 \qquad (9\text{-}101)$$

Equation (3-114) for the Reynolds number less than 10 will not be included here. These equations give the dimensionless time of travel of the drop from its injection. The dimensionless quantity $t\mu/\rho_d D^2$ is tabulated in Table 9-4, starting with zero at $\text{Re} = 10,000$; the dimensionless distance $\rho s'/\rho_d D$ is also tabulated there. These two columns enable one to compute the absolute distance l_d traveled by the drop in completing a fixed fraction of its acceleration. The average value of l_d may be written as

$$l_d = V_a t - \Delta s' \qquad (9\text{-}102)$$

Table 9-4 MOTION OF A DECELERATING DROP

Re	$\rho s'/\rho_d D$	$t\mu/\rho_d D^2$	Re	$\rho s'/\rho_d D$	$t\mu/\rho_d D^2$
0	0	∞	1700	9.002	0.001479
10	0.4058	0.06728	1800	9.175	0.001380
20	0.6629	0.04916	1900	9.339	0.001292
40	1.059	0.03528	2000	9.494	0.001212
60	1.398	0.02838	2250	9.851	0.001044
80	1.704	0.02397	2500	10.17	$9.090(-4)$
100	1.985	0.02083	2750	10.46	$7.988(-4)$
200	3.135	0.01271	3000	10.72	$7.070(-4)$
300	4.015	0.009116	3250	10.96	$6.293(-4)$
400	4.713	0.007047	3500	11.19	$5.627(-4)$
500	5.334	0.005696	3750	11.40	$5.050(-4)$
600	5.855	0.004742	4000	11.59	$4.545(-4)$
700	6.313	0.004026	4250	11.78	$4.099(-4)$
800	6.718	0.003484	4500	11.95	$3.703(-4)$
900	7.074	0.003064	4750	12.11	$3.349(-4)$
1000	7.394	0.002727	5000	12.27	$3.030(-4)$
1100	7.683	0.002452	6000	12.82	$2.020(-4)$
1200	7.946	0.002222	7000	13.29	$1.299(-4)$
1300	8.189	0.002028	8000	13.69	$7.575(-5)$
1400	8.413	0.001861	9000	14.05	$3.367(-5)$
1500	8.622	0.001717	10,000	14.37	0
1600	8.818	0.001591			

which may be written in terms of the tabulated quantities as

$$l_d = \frac{\rho_d D^2}{\mu} V_a \left[\left(\frac{t\mu}{\rho_d D^2} \right)_e - \left(\frac{t\mu}{\rho_d D^2} \right)_0 \right] - \frac{\rho_d D}{\rho} \left[\left(\frac{\rho s'}{\rho_d D} \right)_0 - \left(\frac{\rho s'}{\rho_d D} \right)_e \right] \qquad (9\text{-}103)$$

The subscript e represents the lower end point of the process, at which the drop is held to be traveling close enough to the air velocity for the purpose at hand. This point might be selected at which the Reynolds number is 10 percent of its initial value or at which s' is 10 percent of its initial value. A third alternative is to select $Re_e = 10$, which gives

$$l_d = \frac{\rho_d D^2}{\mu} V_a \left[0.067 - \left(\frac{t\mu}{\rho_d D^2} \right)_0 \right] - \frac{\rho_d D}{\rho} \left[\left(\frac{\rho s'}{\rho_d D} \right)_0 - 0.406 \right] \qquad (9\text{-}104)$$

EXAMPLE 9-9   In Example 9-8, compute the distance traveled by the drop in reaching 10 percent of its initial value of Re.

SOLUTION   Using Eq. (9-103) with the given data gives

$$l_d = 843.9D \left\{ \left[\left(\frac{t\mu}{\rho_d D^2} \right)_e - \left(\frac{t\mu}{\rho_d D^2} \right)_0 \right] Re_0 - \left(\frac{\rho s'}{\rho_d D} \right)_0 + \left(\frac{\rho s'}{\rho_d D} \right)_e \right\}$$

The values of l_d for various drop diameters are given in the following table:

D, mm	l_d, m	D, mm	l_d, m
0.1	0.2587	2.0	28.08
0.3	1.716	2.5	38.84
0.5	3.781	3.0	46.88
1.0	10.71	4.0	65.85
1.5	20.91	5.0	84.51

////

Example 9-9 illustrates how the distance traveled by the drop increases very markedly with increasing drop diameter before reaching the point at which its relative velocity is reduced to 10 percent of original value. The following example shows how a drop of a given diameter approaches the velocity of the airstream.

EXAMPLE 9-10   Using the data of Examples 9-8 and 9-9, determine the distance traveled by the drop as a function of relative drop velocity for a drop diameter of 1 mm.

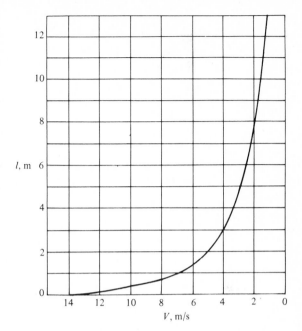

FIGURE 9-16
Distance traveled by a drop as a function of its relative velocity for the data of
Example 9-10.

SOLUTION   The equation in Example 9-9 applies, leading to

$$l = 843.9\left(\frac{t\mu}{\rho_d D^2}\right)_e + 0.8439\left(\frac{\rho s'}{\rho_d D}\right)_e - 8.5411$$

Figure 9-16 shows how l varies with velocity. The values of $(t\mu/\rho_d D^2)_e$ and
$(\rho s'/\rho_d D)_e$ were taken from Table 9-4 at various values of Reynolds number;
the velocity is related to the Reynolds number as $V = \text{Re}/66.67$.   ////

Variation of Efficiency with Drop Velocity

In anticipation of the treatment of multistage scrubbers, it is desirable to know
the efficiency of a single stage under the condition that the drops have not yet
reached the gas velocity at the end of the stage. This matter will be investigated
shortly; in the meantime, the next example explores how the collection efficiency
varies with distance traveled by the drops.

EXAMPLE 9-11   For the data of Example 9-8 and for a drop diameter of 1.0 mm,
calculate and plot the stage efficiency η as a function of the distance l traveled
by the drop from its point of injection.

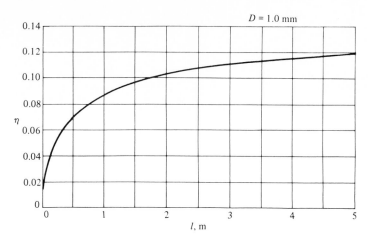

$D = 1.0$ mm

FIGURE 9-17
Collection efficiency as a function of drop travel for the data of Example 9-11.

SOLUTION   From Example 9-8, the initial Reynolds number is 1000, the value of $\sqrt{\rho_p/\rho}\, d/D$ is 0.07115, and the quantity $(\pi/6)\eta\rho Q/\rho_d Q_s$ tabulated in Table 9-3 becomes 6.205η. At the initial point, the value of η is 0.1893; this value assumes that the drop will eventually reach the gas velocity. As the drop travels, the efficiency actually achieved is the difference between that value and the value of η at the particular location of the drop. The values of l are obtained from Example 9-10. Efficiency as a function of distance is plotted in Fig. 9-17.

////

Multistage Spray Scrubber

Figure 9-17 suggests that the bulk of the particle collection achieved by the drop occurs in the first stages of its travel, which suggests an important simplification in dealing with multistage collection. In many and perhaps most cases, we can consider in each stage only the particle collection achieved by the drops injected at the beginning of that stage. Thus, we may expect a conservative estimate for the overall collection efficiency since drops injected at each stage except the last will continue to collect particles in subsequent stages. With this assumption, the collection efficiency can be computed by the method used in Example 9-11; the combined collection efficiency for the scrubber consisting of s stages is given by

$$\eta = 1 - \prod_{i=1}^{s} (1 - \eta_i) \qquad (9\text{-}105)$$

More elaborate formulas can be developed to handle the case where the preceding simplification is not made, but we shall not develop these formulas here.

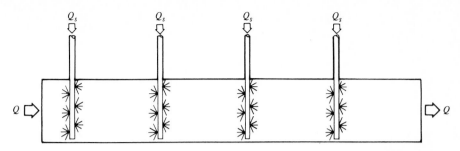

FIGURE 9-18
Four-stage scrubber.

EXAMPLE 9-12 (a) For the data of Examples 9-8 and 9-11, determine the overall collection efficiency of the four-stage scrubber shown in Fig. 9-18. Assume that $Q_s/Q = 0.0001$ in each stage and that the succeeding stages are 2.5 m apart.

(b) How many stages are required to give a collection efficiency of 0.99? How much water flow is required in this case?

(c) What flow rate of water is necessary to obtain an efficiency of 0.99 in a single stage?

SOLUTION (a) From Fig. 9-17, at $l = 2.5$ m, the single-stage efficiency is $\eta_i = 0.108$. Then Eq. (9-105) gives the overall efficiency as

$$\eta = 1 - (1 - \eta_i)^4 = 1 - (1 - 0.108)^4 = 0.367 \qquad Ans.$$

(b) For s stages in series, Eq. (9-105) gives

$$\eta = 1 - (1 - \eta_i)^s$$

Solving this equation for s leads to

$$s = \frac{\ln (1 - \eta)}{\ln (1 - \eta_i)} = \frac{\ln 0.01}{\ln 0.892} = 40 \qquad Ans.$$

The total water flow rate is

$$\frac{Q_s}{Q} = 40(0.0001) = 0.0040 \qquad Ans.$$

(c) From Example 9-8, the quantity $\sqrt{\rho_p/\rho}\, d/D$ is equal to 0.07115 for a drop with a diameter of 1 mm; then Table 9-3 gives

$$\frac{\pi}{6} \frac{\eta \rho Q}{\rho_d Q_s} = 0.7821 \qquad Ans.$$

which gives

$$\frac{Q_s}{Q} = \frac{\pi}{6} \frac{\eta \rho}{0.7821 \rho_d} = \frac{\pi(0.99)(1.185)}{6(0.7821)(1000)} = 0.000785 \qquad Ans. \qquad ////$$

From the preceding example we see that the use of stages in series is much less efficient than the use of a single stage for a given water flow rate. It is also apparent that the use of sufficient water will lead to a theoretically perfect collection efficiency. The amount of water required per unit flow rate of air for $\eta = 1.0$ is given by

$$\frac{Q_s}{Q} = \frac{\pi \eta \rho / 6 \sigma \rho_d}{\pi \eta \rho Q / 6 \sigma \rho_d Q_s} \qquad (9\text{-}106)$$

in which the denominator is obtained from Table 9-3 for the value of the quantity $\sqrt{\rho_p/\rho}\, d/D$ computed for the problem at hand.

Power Requirements

The power required to accelerate the drops can be obtained quickly. This power requirement is in addition to that necessary to overcome friction due to the air flowing through the device; this latter power can be measured by operating the device with the water flow cut off. Equation (2-67) gives the drag force acting on the drop as

$$F = C_D \frac{\pi D^2}{8} \rho V_\infty^{\,2} = C_D \frac{\pi}{8} \frac{\mu^2 \, \mathrm{Re}^2}{\rho} \qquad (9\text{-}107)$$

in which the Reynolds number has been substituted. The work done on each drop is then given by the drag force times the distance moved by the drop relative to the airstream:

$$W_d = \int_0^{S_{max}} F \, ds = \frac{\pi}{8} \frac{\mu^2}{\rho} \int_0^{S_{max}} C_D \, \mathrm{Re}^2 \, ds \qquad (9\text{-}108)$$

The power is then obtained upon multiplying this by the rate at which drops are formed, given by $\dot{N}_d$:

$$\dot{W} = \dot{N}_d W_d = \frac{\pi}{8} \dot{N}_d \frac{\mu^2}{\rho} \int_0^{S_{max}} C_D \, \mathrm{Re}^2 \, ds \qquad (9\text{-}109)$$

Equation (3-80) may be written as

$$s = -\frac{4}{3} D \frac{\rho_d}{\rho} \int_{V_{\infty_0}}^{V_\infty} \frac{dV_\infty}{C_D V_\infty}$$

Differentiating this equation gives

$$ds = -\frac{4}{3} D \frac{\rho_d}{\rho} \frac{1}{C_D V_\infty} dV_\infty = -\frac{4}{3} D \frac{\rho_d}{\rho} \frac{1}{C_D} \frac{d\,\mathrm{Re}}{\mathrm{Re}}$$

Substituting into Eq. (9-109) and performing the integration gives

$$\dot{W} = \frac{\pi}{6} \dot{N}_d \frac{\rho_d}{\rho^2} D\mu^2 \int_0^{\mathrm{Re}_0} \mathrm{Re} \, d\,\mathrm{Re} = \frac{\pi}{12} \dot{N}_d \frac{\rho_d}{\rho^2} \mu^2 D \, \mathrm{Re}_0^2$$

Now, using Eq. (9-87) for $\dot{N}_d$ leads to

$$\dot{W} = \frac{Q_s \rho_d}{2} V_{\infty_0}{}^2$$

We must examine the energy represented by this equation more closely. The energy expended during the acceleration of the drop is equal to the kinetic energy imparted to the drop plus the energy dissipated as friction in the air. The kinetic energy of the drop is represented correctly if its velocity is measured relative to the earth; the preceding equation clearly gives the kinetic energy of the drops. The friction dissipated in the air is given as the integral of force times distance traveled by the air relative to the drop, which leads to the same equation.

Thus our derivation leads to an equation for the power due to the kinetic energy imparted to the drops and as well to an equation for the friction loss in the air during the acceleration of the drops. Since the same equation gives both energies, the total energy required in the acceleration process is twice that evaluated previously:

$$\dot{W} = Q_s \rho_d V_{\infty_0}{}^2 = Q_s \rho_d V_a{}^2 \qquad (9\text{-}110)$$

The pressure drop corresponding to this power is given by

$$\Delta P = \frac{\dot{W}}{Q} = \frac{Q_s}{Q} \rho_d V_a{}^2 \qquad (9\text{-}111)$$

In most devices, the power is actually manifested by a drop in pressure of the airstream in passing through the scrubber. Work is then added to the airstream at some other point to overcome this pressure loss, as given by Eq. (9-111).

EXAMPLE 9-13   Using the data of Example 9-8, how much power is required to accelerate the drops to the air velocity, for an air flow rate of 1.0 m³/s? What pressure drop results in the airstream? Repeat using the water flow rate of Example 9-12c.

SOLUTION   The following data apply:

$$Q = 1.0 \text{ m}^3/\text{s} \qquad \rho_d = 1000 \text{ kg/m}^3$$
$$Q_s = 0.0001 \text{ m}^3/\text{s} \qquad V_a = 15 \text{ m/s}$$

Equations (9-110) and (9-111) give

$$\dot{W} = 0.0001(1000)(15)^2 = 22.5 \text{ W} \qquad Ans.$$

$$\Delta P = \frac{0.0001}{1.0}(1000)(15)^2 = 22.5 \text{ N/m}^2 \qquad Ans.$$

For $Q_s = 0.000785$ m³/s from Example 9-12c, the results are

$$\dot{W} = 0.000785(1000)(15)^2 = 176.6 \text{ W} \qquad Ans.$$
$$\Delta P = 176.6 \text{ N/m}^2 \qquad Ans. \qquad ////$$

9-8   THE CYCLONE SCRUBBER

One aspect of scrubbers is that any type of control device, with a few exceptions, can be converted into a scrubber by the addition of water sprays. In fact, devices not ordinarily thought of as control devices can be made into scrubbers; we saw an example in Sec. 9-7 where a straight length of duct was made into a scrubber by the addition of water sprays. The cyclone collector can be made into a cyclone scrubber by the addition of sprays at some suitable point. Such a device, shown in Fig. 9-19, employs a straight-through cyclone; reverse-flow cyclone scrubbers are not commonly used in practice. Water drops are injected into the flow stream from sprays located along the center of the scrubber. These drops are then given a tangential acceleration by the cyclonic gas motion; the centrifugal force thus applied to the drop accelerates it radially until it reaches the outer shell of the cyclone. During the tangential and radial acceleration process, the drop sweeps up particles. At the wall the drops coalesce into a film which flows down the wall to the sump at the bottom. Small droplets entrained in the air are removed by a simple mist eliminator at the top of the scrubber.

The complexity of analysis of the cyclone scrubber, which can take different forms for different variations in operation of the device, renders a simple and close approximation unlikely. We shall make some plausible assumptions and derive an expression for collection efficiency which, it is hoped, will give meaningful results. No simple equation is likely to accurately correlate experimental data.

Number of Particles Collected

Let us assume cyclonic motion in the region between $R_2/2$ and R_2; Eq. (7-7) gives the tangential velocity for this region:

$$V_\theta = \frac{Q}{Wr \ln R_2/R_1} = 1.443 \frac{Q}{Wr} \qquad (9\text{-}112)$$

The average velocity can be obtained by simply integrating this value of V_θ over the radial distance involved:

$$V_{\theta_{ave}} = \frac{2}{R_2} \int_{R_2/2}^{R_2} V_\theta \, dr = \frac{2Q}{WR_2} \qquad (9\text{-}113)$$

To determine the equation of motion of the drop, start with the forces acting on the drop shown in Fig. 9-20. Let V_1 be the velocity of a drop which is initially projected at velocity $V_{\theta_{ave}}$ into still air as a function of time or position. The relative velocity of the drop is given by

$$V_d = V_{\theta_{ave}} - V_1 \qquad (9\text{-}114)$$

The equation of motion in the r direction is

$$F_c - F_{Dr} = m \frac{dV_r}{dt} \qquad (9\text{-}114a)$$

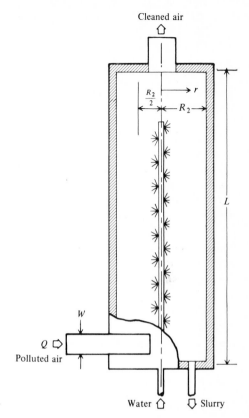

FIGURE 9-19
Cyclone scrubber.

The centrifugal force is given as mV_d^2/r, and the drag force in the radial direction by $C_{Dr} A V_r^2$; using Eq. (9-114) and the value of the mass of the drop, Eq. (9-114a) becomes

$$\frac{d^2r}{dt^2} + \frac{3}{4} C_{Dr} \frac{\rho}{\rho_d D} \left(\frac{dr}{dt}\right)^2 = \frac{1}{r} (V_{\theta_{ave}} - V_1)^2 \qquad (9\text{-}115)$$

in which V_r has been written as dr/dt. Equation (3-7) may be written as

$$\frac{dV_1}{dt} + \frac{3}{4} C_{D1} \frac{\rho}{\rho_d D} V_1^2 = 0 \qquad (9\text{-}116)$$

Solving this equation for t gives

$$t = -\int_{V_{\theta_{ave}}}^{V_1} \frac{dV_1}{3C_{D1}\rho V_1^2/4\rho_d D} \qquad (9\text{-}117)$$

What we are seeking to do at this point is to write Eq. (9-115) in terms of V_1 rather than t, so that the value of V_1 which occurs when the drop strikes the wall of the scrubber can be estimated. This value can be used in conjunction

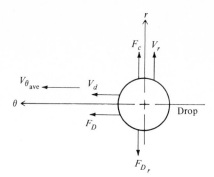

FIGURE 9-20
Velocities and forces acting on a drop in
a cyclone scrubber.

with Table 9-3 to estimate the collection efficiency of the drop. Some rather gross assumptions are made here; let us assume that the Reynolds number in the tangential direction exceeds 500 so that C_{D1} can be taken as the constant value 0.44, as given by Eq. (2-75). Later on, we shall make a similar assumption in the radial direction. Equation (9-117) then becomes

$$t = 3.03 \frac{\rho_d D}{\rho} \left(\frac{1}{V_1} - \frac{1}{V_{\theta_{ave}}} \right) \qquad (9\text{-}118)$$

which can be solved for V_1 and for dV_1/dt as

$$V_1 = \frac{1}{1/V_{\theta_{ave}} + \rho t/3.03\rho_d D} \qquad (9\text{-}119)$$

$$\frac{dV_1}{dt} = -\frac{\rho V_1^2}{3.03\rho_d D} \qquad (9\text{-}120)$$

From Eq. (9-120), the expression for dr/dt is

$$\frac{dr}{dt} = \frac{dr}{dV_1} \frac{dV_1}{dt} = -\frac{\rho V_1^2}{3.03\rho_d D} \frac{dr}{dt} \qquad (9\text{-}121)$$

The second derivative becomes

$$\frac{d^2 r}{dt^2} = \frac{d}{dV_1} \frac{dr}{dt} \frac{dV_1}{dt} = \left(\frac{\rho}{3.03\rho_d D} \right)^2 \left(V_1^4 \frac{d^2 r}{dV_1^2} + 2V_1^3 \frac{dr}{dV_1} \right) \qquad (9\text{-}122)$$

When Eqs. (9-121) and (9-122) are substituted into Eq. (9-115), the differential equation of motion becomes

$$\frac{d^2 r}{dV_1^2} + \frac{2}{V_1} \frac{dr}{dV_1} + \frac{3}{4} C_{Dr} \frac{\rho}{\rho_d D} \left(\frac{dr}{dV_1} \right)^2 = \left(\frac{3.03\rho_d D}{\rho} \right)^2 \frac{1}{r} \frac{(V_{\theta_{ave}} - V_1)^2}{V_1^4} \qquad (9\text{-}123)$$

which can be placed in a nondimensional form if

$$\mathrm{Re}_1 = \frac{V_1 D}{\nu} \qquad \mathrm{Re}_0 = \frac{V_{\theta_{ave}} D}{\nu} \qquad \xi = \frac{\mathrm{Re}_1}{\mathrm{Re}_0} \qquad \zeta = \frac{\rho r}{\rho_d D} \qquad (9\text{-}124)$$

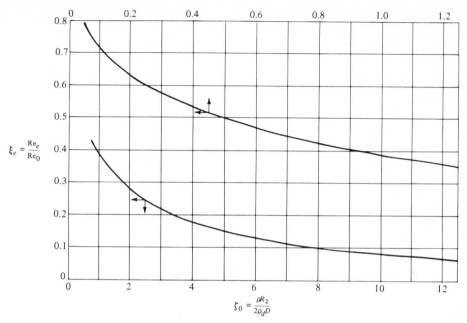

FIGURE 9-21

Nondimensional terminal Reynolds number as a function of ζ_0.

Then

$$\frac{d^2\zeta}{d\xi^2} + \frac{2}{\xi}\frac{d\zeta}{d\xi} + 0.33\left(\frac{d\zeta}{d\xi}\right)^2 = \frac{9.18}{\zeta}\frac{(1-\xi)^2}{\xi^4} \qquad (9\text{-}125)$$

in which the value of C_{Dr} is taken to be 0.44 as in the case of C_{D1}.

At the start of the motion at $r = R_2/2$, the velocity V_1 is equal to the air velocity $V_{\theta_{ave}}$. From Eq. (9-124), this boundary condition becomes

$$\zeta = \zeta_0 = \frac{\rho R_2}{2\rho_a D} \qquad \text{when } \xi = 1 \qquad (9\text{-}126)$$

At the other end of the motion, when the drop strikes the wall, the condition is

$$\xi = \xi_e = \frac{\text{Re}_e}{\text{Re}_0} \qquad \text{when } \zeta = \zeta_e = 2\zeta_0 \qquad (9\text{-}127)$$

The quantity ξ_e is to be solved for, and from this the Reynolds number Re_e at the end of the motion can be obtained. This quantity can be used in Table 9-3 to represent the end of the sweeping motion of the drop. Figure 9-21 shows a curve of ξ_e as a function of ζ_0. An example shows how this curve is used in conjunction with Table 9-3.

EXAMPLE 9-14   A cyclone scrubber treats 10 m^3/s of air. The radius of the scrubber is 2.0 m, and the width W is 0.5 m; the scrubbing fluid is water with a drop diameter of 1 mm. Determine the value of Re_e and the value of $\Delta \dot{N}\rho/C_{nv}\dot{N}_d\rho_d D^3\sigma$ from Table 9-3, assuming that $\rho_p = 1500$ kg/m^3 and $d = 2$ μm.

SOLUTION   Equation (9-113) gives $V_{\theta_{ave}}$ as

$$V_{\theta_{ave}} = \frac{2Q}{WR_2} = \frac{2(10)}{0.5(2)} = 20 \text{ m/s}$$

The values of Re_0 and ζ_0 are computed as

$$Re_0 = \frac{V_{\theta_{ave}}D}{v} = \frac{20(10^{-3})}{1.5 \times 10^{-5}} = 1333$$

$$\zeta_0 = \frac{\rho R_2}{2\rho_d D} = \frac{1.185(2)}{2(1000)(0.001)} = 1.185$$

Figure 9-21 gives $\xi_e = 0.36$, from which

$$Re_e = 0.36 \, Re_0 = 0.36(1333) = 480 \qquad Ans.$$

The following value is computed:

$$\sqrt{\frac{\rho_p}{\rho}\frac{d}{D}} = 0.07116$$

Then Table 9-3 gives

$$\left(\frac{\Delta \dot{N}\rho}{C_{nv}\sigma\dot{N}_d\rho_d D^3}\right)_0 = 1.0794 \qquad \left(\frac{\Delta \dot{N}\rho}{C_{nv}\sigma\dot{N}_d\rho_d D^3}\right)_e = 0.2667$$

The first of the preceding values is obtained at a Reynolds number of 1333, the second at a Reynolds number of 480. Taking the difference between these two values gives

$$\frac{\Delta \dot{N}\rho}{C_{nv}\sigma\dot{N}_d\rho_d D^3} = 1.0794 - 0.2667 = 0.8127 \qquad Ans.$$

This value represents the number of particles swept up by the drop in traversing a spiral path from $R_2/2$ to the outer wall of the scrubber. ////

Efficiency of the Scrubber

The efficiency of the entire scrubber can be obtained from the preceding evaluation. Figure 9-22 shows the scrubber with a section of height dx indicated at position x. Let Q_s be the rate of water injection. The rate at which drops are formed in the section dx is given by

$$d\dot{N}_d = \frac{6Q_s \, dx/L}{\pi D^3} \qquad (9\text{-}128)$$

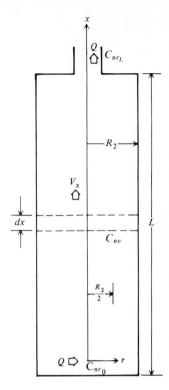

FIGURE 9-22
Element for the determination of collection efficiency of a cyclone scrubber.

The quantity tabulated in Table 9-3 may be written as

$$\frac{\Delta \dot{N} \rho}{C_{nv} \sigma \dot{N}_d \rho_d D^3} = \frac{(d\dot{N})\rho}{C_{nv} \sigma (d\dot{N}_d)\rho_d D^3} \qquad (9\text{-}129)$$

The rate at which drops are collected in the section dx of the scrubber is

$$d\dot{N} = - \frac{d\dot{N}\rho}{C_{nv} \sigma d\dot{N}_d \rho_d D^3} \frac{(dN_d)\sigma C_{nv} \rho_d D^3}{\rho}$$

Substituting Eqs. (9-128) and (9-129) gives

$$d\dot{N} = - \frac{\Delta \dot{N} \rho}{C_{nv} \sigma \dot{N}_d \rho_d D^3} \frac{6}{\pi} \frac{Q_s}{L} \frac{\sigma \rho_d}{\rho} C_{nv} \, dx \qquad (9\text{-}130)$$

The rate of particle collection is related to the rate of change of the number concentration as $d\dot{N} = Q \, dC_{nv}$; then Eq. (9-130) may be rewritten:

$$\frac{dC_{nv}}{C_{nv}} = - \frac{\Delta \dot{N} \rho}{C_{nv} \sigma \dot{N}_d \rho_d D^3} \frac{6}{\pi} \frac{Q_s}{Q} \frac{\sigma \rho_d}{\rho} \frac{dx}{L}$$

The integral of this equation, with $C_{nv} = C_{nv_0}$ at the bottom of the scrubber, is

$$C_{nv} = C_{nv_0} \exp\left(-\frac{\Delta \dot{N} \rho}{C_{nv} \sigma \dot{N}_d \rho_d D^3} \frac{6}{\pi} \frac{Q_s}{Q} \frac{\sigma \rho_d}{\rho} \frac{x}{L}\right) \qquad (9\text{-}131)$$

The collection efficiency is given as

$$\eta = 1 - \frac{C_{nv_L}}{C_{nv_0}} = 1 - \exp\left(-\frac{\Delta \dot{N} \rho}{C_{nv} \sigma \dot{N}_d \rho_d D^3} \frac{6}{\pi} \frac{Q_s}{Q} \frac{\sigma \rho_d}{\rho}\right) \qquad (9\text{-}132)$$

It may be noted that by this derivation the collection efficiency is independent of the length of the scrubber. The next example extends Example 9-14 to consider collection efficiency.

EXAMPLE 9-15   For the scrubber of Example 9-14, determine the water flow rate necessary for a collection efficiency of 0.99 if $\sigma = 1$.

SOLUTION   Equation (9-132) may be solved for Q_s as

$$Q_s = -\frac{\pi}{6} \frac{\rho}{\sigma \rho_p} Q \left(\frac{\Delta \dot{N} \rho}{C_{nv} \sigma \dot{N}_d \rho_d D^3}\right)^{-1} \ln(1 - \eta) \qquad (9\text{-}132a)$$

Substituting the known data, including the result of Example 9-14, gives

$$Q_s = -\frac{\pi}{6} \frac{1.185}{1.0(1000)} (10)(0.8127)^{-1} \ln 0.01 = 0.03516 \text{ m}^3/\text{s} \qquad Ans. \qquad ////$$

Pressure Drop and Power

The pressure drop through the cyclone scrubber can be obtained from Eq. (9-111) if we allow for the total acceleration of the drop:

$$\Delta P = \Delta P_{ns} + \frac{Q_s}{Q} \rho_d V_{\theta_{ave}}^2 \qquad (9\text{-}133)$$

in which ΔP_{ns} is the pressure drop when the spray is turned off. The power required is then given by

$$\dot{W} = Q \Delta P_{ns} + Q_s \rho_d V_{\theta_{ave}}^2 \qquad (9\text{-}134)$$

9-9   THE VENTURI SCRUBBER

The venturi scrubber is a device very similar in principle to the injection scrubber discussed in Sec. 9-7. The primary difference is that the air velocity is speeded up by means of a venturi located in the duct. Water drops are injected into the airstream near the entrance to the maximum-velocity section. These drops scrub the air of its particles as they accelerate to the airspeed; the acceleration is not likely to be complete at the end of the throat, so that collection will

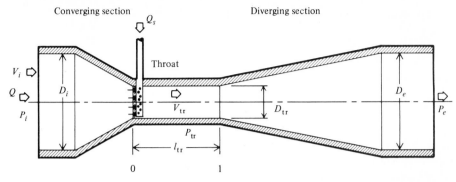

FIGURE 9-23
Cross section of a venturi scrubber.

continue somewhat into the diverging section of the venturi. Figure 9-23 shows a cross-sectional view of a typical venturi scrubber. A collector, such as a louvered collector, cyclone collector, or settling chamber, must be located downstream of the scrubber to collect the drops.

For a properly designed scrubber, the throat length is important; it should be sufficiently long that most of the collection occurs in the throat. If the throat is too short, few particles will be collected in the throat so that much of the potential efficiency of the device will be lost. It would be preferable to employ a larger-diameter, but longer, throat, which would achieve the same collection efficiency with less pressure drop. However, a throat which is longer than necessary will cause an unnecessarily large pressure drop along it.

Collection Efficiency

To analyze the venturi scrubber of Fig. 9-23, assume that the drops are injected at zero velocity at position 0 at the entrance to the throat. At position 1, the relative Reynolds number Re_1, based on the relative velocity of the drop and the air in the throat, can be used in conjunction with Tables 9-3 and 9-4 to determine the collection efficiency. Equation (9-103) may be written for the section 0-1 of length l_{tr} as

$$l_{tr} = \frac{\rho_d D^2}{\mu} V_{tr} \left[\left(\frac{t\mu}{\rho_d D^2} \right)_1 - \left(\frac{t\mu}{\rho_d D^2} \right)_0 \right] + \frac{\rho_d D}{\rho} \left[\left(\frac{\rho s'}{\rho_d D} \right)_1 - \left(\frac{\rho s'}{\rho_d D} \right)_0 \right] \qquad (9\text{-}135)$$

We define the Reynolds number in the throat as

$$Re_{tr} = Re_0 = \frac{\rho D V_{tr}}{\mu} \qquad (9\text{-}136)$$

and the relative Reynolds number at section 1 is given as

$$\text{Re}_1 = \frac{\rho D (V_{tr} - V_{d1})}{\mu} \qquad (9\text{-}137)$$

Equation (9-135) may be written as

$$\frac{\rho l_{tr}}{\rho_d D} = \text{Re}_{tr}\left[\left(\frac{t\mu}{\rho_d D^2}\right)_1 - \left(\frac{t\mu}{\rho_d D^2}\right)_0\right] + \left(\frac{\rho s'}{\rho_d D}\right)_1 - \left(\frac{\rho s'}{\rho_d D}\right)_0 \qquad (9\text{-}138)$$

Equation (9-138) leads to the graph shown in Fig. 9-24 in which the parameter on the curves is the relative Reynolds number at section 1.

Unless the throat is undesirably long, the collection process is not complete at the end of the throat; some additional collection occurs in the diverging section as well. In the diverging section, the drop is accelerating at the same time as the air is slowing down. At some point the two velocities will be equal, and from that point the drop will decelerate with the airstream. It is very difficult to determine the exact point at which the two velocities are equal. As an approximation to the amount of collection occurring in the diverging section, let us assume that the drop ends its collection effectiveness at the point where its relative Reynolds number, based on constant air velocity, is half of Re_1. The value of $\text{Re}_1/2$ is shown in parentheses as a parameter in Fig. 9-24. Table 9-3 is then employed to obtain the difference between the values of

$$\frac{\pi}{6} \frac{\eta \rho Q}{\sigma \rho_d Q_s}$$

between Re_0, which is equal to Re_{tr}, and Re_1. From this difference, the collection efficiency or water flow rate, as required, may be computed.

EXAMPLE 9-16   A venturi scrubber has a throat velocity of 100 m/s and a throat length of 50 cm. The air flow rate through the venturi is 5.0 m^3/s, the particle diameter is 2.0 μm, the particle density is 1500 kg/m^3, and the drop diameter is 1.0 mm. Compute the water flow rate required for a collection efficiency of 1.0.

SOLUTION   The throat Reynolds number, nondimensional throat length, and collection parameter $\sqrt{\rho_p/\rho}\, d/D$ are computed to be

$$\text{Re}_{tr} = \frac{1.185(100)(0.001)}{1.8 \times 10^{-5}} = 6583$$

$$\frac{\rho l_{tr}}{\rho_d D} = \frac{1.185(0.5)}{100(0.001)} = 0.5925$$

$$\sqrt{\frac{\rho_p}{\rho}}\frac{d}{D} = 0.07116$$

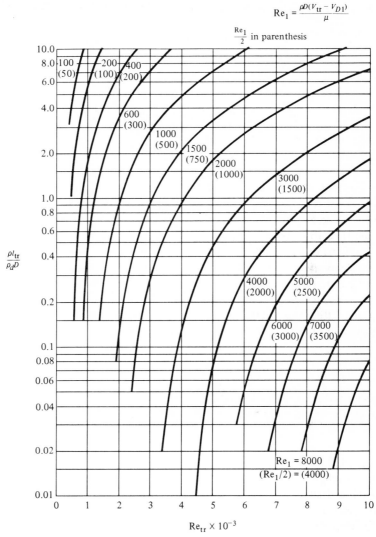

$$\mathrm{Re}_1 = \frac{\rho D(V_{\mathrm{tr}} - V_{D1})}{\mu}$$

$\frac{\mathrm{Re}_1}{2}$ in parenthesis

100
(50)
200
(100)
400
(200)
600
(300)
1000
(500)
1500
(750)
2000
(1000)
3000
(1500)
4000
(2000)
5000
(2500)
6000
(3000)
7000
(3500)
$\mathrm{Re}_1 = 8000$
$(\mathrm{Re}_1/2) = (4000)$

$\frac{\rho l_{\mathrm{tr}}}{\rho_d D}$

$\mathrm{Re}_{\mathrm{tr}} \times 10^{-3}$

FIGURE 9-24
Reynolds number at exit from the throat of a venturi scrubber.

Using the first two of these computed values, Fig. 9-24 gives a value of $\mathrm{Re}_1/2 = 1900$. From Table 9-3, we obtain

$$\frac{\pi}{6}\left(\frac{\eta\rho Q}{\sigma\rho_p Q_s}\right)_0 = 3.519$$

$$\frac{\pi}{6}\left(\frac{\eta\rho Q}{\sigma\rho_d Q_s}\right)_1 = 1.511$$

Using the difference between these two values 2.008 and solving for Q_s gives

$$Q_s = \frac{\pi \eta \rho Q}{6 \sigma \rho_d} \frac{1}{2.008} = 0.00154 \text{ m}^3/\text{s} \qquad Ans. \qquad ////$$

Pressure Drop and Power Loss

The pressure drop across the venturi and the power loss associated with this pressure drop can be evaluated from the following equations:

$$\Delta P = \frac{P_i - P_e}{P_i - P_{tr}} (P_i - P_{tr}) + \frac{Q_s}{Q} \rho_d V_{tr}^2 \qquad (9\text{-}139)$$

$$\dot{W} = Q \frac{P_i - P_e}{P_i - P_{tr}} (P_i - P_{tr}) + Q_s \rho_d V_{tr}^2 \qquad (9\text{-}140)$$

The quantity $(P_i - P_e)/(P_i - P_{tr})$ in Eqs. (9-139) and (9-140), which is plotted in Fig. (9-25), is obtained from values measured in conjunction with venturi meters and is presented in Ref. [7]. The pressure drop and work associated with the water drops is included also in Eqs. (9-139) and (9-140).

The quantity $P_i - P_{tr}$, which appears in Eqs. (9-139) and (9-140), must be evaluated before those equations can be used. For the case of incompressible flow, Eq. (2-120) may be solved for $P_i - P_{tr}$:

$$P_i - P_{tr} = \frac{\rho V_i^2 [1 - (D_{tr}/D_i)^4]}{2 c_d^2 (D_{tr}/D_i)^4} \qquad (9\text{-}141)$$

The discharge coefficient c_d in Eq. (9-141), though very important in solving for the flow rate, is of minor effect when computing the pressure drop. A representative value of 0.985 will be used here, giving

$$P_i - P_{tr} = 0.515 \frac{\rho V_i^2 [1 - (D_{tr}/D_i)^4]}{(D_{tr}/D_i)^4} \qquad (9\text{-}142)$$

This equation is limited to a throat velocity of about 100 m/s; for higher velocities, a compressible flow formula must be used. The throat velocity may be readily computed from the continuity equation.

Compressible Flow Effects

For compressible flow, Eq. (2-125) may be employed. Upon substituting for throat area, using the perfect-gas law and the value of k for air and using the discharge coefficient of 0.985 as before, this equation may be written in the following form:

$$\left(1 + 0.147 \frac{\rho_i V_i^2}{P_i}\right)\left(\frac{P_{tr}}{P_i}\right)^{2/k} - \left(\frac{P_{tr}}{P_i}\right)^{(k+1)/k} = 0.147 \frac{\rho_i V_i^2}{P_i} \left(\frac{D_i}{D_{tr}}\right)^4 \qquad (9\text{-}143)$$

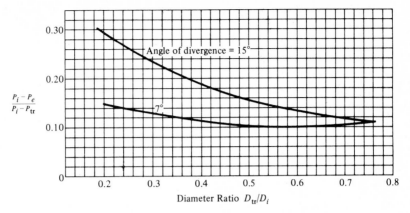

FIGURE 9-25
Pressure drop through a venturi. (*From ASME Research Committee oh Fluid Meters, "Flowmeter Computation Handbook," p. 161, The American Society of Mechanical Engineers, New York, 1961. Used by permission.*)

From this point, a trial-and-error solution is in order; once P_{tr} has been found, it may be used in Eqs. (9-139) and (9-140). Equation (2-123) gives

$$\rho_{tr} = \rho_i \left(\frac{P_{tr}}{P_i}\right)^{1/k} \qquad (9\text{-}144)$$

The throat velocity may be found from the continuity equation as

$$V_{tr} = \frac{\dot{m}}{\rho_{tr}\,\pi D_{tr}{}^2/4} = \frac{4\dot{m}}{\pi\rho_i\,D_{tr}{}^2}\left(\frac{P_i}{P_{tr}}\right)^{1/k} \qquad (9\text{-}145)$$

Note that in Eqs. (9-143) to (9-145) the ratio of throat diameter to inlet diameter appears; for noncircular venturis this ratio may be replaced by the square root of the corresponding area ratio. For a given diameter ratio and given inlet conditions, there is a maximum flow rate through the venturi, which corresponds to sonic velocity at the throat. A throat velocity this high is probably undesirable; for a given diameter ratio, the following table shows the maximum inlet velocity for standard air at the inlet:

D_{tr}/D_i	V_i, m/s
0.9	196
0.8	142
0.7	104
0.6	74
0.5	51
0.4	32
0.3	18
0.2	8
0.15	5

The dividing line which separates the use of the incompressible and compressible flow formulas is located at roughly three-tenths of the values listed in this table.

It may be observed that so far no mention has been made of the effect which throat length has on pressure drop. The values presented are based on a standard venturi flowmeter for which the throat length l_{tr} is equal to the throat diameter D_{tr}. For precise calculations, any extra throat length above this value can be treated as flow through a duct having a diameter equal to the throat diameter, using the formulas of Sec. 2-3.

EXAMPLE 9-17   Assume that the venturi of Example 9-16 has an inlet velocity of 20 m/s. Determine the pressure drop and power loss in this case. Assume a 10° divergence angle for the venturi.

SOLUTION   For a throat velocity of 100 m/s, assume incompressible flow; Eq. (9-142) then gives

$$P_i - P_{tr} = \frac{0.515(1.185)(20)^2[1 - (D_{tr}/D_i)^4]}{(D_{tr}/D_i)^4}$$

The area ratio is obtained as

$$\frac{A_{tr}}{A_i} = \frac{V_i}{V_{tr}} = 0.2$$

The diameter ratio is the square root of the area ratio

$$\frac{D_{tr}}{D_i} = 0.447$$

Then $P_i - P_{tr}$ is

$$P_i - P_{tr} = 5859 \text{ N/m}^2$$

From Fig. 9-25

$$\frac{P_i - P_e}{P_i - P_{tr}} = 0.13$$

The pressure drop is given by Eq. (9-139):

$$\Delta P = \left(\frac{P_i - P_e}{P_i - P_{tr}}\right)(P_i - P_{tr}) + \frac{Q_s}{Q}\rho_d V_{tr}^2$$

$$= 0.13(5859) + \frac{0.00154}{5.0}(1000)(100)^2$$

$$= 762 + 3080 = 3842 \text{ N/m}^2 \qquad Ans.$$

Notice the relative magnitudes of the venturi loss and drop acceleration loss. The power is

$$\dot{W} = Q \, \Delta P = 5.0(3842) = 19.2 \text{ kW} \qquad Ans. \qquad ////$$

The scrubber types which were considered in this chapter are those most amenable to analysis. Several other types are in wide use, such as the packed-bed scrubber, impingement-plate scrubber, and submerged-orifice scrubber. For brief descriptions of these and other types of scrubbers, see Ross [8].

REFERENCES

1 Milne-Thomson, L. M.: "Theoretical Hydrodynamics," 4th ed., p. 464, The Macmillan Company, New York, 1960.
2 Kays, W. M.: "Convective Heat and Mass Transfer," p. 92, McGraw-Hill Book Company, New York, 1966.
3 Lee, J. F., F. W. Sears, and D. L. Turcotte: "Statistical Thermodynamics," pp. 304–309, Addison-Wesley Publishing Company, Inc., Reading, Mass., 1963.
4 Fuchs, N. A.: "The Mechanics of Aerosols," pp. 181–287, Pergamon Press, London, 1964.
5 Davies, C. N.: "Air Filtration," p. 16, Academic Press, Inc., New York, 1973.
6 Perry, J. H.: "Chemical Engineers' Handbook," pp. 18-59–18-68, McGraw-Hill Book Company, New York, 1963.
7 ASME Research Committee on Fluid Meters: "Flowmeter Computation Handbook," p. 161, The American Society of Mechanical Engineers, New York, 1961.
8 Ross, R. D.: "Air Pollution and Industry," pp. 375–397, Van Nostrand Reinhold, New York, 1972.

PROBLEMS

9-1 Show that Eq. (9-9) represents an approximate value for the equation preceding it.
9-2 Show that Eqs. (9-4) and (9-5) satisfy the boundary conditions at infinity.
9-3 Show that Eqs. (9-4) and (9-5) satisfy the continuity equation and the differential equations of motion.
9-4 Obtain Eq. (9-29) from Eq. (9-28).
9-5 Verify Eqs. (9-33) and (9-34).
9-6 Derive Eq. (9-39) from Eq. (9-37).
9-7 Derive Eq. (9-42) from Eq. (9-41) using Eq. (9-40).
9-8 Derive Eq. (9-46) from Eq. (9-45).
9-9 A drop of 1.5 mm in diameter moves through standard air with a velocity of 2.0 m/s. The air contains particles with densities of 1200 kg/m³ and diameters of 3.0 μm. If the attachment coefficient is 0.5, calculate the expected collection efficiency.
9-10 A drop 0.2 mm in diameter moves with a velocity of 30 m/s through standard air. The air contains particles having densities of 1500 kg/m³ and diameters of 0.4 μm. Assuming $\sigma = 0.6$, calculate the Cunningham correction factor C and compute the collection efficiency of the drop.

9-11 Derive Eq. (9-48).

9-12 Evaluate the integral in Eq. (9-57).

9-13 Evaluate the integral in Eq. (9-60).

9-14 In the derivation of Eq. (9-68) it was assumed that the concentration boundary-layer thickness was a constant fraction β of the velocity boundary-layer thickness. A similar derivation can be performed if the concentration boundary-layer thickness δ_n is assumed to be constant over the front face of the sphere. Making this assumption, show that the following equation for collection efficiency results:

$$\eta_{dd} = \frac{3.21\sigma}{Sc^{2/3}} \sqrt{\frac{v}{V_\infty D}}$$

and that the value of β at $\theta = \pi/2$ is given by

$$\beta = \frac{0.954}{Sc^{1/3}}$$

9-15 Starting with Eq. (9-68), derive an asymptotic expression for η_{dd} valid for small particle diameters.

9-16 Starting with Eq. (9-68), derive an asymptotic expression for η_{dd} valid for large particle diameters.

9-17 A drop of water 1.0 mm in diameter moves through air at a relative velocity of 2.5 m/s. What collection efficiency by diffusion can be expected with particles having 0.2 μm diameter? Assume $\sigma = 0.8$.

9-18 For a particle size of 0.1 μm and a drop diameter of 0.75 mm, determine the collection efficiency by diffusion for a range of velocity from 1.0 m/s to 30 m/s. Assume $\sigma = 0.7$.

9-19 For a particle size of 0.1 μm and a drop velocity of 5 m/s, determine the collection efficiency by diffusion for a range of drop diameter from 0.1 to 2 mm. Assume $\sigma = 0.7$.

9-20 Particles whose diameter is 0.05 μm are to be collected by diffusion onto a water drop with a collection efficiency of 0.01. Find a combination of drop diameter and velocity which will accomplish this. Assume $\sigma = 1$.

9-21 Show that Eq. (9-70) can also be derived if the assumption is made that one effect acts on the number of particles approaching the drop and the second effect then acts on the number of particles not collected by the first effect.

9-22 A drop of water has a diameter of 0.5 mm and moves through standard air at a velocity of 5 m/s. If $\sigma = 0.6$, what will be the collection efficiency for particles having a diameter of 0.25 μm and a density of 1500 kg/m^3?

9-23 In a scrubber, the combined collection efficiency for a single drop of 10^{-3}; the particles will encounter 100 drops. What will be the overall collection efficiency?

9-24 Repeat Prob. 9-23 for 1000 drops; for 10,000 drops.

9-25 If $\eta_d = 0.1$ in a scrubber, compute and plot η as a function of n.

9-26 In a scrubber, $\eta_d = 0.05$ and $n = 25$. What overall collection efficiency is expected?

9-27 In a scrubber, $\eta_d = 0.01$ while the overall collection efficiency η must equal 0.99. What value of n is required?

9-28 Verify the empirical equations given with Table 9-1 by computing values of y_2/D and comparing them with the tabulated values.

9-29 A spray chamber treats 50 m³/s of standard air containing particles whose density is 2200 kg/m³; the chamber has a diameter of 5 m and a length of 10 m. The spray fluid is water flowing at the rate of 1 m³/s, with a droplet size of 3 mm. Assume $C = 1$ and $\sigma = 1$, and compute and plot the collection efficiency as a function of particle diameter.

9-30 If ΔP_{ns} is 125 N/m² in Prob. 9-29, determine the air pumping power required.

9-31 A spray chamber handles 2.5 m³/s of standard air containing particles whose density is 900 kg/m³. The spray fluid is water; its volumetric flow rate is 1.5 percent that of the air. The chamber dimensions are 1.0 m in diameter by 4.0 m high. For a drop size of 2 mm and taking C and σ as unity, determine the collection efficiency as a function of particle size. Plot the results.

9-32 For ΔP_{ns} equal to 175 N/m² in Prob. 9-31, determine the air pumping power required.

9-33 A spray chamber is to be designed to treat standard air and to achieve a collection efficiency of 0.995 for particles having diameters of 1.5 μm and densities of 2000 kg/m³. Assume a drop diameter of 2.5 mm, take C and σ as unity and determine suitable values of D_{sc} and L. Take a water flow rate of 0.08 m³/s and an air flow rate of 8.0 m³/s.

9-34 The spray chamber of Prob. 9-29 uses instead of pure water a slurry consisting of the collected drops from the bottom of the chamber mixed with 20 percent of fresh water with 20 percent of the slurry discarded. Assume $C_{mv} = 0.01$ kg/m³ for the polluted air.
(a) Determine the density of the slurry used as spray fluid.
(b) Compute and plot the collection efficiency as a function of particle diameter.

9-35 Repeat Prob. 9-34 if 0.1 percent of the slurry is continuously replaced.

9-36 A drop of water of 0.5 mm diameter is accelerated from zero velocity to 25 m/s in an airstream containing 10^{13} particles/m³ whose diameters are 0.5 μm and whose densities are 2200 kg/m³. If $\sigma = 0.8$, how many particles will be swept up by the drop?

9-37 An injection scrubber is to be designed using water drops of 0.75 mm diameter. The gas-flow rate is 10 m³/s containing particles with densities of 1950 kg/m³. Assume the gas to be standard air, assume $\sigma = 0.9$, and determine the water flow rate for complete collection of particles of diameter d for a range of values of d from 5.0 μm down to 0.1 μm.

9-38 Determine the best drop size to use in an injection scrubber to be used for collecting particles of 0.75 μm diameter and 2000 kg/m³ density with an airspeed of 10 m/s. Assume $\sigma = 1.0$.

9-39 Repeat Prob. 9-38 for airspeeds of 5 m/s, 20 m/s, and 30 m/s.

9-40 An injection scrubber treats a stream of 30 m³/s of standard air containing particles with diameters of 0.5 μm and densities of 1300 kg/m³. The air velocity is 17 m/s, and the water flow rate is 0.04 m³/s. If $\sigma = 0.85$, what collection efficiency is expected for 1.5 mm drops?

9-41 In Prob. 9-40, if the scrubber consists of a square duct, what will be its dimensions? How long should it be if its collection efficiency is 0.9 times its theoretical maximum?

9-42 In Probs. 9-40 and 9-41, how much power is required to operate the scrubber? Include the friction loss for flow through the scrubber duct.

9-43 An eight-stage scrubber operates on a stream of standard air having a flow rate of 5 m³/s and a velocity of 7.5 m/s, containing particles with densities of 3000 kg/m³. It is to have 0.99 efficiency for 1.0-μm particles. The distance between stages is 3 m. Assume $\sigma = 0.75$, and determine the water flow rate into each stage and the total water flow rate. Take $D = 1.0$ mm.

9-44 In Prob. 9-43, determine the size of the scrubber if its cross section is rectangular with an aspect ratio of 2.0. Compute the total pumping power and pressure drop along the scrubber, including the duct loss.

9-45 In Probs. 9-43 and 9-44, determine the water requirement, length, and total power and pressure drop across the scrubber if a single-stage injection scrubber is used.

9-46 For Prob. 9-37, compute and plot the power requirement for particle sizes in the range indicated.

9-47 For an injection scrubber treating 10 m^3/s of standard air containing 1-μm particles having densities of 1200 kg/m^3, compute the water flow rate and pumping power required for complete collection of the particles. Assume an air velocity of 10 m/s and a value of $\sigma = 0.9$, and compute the quantities for varying values of drop diameter D in the range from 0.25 to 10 mm. Include the pressure drop for flow in the duct in computing the power requirement. Assume a square duct.

9-48 Consider the scrubber of Example 9-14 with a water flow rate of 0.01 m^3/s. Compute and plot the collection efficiency as a function of particle diameter over the range from 0.5 to 10 μm. Assume $\sigma = 1$.

9-49 For the scrubber of Example 9-14 with a water flow rate of 0.02 m^3/s, compute and plot the collection efficiency as a function of drop size over the range from 0.5 to 10 mm. Assume $\sigma = 1$.

9-50 A cyclone scrubber treats 100 m^3/s of standard air containing particles whose diameters and densities are 1.0 μm and 2000 kg/m^3, respectively. The radius of the cyclone is 4.0 m, the width W of the inlet is 1.0 m, and the length of the cyclone is 10 m. What will be the collection efficiency if the water flow rate is 1.0 m^3/s? What water flow rate is required to produce a collection efficiency of 0.99? Assume $\sigma = 1$ and $D = 1.0$ mm.

9-51 For the scrubber of Example 9-14 and a water flow rate of 0.01 m^3/s, compute and plot the collection efficiency as a function of the width W over the range from 0.1 to 1.0 m. Assume $\sigma = 1$.

9-52 A cyclone scrubber is to be designed to obtain a collection efficiency of 0.995 on a flow of 25 m^3/s of standard air containing particles of 1.0-μm diameter and 1200 kg/m^3 density. Assume $\sigma = 1$, and make a suitable design for the data given.

9-53 For the two cases of Prob. 9-50, compute the total pressure drop and power required if ΔP_{ns} is equal to 1000 N/m^2.

9-54 A venturi scrubber is to be used to treat 150 m^3/s of standard air which travels through a duct 3 m in diameter. The scrubber must collect all particles whose diameter is 1 μm or larger; the particle density is 2000 kg/m^3. Assume a throat diameter of 1 m and a droplet diameter of 0.2 mm. Calculate the water flow rate required. Assume a throat length of 2.0 m, and assume $\sigma = 1$.

9-55 Compute the power required in Prob. 9-54. Assume a divergence angle of 5°.

9-56 A venturi scrubber is to treat a stream of 35 m^3/s of standard air containing 1.5-μm particles with densities of 1600 kg/m^3. The inlet velocity is 12 m/s, the duct and venturi have circular cross sections, the throat diameter is half the inlet duct diameter, and the throat length is twice its diameter. Compute the water flow rate required as a function of drop diameter over the range from 200 μm to 2 mm. Assume $\sigma = 1$.

9-57 A venturi scrubber treating standard air containing particles with densities of 3000 kg/m^3 has an inlet velocity of 18 m/s, an inlet diameter of 0.5 m, a throat diameter of 20 cm, and a throat length of 50 cm. The drop diameter is 0.5 mm, $\sigma = 1$, and $Q_s/Q = 0.0003$. Compute and plot the collection efficiency as a function of particle size over the range from 0.5 to 5 μm, neglecting the Cunningham correction factor.

9-58 Compute the power consumed in Prob. 9-56 for a divergence angle of 15°.

9-59 Compute the power used in Prob. 9-57. Assume a divergence angle of 8°.

9-60 A venturi scrubber is to be used to remove all the 1.5-μm particles from a stream of 75 m^3/s of standard air. The particle density is 2200 kg/m^3. The inlet duct diameter is 2 m. For each of the following cases compute the water flow rate and the power required, assuming a drop diameter of 0.5 mm:

(a) The throat diameter is 1 m and its length is 2 m.

(b) The throat diameter and length are 1.5 and 4 m, respectively.

(c) The throat diameter is 0.7 m, and its length is 2 m.

Assume a divergence angle of 12°.

9-61 A venturi scrubber is located in a rectangular duct having dimensions of 1.0 by 2.8 m. The venturi throat is circular with a diameter of 0.8 m and a length of 2.2 m. The inlet air velocity is 17 m/s (standard air), the air contains 1.5-μm particles of 1400-kg/m^3 density, all of which must be collected. The water-drop diameter is 0.75 mm, and $\sigma = 1$. The venturi divergence angle is 15°. Compute the water flow rate and the power loss for the scrubber.

10

FILTERS

The final collection mechanism for particulates which will be considered in this book is that of filters. Inside the filter, cylindrical fibers are interspersed in the flow path of the air and the particles are collected onto these fibers by the mechanisms of interception, inertial impaction, and diffusion. Filters can be classified as one of two types, based on the way in which the fibers are held in place. In the first type, the *packed filter*, the fibers are loosely packed into a substantial volume, presenting a fairly long path along which the air must pass on its way through the filter. This type of filter has wide use in air-conditioning applications and other applications where the dust loading, or particle-number concentration, is relatively small. In the second type of filter, called the *single-layer filter*, fibers are woven into a thin layer of cloth, for example. Bag filters are the typical example of this type of filter; filter paper also falls into this category. Figure 10-1a shows a packed filter, while Fig. 10-1b shows a single-layer filter.

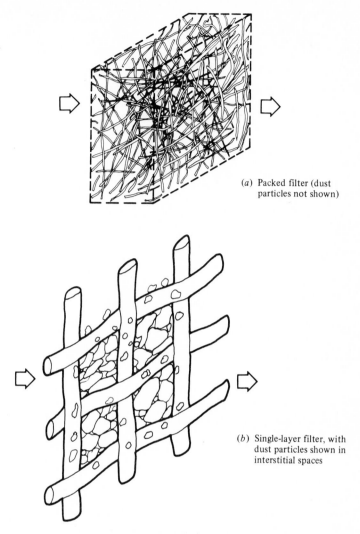

(*a*) Packed filter (dust particles not shown)

(*b*) Single-layer filter, with dust particles shown in interstitial spaces

FIGURE 10-1
Filter element in packed and single-layer filters.

10-1 1NTERCEPTION AND INERTIAL IMPACTION ON CYLINDRICAL FIBERS

To analyze mathematically the collection process in a packed filter, we shall treat the flow around a long cylinder held normal to the flow. This form of analysis is of limited usefulness for a single-layer filter since the filtering mechanism in this case involves the flow of the air through a tortuous passage between particles already collected.

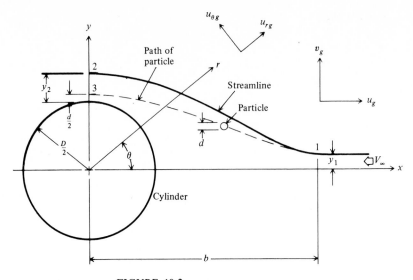

FIGURE 10-2
Collection on a cylinder by interception and inertial impaction.

Figure 10-2 shows a fiber of diameter D held normal to the flow of air having velocity V_∞ at a distance b upstream from the center of the cylinder. The distance b is related to the packing density of the filter and to the size of the fibers. Let us first define the packing density c as the ratio of the volume of the fibers in a filter per unit volume of the filter. If L_f represents the total length of fiber in unit volume of filter, then

$$c = \frac{\pi D^2 L_f}{4} \qquad (10\text{-}1)$$

Let the volume of a cylinder of radius b and length L_f equal the unit volume of the filter, which gives for b

$$b = \sqrt{\frac{1}{\pi L_f}} = \frac{1}{\sqrt{4c/D^2}} = \frac{D}{2\sqrt{c}} \qquad (10\text{-}2)$$

The cylinder of radius b represents the space assigned to the flow around the cylinder of radius $D/2$; we assume that on the surface of the outer cylinder the flow is undisturbed by the presence of the inner cylinder.

Next, let us examine the flow field around the cylinder. As usual, the Reynolds number characterizes the nature of the flow; here the Reynolds number is defined as

$$\text{Re} = \frac{V_\infty \rho D}{\mu} \qquad (10\text{-}3)$$

The velocity V_∞ is the velocity inside the filter and is greater than the velocity of the air approaching the filter due to the partial blocking of flow by the fibers. The value of V_∞ is obtained from

$$V_\infty = \frac{Q}{A_f(1 - c)} \qquad (10\text{-}4)$$

in which A_f is the face, or cross-sectional, area of the filter.

A suitable choice of equations for the flow field inside the filter is not a simple one. Davies [1, pp. 53–54] discusses this question in detail. The nature of the boundary condition at $r = b$ is in question at the present time. Moreover, the range of Reynolds number for practical application of filters extends from that for purely viscous flow, that is, Stokes' flow, into the transition region but not to such high Reynolds numbers that potential flow is applicable. Davies [1, pp. 53–54] has concluded that the Kuwabara flow solution gives closest agreement to experimental results and makes the most sense to use. We shall follow his judgment in this matter even though there are several other flow solutions which could be used. Our solution will be only a somewhat crude approximation, using the Kuwabara flow solution; as an approximation, it will be analogous to the approximation used in Chap. 9 for flow around spheres in the Reynolds number range suitable for scrubbers.

The Kuwabara flow solution, as given by Davies [1, pp. 53–54] is expressed in terms of the stream function ψ as

$$\psi = -\frac{V_\infty r}{2\,\mathrm{Ku}}\left[2\ln\frac{2r}{D} - 1 + c + \frac{D^2}{4r^2}\left(1 - \frac{c}{2}\right) - \frac{2cr^2}{D^3}\right]\sin\theta \qquad (10\text{-}5)$$

The radial and tangential velocity components are given in terms of the stream function as

$$u_{rg} = \frac{1}{r}\frac{\partial\psi}{\partial\theta} \qquad u_{\theta g} = -\frac{\partial\psi}{\partial r} \qquad (10\text{-}6)$$

The quantity Ku in Eq. (10-5) is given as

$$\mathrm{Ku} = c - \frac{3}{4} - \frac{c^2}{4} - \frac{1}{2}\ln c \qquad (10\text{-}7)$$

Using Eqs. (10-5) and (10-6), the radial and tangential velocity components become

$$u_{rg} = -\frac{V_\infty}{2\,\mathrm{Ku}}\left[2\ln\frac{2r}{D} - 1 + c + \frac{D^2}{4r^2}\left(1 - \frac{c}{2}\right) - \frac{2cr^2}{D^2}\right]\cos\theta \qquad (10\text{-}8)$$

$$u_{\theta g} = \frac{V_\infty}{2\,\mathrm{Ku}}\left[2\ln\frac{2r}{D} + 1 + c - \frac{D^2}{4r^2}\left(1 - \frac{c}{2}\right) - \frac{6cr^2}{D^2}\right]\sin\theta \qquad (10\text{-}9)$$

The x and y components of velocity can be found from these using the following equations:

$$u_g = u_{r_g} \cos \theta - u_{\theta_g} \sin \theta \qquad (10\text{-}10)$$

$$v_g = u_{r_g} \sin \theta + u_{\theta_g} \cos \theta \qquad (10\text{-}11)$$

Substituting Eqs. (10-8) and (10-9) into Eqs. (10-10) and (10-11) gives for these components

$$u_g = -\frac{V_\infty}{2 \text{ Ku}}\left[\ln \frac{4(x^2 + y^2)}{D^2} + c - 2c\,\frac{x^2 + 3y^2}{D^2} + \frac{y^2 - x^2}{x^2 + y^2}\right.$$
$$\left. + \frac{D^2}{4}\left(1 - \frac{c}{2}\right)\frac{x^2 - y^2}{(x^2 + y^2)^2}\right] \qquad (10\text{-}12)$$

$$v_g = \frac{V_\infty}{2 \text{ Ku}}\left[\frac{2xy}{x^2 + y^2} - \frac{D^2 xy}{2(x^2 + y^2)^2}\left(1 - \frac{c}{2}\right) - \frac{4cxy}{D^2}\right] \qquad (10\text{-}13)$$

In our approximate analysis, let us choose a reasonable average velocity for the region between points 1 and 2 in Fig. 10-2. The value of the stream function for the streamline passing through points 1 and 2 is $-V_\infty y_1$. Apply this at point 2, using Eq. (10-5) and noting that $\theta = \pi/2$ and $r = D/2 + y_2$. Thus y_1 can be expressed in terms of y_2, and in nondimensional form this relation is

$$\frac{2y_1}{D} = \frac{1 + 2y_2/D}{2 \text{ Ku}}\left[2 \ln\left(1 + \frac{2y_2}{D}\right) - (1 - c) + \frac{2 - c}{2}\,\frac{1}{(1 + 2y_2/D)^2}\right.$$
$$\left. - \frac{c}{2}\left(1 + \frac{2y_2}{D}\right)^2\right] \qquad (10\text{-}14)$$

The value of the stream function is also $u_{g_2} y_2$, in which u_{g_2} is the average velocity in the region between point 2 and the surface of the cylinder. Thus

$$u_{g_2} = -\frac{y_1}{y_2}\,V_\infty = -V_\infty\,\frac{2y_1/D}{2y_2/D} \qquad (10\text{-}15)$$

The average value of u_g will be taken as

$$u_{g_{ave}} = \tfrac{1}{2}(u_{g_2} - V_\infty) \qquad (10\text{-}16)$$

Using Eq. (10-15), this becomes

$$u_{g_{ave}} = -\frac{V_\infty}{2}\left(1 + \frac{2y_1/D}{2y_2/D}\right) \qquad (10\text{-}17)$$

The average value of the vertical component of air velocity v_g is obtained from the fact that the gas must travel the vertical distance from point 1 to point 2 in the time during which it travels horizontally between these two points. Thus

$$u_{g_{ave}} = \frac{y_2 + D/2 - y_1}{-b/u_{g_{ave}}} = -u_{g_{ave}}\frac{D}{2b}\left(1 + \frac{2y_2}{D} - \frac{2y_1}{D}\right) \qquad (10\text{-}18)$$

Let us define the following nondimensional quantities:

$$\xi = \frac{2y_1}{D} \qquad (10\text{-}19)$$

$$\zeta = \frac{2y_2}{D} \qquad (10\text{-}20)$$

$$\chi = -\frac{u_{g_{ave}}}{V_\infty} \qquad (10\text{-}21)$$

$$\Omega = \frac{V_\infty}{aD} \qquad (10\text{-}22)$$

Using these, Eq. (10-17) becomes

$$\chi = \frac{1}{2}\left(1 + \frac{\xi}{\zeta}\right) \qquad (10\text{-}23)$$

Using Eq. (10-2) also, Eq. (10-18) may be written

$$\frac{v_{g_{ave}}}{V_\infty} = \sqrt{c}\,\chi(1 + \zeta - \xi) \qquad (10\text{-}24)$$

Equation (10-14) becomes

$$\xi = \frac{1+\zeta}{2\,Ku}\left[2\ln(1+\zeta) + c - 1 + \frac{2-c}{2(1+\zeta)^2} - \frac{c}{2}(1+\zeta)^2\right] \qquad (10\text{-}25)$$

Equation (10-22), defined here for convenience, will be used shortly. Equations (10-23) and (10-25) will be presented in tabular and graphic form after the analysis has been completed.

Up to this point, we have considered only the flow of the air; now let us consider the motion of the particle of diameter d in this flow field. Equations (3-140) and (3-141) may be written in the following way if u_g and v_g are taken as the average values defined previously and if the quantity a is defined as

$$a = \frac{18\mu}{\rho_p d^2 C} \qquad s^{-1} \qquad (10\text{-}26)$$

$$\frac{d^2x}{dt^2} + a\frac{dx}{dt} = au_{g_{ave}} \qquad (10\text{-}27)$$

$$\frac{d^2y}{dt^2} + a\frac{dy}{dt} = av_{g_{ave}} \qquad (10\text{-}28)$$

These are the equations of motion of the particle; they may be readily solved subject to the initial condition that when t is zero,

$$x = b \qquad y = y_1 \qquad \frac{dx}{dt} = u_{g_{ave}} \qquad \frac{dy}{dt} = 0$$

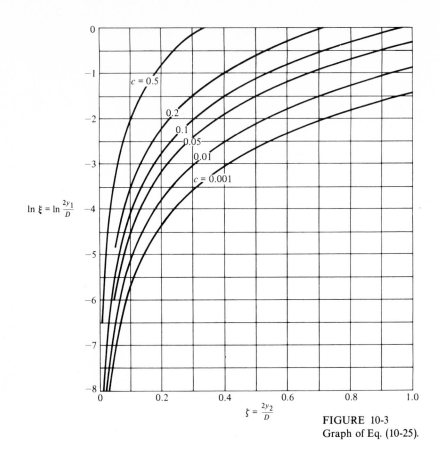

$$\ln \xi = \ln \frac{2y_1}{D}$$

$$\zeta = \frac{2y_2}{D}$$

FIGURE 10-3
Graph of Eq. (10-25).

The solution is

$$x = b + u_{g_{ave}} t \qquad (10\text{-}29)$$

$$y = y_1 - \frac{v_{g_{ave}}}{a}\left(1 - e^{-at}\right) + v_{g_{ave}} t \qquad (10\text{-}30)$$

Other boundary conditions are preferable, but lead to mathematically intractable results.

From Fig. 10-2, it will be noted that when $x = 0$, $y = (D + d)/2$, representing the condition in which the particle is just captured by the cylinder. Equation (10-29) gives the time of travel as

$$t_2 = -\frac{b}{u_{g_{ave}}} \qquad (10\text{-}31)$$

Equation (10-30) then gives y_1 as

$$y_1 = \frac{D + d}{2} + \frac{v_{g_{ave}}}{a}\left(1 - e^{ab/u_{g_{ave}}}\right) + \frac{v_{g_{ave}} b}{u_{g_{ave}}} \qquad (10\text{-}32)$$

which may be written in nondimensional form using Eqs. (10-19) to (10-22):

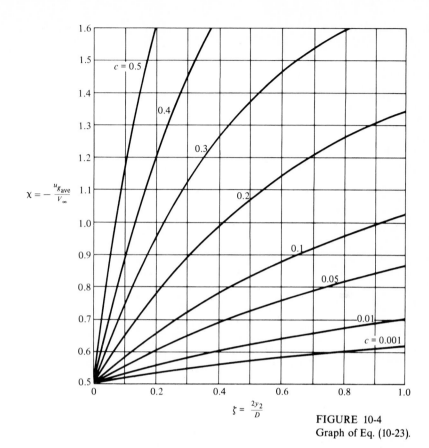

$$\chi = -\frac{u_{g_{\text{ave}}}}{V_\infty}$$

$$\zeta = \frac{2y_2}{D}$$

FIGURE 10-4
Graph of Eq. (10-23).

$$\xi = 1 + \frac{d}{D} + 2\Omega \frac{v_{g_{\text{ave}}}}{V_\infty}\left[1 - \exp\left(-\frac{1}{2\sqrt{c\Omega\chi}}\right)\right] - \frac{v_{g_{\text{ave}}}/V_\infty}{\sqrt{c\chi}} \qquad (10\text{-}33)$$

Substituting Eq. (10-24) and solving for ζ, we have

$$\xi = \frac{d/D + 2\sqrt{c\Omega\chi}(1 - \xi)\left[1 - \exp\left(-\frac{1}{2\sqrt{c\Omega\chi}}\right)\right]}{1 - 2\sqrt{c\Omega\chi}\left[1 - \exp\left(-\frac{1}{2\sqrt{c\Omega\chi}}\right)\right]} \qquad (10\text{-}34)$$

Equation (10-34) can be solved iteratively for ζ using Eq. (10-25) for ξ and Eq. (10-23) for χ. The collection efficiency is given from Fig. 10-2 as

$$\eta_{fi} = \frac{\sigma y_1}{D/2} = \sigma\xi \qquad (10\text{-}35)$$

The iterative solution proceeds as follows: First, estimate a likely value of ζ. Using this value, a value of ξ is found from Fig. 10-3 (or Table 10-1 can also be used); Fig. 10-3 and Table 10-1 are based on Eq. (10-25). Using the assumed value of ζ and the value of ξ just obtained, a value of χ can be found from Fig. 10-4, or from Eq. (10-23). The value of ξ and that of χ are then used in

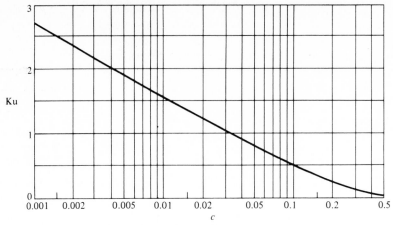

FIGURE 10-5
The Kuwabara parameter Ku as a function of c, Eq. (10-7).

Table 10-1 ξ **AS A FUNCTION OF** ζ **AND** c

$\zeta \backslash c$	0.001	0.002	0.005	0.0075	0.01	0.02
0.01	0.0000367	0.000042	0.0000519	0.0000579	0.0000629	0.0000794
0.02	0.0001458	0.000167	0.0002063	0.0002299	0.0002500	0.0003155
0.05	0.0008938	0.001024	0.001265	0.001409	0.001532	0.001933
0.10	0.003466	0.003971	0.004904	0.005464	0.005941	0.007493
0.15	0.007573	0.008675	0.01071	0.01194	0.01298	0.01636
0.20	0.01309	0.01499	0.01851	0.02063	0.02242	0.02826
0.25	0.01991	0.02281	0.02816	0.03137	0.03410	0.04296
0.30	0.02795	0.03201	0.03951	0.04402	0.04784	0.06024
0.40	0.04733	0.05421	0.06690	0.07451	0.08097	0.1019
0.50	0.07071	0.08097	0.09991	0.1113	0.1209	0.1520
0.60	0.09764	0.1118	0.1379	0.1536	0.1668	0.2095
0.70	0.1278	0.1463	0.1804	0.2008	0.2181	0.2738
0.80	0.1608	0.1841	0.2271	0.2527	0.2744	0.3441
0.90	0.1966	0.2250	0.2774	0.3087	0.3351	0.4199
1.00	0.2348	0.2688	0.3312	0.3685	0.3999	0.5006

$\zeta \backslash c$	0.05	0.10	0.15	0.20	0.25	0.30
0.01	0.0001183	0.0001791	0.000246	0.000324	0.000419	0.000535
0.02	0.0004701	0.0007112	0.000976	0.001286	0.001661	0.002122
0.05	0.002879	0.004350	0.005964	0.007846	0.01011	0.01289
0.10	0.01115	0.01681	0.02299	0.03017	0.03877	0.04927
0.15	0.02431	0.03658	0.04992	0.06533	0.08370	0.1060
0.20	0.04194	0.06299	0.08575	0.1119	0.1429	0.1803.
0.25	0.06368	0.09543	0.1296	0.1687	0.2147	0.2698
0.30	0.08920	0.1339	0.1807	0.2345	0.2975	0.3724
0.40	0.1505	0.2240	0.3019	0.3894	0.4905	0.6087
0.50	0.2239	0.3318	0.4446	0.5699	0.7123	0.8757
0.60	0.3079	0.4541	0.6050	0.7702	0.9546	1.162
0.70	0.4013	0.5887	0.7798	0.9854	1.210	1.456
0.80	0.5029	0.7340	0.9661	1.211	1.474	1.751
0.90	0.6120	0.8882	1.1614	1.444	1.789	2.037
1.00	0.7276	1.0501	1.364	1.681	2.000	2.307

Eq. (10-34) to solve for a closer approximation for ζ. This procedure is then repeated until subsequent iterations produce negligible change in ζ. This iterative procedure converges quite rapidly and is illustrated in the following example. Figure 10-5 shows the variation of the Kuwabara parameter Ku with packing density c, as given in Eq. (10-7).

EXAMPLE 10-1 Evaluate η_{fi} for the case where

$$V_\infty = 0.2 \text{ m/s} \qquad c = 0.05$$
$$d = 0.4 \ \mu m \qquad \rho_p = 1000 \text{ kg/m}^3$$
$$D = 4.0 \ \mu m$$

Assume standard air, and note that the Cunningham correction factor $C = 1.42$ from Table 3-1.

SOLUTION   The quantity a is given by Eq. (10-26) as

$$a = \frac{18(1.84 \times 10^{-5})}{1000(0.4 \times 10^{-6})^2(1.42)} = 1.46 \times 10^6 \text{ s}^{-1}$$

The value $2\sqrt{c} = 0.4472$, $d/D = 0.1$, and the parameter Ω is

$$\Omega = \frac{V_\infty}{aD} = \frac{0.2}{1.46 \times 10^6(4 \times 10^{-6})} = 0.03425$$

Equation (10-34) simplifies to

$$\zeta = \frac{0.1 + 0.01532\chi(1 - \xi)(1 - e^{-65.3/\chi})}{1 - 0.01532\chi(1 - e^{-65.3/\chi})}$$

To begin the iterative process, try $\zeta = 0.10$. Figure 10-3 gives

$$\ln \xi = -4.5 \qquad \xi = 0.0111$$

Figure 10-4 gives $\chi = 0.55$. Then the preceding equation yields

$$\zeta = \frac{0.1 + 0.01532(0.55)(1 - 0.0111)(1 - e^{-65.3/0.55})}{1 - 0.01532(0.55)(1 - e^{-65.3/0.55})} = 0.1093$$

The exponential contributes negligibly in the preceding equation; for subsequent iterations we may neglect it. For the next iteration

$$\ln \xi = -4.3 \qquad \xi = 0.0136 \qquad \chi = 0.56$$

$$\zeta = \frac{0.1 + 0.01532(0.56)(1 - 0.0136)}{1 - 0.01532(0.56)} = 0.1094$$

The resulting value $\xi = 0.0136$. The collection efficiency due to interception and inertial impaction is then

$$\eta_{fi} = \xi = 0.0136 \qquad Ans.$$

The value of efficiency due to interception alone is given by $\zeta = d/D = 0.1$, which gives, from Fig. 10-3,

$$\eta_{fi} = 0.0111 \qquad ////$$

10-2   DIFFUSION TO CYLINDRICAL FIBERS

The process of diffusion operates effectively with small particles, which are deposited onto cylinders in much the same way as they are deposited onto spheres. Although diffusion collection is of little consequence for particles larger than 0.5 μm in diameter, it becomes the dominant mode of collection for particles smaller than about 0.2 μm in diameter. In this section, the collection mechanism of diffusion is examined.

Figure 10-6 shows a cylindrical fiber of length L_f with a concentration boundary layer formed around it. The boundary layer thickness δ_n is least at the leading stagnation point on the cylinder, increasing gradually on the front half of the cylinder and much more rapidly around the rear portion of the cylinder. The boundary layer becomes infinitely thick at the rear stagnation point. Inside the boundary layer, diffusion causes particles to bombard the cylinder and upon striking remain attached to it. The effective particle concentration C_{nv} is zero at the surface of the cylinder and is very close to C_{nv_∞} at the edge of the boundary layer. Figure 10-6 also shows a portion of the boundary layer of length $D/2\,d\theta$, as well as the velocity components and particle flow rates across the four sides of the element. The particle flow rate $d\dot{N}$ represents the rate at which particles are removed from the boundary layer by attachment to the cylinder. The other particle flow rates are due to convection as the air flows across the boundaries of the control volume.

An application of the principle of conservation of number of particles to the control volume of Fig. 10-6 shows that

$$\dot{N}_2 - \dot{N}_1 + d\dot{N} + d\dot{N}_3 - d\dot{N}_4 = 0 \qquad (10\text{-}36)$$

Equation (9-49) gives for $d\dot{N}$

$$d\dot{N} = \mathscr{D}\left(\frac{\partial C_{nv}}{\partial r}\right)_{r=D/2} \frac{DL_f\,d\theta}{2} \qquad (10\text{-}37)$$

in which $\mathscr{D}$ is given by Eq. (9-50):

$$\mathscr{D} = \frac{kTC}{3\pi d\mu} \qquad (10\text{-}38)$$

In this analysis, the Kuwabara flow solution, Eqs. (10-8) and (10-9), are used for the air velocity around the cylinder. The particle concentration in the boundary layer will be approximated by the cubic equation of Eq. (9-51), which may be written

$$C_{nv} = C_{nv_\infty}\left[\frac{3}{2}\frac{y}{\delta_n} - \frac{1}{2}\left(\frac{y}{\delta_n}\right)^3\right] \qquad (10\text{-}39)$$

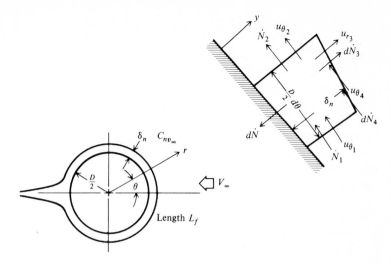

FIGURE 10-6
Flow around a cylindrical fiber including the concentration boundary layer.

in which y is defined as $r - D/2$. Using Eq. (10-39), Eq. (10-37) becomes

$$d\dot{N} = \frac{3}{4}\frac{DL_f C_{nv_\infty}}{\delta_n}\,d\theta \qquad (10\text{-}40)$$

The remaining quantities in Eq. (10-36) may be written

$$\dot{N}_2 - \dot{N}_1 = L_f\frac{d}{d\theta}\int_0^{\delta_n} C_{nv}u_\theta\,dy\,d\theta \qquad (10\text{-}41)$$

$$d\dot{N}_3 = u_r(\delta_n)C_{nv_\infty}L_f\frac{D}{2}\,d\theta \qquad (10\text{-}42)$$

$$d\dot{N}_4 = u_\theta(\delta_n)C_{nv_\infty}L_f\frac{d\delta_n}{d\theta}\,d\theta \qquad (10\text{-}43)$$

Then Eq. (10-36) becomes

$$\frac{d}{d\theta}\int_0^{\delta_n} C_{nv}u_\theta\,dy + \frac{3}{4}\frac{D\mathscr{D}C_{nv_\infty}}{\delta_n} + u_r(\delta_n)C_{nv_\infty}\frac{D}{2} - u_\theta(\delta_n)C_{nv_\infty}\frac{d\delta_n}{d\theta} = 0 \qquad (10\text{-}44)$$

We shall assume that the concentration boundary layer is quite thin; with this assumption, Eq. (10-9) may be approximated as

$$u_\theta = \left(\frac{\partial u_\theta}{\partial y}\right)_{y=0}y = \frac{4V_\infty}{\mathrm{Ku}}(1-c)\frac{y}{D}\sin\theta \qquad (10\text{-}45)$$

while Eq. (10-8) may be approximated similarly as

$$u_r = \left(\frac{\partial u_r}{\partial y}\right)_{y=0} y = -\frac{4V_\infty}{\text{Ku}}(1-c)\frac{y^2}{D^2}\cos\theta \qquad (10\text{-}46)$$

When these and Eq. (10-39) are substituted into Eq. (10-44) and certain simplifications are made, then

$$\delta_n \frac{d}{d\theta}(\delta_n{}^2 \sin\theta) - \frac{5}{4}\delta_n{}^3 \cos\theta - \frac{5}{2}\delta_n{}^2 \frac{d\delta_n}{d\theta}\sin\theta = -\frac{15}{32}\frac{D^2\mathscr{D}\,\text{Ku}}{V_\infty(1-c)} \qquad (10\text{-}47)$$

Equation (10-47) may be further simplified to

$$\frac{d\delta_n{}^3}{d\theta} + \frac{3}{2}\delta_n{}^3 \cot\theta = \frac{45}{16}\frac{D^2\mathscr{D}\,\text{Ku}}{V_\infty(1-c)}\csc\theta \qquad (10\text{-}48)$$

which may be solved using the integrating factor $(\sin\theta)^{3/2}$, giving

$$\delta_n{}^3 = \frac{45}{16}\frac{D^2\mathscr{D}\,\text{Ku}}{V_\infty(1-c)}\frac{1}{(\sin\theta)^{3/2}}\int_0^\theta \sqrt{\sin\theta}\,d\theta \qquad (10\text{-}49)$$

When Eq. (10-49) is substituted into Eq. (10-40), the result is

$$\frac{d\dot N}{d\theta} = \frac{3}{4}\left(\frac{16}{45}\right)^{1/3}\left[\frac{D\mathscr{D}^2 V_\infty(1-c)}{\text{Ku}}\right]^{1/3} L_f C_{nv_\infty}\frac{\sqrt{\sin\theta}}{(\int_0^\theta \sqrt{\sin\theta}\,d\theta)^{1/3}} \qquad (10\text{-}50)$$

Upon integrating Eq. (10-50) from 0 to π, the rate at which particles are removed from the boundary layer is

$$\dot N = \frac{3}{4}\left(\frac{16}{45}\right)^{1/3}\left[\frac{D\mathscr{D}^2 V_\infty(1-c)}{\text{Ku}}\right]^{1/3} L_f C_{nv_\infty}\int_0^\pi \frac{\sqrt{\sin\theta}\,d\theta}{(\int_0^\theta \sqrt{\sin\theta'}\,d\theta')^{1/3}} \qquad (10\text{-}51)$$

The collection efficiency of the cylinder by diffusion is given by the following equation, where use is also made of the attachment coefficient:

$$\eta_{fd} = \frac{\sigma\dot N}{DL_f V_\infty C_{nv_\infty}/2} \qquad (10\text{-}52)$$

Using Eq. (10-51), Eq. (10-52) becomes

$$\eta_{fd} = \frac{3}{2}\left(\frac{16}{45}\right)^{1/3}\sigma\left[\frac{\mathscr{D}^2(1-c)}{\text{Ku}\,D^2 V_\infty{}^2}\right]^{1/3}\int_0^\pi \frac{\sqrt{\sin\theta}\,d\theta}{(\int_0^\theta \sqrt{\sin\theta'}\,d\theta')^{1/3}} \qquad (10\text{-}53)$$

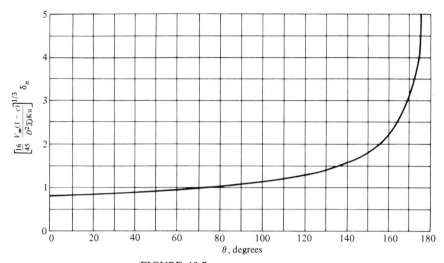

FIGURE 10-7
Nondimensional concentration boundary layer around a cylinder.

The value of the integral in Eq. (10-53) is 2.686; the equation then becomes

$$\eta_{fd} = 2.854\sigma \left[\frac{\mathscr{D}^2(1-c)}{\text{Ku } D^2 V_\infty^{\ 2}} \right]^{1/3} \qquad (10\text{-}54)$$

Figure 10-7 shows the variation of boundary-layer thickness around the cylinder as obtained from the numerical integration of Eq. (10-49). The limiting value of the nondimensional boundary-layer thickness as θ approaches zero is 0.874.

EXAMPLE 10-2   Compute the collection efficiency of a 2-μm-diameter fiber in a stream of standard air due to diffusion only. The velocity V_∞ is 0.1 m/s, and the packing density c is 0.1. Let the particle diameter range from 0.5 to 0.001 μm. Assume $\sigma = 1$.

SOLUTION   Combining Eqs. (10-38) and (10-54) gives

$$\eta_{fd} = 0.6396\left(\frac{1-c}{\text{Ku } D^2 V_\infty^{\ 2}}\right)^{1/3}\left(\frac{kTC}{d\mu}\right)^{2/3} \qquad (10\text{-}54a)$$

Figure 10-5 or Eq. (10-7) gives Ku = 0.499.   Then

$$\eta_{fd} = 8.384 \times 10^{-7}\left(\frac{C}{d}\right)^{2/3}$$

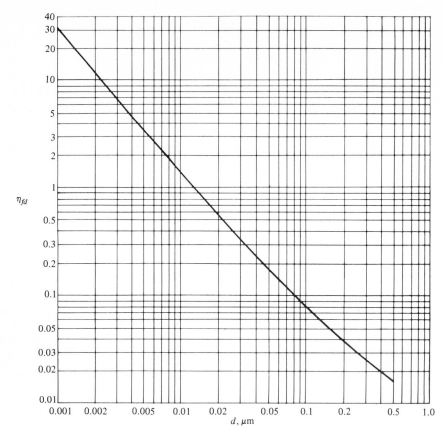

FIGURE 10-8
Collection efficiency of a fiber due to diffusion for the data of Example 10-2.

Obtaining the Cunningham correction factor C from Table 3-1 or Eq. (3-55) and computing η_{fd} for various values of d gives the curve plotted in Fig. 10-8. Notice that the collection efficiency of the fiber can become quite large for small particle diameters. That the efficiency can exceed unity is no cause for concern since this merely means that the fiber can collect particles from a region extending beyond diameter b. ////

10-3 COMBINED MODES OF COLLECTION

The collection efficiency for inertial impaction and interception can be combined with that for diffusion to obtain a combined collection efficiency for the fiber. The result is very similar to that obtained in Chap. 9 for combined modes of collection by a drop. In fact, Eq. (9-70) may be adapted directly to the present situation:

$$\eta_f = \eta_{fi} + \eta_{fd} - \eta_{fi}\eta_{fd} \qquad (10\text{-}55)$$

The following example shows how the combined collection efficiency varies with particle diameter.

EXAMPLE 10-3   Using the data for Example 10-2, with $\rho_p = 1500$ kg/m^3. compute the collection efficiency for the fiber due to inertial impaction and interception combined with that due to diffusion. Let the particle diameter range from 0.001 to 10 μm.

SOLUTION   Example 10-2 gives the efficiency due to diffusion; all that is necessary is to extend the range of calculations to higher particle diameters. The solution for inertial impaction and interception will have to be used for this case. From the given data we have

$$a = \frac{2.208 \times 10^{-7}}{d^2C} \qquad \Omega = \frac{50{,}000}{a}$$

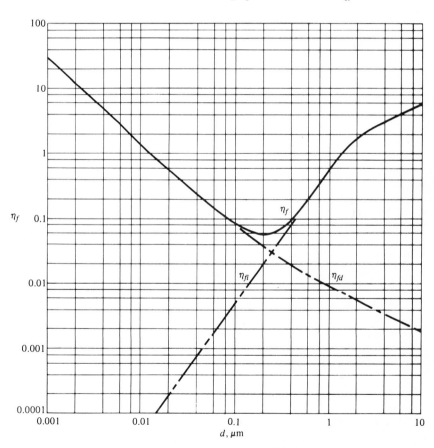

FIGURE 10-9
Efficiency curves for Example 10-3.

Equation (10-34) gives

$$\zeta = \frac{d/2 \times 10^{-6} + 0.632\Omega\chi(1 - \xi)(1 - e^{-1.582/\Omega\chi})}{1 - 0.632\Omega\chi(1 - e^{-1.582/\Omega\chi})}$$

Equations (10-25) and (10-23) give

$$\xi = \frac{1 + \zeta}{0.998}\left[\ln(1 + \zeta)^2 - 0.9 + \frac{0.95}{(1 + \zeta)^2} - 0.05(1 + \zeta)^2\right]$$

$$\chi = 0.5\left(1 + \frac{\xi}{\zeta}\right)$$

These equations can be solved iteratively or by trial and error for the larger diameters.

Equation (10-55) then gives the combined collection efficiency; these efficiencies are plotted in Fig. 10-9. Note that from Eq. (10-2)

$$b = \frac{D}{2\sqrt{c}} = \frac{2}{0.632} = 3.16 \ \mu m$$

which suggests that for particle diameters greater than 2 or 3 μm the calculated results have little meaning.　　　　　////

10-4   COLLECTION EFFICIENCY FOR A FILTER BED

A packed filter, or filter bed, consists of a volume of fibers compressed so that they occupy a relative volume c of the filter. Figure 10-10 shows a filter having dimensions of W by H in the plane normal to the air-flow direction and length L in the air-flow direction. The filtering action takes place continuously through the thickness of the filter as each fiber collects a fraction η_f of the particles passing by it. At this point, we shall derive an equation for the collection efficiency of the entire filter if the individual fiber efficiency is known.

Figure 10-10 also shows a section of the filter having dimensions of W by H by thickness dx located at an arbitrary position x inside the filter. We assume that the filter is plane, has uniform packing density c and fiber diameter D, and that the fiber efficiency is also uniform. By Eq. (10-4)

$$V_\infty = \frac{Q}{A_f(1 - c)} = \frac{V_0}{1 - c}$$

in which A_f is the face area of the filter given by WH. The flow rate of particles past any section at position x is given by

$$\dot{N}(x) = C_{nv}(x)Q$$

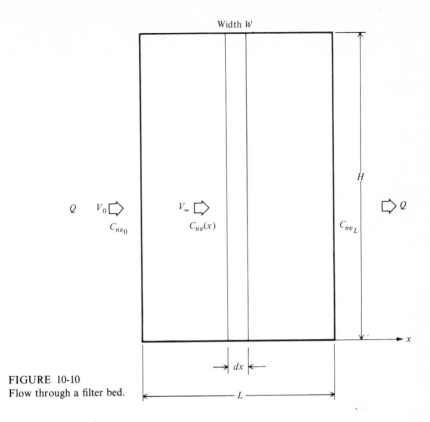

FIGURE 10-10
Flow through a filter bed.

The rate at which particles are removed from the stream in the section of thickness dx is given by

$$d\dot{N} = -\eta_f C_{nv}(x) V_\infty DL_f WH\, dx$$

in which L_f is the total length of fiber in unit volume of the filter. Combining the preceding three equations gives

$$\frac{dC_{nv}}{C_{nv}} = -\eta_f \frac{DL_f\, dx}{1-c}$$

Using Eq. (10-1) then leads to

$$\frac{dC_{nv}}{C_{nv}} = -\frac{4c\eta_f\, dx}{\pi(1-c)D} \qquad (10\text{-}56)$$

Upon integration between the limits of 0 and L for x and between the corresponding limits of C_{nv_0} and C_{nv_L} for C_{nv}, the collection efficiency is given as

$$\eta = 1 - \frac{C_{nv_L}}{C_{nv_0}} = 1 - \exp\left[-\frac{4c\eta_f L}{\pi D(1-c)}\right] \qquad (10\text{-}57)$$

The previous derivation is limited to a plane filter. A cylindrical or a spherical filter could be analyzed in a similar manner if it could be assumed that η_f is constant in the flow direction. For flow in the radial direction, this assumption is not valid and there is no simple way to analyze such filters. Of course, a cylindrical filter with flow in the axial direction is simply a special case of a plane filter, and Eq. (10-57) applies.

EXAMPLE 10-4   Use the data of Examples 10-2 and 10-3; choose the point of minimum η_f in Fig. 10-9. Determine the dimensions of a filter handling 1.0 m³/s of standard air and giving overall collection efficiencies ranging from 0.9 to 0.99999. Assume $W = H$.

SOLUTION   The data of the problem are

$$D = 2.0 \ \mu\text{m} \qquad \rho_p = 1{,}500 \ \text{kg/m}^3$$
$$V_\infty = 0.1 \ \text{m/s} \qquad d = 0.2 \ \mu\text{m}$$
$$c = 0.1 \qquad \eta_f = 0.08$$
$$\sigma = 1.0$$

The face area is computed to be

$$A_f = \frac{Q}{(1 - c)V_\infty} = \frac{1.0}{0.9(0.1)} = 11.11 \ \text{m}^2$$

which gives, for $W = H$,

$$W = H = \sqrt{A_f} = \sqrt{11.11} = 3.33 \ \text{m} \qquad Ans.$$

Equation (10-57) can be solved for L giving

$$L = -\frac{\pi D(1 - c)}{4c\eta_f} \ln(1 - \eta) = -0.0001767 \ln(1 - \eta)$$

The results are summarized in the following table:

η	L, m
0.9	0.000407
0.95	0.000529
0.99	0.000814
0.995	0.000936
0.999	0.00122
0.9995	0.00134
0.9999	0.00163
0.99995	0.00175
0.99999	0.00203

////

The collection efficiency of a filter can be readily enhanced by electrical effects. If either the particles or the fibers, but not both, are charged, collection is improved by virtue of the image charges induced in the uncharged medium. The uncharged medium becomes a dipole, which attracts or is attracted to the charged medium. If the particles and fibers are both charged but with opposite polarity, the collection is improved even more; however, if the particles and fibers are charged with like polarity, the collection efficiency is greatly reduced. In normal use, no attempt is made to charge the particles. The fibers can be charged in the following manner: Small particles of a suitable resin are inserted into the filter, where they attach to the fibers. Charges are then imparted by mechanical means, as occur naturally during the carding process. If the resin has a suitably high electrical resistivity, the charges can remain in place for years. No attempt will be made here to quantify the improvement in collection efficiency possible by charging the filter.

10-5 PRESSURE DROP IN A PACKED FILTER

The pressure drop through a packed filter is an important parameter; if the filter is clean, this pressure drop is amenable to calculation. As the filter collects particles and the interstitial spaces partially fill with the collected material, the pressure drop increases from the minimum value for the clean filter. When the pressure drop becomes too great, the filter must be taken out of service and either cleaned and returned to service or replaced. We shall not attempt to derive an equation for pressure drop in this section; rather, suitable equations will be quoted from the literature.

Davies [1, pp. 40–41], discusses pressure drop through filters in some detail. The following equation, derived from Kuwabara flow theory, is selected from his book as being a theoretical equation which appears to give good agreement with experiments conducted on clean filters:

$$\frac{\Delta P A_f D^2}{\mu Q L} = \frac{16c}{\text{Ku}} \quad (10\text{-}58)$$

in which

$$\text{Ku} = -\frac{1}{2} \ln c - \frac{3}{4} + c - \frac{c^2}{4}$$

as before. Equation (10-58) may be written

$$\Delta P = \frac{16c}{\text{Ku}} \frac{\mu Q L}{A_f D^2} = \frac{16c}{\text{Ku}} \frac{\mu L V_0}{D^2} \quad (10\text{-}59)$$

Equation (10-58) is plotted in Fig. 10-11.

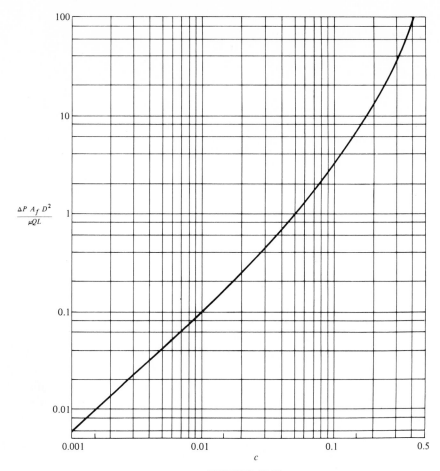

FIGURE 10-11
Nondimensional pressure drop in a packed filter.

If the filter is to operate at low pressure, as in a high-altitude atmospheric sampler, Eq. (10-59) may be modified to read

$$\Delta P = \frac{\mu Q L}{A_f D^2} \frac{16c(1 + 4\lambda/D)}{\mathrm{Ku} + 2\lambda(-\ln c + c^2/2 - \frac{1}{2})/D} \qquad (10\text{-}60)$$

in which λ is the mean-free path of the air molecules given by Eqs. (3-56) and (3-57) and having the value of 0.0667 μm for standard air. As usual, the power expended due to the flow of air through the filter is given by

$$\dot{W} = Q \, \Delta P$$

EXAMPLE 10-5   Use the data of Example 10-2, with a filter thickness of 2 mm, and estimate the pressure drop when the filter is clean.

SOLUTION   The approach velocity is $V_0 = (1 - c)V_\infty = 0.9(0.1) = 0.09$ m/s. Figure 10-11 gives

$$\frac{\Delta P A_f D^2}{\mu Q L} = 3.2$$

Then ΔP can be calculated as

$$\Delta P = \frac{3.2\mu Q L}{A_f D^2} = \frac{3.2\mu V_0 L}{D^2}$$

$$\Delta P = \frac{3.2(1.84 \times 10^{-5})(0.09)(0.002)}{(2 \times 10^{-6})^2} = 2650 \text{ N/m}^2 \qquad \textit{Ans.} \qquad ////$$

10-6   COLLECTION EFFICIENCY AND PRESSURE DROP FOR A SINGLE-LAYER FILTER

By far, the greatest use of filters in industrial air pollution control is in the form of single-layer, or cloth, filters. Such a filter consists of a single layer of cloth woven from natural, synthetic, metal, or glass fibers. When the filter element is first installed and has not been exposed to dust, its collection efficiency is rather poor. As the filter operates, it collects particles in the interstitial spaces between the fibers; these particles improve the collection efficiency greatly. During the rest of the useful life of the filter, this interstitial dust layer will remain in place. Filters containing such an interstitial dust layer and with no holes present other than the pores give excellent collection efficiencies when operating at the low air velocities typical of single-layer-filter applications. Typical collection efficiencies range from 0.97 to 0.999 using air velocities of the order of 1 to 2 cm/s. Even so, in a large installation there is usually one or more filter elements in need of replacement or repair at any given time so that the maximum efficiency is seldom realized.

Aside from the need for careful maintenance to prevent a deterioration of efficiency, collection efficiency is not a matter of concern in the design or selection of single-layer filters. The efficiency is high enough for most applications, and those applications requiring higher efficiencies must use another type of device. The important considerations in the evaluation of single-layer filters are the pressure drop as the air flows through the filter, the first cost of the device, and the cost and frequency of replacement of the filter elements.

As the filter continues to operate, a dust cake forms on the upstream side of the filter. This dust cake improves the collection efficiency slightly and increases the pressure drop greatly. Eventually, as the dust cake becomes thicker and thicker, the pressure drop becomes so high that the filter must be withdrawn

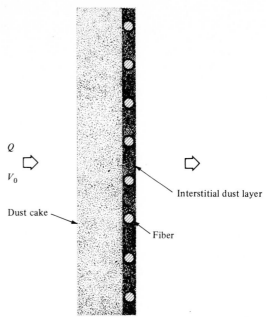

FIGURE 10-12
Single-layer filter with interstitial dust
layer and with dust cake attached.

from service and the dust cake removed. A variety of methods have been devised
for the removal of the dust cake; most of these involve either shaking the dust
loose from the filter or blowing the dust cake away by a reverse flow of air. When
the dust cake has been removed, the filter is placed back in service and the
cycle is repeated.

Figure 10-12 shows a section of the filter with an interstitial dust layer and
a dust cake present. Equation (10-59) shows that the pressure drop in a filter
increases linearly with the air velocity and with the thickness of the filter. This
conclusion holds for the single-layer filter and dust cake; if the remaining terms
are taken as constants characteristic of the filter material and weave, dust
composition and size distribution, and filtering velocity, then the following
equation for pressure drop may be written:

$$\Delta P = (K_1 + K_2 C_{ma})V_0 = (K_1 + K_2 C_{ma})\frac{Q}{A_f} \quad (10\text{-}61)$$

In Eq. (10-61) C_{ma} is the mass-area concentration in units of kilograms per
square meter and represents the mass of dust present in the dust cake attached
to unit area of filter. Clearly, C_{ma} is proportional to the thickness of the
dust cake. The constants K_1 and K_2, in units of newton seconds per cubic meter
(or kilograms per square meter per second) and (second)$^{-1}$, respectively, must be
determined for the individual situation. Hemeon [2] lists some values of the

constants (for which he uses the symbols K_0, in place of K_1, and K_d, in place of K_2); his values do not represent a wide segment of industrial activity and show much variation within a single type of activity. No table of values of K_1 and K_2 will be provided here; readers will have to estimate the appropriate values for their situations. Typical values of these constants appear to lie in the following ranges:

$$12{,}000 < K_1 < 120{,}000 \text{ N} \cdot \text{s/m}^3$$
$$10{,}000 < K_2 < 130{,}000 \text{ s}^{-1}$$

The thickness of the dust cake increases with time; its growth can be predicted if the flow rate and dust concentration in the airstream are known. The result is

$$C_{ma} = \frac{Q C_{mv} t}{A_f} = V_0 C_{mv} t \qquad (10\text{-}62)$$

Here C_{mv} is the mass-volume concentration of the airstream and t is the time since the filter was last cleaned. The mass-area concentration can also be found as the ratio of the mass collected when the filter is cleaned to the area of filter.

The values of K_1 and K_2 can be estimated either from an existing installation or from a portable filter attached to the dust in question. In either case, the installation should be instrumented to read pressure drop, flow rate, and mass of material dislodged upon removal of the dust cake. Then K_1 is computed from Eq. (10-61) when the data are taken immediately after cleaning; next K_2 is computed from the data taken just before cleaning after K_1 and C_{ma} have been computed. To compute C_{ma}, the mass of dust dislodged from the filter is divided by the filter area. Care must be taken to ensure that the weight of dust measured is due solely to dislodgment from the filter; most filters also collect dust by centrifugal action before the airstream enters the filter. Usually, the dust collected by both mechanisms is stored in the same hopper.

EXAMPLE 10-6 A filter has $K_1 = 30{,}000 \text{ N} \cdot \text{s/m}^3$ and $K_2 = 75{,}000 \text{ s}^{-1}$. The filter area is 8000 m^2, Q is 120 m^3/s, and C_{mv} is 0.02 kg/m^3. What is the pressure drop through the filter immediately after cleaning and after 3 h of operation?

SOLUTION At zero time following cleaning, $C_{ma} = 0$; Eq. (10-61) gives

$$\Delta P = \frac{K_1 Q}{A_f} = \frac{30{,}000(120)}{8000} = 450 \text{ N/m}^2 \qquad Ans.$$

After 3 h, or 10,800 s, by Eq. (10-62)

$$C_{ma} = \frac{Q C_{mv} t}{A_f} = \frac{120(0.02)(10{,}800)}{8000} = 3.24 \text{ kg/m}^2$$

Then, by Eq. (10-61) again,

$$\Delta P = (K_1 + K_2 C_{ma}) \frac{Q}{A_f}$$

$$= [30,000 + 75,000(3.24)] \frac{120}{8000} = 4095 \text{ N/m}^2 \qquad Ans. \qquad ////$$

EXAMPLE 10-7   A filter has 1000 m² of face area and treats 10 m³/s of air carrying a dust with a concentration C_{mv} equal to 0.001 kg/m³. Assume $K_1 = 20,000$ N·s/m³ and $K_2 = 25,000$ s⁻¹. If the filter must be cleaned when $\Delta P = 2000$ N/m², after what period of time must cleaning occur?

SOLUTION   Substitute Eq. (10-62) into Eq. (10-61) and solve for t, giving

$$t = \frac{A_f}{K_2 Q C_{mv}} \left(\frac{\Delta P A_f}{Q} - K_1 \right) \qquad (10\text{-}62a)$$

Then

$$t = \frac{1000}{25,000(10)(0.001)} \left[\frac{2000(1000)}{10} - 20,000 \right] \frac{1}{3600} = 200 \text{ h} \qquad Ans. \qquad ////$$

EXAMPLE 10-8   In a test for measuring K_1 and K_2, the following data were obtained:

ΔP after cleaning	400 N/m²
ΔP before cleaning	2100 N/m²
Flow rate	0.5 m³/s
Mass collected	55 kg
Filter area	40 m²

Determine K_1 and K_2.

SOLUTION   Just after cleaning, when $C_{ma} = 0$, Eq. (10-61) gives

$$K_1 = \frac{\Delta P A_f}{Q} = \frac{400(40)}{0.5} = 32,000 \text{ N·s/m}^3 \qquad Ans.$$

Just before cleaning, $C_{ma} = m/A_f = 55/40 = 1.375$ kg/m². Solving Eq. (10-61) for K_2 gives

$$K_2 = \frac{\Delta P A_f/Q - K_1}{C_{ma}} = \frac{2100(40)/0.5 - 32,000}{1.375} = 98,909 \text{ s}^{-1} \qquad Ans. \qquad ////$$

10-7   BAG FILTERS AND BAGHOUSES

The bag filter most commonly used is in the shape of a long cylinder, as shown in Fig. 10-13, where the filter is clamped around a sleeve at the bottom and around a cap at the top. The air enters at the bottom, flows through the filter

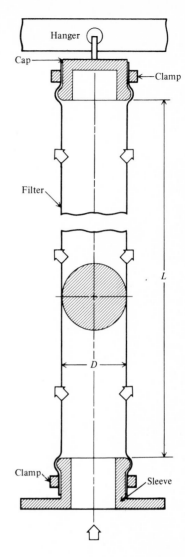

FIGURE 10-13
Bag-type filter element clamped around a
sleeve at the bottom and around a cap at
the top.

along its sides, leaving the dust to form a cake on the inside of the filter, and
flows outside the filter to the exit duct. This method of suspending the filter at
the top is not the only method in use; frequently, the filter is closed at the top,
with a hook or strap sewed to it, and the filter is suspended by this hook or
strap. Many other variations of the shape and method of mounting the filter
are available. Danielson [3] provides an excellent discussion of the various details
of construction and mounting of filters.

For the cylindrical bag filter shown in Fig. 10-13, the filtering area is given by

$$A_f = \pi n D L$$

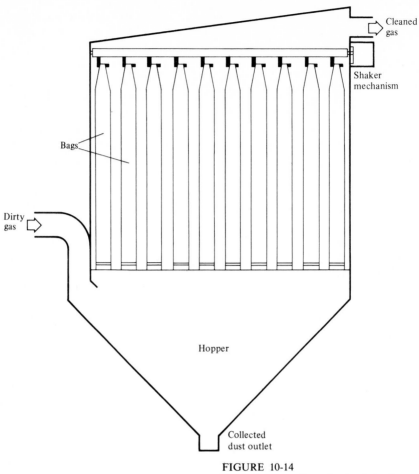

FIGURE 10-14
Vertical cross section of a simple baghouse.

Generally, bags are made in diameters ranging from 5 to 12 in (12.5 to 30 cm) with lengths from 6 to 20 ft (2 to 6 m). For satisfactory performance, the length to diameter ratio should not usually exceed 20 : 1. Thus a bag 15 cm in diameter could be as much as 3.0 m long.

The material from which the filter is woven depends on the temperature of the gas and its degree of acidity or alkalinity, as well as the desired life of the filter element. The elements need to be replaced every few months or every few years, depending upon the severity of service and upon the relative cost and relative life of elements made from different materials. As listed by Danielson [3], the common materials used for bag filters, listed in order of ability to withstand increasingly higher temperatures, are cotton, Dynel, wool, nylon, Orlon, Dacron, and glass. For details, the reader should consult Ref. [3] or some other source.

Figure 10-14 shows a vertical cross section through a simple baghouse.

Here the top end of each bag is sewed to a strap which is fastened to a hanger. The hanger is attached to the shaker mechanism. The gas to be cleaned enters at the left, flows downward through the top of the hopper, then flows into the bags at the bottom, through the sides of the bag filters, and then flows out at the top right side of the baghouse. Many of the larger dust particles are separated from the air by centrifugal action directly into the hopper, which relieves the filters of much of the burden of collecting the large particles and requires the filters to be cleaned less often. When the dust cake builds to a sufficient thickness, as evidenced by a sufficiently large pressure drop across the baghouse, the bags must be cleaned. First the gas flow is shut off, and then the shaker mechanism is actuated for a short time until the cake is removed. It is important that the gas flow be cut off completely, as by turning off the blower or by closing a tightly sealed damper; even a very small flow of gas can destroy the cleaning effectiveness.

Large baghouses are built with several compartments, so that one compartment can be closed for cleaning as needed. The remaining compartments must be sufficiently large to handle the cleaning load. In fact, it is best to design the baghouse so that two compartments can be off-line when needed, one for cleaning and one for repair or replacement of bags. The bags should be arranged within the compartment in such a way as to utilize the space effectively and yet provide access to each individual bag for replacement. This is done by placing the bags close together, about 5 cm apart, while leaving aisles between every fourth to eighth row of bags. The aisles should be wide enough to provide access to the bags; $\frac{1}{3}$ to $\frac{2}{3}$ m is typical. Although staggering the bags allows more bags to be installed in a compartment of given size, this is not a good plan because the staggered rows of bags are not easily accessible. Bags can be replaced one by one as they become defective, or all bags in a single compartment can be replaced at one time. It must be decided which method is more economical for a given installation.

10-8 CLEANING CYCLES FOR BAGHOUSES

The cleaning cycle for a single-compartment baghouse is easily computed; Example 10-7 shows how this computation is made. For a multiple-compartment baghouse, however, the procedure is more involved. In this section, equations are derived to determine the cleaning interval for a multiple-compartment baghouse. These computations are mainly for design purposes; in practice, the baghouse will be cleaned according to a set sequence whenever the pressure drop across the filter reaches a certain preset value.

In this derivation, we assume a baghouse having n_c compartments and with a total flow rate Q, fixed by the process which the baghouse serves, distributed among the different compartments. During normal use, n_c is either the number of compartments in the baghouse or perhaps one less if allowance is made for a compartment that is out of service due to repair. During the cleaning part of the

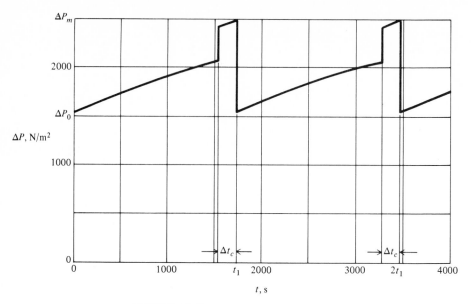

FIGURE 10-15
Pressure drop across a baghouse during two consecutive cleaning cycles.

cycle, $n_c - 1$ compartments are active. The pressure drop is ΔP, and its maximum value is ΔP_m; this value is to be reached just as the newly cleaned compartment is activated. The weight of dust cake on each filter is given by C_{ma_i}, where the subscript i represents the ith compartment. We assume that all bags in a given compartment behave identically. The flow rate through the ith compartment is Q_i, and the filter area in each compartment is A_{fi}, which is the same for all compartments. The analysis begins at the time when compartment 1 has just been cleaned and is reactivated. Let t_1 be the length of the cleaning cycle, that is, the time period between the start of one cleaning process and the start of the next cleaning process. Also, let Δt_c be the length of the cleaning process.

Figure 10-15 shows the variation with time of pressure drop ΔP across the baghouse during two cleaning cycles. This pressure drop increases abruptly when one compartment is removed for cleaning; it drops abruptly when the newly cleaned compartment is reactivated. Figure 10-16 shows how the weight of the dust cake C_{ma_i} increases during successive cleaning cycles for the different compartments. As the ith compartment is cleaned, the value of C_{ma_i} drops to zero. Figure 10-17 shows the flow rates through the different compartments during one cleaning cycle. The flow rate drops through the newly cleaned compartment while increasing through the others. When the fifth compartment is removed for cleaning, the flow rate increases abruptly through the others. (All these figures are drawn for a five-compartment baghouse.)

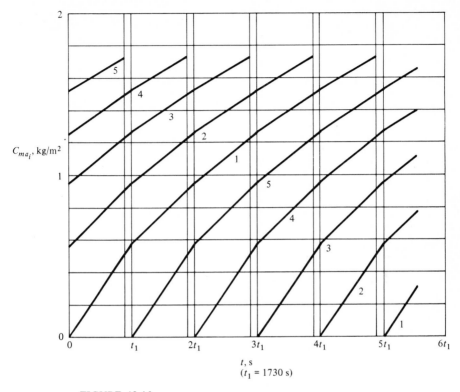

C_{ma_i}, kg/m²

t, s
($t_1 = 1730$ s)

FIGURE 10-16
Weight of dust cake on filters in the different compartments in a five-compart-ment baghouse during five consecutive cleaning cycles.

Equations (10-61) and (10-62) form the basis for the analysis. These may be written for each compartment as

$$\Delta P = (K_1 + K_2 C_{ma_i}) \frac{Q_i}{A_{fi}} \qquad (10\text{-}63)$$

$$C_{ma_i} = \frac{Q_i C_{mv} t}{A_{fi}} \qquad (10\text{-}64)$$

The total flow rate is equal to the sum of the flow rates through each compart-ment and can be expressed by the following equations:

$$Q = \sum_{i=1}^{n_c} Q_i \qquad 0 < t < t_1 - \Delta t_c \qquad (10\text{-}65)$$

$$Q = \sum_{i=1}^{n_c - 1} Q_i \qquad t_1 - \Delta t_c < t < t_1 \qquad (10\text{-}66)$$

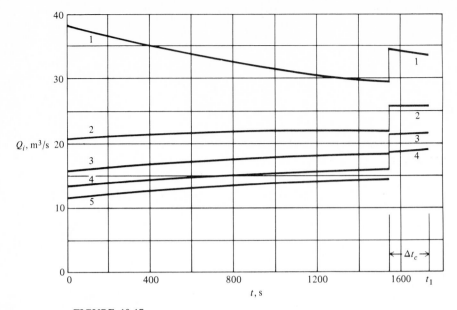

FIGURE 10-17
Flow rate through each compartment in a five-compartment baghouse during one cleaning cycle.

The weight of dust cake on the filters of the ith compartment is given by the integral of Eq. (10-64):

$$C_{ma_i} = C_{ma_{i0}} + \int_0^t \frac{Q_i C_{mv}}{A_{fi}} dt \quad (10\text{-}67)$$

Taking the derivative of Eq. (10-67) and using Eq. (10-63),

$$\frac{dC_{ma_i}}{dt} = \frac{C_{mv}}{A_{fi}} Q_i = \frac{C_{mv} \Delta P}{K_1 + K_2 C_{ma_i}} \quad (10\text{-}68)$$

Equation (10-68) may be integrated and rewritten as

$$K_1(C_{ma_i} - C_{ma_{i0}}) + \frac{K_2}{2}(C_{ma_i}{}^2 - C_{ma_{i0}}{}^2) = C_{mv} \int_0^t \Delta P \, dt \quad (10\text{-}69)$$

Equation (10-69) can be solved for C_{ma_i}; the positive sign is used in the quadratic equation, and, in addition, the quantity ϕ is defined as

$$\phi = \frac{2K_2 C_{mv}}{K_1{}^2} \int_0^t \Delta P \, dt \quad (10\text{-}70)$$

The resulting equation for C_{ma_i} is

$$C_{ma_i} = \frac{K_1}{K_2}\left\{-1 + \left[\left(1 + \frac{K_2}{K_1}C_{ma_{i0}}\right)^2 + \phi\right]^{1/2}\right\} \qquad (10\text{-}71)$$

Equation (10-71) may be written as

$$(K_1 + K_2 C_{ma_i})^2 = K_1{}^2\left[\left(1 + \frac{K_2}{K_1}C_{ma_{i0}}\right)^2 + \phi\right]$$

If we define ϕ_1 as the value of ϕ defined by Eq. (10-70) when t is equal to t_1, the time at the completion of the cleaning cycle, then this equation becomes

$$(K_1 + K_2 C_{ma_{i1}})^2 = K_1{}^2\left[\left(1 + \frac{K_2}{K_1}C_{ma_{i0}}\right)^2 + \phi_1\right] \qquad (10\text{-}72)$$

At the start of the new cleaning cycle, the state of cleanliness of the compartments has shifted cyclicly by one position, so that compartment 1 occupies the position of compartment 2 at the beginning of the previous cycle, and so forth, as expressed by the following relations when $t = t_1$:

$$C_{ma_{i1}} = C_{ma_{(i+1)0}} \qquad i = 1, 2, \ldots, n_c - 1$$
$$C_{ma, n_c 1} = C_{ma_{10}} = 0 \qquad\qquad (10\text{-}73)$$

in which we have assumed that compartment 1 starts out with no dust cake. Also, $C_{ma_{i1}}$ is the weight of dust cake on the filter elements in the ith compartment at the end of the cleaning cycle.

Combining Eqs. (10-72) and (10-73) gives

$$(K_1 + K_2 C_{ma_{(i+1)0}})^2 = (K_1 + K_2 C_{ma_{i0}})^2 + K_1{}^2\phi_1 \qquad (10\text{-}74)$$

which may be expanded as follows:

$$(K_1 + K_2 C_{ma_{20}})^2 = K_1{}^2 + K_1{}^2\phi_1$$
$$(K_1 + K_2 C_{ma_{30}})^2 = (K_1 + K_2 C_{ma_{20}})^2 + K_1{}^2\phi_1 = K_1{}^2 + 2K_1{}^2\phi_1$$
$$\cdots\cdots\cdots\cdots\cdots\cdots\cdots\cdots\cdots\cdots\cdots\cdots\cdots\cdots\cdots\cdots\cdots\cdots\cdots\cdots$$

$$(K_1 + K_2 C_{ma, n_c 0})^2 = K_1{}^2 + (n_c - 1)K_1{}^2\phi_1$$

Then we see that for the ith compartment

$$K_1 + K_2 C_{ma_{i0}} = K_1\sqrt{1 + (i - 1)\phi_1}$$

Solving for $C_{ma_{i0}}$,

$$C_{ma_{i0}} = -\frac{K_1}{K_2} + \frac{K_1}{K_2}\sqrt{1 + (i - 1)\phi_1} \qquad (10\text{-}75)$$

Substituting Eq. (10-75) into Eq. (10-71) gives for C_{ma_i}

$$C_{ma_i} = \frac{K_1}{K_2}[\sqrt{1 + \phi + (i - 1)\phi_1} - 1] \qquad (10\text{-}76)$$

We substitute Eq. (10-76) into Eq. (10-63), giving

$$Q_i = \frac{A_{fi}\,\Delta P}{K_1\sqrt{1 + \phi + (i-1)\phi_1}} \tag{10-77}$$

Then substituting Eq. (10-77) into Eqs. (10-65) and (10-66) gives

$$Q = \sum_{i=1}^{n_c} \frac{A_{fi}\,\Delta P}{K_1\sqrt{1 + \phi + (i-1)\phi_1}} \qquad 0 < t < t_1 - \Delta t_c$$

$$Q = \sum_{i=1}^{n_c - 1} \frac{A_{fi}\,\Delta P}{K_1\sqrt{1 + \phi + (i-1)\phi_1}} \qquad t_1 - \Delta t_c < t < t_1 \tag{10-78}$$

Evaluating the second of Eqs. (10-78) when $t = t_1$, $\phi = \phi_1$, and $\Delta P = \Delta P_m$, we have

$$\frac{QK_1}{A_{fi}\,\Delta P_m} = \sum_{i=1}^{n_c - 1} \frac{1}{\sqrt{1 + i\phi_1}} \tag{10-79}$$

This function has been evaluated for various values of ϕ_1 and n_c; Fig. 10-18 enables ϕ_1 to be determined. For the case in which ϕ_1 is greater than 10, Eq. (10-79) may be approximated as

$$\phi_1 \approx \frac{\left(\sum_{i=1}^{n_c - 1} 1/\sqrt{i}\right)^2}{(QK_1/A_{fi}\,\Delta P_m)^2} \tag{10-80}$$

The numerator for Eq. (10-80) has been evaluated as shown in Table 10-2.

Equation (10-70) may be solved for ΔP by differentiation, giving

$$\Delta P = \frac{K_1^2}{2K_2 C_{mv}}\frac{d\phi}{dt} \tag{10-81}$$

Equations (10-78) may be integrated, giving

$$Q_t = \frac{A_{fi}}{K_1} \sum_{i=1}^{n_c} \int_0^t \frac{\Delta P\,dt}{\sqrt{1 + \phi + (i-1)\phi_1}}$$

$$Q(t - t_1 + \Delta t_c) = \frac{A_{fi}}{K_1} \sum_{i=1}^{n_c - 1} \int_{t_1 - \Delta t_c}^{t_1} \frac{\Delta P\,dt}{\sqrt{1 + \phi + (i-1)\phi_1}}$$

When Eq. (10-81) is substituted into these equations, the result is

$$t = \frac{A_{fi} K_1}{2K_2 C_{mv} Q} \sum_{i=1}^{n_c} \int_0^{\phi} \frac{d\phi}{\sqrt{1 + \phi + (i-1)\phi_1}} \qquad 0 < \phi < \phi_1' \tag{10-82}$$

$$t = t_1 - \Delta t_c + \frac{A_{fi} K_1}{2K_2 C_{mv} Q} \sum_{i=1}^{n_c - 1} \int_{\phi_1'}^{\phi} \frac{d\phi}{\sqrt{1 + \phi + (i-1)\phi_1}} \qquad \phi_1' < \phi < \phi_1 \tag{10-83}$$

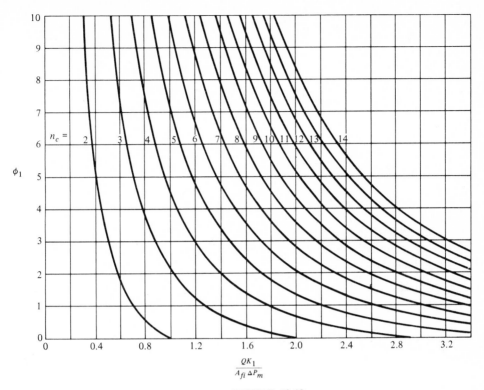

FIGURE 10-18
Graph of ϕ_1 for a baghouse having n_c compartments.

Table 10-2   NUMERATOR OF EQ. (10-80)

n_c	$\left(\sum_{i=1}^{n_c-1} 1/\sqrt{i}\right)^2$
2	1.000
3	2.914
4	5.219
5	7.753
6	10.44
7	13.25
8	16.14
9	19.11
10	22.13
11	25.21
12	28.33
13	31.49
14	34.67

When Eqs. (10-82) and (10-83) have been integrated, the result is

$$t = \frac{A_{fi} K_1}{K_2 C_{mv} Q} \sum_{i=1}^{n_c} [\sqrt{1 + \phi + (i-1)\phi} \quad \sqrt{1 + (i-1)\phi_1}]$$

$$0 < \phi < \phi_1' \quad (10\text{-}84)$$

$$t = t_1 - \Delta t_c + \frac{A_{fi} K_1}{K_2 C_{mv} Q} \sum_{i=1}^{n_c - 1} [\sqrt{1 + \phi + (i-1)\phi_1} - \sqrt{1 + \phi_1' + (i-1)\phi_1}]$$

$$\phi_1' < \phi < \phi_1 \quad (10\text{-}85)$$

In Eqs. (10-83) to (10-85) ϕ_1' is given by

$$\phi_1' = \frac{2K_2 C_{mv}}{K_1^2} \int_0^{t_1 - \Delta t_c} \Delta P \, dt \quad (10\text{-}86)$$

When Eq. (10-84) is evaluated at $t_1 - \Delta t_c$, for which $\phi = \phi_1'$, when the result is substituted into Eq. (10-85), and when that equation is evaluated at t_1, for which $\phi = \phi_1$, the final equation is

$$t_1 = \frac{A_{fi} K_1}{K_2 C_{mv} Q} [\sqrt{1 + \phi_1' + (n_c - 1)\phi_1} - 1] \quad (10\text{-}87)$$

The quantity ϕ_1' must still be determined. If Eq. (10-85) is evaluated when $t = t_1$ and $\phi = \phi_1$, the result is

$$\frac{K_2 C_{mv} Q \, \Delta t_c}{A_{fi} K_1} = \sum_{i=1}^{n_c - 1} \{(1 + i\phi_1)^{1/2} - [1 + i\phi_1 - (\phi_1 - \phi_1')]^{1/2}\}$$

The second radical in the preceding equation may be expanded by the binomial theorem and higher terms neglected, giving

$$\frac{K_2 C_{mv} Q \, \Delta t_c}{A_{fi} K_1} \approx \frac{1}{2} (\phi_1 - \phi_1') \sum_{i=1}^{n_c - 1} (1 + i\phi_1)^{-1/2}$$

Combining this equation with Eq. (10-79) gives

$$\phi_1' \approx \phi_1 - \frac{2K_2 C_{mv} \Delta P_m \, \Delta t_c}{K_1^2} \quad (10\text{-}88)$$

When Eqs. (10-84) and (10-85) are substituted into Eq. (10-81), the following equations for pressure drop are obtained:

$$\Delta P = \frac{K_1 Q / A_{fi}}{\sum_{i=1}^{n_c} [1 + \phi + (i-1)\phi_1]^{-1/2}} \quad 0 < \phi < \phi_1'$$

$$\Delta P = \frac{K_1 Q / A_{fi}}{\sum_{i=1}^{n_c - 1} [1 + \phi + (i-1)\phi_1]^{-1/2}} \quad \phi_1' < \phi < \phi_1$$

$$(10\text{-}89)$$

In the practical evaluation of baghouse cleaning cycles, first the quantity ϕ_1 is determined from Figure 10-18 and then ϕ_1' is computed from Eq. (10-88).

Next the length of the cleaning cycle t_1 is computed from Eq. (10-87). Although t_1 is the quantity of primary interest, further data on the baghouse performance can then be obtained from Eqs. (10-84), (10-85), (10-75), (10-76), (10-89), and (10-63), applied in that order. Figures 10-15 to 10-17 were drawn from computations performed in this manner using the data of the following example.

EXAMPLE 10-9   A baghouse has five compartments, with 1000 m² of filtering area in each compartment. The total flow rate is 100 m³/s, and C_{mv} is 0.01 kg/m³. Take ΔP_m as 2500 N/m², K_1 as 40,000 N · s/m³, K_2 as 60,000 s⁻¹, and Δt_c as 3.0 min. Determine t_1 and compute and plot curves of ΔP as a function of time, C_{ma} as a function of time for each compartment, and Q_i as a function of time for each compartment.

SOLUTION   From the given data, we may compute

$$\frac{QK_1}{A_{fi}\Delta P_m} = \frac{100(40,000)}{1000(2500)} = 1.60$$

From Fig. 10-18, ϕ_1 is read as 2.45. From Eq. (10-88), ϕ_1' is calculated as

$$\phi_1' = 2.45 - \frac{2(60,000)(0.01)(2500)(180)}{(40,000)^2} = 2.11$$

Equation (10-87) gives t_1:

$$t_1 = \frac{1000(40,000)}{60,000(0.01)(100)} [\sqrt{1 + 2.11 + 4(2.45)} - 1]$$

$$= 1279 \text{ s}$$

$$= 28.8 \text{ min} \qquad Ans.$$

Equation (10-75) enables $C_{ma_{i0}}$ to be calculated, giving

i	$C_{ma_{i0}}$, kg/m²
1	0
2	0.572
3	0.953
4	1.26
5	1.52

Next, various values of ϕ are selected and converted to time by Eqs. (10-84) and (10-85). Equation (10-76) gives C_{ma_1} to C_{ma_5} as functions of ϕ and hence of time. Equations (10-89) give ΔP as a function of ϕ, while Eq. (10-63) can be solved for Q_i to give Q_1 to Q_5 as functions of time. The pressure-drop variation is plotted in Fig. 10-15. Figure 10-16 shows how the dust cake builds up on the filter elements, and Fig. 10-17 shows the manner in which the flow rates through the various compartments change during the cleaning cycle.   ////

REFERENCES

1 Davies, C. N.: "Air Filtration," Academic Press, Inc., New York, 1973.
2 Hemeon, W. C. L.: "Plant and Process Ventilation," 2d ed., pp. 437–448, Industrial Press, Inc., New York, 1963.
3 Danielson, J. A., ed.: Air Pollution Engineering Manual, pp. 106–135, *Public Health Service Publ. No. 999-AP-40*, National Center for Air Pollution Control, Cincinnati, Ohio, 1967. (Available from the U.S. Government Printing Office, Washington, D.C.)

PROBLEMS

10-1 Derive Eqs. (10-8) and (10-9) from Eq. (10-5).
10-2 Obtain Eqs. (10-12) and (10-13) from Eqs. (10-8) to (10-11).
10-3 Derive Eqs. (10-14) and (10-25).
10-4 Derive Eqs. (10-29) and (10-30).
10-5 Derive Eqs. (10-32) to (10-34).
10-6 Determine the collection efficiency of a 6.0-μm-diameter fiber acting on particles 1.0 μm in diameter and with densities of 3000 kg/m^3. The filter packing density is 0.15, and the flow velocity approaching the filter is 1.0 m/s. Assume standard air. Consider only inertial impaction and interception.
10-7 For a packed filter having $c = 0.10$, $V_\infty = 0.5$ m/s, and $D = 4.0$ μm, the particle diameter ranges from 0.1 to 2.0 μm. For a particle density of 1000 kg/m^3, compute and plot the collection efficiency as a function of particle diameter. Assume standard air.
10-8 For the data of Example 10-1, allow the velocity V_∞ to vary from 0.1 to 2.0 m/s. Compute and plot the Reynolds number and the collection efficiency due to interception and inertial impaction as a function of V_∞.
10-9 For the data of Example 10-1, compute and plot the collection efficiency due to interception and inertial impaction as the fiber diameter varies from 1.0 to 20 μm.
10-10 Repeat Prob. 10-6 for a particle 20 μm in diameter.
10-11 Using the data of Example 10-2, compute and plot the concentration boundary-layer thickness around the cylinder.
10-12 Compute and plot the collection efficiency by diffusion only for particles of 0.1 μm diameter for $V_\infty = 0.4$ m/s, $c = 0.05$, standard air, and $\sigma = 1$. Let the fiber diameter vary from 0.5 to 10 μm.
10-13 Compute and plot the collection efficiency by diffusion only if $D = 5$ μm, $d = 0.05$ μm, and $c = 0.03$ for standard air if the velocity V_∞ varies from 0.05 to 1.5 m/s. Assume $\sigma = 1$.
10-14 Consider a filter having a packing density of 0.04 and an approach velocity of 0.55 m/s of standard air. The filter diameter is 7 μm; the particle diameter and density are 0.4 μm and 1500 kg/m^3, respectively. Compute the collection efficiency of the fiber by inertial impaction and interception and by diffusion. Assume $\sigma = 1$.
10-15 Compute the collection efficiency by diffusion only for a fiber in a filter if $V_\infty = 0.3$ m/s, $D = 2.5$ μm, $c = 0.06$, $d = 0.0075$ μm, and the fluid is air at 25,000 N/m^2 pressure and 500 K temperature. Assume $\sigma = 0.85$.
10-16 A filter having $c = 0.15$ and $V_\infty = 0.30$ m/s operates with standard air containing particles with densities of 1100 kg/m^3 and diameters of 0.2 μm. The fiber diameter

is 4 μm. Determine the combined collection efficiency for a single fiber, assuming that $\sigma = 1$.

10-17 Repeat Prob. 10-16 for the following data: $c = 0.07$, $V_{\infty} = 0.50$ m/s, $\rho_p = 1500$ kg/m^3, $D = 5$ μm, and $d = 0.3$ μm.

10-18 Using the data of Prob. 10-16, determine the thickness of a filter having a collection efficiency of 0.999.

10-19 Using the data of Prob. 10-17, what collection efficiency will be obtained from a filter 2.0 cm thick?

10-20 Derive an equation for the collection efficiency of a cylindrical filter, with flow in the radial direction, if the fiber collection efficiency η_f can be assumed constant. Do the derivation both for radial outflow and radial inflow.

10-21 Repeat Prob. 10-20 for the case of a sphere.

10-22 A filter for use in trapping radioactive particles must provide an overall collection efficiency of at least 0.99999 for particles of any size. The flow rate is 1.0 m^3/s, $\rho_p = 1200$ kg/m^3, $c = 0.2$, $D = 4$ μm, and the fluid is standard air. If the width and height of the filter are 1.5 and 1.0 m, respectively, how thick must the filter be? Assume $\sigma = 1.0$.

10-23 For a given application, a filter 5 cm thick provides a collection efficiency of 0.99. How thick should the filter be to provide an efficiency of 0.99999?

10-24 A filter is to collect 0.25-μm particles with densities of 1000 kg/m^3 in a stream of 2.5 m^3/s of standard air. Take $D = 3.0$ μm, $c = 0.15$, and $\sigma = 1.0$. Estimate the dimensions of a filter having a thickness of 2.0 cm and an efficiency of 0.99.

10-25 A filter is to collect 1.0-μm particles with densities of 1100 kg/m^3. The flow rate of standard air through the filter is 2.0 m^3/s, $c = 0.01$, $D = 10$ μm, $\sigma = 1$, and the filter has dimensions of 0.8 by 0.6 m by 1.0 cm thick. What collection efficiency is expected?

10-26 Estimate the pressure drop for the filter of Prob. 10-18 when the filter is clean.

10-27 Estimate the pressure drop for the filter of Prob. 10-19 when the filter is clean.

10-28 What pressure drop and power loss are expected for the filter of Prob. 10-22?

10-29 Estimate the pressure drop and power loss for the filter of Prob. 10-24 when clean.

10-30 Estimate the pressure drop and power loss for the filter of Prob. 10-25 when clean.

10-31 Using Eq. (10-59), derive an equation for the pressure drop in a cylindrical filter with flow in the radial direction.

10-32 Repeat Prob. 10-31 for a spherical filter.

10-33 Derive an equation for the thickness of the dust cake in terms of C_{ma}, the density of the dust, and the porosity $(1 - c)$ of the dust cake.

10-34 Derive equations for K_1 and K_2 in terms of the quantities in Eq. (10-59) if that equation is assumed to apply to the dust cake and filter with its interstitial layer. (Note that in reality this is an unlikely assumption.)

10-35 A filter has 30,000 m^2 of surface area and serves a stream of 275 m^3/s of standard air, containing a concentration of 0.025 kg/m^3 of dust particles. If $K_1 = 100,000$ N $\cdot$ s/m^3 and $K_2 = 50,000$ s^{-1}, determine the pressure drop immediately after cleaning, and compute and plot pressure drop as a function of time thereafter.

10-36 For the data of Prob. 10-35, compute and plot power loss as a function of time.

10-37 A filter is to be designed to treat 25 m^3/s of standard air, with $C_{mv} = 0.008$ kg/m^3. For this dust and for the type of filter proposed, it has been found that $K_1 = 40,000$ N $\cdot$ s/m^3 and $K_2 = 65,000$ s^{-1}. For $V_0 = 1.0$ cm/s, how much surface area of filter is required and what will be the pressure drop immediately after cleaning?

10-38 For the filter of Prob. 10-37, how many hours of service will the filter provide before it must be cleaned, if the maximum allowable pressure drop is 3000 N/m²?

10-39 A filter is to be designed to treat 100 m³/s of standard air having $C_{mv} = 0.01$ kg/m³. It is known that $K_1 = 70,000$ N·s/m³ and $K_2 = 50,000$ s⁻¹. The housing and support structure for the filter costs \$100/m² of filter surface and lasts 20 years; the filter cloth costs \$20/m² of filter surface and lasts 3 years; the capitalization rate is 0.15 for 20 years and 0.5 for 3 years. Power costs \$.012/kWh, and the blower is 0.75 efficient. Determine the filter area, face velocity, and annual cost if these are selected to give minimum annual cost. The filter is cleaned every 12 h and operates continuously except for brief cleaning intervals.

10-40 For the data of Prob. 10-38, what is the maximum power loss?

10-41 A filter is to treat 50 m³/s of standard air with $C_{mv} = 0.03$ kg/m³. It is known that K_1 and K_2 are 50,000 N·s/m³ and 50,000 s⁻¹, respectively. The maximum allowable pressure drop is 4000 N/m², and the filter must operate for 8 h between cleanings. If no further restriction on face velocity is needed, determine the filter area required. Determine also the maximum power loss.

10-42 In a measurement to determine K_1 and K_2, a small portable filter unit was installed. The measured data show that Q was 1.0 m³/s, ΔP was 2500 N/m² just before the filter was cleaned, and ΔP was 700 N/m² immediately after cleaning; 100 kg of dust was removed during cleaning. If the filter area was 75 m², what values of K_1 and K_2 were found?

10-43 In a single-compartment baghouse, there are 1000 bags, each 20 cm in diameter by 3.5 m long. If the bags are cylindrical, compute the filter area of the baghouse.

10-44 A baghouse consists of 1000 m² of filter area. Replacement cotton filters cost \$5/m² and must be replaced every 3 mo. Nylon filters cost \$15/m² and must be replaced every 10 mo. The installation cost for the cotton bags is \$.50/m²; that for the nylon bags is \$.75/m². Determine the annual cost for each bag material.

10-45 Derive Eq. (10-71) from Eqs. (10-69) and (10-70).

10-46 Derive Eq. (10-87) from previous equations.

10-47 In place of Eq. (10-88), derive a closer approximation by retaining an extra term in the binomial expansion of the radical in the second preceding equation.

10-48 A baghouse has four compartments and handles air at the rate of 50 m³/s having C_{mv} equal to 0.006 kg/m³. Each compartment has 500 m² of filtering surface area, $K_1 = 12,000$ N·s/m³, and $K_2 = 15,000$ s⁻¹. If Δt_c is 150 s and the maximum pressure drop is 500 N/m², determine t_1.

10-49 A baghouse has eight compartments, each having a filter surface area of 3000 m² and for which $K_1 = 20,000$ N·s/m³ and $K_2 = 15,000$ s⁻¹. The total flow rate is 360 m³/s, and C_{mv} is 0.01 kg/m³. For a cleaning time Δt_c of 100 s and for a maximum pressure drop ΔP_m of 1000 N/m², determine the length of the cleaning cycle t_1.

10-50 For the baghouse of Prob. 10-49, recompute the length of the cleaning cycle if one compartment is taken out of service for repair.

ABSORPTION DEVICES

The collection of particles by scrubbers was discussed in Chap. 9. Scrubbers are also valuable for removing gaseous pollutants from an airstream. Absorption columns and towers are used for this purpose as well. In this chapter, we shall consider the use of scrubbers, columns, and towers for the removal of gaseous pollutants. Sections 11-1 to 11-3 deal with the collection of gases by moving drops; Sec. 11-4 deals with absorption columns and towers.

The basic collection mechanism is diffusion. One constituent of a mixture diffuses through another in response to a concentration gradient, with concentration decreasing in the direction of flow. This is expressed by Eq. (9-49) as

$$\frac{\dot{m}}{A} = \mathscr{D}\,\frac{\partial C_{mv}}{\partial r} \qquad (11\text{-}1)$$

in which $\mathscr{D}$ is the diffusion coefficient. Consider the spherical shell of thickness dr shown in Fig. 11-1. The mass flow rates by diffusion into and out of the shell are given by

$$\dot{m}_1 = -\left(4\pi r^2 \mathscr{D}\,\frac{\partial C_{mv}}{\partial r}\right)_r,$$

$$\dot{m}_2 = -\left(4\pi r^2 \mathscr{D}\,\frac{\partial C_{mv}}{\partial r}\right)_{r+dr}$$

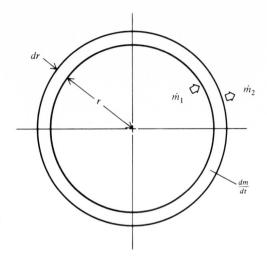

FIGURE 11-1
Spherical shell of thickness dr with entering and leaving mass flow rates due to diffusion.

The rate of accumulation of mass within the shell is given by

$$\frac{dm}{dt} = 4\pi r^2 \, dr \, \frac{\partial C_{mv}}{\partial t}$$

The mass flow rate into the shell must equal the sum of the mass flow rate out of the shell plus the rate of mass accumulation within the shell, or $\dot{m}_1 = \dot{m}_2 + dm/dt$. When the preceding equations are substituted into this equation and the difference terms expressed as a derivative, we have the following partial differential equation:

$$\frac{1}{r^2} \frac{\partial}{\partial r} \left(r^2 \frac{\partial C_{mv}}{\partial r} \right) = \frac{1}{\mathscr{D}} \frac{\partial C_{mv}}{\partial t} \qquad (11\text{-}2)$$

Equations (11-1) and (11-2) are the basic equations for diffusion in a spherical geometry; they must be solved subject to the appropriate boundary conditions for the particular problem.

11-1 ABSORPTION OF GASES BY MOVING DROPS

Consider a drop moving through a mixture of air and a gaseous pollutant. The velocity of the air relative to the drop is V_∞ at a substantial distance from the drop; the concentration of the pollutant gas is C_{mv_∞} at that point. A concentration boundary layer will be formed next to the surface of the drop, across which the concentration will change from essentially C_{mv_∞} to a suitable value C_{mv_0} at the surface of the drop. The value of C_{mv_0}, unknown at present, will be discussed in Sec. 11-2. Figure 11-2 shows the drop and the concentration boundary layer.

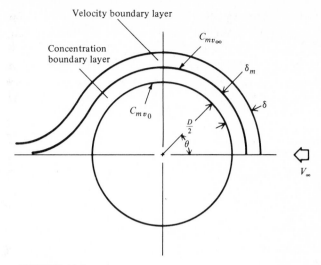

FIGURE 11-2
Concentration and velocity boundary layers on a spherical drop.

An approximation to the variation of concentration across the boundary layer is given by the following equation, which is similar to Eq. (9-51):

$$C_{mv} - C_{mv_0} = (C_{mv_\infty} - C_{mv_0})\left[\frac{3}{2}\frac{y}{\delta_m} - \frac{1}{2}\left(\frac{y}{\delta_m}\right)^3\right] \qquad (11\text{-}3)$$

If we assume that the concentration boundary-layer thickness δ_m is proportional to that of the velocity boundary layer δ, that is,

$$\delta_m = \beta\delta \qquad (11\text{-}4)$$

then Eq. (11-3) becomes

$$C_{mv} = C_{mv_0} + (C_{mv_\infty} - C_{mv_0})\left[\frac{3}{2\beta}\frac{y}{\delta} - \frac{1}{2\beta^3}\left(\frac{y}{\delta}\right)^3\right] \qquad (11\text{-}5)$$

From this equation

$$\left(\frac{\partial C_{mv}}{\partial y}\right)_{y=0} = (C_{mv_\infty} - C_{mv_0})\frac{3}{2\beta\delta} \qquad (11\text{-}6)$$

Equation (11-1) may be written for the surface of the drop:

$$\frac{\dot{m}}{A} = \mathscr{D}\left(\frac{\partial C_{mv}}{\partial y}\right)_{y=0}, \qquad (11\text{-}7)$$

Using Eq. (11-6) gives

$$\frac{\dot{m}}{A} = \frac{3\mathscr{D}}{2\beta\delta}(C_{mv_\infty} - C_{mv_0}) \qquad (11\text{-}8)$$

Take an element of area on the surface of the sphere given by $2\pi(D/2)\sin\theta\,(D/2)\,d\theta$; the mass flow rate across this area is

$$dm = \frac{3\pi\mathscr{D}D^2}{4\beta\delta}\sin\theta\,d\theta(C_{mv_\infty} - C_{mv_0}) \qquad (11\text{-}9)$$

The mass flow rate over the front surface of the sphere is given by

$$\dot{m} = \frac{3\pi\mathscr{D}D^2}{4\beta}(C_{mv_\infty} - C_{mv_0})\int_0^{\pi/2}\frac{\sin\theta\,d\theta}{\delta} \qquad (11\text{-}10)$$

The integral in Eq. (11-10) has been performed in Eq. (9-58); the result is

$$\dot{m} = 1.783\,\frac{D^2\mathscr{D}(C_{mv_\infty} - C_{mv_0})}{\beta}\sqrt{\frac{V_\infty}{\nu D}} \qquad (11\text{-}11)$$

The mass flow rate given by Eq. (11-11) is equal to the difference between the mass flowing past the section at $\theta = \pi/2$ if C_{mv} at that section is equal to C_{mv_∞} and the mass flowing past that section for the value of C_{mv} given by Eq. (11-3). This is expressed as

$$\dot{m} = C_{mv_\infty}\int_0^{\delta_m}2\pi\,\frac{D}{2}\,u\,dy - \int_0^{\delta_m}C_{mv}\,2\pi\,\frac{D}{2}\,u\,dy$$

When this equation is integrated, Eq. (11-11) substituted for $\dot{m}$, and the resulting equation solved for $\nu/\mathscr{D}$, then

$$\frac{\nu}{\mathscr{D}} = \frac{1.288}{\beta^3(1 - 0.07143\beta)} \qquad (11\text{-}12)$$

If $\beta < 1$, Eq. (11-12) may be solved approximately for β as

$$\beta \approx \frac{1.088}{(\nu/\mathscr{D})^{1/3}} = \frac{1.088}{Sc^{1/3}} \qquad (11\text{-}13)$$

Equation (11-11) becomes

$$\dot{m} = 1.64D^2\mathscr{D}(C_{mv_\infty} - C_{mv_0})\,Sc^{1/3}\sqrt{\frac{V_\infty}{\nu D}} \qquad (11\text{-}14)$$

in which the Schmidt number Sc is defined as

$$Sc = \frac{\nu}{\mathscr{D}} \qquad (11\text{-}15)$$

The mass transfer coefficient h_m is given as

$$h_m = \frac{\dot{m}}{\pi D^2(C_{mv_\infty} - C_{mv_0})}\qquad \text{m/s} \qquad (11\text{-}16)$$

The Sherwood number Sh is defined as

$$Sh = \frac{h_m D}{\mathscr{D}} \qquad (11\text{-}17)$$

Combining Eqs. (11-14), (11-16), and (11-17) gives

$$Sh = 0.522 \, Sc^{1/3} \sqrt{\frac{V_\infty D}{\nu}} = 0.522 \, Sc^{1/3} \, Re^{1/2} \qquad (11\text{-}18)$$

The constant in Eq. (11-18) takes into account only the mass transfer to the front half of the drop; the rear half is very difficult to predict due to the wake effects.

An empirical equation for mass transfer to a sphere is given by Geankoplis [1]:

$$Sh = 2 + 0.552 \, Sc^{1/3} \, Re^{0.53} \qquad (11\text{-}19)$$

A comparison between Eqs. (11-18) and (11-19), noting the different exponent for the Reynolds number, suggests that a coefficient of 0.75 would be appropriate for Eq. (11-18):

$$Sh = 0.75 \, Sc^{1/3} \, Re^{1/2} \qquad (11\text{-}20)$$

Using the modified coefficient of Eq. (11-20), Eq. (11-14) becomes

$$\dot{m} = 2.36 D^2 \mathscr{D} (C_{mv_\infty} - C_{mv_0}) \, Sc^{1/3} \sqrt{\frac{V_\infty}{\nu D}} \qquad (11\text{-}21)$$

From Eq. (11-16)

$$h_m = 0.75 \mathscr{D} \, Sc^{1/3} \, Re^{1/2} \qquad (11\text{-}22)$$

The collection efficiency of the drop is the ratio of the mass flow rate into the drop to the rate at which mass of the pollutant flows through a circle whose diameter is that of the drop. This latter rate is

$$\dot{m}_0 = \frac{\pi}{4} D^2 V_\infty C_{mv_\infty} \qquad (11\text{-}23)$$

The collection efficiency is then given by

$$\eta_d = 3 \mathscr{D} \frac{C_{mv_\infty} - C_{mv_0}}{C_{mv_\infty}} \frac{1}{\sqrt{\nu D V_\infty}} \, Sc^{1/3}$$

Since $\mathscr{D} = \nu/Sc$, this equation becomes

$$\eta_d = 3 \frac{C_{mv_\infty} - C_{mv_0}}{C_{mv_\infty}} \frac{1}{Sc^{2/3} \, Re^{1/2}} \qquad (11\text{-}24)$$

Equation (11-24) can be written in terms of the attachment coefficient σ, which was first introduced in Eq. (9-1):

$$\eta_d = \frac{3\sigma}{Sc^{2/3} \, Re^{1/2}} \qquad (11\text{-}25)$$

in which

$$\sigma = \frac{C_{mv_\infty} - C_{mv_0}}{C_{mv_\infty}} \qquad (11\text{-}26)$$

These equations can be used if the value of C_{mv_0} is known, which will be the topic of the next section. If C_{mv_0} can be assumed constant during the motion of the drop, then Eq. (9-71) is valid and the methods of Chap. 9 for the various types of scrubbers may be applied directly. However, if C_{mv_0} is allowed to vary during the scrubbing process, then an additional derivation is necessary.

EXAMPLE 11-1   Ammonia is to be removed from standard air in a scrubber. Assume $C_{mv_\infty} = 0.0007$ and $C_{mv_0} = 0.0001$. Determine the Sherwood number and collection efficiency for the drop. The drop is 1 mm in diameter and has a velocity of 75 m/s relative to the air.

SOLUTION   The Reynolds number is computed to be 4839. From Table 11-1, the diffusion coefficient is 2.2×10^{-5} m²/s for ammonia diffusing in air; the Schmidt number is calculated as

$$Sc = \frac{\nu}{\mathscr{D}} = \frac{1.55 \times 10^{-5}}{2.2 \times 10^{-5}} = 0.705$$

The Sherwood number is computed from Eq. (11-20) as

$$Sh = 0.75\, Sc^{1/3}\, Re^{1/2} = 0.75(0.705)^{1/3}(4839)^{1/2} = 46.4 \qquad Ans.$$

Equation (11-26) gives σ as $(0.0007 - 0.0001)/0.0007 = 0.857$; Eq. (11-25) then gives for the efficiency

$$\eta_d = \frac{3\sigma}{Sc^{2/3}\, Re^{1/2}} = \frac{3(0.857)}{(0.705)^{2/3}(4839)^{1/2}} = 0.0467 \qquad Ans. \qquad ////$$

11-2 HENRY'S LAW AND DIFFUSION INSIDE THE DROP

The value of C_{mv_0} in Eq. (11-26) is related to the amount of pollutant substance absorbed in the drop and the way in which the pollutant is distributed within the drop at the time considered. The relation between C_{mv_0} and the amount of pollutant inside the drop is based on Henry's law, which we shall now discuss.

Henry's Law

Suppose that a drop of liquid (the solvent) is placed in contact with a gas or vapor (the solute). For many, but not all, combinations of gas and liquid, the gas will go into solution in the liquid. If conditions are held constant long enough, the solution will equilibrate and a certain amount of solute will be dissolved in the solvent. For a given combination of solute and solvent and for a given size of drop, the amount of solute contained in the equilibrium solution will increase as the pressure or partial pressure of the solute gas surrounding the drop increases. For sufficiently dilute solutions, this increase is linear with pressure, a relationship known as *Henry's law*. Henry's law is usually expressed as

$$P_p = HC'_n \quad (11\text{-}27)$$

in which P_p is the partial pressure of the pollutant gas or vapor surrounding the drop, H is known as *Henry's law constant*, and C'_n is the mole fraction of the pollutant substance in the drop; the prime indicates a property of the liquid inside the drop.

Although Henry's law is valid only for sufficiently dilute solutions, its range of validity frequently extends up to partial pressures of 1 atm or more. We shall normally deal with partial pressures lower than this and therefore shall assume Henry's law to hold. For operation at about 1 atm pressure, as is usual in air pollution work, with partial pressures of the pollutant below this, it is expected that if Henry's law does not hold, the drop will be too rich in pollutant to be effective. The usual solvent is water, and our notation will reflect this, but on occasion other solvents may be used.

When the relations of Sec. 4-2 are employed, Eq. (11-27) may be written in several alternative forms:

$$C_{mv} = \frac{M_p H C'_n}{R_u T} \quad (11\text{-}28)$$

$$C_{mv} = \frac{H M_w}{R_u T} \frac{C'_m}{1 + (M_w/M_p - 1)C'_m} \quad (11\text{-}29)$$

in which M_p and M_w are the molecular weights of the pollutant and the solvent, respectively, R_u is the universal gas constant, C'_m is the mass fraction of pollutant in the drop, and C_{mv} is the mass-volume concentration of the pollutant vapor in the air. Using the relation

$$C'_{mv} = \rho_d C'_m \quad (11\text{-}30)$$

in which ρ_d is the density of the solution in the drop, which is closely equal to the density of the solvent, Eq. (11-29) becomes

$$C_{mv} = \frac{H M_w}{R_u T} \frac{C'_{mv}}{\rho_d + (M_w/M_p - 1)C'_{mv}} \quad (11\text{-}31)$$

Either Eq. (11-30) or (11-31) can be used as a boundary condition at the surface of the drop, depending on the manner in which concentration within the drop is expressed.

For calculation, we shall need to know values of Henry's law constant. These values are available to a limited extent in the literature (Refs. [2] and [3]). Henry's law constant can sometimes be found from solubility data, which may be expressed in several ways provided Henry's law holds up to a partial pressure of 1 atm. Three forms of solubility are: α is the ratio of the partial volume of the gas, reduced to 0°C and 1.0 atm, dissolved in unit volume of solvent when the partial pressure of the gas is 1.0 atm; β is the same as α except that the partial volume of the gas is measured at 1.0 atm and at the temperature of the solvent; δ is the mass of dissolved gas per unit mass of solvent when the gas is at 1.0 atm. Henry's-law constant can be related to α, β, and δ as

$$H \approx \frac{R_u T^\circ \rho_w}{M_w \alpha} = \frac{1.261 \times 10^8}{\alpha} \qquad (11\text{-}32)$$

$$H \approx \frac{R_u T \rho_w}{M_w \beta} = \frac{461{,}906T}{\beta} \qquad (11\text{-}33)$$

$$H \approx \frac{P^\circ M_p}{M_w \delta} = \frac{5629 M_p}{\delta} \qquad (11\text{-}34)$$

in which $T^\circ = 273$ K and $P^\circ = 1.0$ atm; α, β, and δ are dimensionless, and should also be unitless in the above equations.

Table 11-1 lists values of Henry's law constant for various gases and vapors dissolved in water. The diffusion coefficients are also shown for diffusion in air and in water.

EXAMPLE 11-2 A drop of water is used to scrub SO_2 from standard air. If C_{mv_∞} is 0.0002 kg/m^3, determine σ as a function of C_m for the SO_2 dissolved in the water.

SOLUTION From Table 11-1, the Henry's law constant H is found as 4.85×10^6 N/m^2. Equation (11-29) gives for C_{mv_0}

$$C_{mv_0} = \frac{4.85 \times 10^6 (18)}{8314(298)} \frac{C'_m}{1 + (\frac{18}{64} - 1)C'_m} = \frac{35.24 C'_m}{1 - 0.7188 C'_m}$$

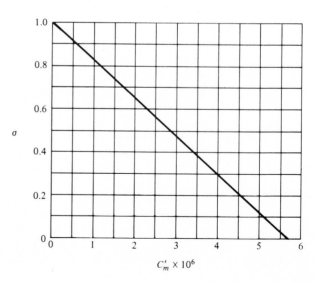

FIGURE 11-3
Variation of σ with C'_m from Example 11-2.

Equation (11-26) then gives for σ

$$\sigma = 1 - \frac{C_{mv_0}}{C_{mv_\infty}}$$

The curve for σ as a function of C'_m is plotted in Fig. 11-3. Note that σ quickly becomes negative as the amount of SO_2 dissolved in the water increases. This type of scrubbing system requires that the water be almost pure. ////

Table 11-1 HENRY'S LAW CONSTANT AND DIFFUSION COEFFICIENTS FOR VARIOUS GASES AND VAPORS

Substance	Formula	Molecular weight	Henry's law constant,* N/m²	Diffusion coefficient	
				Air,† 10^{-5} m²/s	Water,‡ 10^{-9} m²/s
Acetylene	C_2H_2	26.0	1.35×10^8	1.7	2.0
Air		29.0	7.29×10^9		2.0
Ammonia	NH_3	17.0	303,000	2.2	2.0
Argon	Ar	79.8	3.61×10^9	1.7	1.4
Bromine	Br_2	159.8	7.47×10^6	1.0	1.3
Carbon dioxide	CO_2	44.0	1.65×10^8	1.5	2.0
Carbon monoxide	CO	28.0	5.87×10^9	2.0	2.0
Carbonyl sulfide	COS	60.1	2.63×10^8	1.3	1.5
Chlorine	Cl_2	70.9	6.82×10^7	1.2	1.5
Ethane	C_2H_6	30.0	3.07×10^9	1.5	1.4
Ethylene	C_2H_4	28.0	1.16×10^9	1.6	1.5
Gasoline			7.33×10^9		0.7
Helium	He	4.0	1.27×10^{10}	7.0	6.0
Hydrogen	H_2	2.0	7.07×10^9	7.5	4.8
Hydrogen cyanide	HCN	27.1	6.4×10^5	1.5	1.8
Hydrogen sulfide	H_2S	34.1	5.52×10^7	1.7	1.6
Krypton	Kr	83.8	2.20×10^9	1.5	2.0
Methane	CH_4	16.0	4.19×10^9	2.2	1.8
Methyl chloride	CH_3Cl	50.5	1.49×10^5	1.3	1.5
Neon	Ne	20.2	8.4×10^9	3.2	3.0
Nitric oxide	NO	30.0	2.91×10^9	2.0	2.4
Nitrogen	N_2	28.0	8.40×10^9	2.0	1.9
Nitrous oxide	N_2O	44.0	2.27×10^8	1.5	1.8
Oxygen	O_2	32.0	4.44×10^9	2.0	2.5
Ozone	O_3	48.0	4.64×10^8		2.0
Phosphine	PH_3	34.0	5.71×10^9	1.6	
Propylene	C_3H_6	42.0	5.73×10^8		1.1
Sulfur dioxide	SO_2	64.1	4.85×10^6	1.3	1.7
Water	H_2O	18.0		2.6	
Xenon	Xe	131.3	1.28×10^9	1.3	1.7

* These values were computed from: National Research Council, "International Critical Tables," vol. 3, pp. 255–261, McGraw-Hill Book Company, New York, 1928.

† These values were computed by approximate methods.

‡ These values were taken from: R. C. Reid and T. K. Sherwood, "The Properties of Gases and Liquids, Their Estimation and Correlation," pp. 554–555, McGraw-Hill Book Company, New York, 1966; R. H. Perry, C. H. Chilton, and S. D. Kirkpatrick, eds., "Chemical Engineers Handbook," 4th ed., pp. 14–22 and 14–23, McGraw-Hill Book Company, New York, 1963; and from values computed by approximate methods.

Effect of the Distribution of Concentration within the Drop

When the drop is first injected into the airstream, its concentration of pollutant is uniformly distributed. As the scrubbing process proceeds, however, and pollutant is absorbed in the drop, the concentration becomes higher at the surface than at points inside the drop. The effect is to retard the absorption of the pollutant as the process continues.

It is possible to derive an equation for the concentration within the drop as a function of radial position and time under the boundary condition which exists at the surface of the drop. This boundary condition is approximately expressed as

$$\frac{\partial C'_{mv}}{\partial r}\left(\frac{D}{2}, t\right) = 0.75 \frac{\mathscr{D}}{\mathscr{D}'} \, \text{Sc}^{1/3} \sqrt{\frac{V_\infty}{vD}} \left[C_{mv\infty} - \frac{HM_w}{R_u T \rho_d} C'_{mv}\left(\frac{D}{2}, t\right)\right] \quad (11\text{-}35)$$

A simpler and sufficiently accurate solution is obtained if the concentration at the surface of the drop is assumed to reach immediately the value corresponding to C_{mv_s}, as expressed by Henry's law, Eq. (11-31). The solution is then adjusted slightly to allow for the mass flow rate into the drop as given by Eq. (11-21). The details of this solution are omitted here; the general solution is

$$C'_{mv}(r, t) = C'_{mv_s} + \frac{D}{\pi}(C'_{mv_s} - C'_{mv_i})\frac{1}{r}\sum_{n=1}^{\infty}\frac{(-1)^n \sin 2n\pi r/D}{n} \exp\left(-4\pi^2 n^2 \mathscr{D}'\frac{t}{d^2}\right)$$

$$(11\text{-}36)$$

In Eq. (11-36), C'_{mv_s} is the concentration inside the drop at its surface. Combining Eqs. (11-1), (11-21), (11-31), and (11-36) and solving for C_{mv_0} gives

$$\frac{C_{mv_0}}{C_{mv_s}} = \frac{\mathscr{D} \, \text{Sc}^{1/3} \, \text{Re}^{1/2} + 5.32\mathscr{D}'(C'_{mv_i}/C_{mv\infty})\sum_{n=1}^{\infty} \exp\left(-4\pi^2 n^2 \mathscr{D}' \dfrac{t}{D^2}\right)}{\mathscr{D} \, \text{Sc}^{1/3} \, \text{Re}^{1/2} + 5.32\mathscr{D}' \dfrac{R_u T \rho_d}{HM_w}\sum_{n=1}^{\infty} \exp\left(-4\pi^2 n^2 \mathscr{D}' \dfrac{t}{D^2}\right)} \quad (11\text{-}37)$$

Let us substitute Eq. (11-37) into Eq. (11-26), after first defining Fo′ and ϕ(Fo′):

$$\text{Fo}' = \frac{\mathscr{D}'t}{D^2} \quad (11\text{-}38)$$

$$\phi(\text{Fo}') = 5.32\sum_{n=1}^{\infty} e^{-4\pi^2 n^2 \, \text{Fo}'} \quad (11\text{-}39)$$

The resulting equation for σ is

$$\sigma = 1 - \frac{\mathscr{D} \, \text{Sc}^{1/3} \, \text{Re}^{1/2} + \mathscr{D}'(C'_{mv_i}/C_{mv\infty})\phi(\text{Fo}')}{\mathscr{D} \, \text{Sc}^{1/3} \, \text{Re}^{1/2} + \mathscr{D}'(R_u T \rho_d/HM_w)\phi(\text{Fo}')} \quad (11\text{-}40)$$

Table 11-2 (p. 474) gives values of ϕ(Fo′) as a function of Fo′. The quantity Fo′ is a mass-transfer Fourier number for the region inside the drop.

EXAMPLE 11-3　A drop of water is used to scrub SO_2 from standard air. Assume that C'_{mv_i} is zero, that V_∞ is 75 m/s, and that $C_{mv\infty}$ is 0.0002 kg/m^3.

Compute and plot σ and η_d as functions of time for diameters of 0.1, 0.5, 1, and 2 mm.

SOLUTION   From Table 11-1, $\mathscr{D} = 1.3 \times 10^{-5}$ m²/s, $\mathscr{D}' = 1.7 \times 10^{-9}$ m²/s, and $H = 4.85 \times 10^6$ N/m². Then

$$\text{Sc} = \frac{\nu}{\mathscr{D}} = \frac{1.55 \times 10^{-5}}{1.3 \times 10^{-5}} = 1.19$$

$$\text{Re} = \frac{V_\infty D}{\nu} = \frac{75}{1.55 \times 10^{-5}} D = 4.84 \times 10^6 D$$

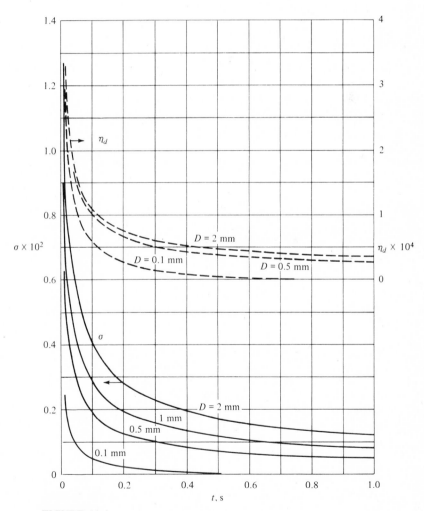

FIGURE 11-4
Variation of attachment coefficient σ and collection efficiency η_d with time for the data of Example 11-3.

From Eq. (11-38), $Fo' = \mathscr{D}'t/D^2 = 1.7 \times 10^{-9}t/D^2$. Equation (11-39) or Table 11-2 gives $\phi(Fo')$. After evaluating the various coefficients, Eq. (11-40) becomes

$$\sigma = 1 - \frac{1}{1 + 0.003497\phi\,(Fo')/\sqrt{Re}}$$

Equation (11-25) gives η_d as

$$\eta_d = \frac{3\sigma}{Sc^{2/3}\,Re^{1/2}} = 2.672\,\frac{\sigma}{\sqrt{Re}}$$

The preceding equations for σ and η_d have been evaluated, and the results are plotted in Fig. 11-4 (p. 473).   ////

Effect of Chemical Reaction within the Drop

Suppose a chemical reaction takes place within the drop in which the dissolved pollutant reacts with a reagent in aqueous solution to form certain products which then remain in aqueous solution themselves. Although the reaction can be either fast or slow, we shall consider only the case of a fast reaction. For such a fast reaction, there will be a zone near the surface of the drop throughout which the reagent will be expended; there will also be a reaction front at which the reaction is occurring at any given time. This situation is shown in Fig. 11-5, where the radius of the reaction front is r_0.

The pollutant diffuses across the reagent-free zone from the surface of the drop to the reaction front. The reagent diffuses from the interior of the drop toward the reaction front; at radius r_1 the reagent has essentially its initial concentration within the drop. At the reaction front the concentrations of both

Table 11-2 VALUES OF $\phi(Fo')$ AS DEFINED BY EQ. (11-39)

Fo'	$\phi(Fo')$	Fo'	$\phi(Fo')$
0.00001	234.6	0.007	6.309
0.00005	103.5	0.008	5.729
0.0001	72.38	0.009	5.250
0.0005	30.90	0.01	4.844
0.0006	27.97	0.02	2.646
0.0007	25.70	0.03	1.674
0.0008	23.87	0.04	1.106
0.0009	22.35	0.05	0.7410
0.001	21.07	0.06	0.4984
0.002	14.12	0.07	0.3356
0.003	11.04	0.08	0.2261
0.004	9.204	0.09	0.1524
0.005	7.952	0.1	0.1027
0.006	7.027	0.2	0.001981

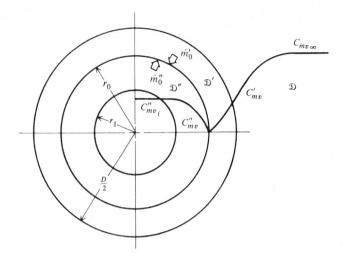

FIGURE 11-5
Quasi-steady concentration distribution in a drop with chemical reaction occurring.

the pollutant and the reagent are zero. The chemical reaction occurring at the front follows the general equation

$$\iota'R' + \iota''R'' \rightarrow \iota_3 R_3 + \iota_4 R_4 \qquad (11\text{-}41)$$

in which R' is the pollutant, R'' is the reagent, and R_3 and R_4 are the products formed. The coefficients ι', ι'', ι_3, and ι_4 are the relative number of moles of these respective constituents entering into the reaction. We shall be interested only in the left side of Eq. (11-41), and the ratio of the mole numbers is of particular importance; this ratio ι is defined as

$$\iota = \frac{\iota'}{\iota''} \qquad (11\text{-}42)$$

Let us predict the effect which the chemical reaction has upon the collection efficiency of the drop. Our analysis will be an approximate one based on a quasi-steady state. Assume that at any given time the steady concentration distribution holds even though that distribution changes with time. Since we shall be dealing with a steady distribution, Eq. (11-2) may be written

$$\frac{\partial^2 C_{mv}}{\partial r^2} + \frac{2}{r}\frac{\partial C_{mv}}{\partial r} = 0 \qquad (11\text{-}43)$$

The solution to this differential equation is of the form

$$C_{mv} = C_1 + \frac{C_2}{r} \qquad (11\text{-}44)$$

and applies to both the pollutant C'_{mv} and the reagent C''_{mv}.

Equation (11-35) is one boundary condition satisfied by the two concentrations. The other three boundary conditions are

$$C'_{mv}(r_0, t) = 0 \qquad C''_{mv}(r_0, t) = 0 \qquad C''_{mv}(r_1, t) = C''_{mv_i} \qquad (11\text{-}45)$$

An additional relation occurs at the reaction front. The ratio of the mass flow rate of pollutant to that of the reagent at the front is given as

$$\frac{\dot{m}'_0}{\dot{m}''_0} = \frac{\iota' M'}{\iota'' M''} = \iota \frac{M'}{M''} \qquad (11\text{-}46)$$

in which Eq. (11-42) has been used. Applying Eq. (11-1) to the pollutant and reagent gives

$$\dot{m}'_0 = 4\pi r_0{}^2 \mathscr{D}' \frac{\partial C'_{mv}}{\partial r}(r_0, t) \qquad (11\text{-}47)$$

$$\dot{m}''_0 = -4\pi r_0{}^2 \mathscr{D}'' \frac{\partial C''_{mv}}{\partial r}(r_0, t) \qquad (11\text{-}48)$$

Substituting Eqs. (11-47) and (11-48) into Eq. (11-46) gives

$$\frac{\partial C'_{mv}}{\partial r}(r_0, t) = -\iota \frac{M'}{M''} \frac{\mathscr{D}''}{\mathscr{D}'} \frac{\partial C''_{mv}}{\partial r}(r_0, t) \qquad (11\text{-}49)$$

The concentrations C'_{mv} and C''_{mv} can be evaluated by applying Eqs. (11-35) and (11-45) to Eq. (11-44), resulting in

$$C'_{mv} = \frac{0.75(\mathscr{D}/\mathscr{D}')\,\mathrm{Sc}^{1/3}\sqrt{V_\infty/vD}\,C_{mv_\infty}(1 - r_0/r)}{4r_0/D^2 + 0.75(\mathscr{D}/\mathscr{D}')\,\mathrm{Sc}^{1/3}\sqrt{V_\infty/vD}\,(HM_w/R_u T\rho_d)(1 - 2r_0/D)} \qquad (11\text{-}50)$$

$$C''_{mv} = C''_{mv_i}\frac{1 - r_0/r}{1 - r_0/r_1} \qquad (11\text{-}51)$$

Equations (11-50) and (11-51) can be substituted back into Eq. (11-49) and the result solved for r_0 in terms of r_1. This equation for r_0 can then be substituted into Eq. (11-49) and the result substituted into Eq. (11-47) to give

$$\dot{m}' = \frac{4\pi \mathscr{D}'(\alpha C_{mv_\infty} + \beta \alpha \alpha_H C''_{mv_i})r_1}{4r_1/D^2 + \alpha \alpha_H(1 - 2r_1/D)} = \dot{m}'_{cr} \qquad (11\text{-}52)$$

in which

$$\alpha = 0.75 \frac{\mathscr{D}}{\mathscr{D}'}\,\mathrm{Sc}^{1/3}\sqrt{\frac{V_\infty}{vD}} \qquad \alpha_H = \frac{HM_w}{R_u T\rho_d} \qquad \beta = \iota \frac{M'}{M''} \frac{\mathscr{D}''}{\mathscr{D}'} \qquad (11\text{-}53)$$

and in which $\dot{m}'_{cr}$ is the mass flow rate of pollutant for the case in which a chemical reaction occurs.

Let us compare this with a corresponding case in which no chemical reaction takes place. This situation is shown in Fig. 11-6; the value of C'_{mv_i} is taken as zero, which is in closest agreement to the case of a chemical

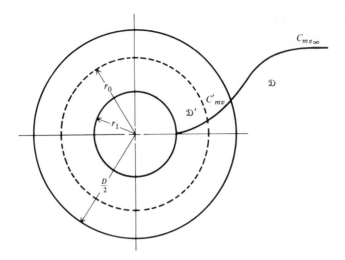

FIGURE 11-6
Quasi-steady concentration distribution in a drop without chemical reaction.

reaction occurring. The mass flow rate of pollutant for this case is given by the following equation:

$$\dot{m}'_{ncr} = \frac{4\pi r_1 \alpha C_{mv_\infty} \mathscr{D}'}{4r_1/D^2 + \alpha\alpha_H(1 - 2r_1/D)} \qquad (11\text{-}54)$$

From Eq. (11-26) we may write for the attachment coefficients σ_{cr} in the case of a chemical reaction and σ_{ncr} in the case of no chemical reaction:

$$\frac{\sigma_{cr}}{\sigma_{ncr}} = \frac{C_{mv} - (C_{mv_0})_{cr}}{C_{mv} - (C_{mv_0})_{ncr}} = \frac{\dot{m}'_{cr}}{\dot{m}'_{ncr}} \qquad (11\text{-}55)$$

Substituting Eqs. (11-52) and (11-54) into Eq. (11-55) gives

$$\frac{\sigma_{cr}}{\sigma_{ncr}} = 1 + \beta\alpha_H \frac{C''_{mv_i}}{C_{mv_\infty}} \qquad (11\text{-}56)$$

Substituting Eqs. (11-53) gives

$$\frac{\sigma_{cr}}{\sigma_{ncr}} = 1 + \iota\frac{M'}{M''}\frac{\mathscr{D}''}{\mathscr{D}'}\frac{HM_w}{R_u T \rho_d}\frac{C''_{mv_i}}{C_{mv_\infty}} \qquad (11\text{-}57)$$

This equation can be solved for σ_{cr}. Equation (11-40) gives σ_{ncr}, after which the collection efficiency η_d can be obtained from Eq. (11-25).

For very short times, the value of σ_{cr} given by Eq. (11-57) can exceed 1.0. This is impossible and indicates that the solution is not valid for such short times. It is doubtful that any problem will arise of this nature, but if it should, a simple solution is to set σ equal to 1.0 in all cases where Eq. (11-57) indicates a value of $\sigma > 1.0$.

EXAMPLE 11-4   Use the data of Example 11-3 except that potassium sulfide is present in the water at an initial concentration of 0.1 kg/m^3. The chemical reaction involved is

$$SO_2 + K_2SO_3 + H_2O \rightarrow 2KHSO_3$$

Assume the diffusion coefficient $\mathscr{D}'' = 1.2 \times 10^{-9} \text{ m}^2/\text{s}$ for potassium sulfide in water. Determine how σ and η_d are related to the respective values from Example 11-3 when no chemical reaction is present.

SOLUTION   From the chemical equation, $\iota = 1$. The molecular weights are $M_w = 18$, $M' = 64$, and $M'' = 158$. The diffusion coefficients are $\mathscr{D}' = 1.7 \times 10^{-9}$ m^2/s and $\mathscr{D}'' = 1.2 \times 10^{-9} \text{ m}^2/\text{s}$. Equation (11-57) gives

$$\frac{\sigma_{cr}}{\sigma_{ncr}} = 1 + 1.0 \frac{64}{158} \frac{1.2 \times 10^{-9}}{1.7 \times 10^{-9}} \frac{4.85 \times 10^6 (18)}{8314(298)(1000)} \frac{0.1}{0.0002} = 6.04$$

Thus

$$\sigma_{cr} = 6.04 \sigma_{ncr} \qquad \eta_{d_{cr}} = 6.04 \eta_{d_{ncr}} \qquad ////$$

11-3   GAS SCRUBBERS

As we have seen in Chap. 9, in a scrubber a large number of liquid drops are injected into the airstream. As the air flows around these drops, absorption of the pollutant onto the drop surface takes place. In analyzing the performance of a scrubber, the behavior of a large number of drops, acting together and in sequence, must be considered. Much of this analysis has already been performed in Chap. 9, and we may use the equations of that chapter along with the equations for the collection efficiency of an individual drop developed in this chapter.

When a parcel of air encounters a series of n drops, one after the other, the resulting overall collection efficiency is given by Eq. (9-71):

$$\eta = 1 - (1 - \eta_d)^n \qquad (11\text{-}58)$$

This equation is just as useful for gaseous pollutants as for particulates. Equations (9-89) and (9-90) for pressure drop and power requirement developed in Chap. 9 apply to the respective scrubber types when used for gaseous pollutants; these equations will not be repeated in this chapter.

Spray Chambers

The spray chamber of Sec. 9-6 can be used to remove gaseous pollutants, although its overall collection efficiency is expected to be rather low. Equations (9-84) and (9-85) apply:

$$n = \frac{1.5 \dot{m}_s L}{\rho_d D} \left[\frac{1}{Q} + \frac{1}{A_{sc}(V_\infty - V_a)} \right] \qquad (11\text{-}59)$$

$$n = \frac{0.0015 \dot{m}_s L}{D} \left[\frac{1}{Q} + \frac{1}{A_{sc}(V_\infty - V_a)} \right] \qquad (11\text{-}60)$$

Equation (11-60) is valid if water is the scrubbing fluid; it is also a good approximation for dilute aqueous solutions.

EXAMPLE 11-5 A spray chamber is to be used to remove SO_2 from standard air. The pollutant concentration C_{mv_∞} is 0.001 kg/m³, while the drops are formed from initially pure water. The drop diameter is 1 mm, the air flow rate is 9.5 m³/s, and the spray chamber is 2 m in diameter and 6 m long. Neglect the increase in concentration of SO_2 in the drops. Compute the overall collection efficiency of the spray chamber for a range of water flow rates.

SOLUTION The cross-sectional area of the spray chamber is 3.14 m², and the air velocity is 3.02 m/s. Equations (3-51) and (3-29) give the terminal velocity of the drop, which is equal to V_∞, as 4.33 m/s, corresponding to a Reynolds number $Re_t = 280$. Equation (11-60) gives

$$n = \frac{0.0015(6)\dot{m}_s}{0.001}\left[\frac{1}{9.5} + \frac{1}{3.14(4.33 - 3.02)}\right] = 3.13\dot{m}_s$$

In terms of the volumetric flow rate of water, this is

$$n = 3130Q_s$$

Using Sc = 1.19 from Example 11-3, Eq. (11-25) gives

$$\eta_d = \frac{3(1.0)}{(1.19)^{2/3}\sqrt{280}} = 0.160$$

in which σ has been set equal to 1.0 in accordance with the assumption that the concentration of SO_2 in the drop remains negligibly small. Equation (11-58) then gives

$$\eta = 1 - (0.84)^n$$

Some calculations give the following results:

Q_s, m³/s	η
0.001	0.421
0.005	0.935
0.01	0.996
0.02	0.99998

////

The results of Example 11-5 look very favorable—indeed too favorable. The neglect of the increase of concentration at the surface of the drop as the drop absorbs SO_2 is at fault. In the next example, this buildup of SO_2 at the drop surface is taken into account, with very different results.

EXAMPLE 11-6 Repeat Example 11-5 but take into account the increase in concentration of the SO_2 within the drop during its fall.

SOLUTION The drop efficiency η_d decreases continuously as the drop falls; Fig. 11-4 shows that the efficiency drops rapidly at first but then levels off. In light of this behavior, we shall compute the efficiency at the end of the drop's fall and assume that this efficiency applies throughout the fall. The time of drop exposure is computed as

$$t = \frac{L}{V_\infty - V_a} = \frac{6.0}{4.33 - 3.02} = 4.58 \text{ s}$$

The Fourier number Fo' becomes

$$\text{Fo}' = \frac{\mathscr{D}'t}{D^2} = \frac{1.7 \times 10^{-9}(4.85)}{(0.001)^2} = 0.00825$$

Table 11-2 then gives $\phi(Fo')$ as 5.609. The Schmidt number, from Example 11-3, is 1.19; and the Reynolds number, from Example 11-5, is 280.

Equation (11-40) then gives

$$\sigma = 1 - \frac{\mathscr{D} \, \text{Sc}^{1/3} \, \text{Re}^{1/2}}{\mathscr{D} \, \text{Sc}^{1/3} \, \text{Re}^{1/2} + \mathscr{D}'(R_u T \rho_d / H M_w)\phi(Fo')}$$

$$= 1 - \frac{0.0002305}{0.0002305 + 1.7 \times 10^{-9}(8314)(298)(1000)(5.609)/[4.85 \times 10^6(18)]}$$

$$= 0.001244$$

Equation (11-25) gives

$$\eta_d = \frac{3(0.001244)}{(1.19)^{2/3}\sqrt{280}} = 0.0001987$$

Equation (11-58) gives the overall efficiency as

$$\eta = 1 - 0.9998^n$$

From Example 11-5, $n = 3130Q_s$. The results are

Q_s, m^3/s	η
0.001	0.000626
0.01	0.00624
0.1	0.0607
1.0	0.465
2.0	0.714
4.0	0.918

////

The results of Example 11-6, although not favorable, seem more realistic than those of Example 11-5. The use of chemical reactions or higher-energy devices may give improved efficiency.

Collection by an Accelerating Drop

The collection of particulates by an accelerating drop was treated in Sec. 9-7. The collection of gaseous pollutants by an accelerating drop is handled similarly. Equations (9-91) to (9-94) still apply to the motion of the drop; Table 9-4 gives time and distance as functions of the Reynolds number of the drop. However, Table 9-3 must be replaced by a table which makes use of the proper efficiency expression for the collection of gases. An equation analogous to Eq. (9-95) is

$$\Delta m = \frac{\pi D^2}{4} C_{mv_\infty} \sigma \int_0^{S_{max}} \frac{\eta_d}{\sigma} ds' \qquad (11\text{-}61)$$

Writing $\dot{N}_d$ as the rate of formation of drops, letting $\Delta\dot{m}$ be the rate at which pollutant is removed from the air, and substituting Eq. (11-25) for η_d, we have

$$\Delta\dot{m} = \frac{3\pi D^2}{4} \frac{C_{mv_\infty}\sigma\dot{N}_d}{Sc^{2/3}} \int_0^{S_{max}} \frac{ds'}{\sqrt{Re}} \qquad (11\text{-}62)$$

The efficiency for the stage is given by $\eta = \Delta\dot{m}/C_{mv_\infty}Q$, and Eq. (9-87) gives for the drop formation rate $\dot{N}_d = 6Q_s/\pi D^3$. Using these relations, Eq. (11-62) can be written in the following alternative forms:

$$\frac{\Delta\dot{m}\rho\,Sc^{2/3}}{\sigma\dot{N}_d C_{mv_\infty}\rho_d D^3} = \frac{\eta\rho Q\,Sc^{2/3}}{\sigma\dot{N}_d\rho_d D^3} = \frac{\pi}{6}\frac{\eta\rho Q\,Sc^{2/3}}{\sigma\rho_d Q_s} = \frac{3\pi}{4}\int_0^{Re_{max}}\frac{1}{\sqrt{Re}}d\left(\frac{\rho s'}{\rho_d D}\right) \qquad (11\text{-}63)$$

Equation (11-63) may be integrated numerically in a form similar to Eq. (9-99), giving

$$\left(\frac{\eta\rho Q\,Sc^{2/3}}{\sigma\dot{N}_d\rho_d D^3}\right)_{n+1} = \left(\frac{\eta\rho Q\,Sc^{2/3}}{\sigma\dot{N}_d\rho_d D^3}\right)_n + \frac{3\pi}{8}\left(\frac{1}{\sqrt{Re_{n+1}}}+\frac{1}{\sqrt{Re_n}}\right)\left[\left(\frac{\rho s'}{\rho_d D}\right)_{n+1} - \left(\frac{\rho s'}{\rho_d D}\right)_n\right] \qquad (11\text{-}64)$$

Equation (11-64) is solved using Eqs. (9-91) to (9-94), and the results are tabulated in Table 11-3.

Multistage Spray Scrubber

Equation (9-105) gives the overall collection efficiency for the scrubber shown in Fig. 9-18 in terms of the efficiencies of the individual stages. The efficiency for each stage can be computed using the methods of this chapter. For simple gas scrubbing, the efficiency is the same for each stage, assuming that each is constructed alike. If a chemical reaction is present, then the efficiency must be computed separately for each stage. The next example illustrates the procedure involved if no chemical reaction is present.

EXAMPLE 11-7   Solve Example 9-12 for a scrubber used to remove SO_2 from standard air. Assume C_{mv_∞} is initially 0.001 kg/m³, the drop diameter is 1.0 mm, the air velocity is 5.0 m/s, and the water is pure when injected.

SOLUTION   From Table 11-1, $H = 4.85 \times 10^6$ N/m², $\mathscr{D} = 1.3 \times 10^{-5}$ m²/s, and $\mathscr{D}' = 1.7 \times 10^{-9}$ m²/s. The Reynolds number of the drop at the point of injection is $Re_0 = V_0 D/\nu = 5.0(10^{-3})/1.55 \times 10^{-5} = 323$. The Schmidt number $Sc = 1.19$. The length of the stage is 2.5 m. At the Reynolds number of 323, Table 9-4 gives

$$\left(\frac{t\mu}{\rho_d D^2}\right)_0 = 0.008640 \qquad \left(\frac{\rho s'}{\rho_d D}\right)_0 = 4.18$$

At the end of the stage, the drop will have accelerated to almost the air velocity; the Reynolds number of the drop relative to the air will be small at that point. To evaluate this Reynolds number at the end of the stage, first compute

$$\left(\frac{\rho s'}{\rho_d D}\right)_L = \left(\frac{\rho s'}{\rho_d D}\right)_0 - \frac{\rho L}{\rho_d D} = 4.18 - \frac{1.185(2.5)}{1000(0.001)} = 1.218$$

This value can be found in Table 9-4 at a Reynolds number of 49.4. The value of $(t\mu/\rho_d D^2)_L$ at the end of the stage is 0.03204 from the same table. We

Table 11-3   GASEOUS COLLECTION RATE FOR AN ACCELERATING DROP

Re	$\eta\rho Q\, Sc^{2/3}/\sigma N_d \rho_d D^3$	Re	$\eta\rho Q\, Sc^{2/3}/\sigma N_d \rho_d D^3$
0	0	1700	1.710
10	0.3973	1800	1.714
20	0.5608	1900	1.718
40	0.7391	2000	1.721
60	0.8538	2250	1.729
80	0.9407	2500	1.736
100	1.011	2750	1.741
200	1.242	3000	1.746
300	1.375	3250	1.751
400	1.466	3500	1.755
500	1.533	3750	1.758
600	1.586	4000	1.761
700	1.628	4250	1.764
800	1.643	4500	1.767
900	1.655	4750	1.769
1000	1.665	5000	1.771
1100	1.674	6000	1.779
1200	1.682	7000	1.785
1300	1.689	8000	1.789
1400	1.695	9000	1.793
1500	1.701	10,000	1.796
1600	1.706		

then solve for t_L, the time required for the drop to traverse the length of the stage during its acceleration process:

$$t_L = (0.03204 - 0.00864)\frac{1000(10^{-6})}{1.84 \times 10^{-5}} = 1.27 \text{ s}$$

which gives a Fourier number Fo' $= \mathscr{D}'t_L/D^2 = 0.00216$; Table 11-2 shows that $\phi(\text{Fo}')$ is 13.63.

Equation (11-40) gives the attachment coefficient σ as

$$\sigma = 1 - \frac{0.00009682}{0.00009682 + 6.58 \times 10^{-7}} = 0.006746$$

Table 11-3 gives the two values of the efficiency parameter as

$$\left(\frac{\eta\rho Q \text{ Sc}^{2/3}}{\sigma \dot{N}_d \rho_d D^3}\right)_0 = 1.396 \qquad \left(\frac{\eta\rho Q \text{ Sc}^{2/3}}{\sigma \dot{N}_d \rho_d D^3}\right)_L = 0.7930$$

Taking the difference between these values and using one of Eqs. (11-63), we have

$$\frac{\pi}{6}\frac{\eta\rho Q \text{ Sc}^{2/3}}{\sigma\rho_d Q_s} = 1.396 - 0.793 = 0.603$$

Solving this equation for η_i,

$$\eta_i = \frac{0.603(6)(0.006746)(1000)Q_s/Q}{1.185(1.19)^{2/3}\pi} = 5.838\frac{Q_s}{Q}$$

(a) For four stages and $Q_s/Q = 0.0001$ in each stage, the overall efficiency is given by Eq. (9-105):

$$\eta = 1 - (1 - \eta_i)^4 = 1 - (1 - 0.000584)^4 = 0.00233 \qquad Ans.$$

(b) To determine the number of stages required to give 0.99 efficiency, solve Eq. (9-105) for s:

$$s = \frac{\ln(1 - \eta)}{\ln(1 - \eta_i)} = \frac{\ln 0.01}{\ln(1 - 0.000584)} = 7883 \qquad Ans.$$

For this many stages, $Q_s/Q = 0.0001(7883) = 0.788.$ $\quad Ans.$

(c) To obtain an efficiency of 0.99 with just one stage, the water flow rate is computed as

$$\eta_i = 0.99 = 5.838\frac{Q_s}{Q}$$

$$\frac{Q_s}{Q} = \frac{0.99}{5.838} = 0.170 \qquad Ans. \qquad ////$$

We see that this scrubber is impracticable; with a chemical reaction it should be feasible, as may be explored in Prob. 11-33 at the end of the chapter.

Cyclone Scrubber

The cyclone scrubber for use with gaseous pollutants is analyzed in a manner very similar to that of Sec. 9-8 for the cyclone scrubber used to collect particulates. For the motion of the drop, Fig. 9-21 applies and is used in the same way as in the first part of Example 9-14. The quantity $\eta\rho Q\,\mathrm{Sc}^{2/3}/\sigma\dot{N}_d\rho_d D^3$ is taken from Table 11-3 as the difference between the entries at Re_0 and at Re_e. Next, let us derive equations for the collection efficiency of the scrubber.

An equation similar to Eq. (9-131) is obtained:

$$\frac{dC_{mv}}{C_{mv}} = -\frac{6}{\pi}\frac{Q_s}{Q}\frac{\sigma\rho_d}{\rho\,\mathrm{Sc}^{2/3}}\frac{\eta\rho Q\,\mathrm{Sc}^{2/3}}{\sigma\dot{N}_d\rho_d D^3}\frac{dx}{L} \qquad (11\text{-}65)$$

Using Eq. (11-57) along with the notation of Eq. (11-53), we have

$$\frac{\sigma_{cr}}{\sigma_{ncr}} = 1 + \beta\alpha_H\frac{C''_{mv_i}}{C_{mv}} \qquad (11\text{-}66)$$

Let us also define

$$\beta_1 = \frac{6}{\pi}\frac{Q_s}{Q}\frac{\sigma_{ncr}\rho_d}{\rho\,\mathrm{Sc}^{2/3}}\frac{\eta\rho Q\,\mathrm{Sc}^{2/3}}{\sigma\dot{N}_d\rho_d D^3} \qquad (11\text{-}67)$$

Equation (11-65) then becomes

$$\frac{dC_{mv}}{C_{mv} + \beta\alpha_H C''_{mv_i}} = -\beta_1\frac{dx}{L} \qquad (11\text{-}68)$$

Equation (11-68) integrates to the following equation if we use the initial condition that when $x = 0$, $C_{mv} = C_{mv_0}$:

$$C_{mv} = -\beta\alpha_H C''_{mv_i} + (C_{mv_0} + \beta\alpha_H C''_{mv_i})e^{-\beta_1 x/L} \qquad (11\text{-}69)$$

If the value of the attachment coefficient σ_{cr} does not exceed unity by the end of the scrubber, at which $x = L$, then Eq. (11-69) yields the efficiency:

$$\eta = \left(1 + i\frac{M'}{M''}\frac{\mathscr{D}''}{\mathscr{D}'}\frac{HM_w}{R_u T\rho_d}\frac{C''_{mv_i}}{C_{mv_0}}\right)\left\{1 - \exp\left[-\frac{6}{\pi}\frac{Q_s}{Q}\frac{\sigma_{ncr}\rho_d}{\rho\,\mathrm{Sc}^{2/3}}\left(\frac{\eta\rho Q\,\mathrm{Sc}^{2/3}}{\sigma\dot{N}_d\rho_d D^3}\right)\right]\right\} \qquad (11\text{-}70)$$

In the event that σ_{cr} exceeds unity at the end of the scrubber, then Eq. (11-70) is not valid. This situation may occur because, as Eq. (11-66) shows, the value of σ_{cr} increases along the scrubber with increasing value of x as the value of C_{mv} drops. Combining Eqs. (11-66) and (11-69) and setting σ_{cr} equal to unity gives

$$\frac{C_{mv_1}}{C''_{mv_i}} = \frac{\beta\alpha_H\sigma_{ncr}}{1 - \sigma_{ncr}} \qquad (11\text{-}71)$$

in which C_{mv_1} is the value of C_{mv} at position x_1 at which $\sigma_{cr} = 1$. Using the initial condition that when $x = x_1$, $C_{mv} = C_{mv_1}$, Eq. (11-68) integrates to

$$\frac{C_{mv}}{C''_{mv_i}} = -\beta\alpha_H\left(1 - \frac{e^{-\beta_1(x-x_1)/L}}{1 - \sigma_{ncr}}\right) \qquad (11\text{-}72)$$

in which Eq. (11-71) has been substituted. Equation (11-69) can be solved for x_1:

$$x_1 = \frac{L}{\beta_1}\ln\left[\left(1 + \frac{C_{mv_0}}{\beta\alpha_H\,C''_{mv_i}}\right)(1 - \sigma_{ncr})\right] \qquad (11\text{-}73)$$

Equation (11-72) gives the collection efficiency; after substitution of Eq. (11-73), the final expression is

$$\eta = 1 + \beta\alpha_H\left[1 - \left(1 + \frac{C_{mv_0}}{\beta\alpha_H\,C''_{mv_i}}\right)e^{-\beta_1}\right]\frac{C''_{mv_i}}{C_{mv_0}} \qquad (11\text{-}74)$$

in which β_1 is given by Eq. (11-67) and α_H and β are given by

$$\alpha_H = \frac{HM_w}{R_u\,T\rho_d} \qquad \beta = \iota\,\frac{M'\,\mathscr{D}''}{M''\,\mathscr{D}'}$$

The detailed application of these equations to a scrubber is illustrated in the following example.

EXAMPLE 11-8   A cyclone scrubber is used to remove SO_2 from 100 m³/s of standard air. The radius of the scrubber is 4.0 m, the width $W = 1.0$ m, and $C_{mv} = 0.001$ kg/m³ at the entrance to the scrubber. The flow rate of spray fluid is 0.5 m³/s, and the drop diameter is 0.7 mm. The fluid consists of water and potassium sulfide with an initial concentration of 10 kg/m³. (The chemical reaction is shown in Example 11-4.) What collection efficiency is expected?

SOLUTION   From Eq. (9-113), $V_{\theta_{ave}} = 2Q/WR_2 = 50$ m/s. The Reynolds number $Re_0 = V_{\theta_{ave}}D/v = 2258$; from Eq. (9-124), $\zeta_0 = \rho R_2/2\rho_d D = 3.39$. Figure 9-21 then shows that $\xi_e = 0.20$, from which $Re_e = \xi_e\,Re_0 = 452$. From Table 9-4, at the respective Reynolds numbers of 2258 and 452, we have

$$\left(\frac{t\mu}{\rho_d D^2}\right)_0 = 0.001040 \qquad \left(\frac{t\mu}{\rho_d D^2}\right)_e = 0.006344$$

The time of travel of the drop from injection until it strikes the outer wall of the scrubber is obtained from the difference between the two preceding values:

$$t = (0.006344 - 0.001040)\frac{\rho_d D^2}{\mu} = 0.005304\frac{1000(7 \times 10^{-4})^2}{1.84 \times 10^{-5}} = 0.1412 \text{ s}$$

The Fourier number $Fo' = \mathscr{D}'t/D^2 = 0.000490$; and, from Table 11-2, $\phi(Fo') = 31.94$. Equation (11-40), based on Re_e, gives $\sigma_{ncr} = 0.005233$ for no chemical reaction.

Next, we enter Table 11-3 at the Reynolds numbers of 2258 and 452 to obtain

$$\left(\frac{\eta \rho Q \, \text{Sc}^{2/3}}{\sigma \dot{N}_d \rho_d D^3}\right)_0 = 1.729 \qquad \left(\frac{\eta \rho Q \, \text{Sc}^{2/3}}{\sigma \dot{N}_d \rho_d D^3}\right)_e = 1.501$$

Then

$$\frac{\eta \rho Q \, \text{Sc}^{2/3}}{\sigma \dot{N}_d \rho_d D^3} = 1.729 - 1.501 = 0.228$$

Equation (11-67) gives β_1:

$$\beta_1 = \frac{6}{\pi} \frac{0.5}{100} \frac{0.005233(1000)}{1.185(1.19)^{2/3}} (0.228) = 0.008562$$

Also,

$$\alpha_H = \frac{4.85 \times 10^6 (18)}{8314(298)(1000)} = 0.03524$$

$$\beta = 1 \frac{64}{158} \frac{1.2 \times 10^{-9}}{1.7 \times 10^{-9}} = 0.2859$$

Equation (11-73) gives x_1:

$$\frac{x_1}{L} = \frac{1}{0.008562} \ln\left[1 + \frac{0.001}{0.2859(0.03524)(10)}\right](1 - 0.005233) = 0.5407$$

Equation (11-74) gives the efficiency:

$$\eta = 1 + 0.2859(0.03524)\left\{1 - \left[1 + \frac{0.001}{0.2859(0.03524)(10)}\right]e^{-0.008562}\right\}\frac{10}{0.001}$$

$$= 0.8675 \qquad\qquad\qquad \text{////}$$

Venturi Scrubber

The venturi scrubber requires no further analysis. The motion of the drops is treated in Sec. 9-9. After computing the throat velocity, assuming either incompressible flow or using the compressible-flow formulas of that section, the initial Reynolds number is given by

$$\text{Re}_0 = \frac{\rho_{tr} D V_{tr}}{\mu} \qquad (11\text{-}75)$$

After computing the distance parameter $\rho_{tr} l_{tr}/\rho_d D$, Fig. 9-24 gives the value of $\text{Re}_1/2$, which is taken as the terminal Reynolds number of the drop motion.

Table 9-4 enables the time t for the drop to reach $\text{Re}_1/2$ to be obtained from the entries at Re_0 and $\text{Re}_1/2$. From this value of t, the Fourier number Fo' is obtained as $\text{Fo}' = \mathscr{D}'t/D^2$. Table 11-2 then gives $\phi(\text{Fo}')$, and Eq. (11-40) gives σ_{ncr}, the attachment coefficient if no chemical reaction is present. If there is a chemical reaction, then Eq. (11-57) may be used to obtain σ_{cr}. Finally,

Table 11-3 may be used to obtain the value of the following parameter, which by Eq. (11-63) may be written in the second form:

$$\frac{\eta \rho Q \text{ Sc}^{2/3}}{\sigma_{cr} \dot{N}_d \rho_d D^3} = \frac{\pi}{6} \frac{\eta \rho Q \text{ Sc}^{2/3}}{\sigma_{cr} \rho_d Q_s}$$

Once this parameter has been obtained, then either η or Q_s may be solved for as needed. (The subscript tr in the preceding equations refers to values in the throat section of the venturi.)

EXAMPLE 11-9   A venturi scrubber is to remove SO_2 from a stream of standard air. The given data are

$$V_i = 10 \text{ m/s} \qquad C_{mv_0} = 0.01 \text{ kg/m}^3$$
$$D_i = 1.0 \text{ m} \qquad D = 0.3 \text{ mm}$$
$$D_{tr} = 0.5 \text{ m} \qquad l_{tr} = 2.0 \text{ m}$$

If the spray fluid is water and if no chemical reaction occurs, determine Q_s for perfect efficiency.

SOLUTION   The flow rate and throat velocity are determined as $Q = \pi D_i^2 V_i/4 = 7.854$ m^3/s and $V_{tr} = 4Q/\pi D_{tr}^2 = 40.0$ m/s. The initial Reynolds number is $\text{Re}_0 = \rho_{tr} D V_{tr}/\mu = 773$, and the quantity $\rho_{tr} l_{tr}/\rho_d D = 7.90$. Figure 9-24 shows that $\text{Re}_1/2 = 50$. Using Table 9-4 at these two values of Reynolds number gives

$$\left(\frac{t\mu}{\rho_d D^2}\right)_0 = 0.003630 \qquad \left(\frac{t\mu}{\rho_d D^2}\right)_1 = 0.03183$$

Taking the difference between these values and solving for t gives

$$t = (0.03183 - 0.003630)\frac{\rho_d D^2}{\mu} = 0.02820 \frac{1000(0.3 \times 10^{-3})^2}{1.84 \times 10^{-5}} = 0.1379 \text{ s}$$

The Fourier number Fo$' = 0.02605$, and $\phi(\text{Fo}') = 12.26$ from Table 11-2. Equation (11-40) gives

$$\sigma = 1 - \frac{0.00009747}{0.00009747 + 5.91 \times 10^{-7}} = 0.006032$$

Table 11-3 gives

$$\left(\frac{\eta \rho Q \text{ Sc}^{2/3}}{\sigma \dot{N}_d \rho_d D^3}\right)_0 = 1.639 \qquad \left(\frac{\eta \rho Q \text{ Sc}^{2/3}}{\sigma \dot{N}_d \rho_d D^3}\right)_1 = 0.7965$$

Taking the difference between these values and using Eq. (11-63) gives

$$\frac{\pi}{6} \frac{\eta \rho Q \text{ Sc}^{2/3}}{\sigma \rho_d Q_s} = 1.639 - 0.7965 = 0.8426$$

Finally, the preceding equation is solved for Q_s, giving

$$Q_s = \frac{\pi(1)(1.185)(7.854)(1.19)^{2/3}}{6(0.8426)(0.006032)(1000)} = 1.077 \ \text{m}^3/\text{s} \qquad Ans. \qquad ////$$

EXAMPLE 11-10 Repeat Example 11-9 but assume that potassium sulfide is present in the water with an initial concentration of 2 kg/m^3.

SOLUTION The chemical reaction is given in Example 11-4. Equation (11-57) gives the attachment coefficient for the case of a chemical reaction:

$$\sigma_{cr} = 0.006032\left[1 + 1\frac{64}{158}\frac{1.2 \times 10^{-9}}{1.7 \times 10^{-9}}\frac{4.85 \times 10^6(18)}{8314(298)(1000)}\frac{2}{0.01}\right] = 0.0182$$

From Example 11-9

$$\frac{\pi}{6}\frac{\eta\rho Q \ \text{Sc}^{2/3}}{\sigma_{cr}\rho_d Q_s} = 0.8426$$

Solving this equation for Q_s gives

$$Q_s = \frac{\pi(1)(1.185)(7.854)(1.19)^{2/3}}{6(0.8426)(0.0182)(1000)} = 0.357 \ \text{m}^3/\text{s} \qquad Ans.$$

This analysis should be conservative inasmuch as the value of σ_{cr} is computed at the start of the process and should become greater as collection proceeds; this increase is ignored in the analysis. ////

11-4 ABSORPTION TOWERS

Absorption columns or towers are classified into two types: packed or plate towers. In either case, special packing material or inserts are included inside the tower to facilitate the mutual interaction of the liquid and the gas.

Packed Towers

A packed tower consists of a vertical cylindrical shell which is mostly filled with a suitable packing material. An example is shown in Fig. 11-7 in which Raschig rings are used as the packing material. The packing may be placed randomly inside the shell, as is done in Fig. 11-7, or stacked packings may be used. Packing elements come in a variety of shapes, sizes, and materials. Random packing elements include gravel, Raschig rings, Berl saddles, and Intalox saddles; stacked packing elements include Raschig rings, spiral rings, expanded-metal lath, wood grids, and drip-point grids. Treybal [4, chap. 6] presents a thorough discussion of the construction of absorption towers, as well as many of the technical details involved in their design and construction. In this section, we shall be

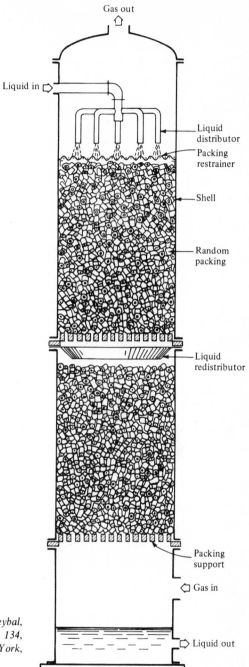

FIGURE 11-7
Packed tower. (*From R. E. Treybal,
"Mass-Transfer Operations," p. 134,
McGraw-Hill Book Company, New York,
1955. Used by permission.*)

concerned with the size of an absorption tower necessary to achieve a given collection efficiency and with the pressure drop and power requirements of the flow through the tower.

Number of Transfer Units

At this point, let us begin to analyze the action of an absorption tower such as the one in Fig. 11-7. We shall assume counterflow operation, as shown in the figure, in which the liquid flows from top to bottom under the action of gravity while the gas flows from bottom to top. We shall assume dilute solutions in both the gas and liquid phases, so that Henry's law will hold. This is given by Eq. (11-27) and may be written as

$$C_{n_e} = \frac{H}{P} C'_n \qquad (11\text{-}76)$$

in which the subscript e means that C_n applies to an equilibrium state between the gas and the liquid. Of course, the tower does not operate in equilibrium since if that were the case, an infinitely long tower would be required. Instead, a concentration difference $C_n - C_{n_e}$ serves as a driving force for mass transfer from the gas phase to the liquid phase.

Figure 11-8 shows a schematic diagram of a tower with a section located at x and an element of height dx shown. The height of the tower is L, and its cross-sectional area is A. The mass flow rates and concentrations of the two streams are shown. A mass balance of the pollutant on the lower section of the tower gives the following equation:

$$C_n = C_{n_1} - \frac{\dot{m}_l M_a}{\dot{m}_a M_l} (C'_{n_2} - C'_n) \qquad (11\text{-}77)$$

This equation, along with the equilibrium condition, Eq. (11-76), is shown in Fig. 11-9. The difference in height between the operating line and the equilibrium line is the driving force for mass transfer.

Next, consider the differential element of height dx and cross-sectional area A, as shown in Fig. 11-8. In this element, mass is transferred from the gas stream to the liquid stream in an amount which is somehow proportional to the area of contact between the two streams, as well as to the difference in concentrations between the operating line and the equilibrium line. This area of contact is not amenable to calculation but will be determined from empirical data. We may write across the element

$$\frac{1}{M_a} d(\dot{m}_a C_n) = \frac{1}{M_l} d(\dot{m}_l C'_n) = -KA(C_n - C_{n_e}) \, dx$$

If we assume that $\dot{m}_a$ and $\dot{m}_l$ are constant, then this equation may be written

$$\frac{dC_n}{C_n - HC'_n/P} = -\frac{KAM_a}{\dot{m}_a} dx \qquad (11\text{-}78)$$

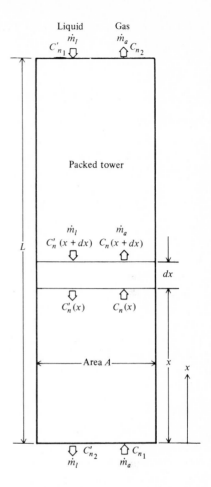

FIGURE 11-8
Schematic diagram of a packed tower for
purposes of analysis.

When Eq. (11-77) is substituted into Eq. (11-78), we obtain

$$\frac{dC_n}{C_n(1 - H\dot{m}_a M_l/P\dot{m}_l M_a) + H(\dot{m}_a M_l C_{n_1}/\dot{m}_l M_a - C'_{n_2})/P} = -\frac{KAM_a}{\dot{m}_a}dx \quad (11\text{-}79)$$

Equation (11-79) may be integrated between the two ends of the absorber to give

$$\ln \frac{(1 - H\dot{m}_a M_l/P\dot{m}_l M_a)C_{n_1} + H(\dot{m}_a M_l C_{n_2}/\dot{m}_l M_a - C'_{n_1})/P}{C_{n_2} - HC'_{n_1}/P}$$

$$= \frac{KALM_a}{\dot{m}_a}\left(1 - \frac{H\dot{m}_a M_l}{P\dot{m}_l M_a}\right) \quad (11\text{-}80)$$

Next, we define the transfer unit as the height of tower required to cause a concentration change of the gas equal to the average concentration driving force $C_n - C_{n_e}$. It follows that the number of transfer units (NTU) in the height L

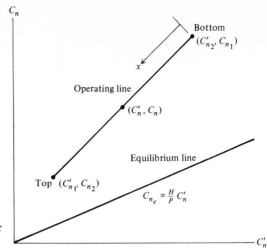

FIGURE 11-9
Operating line and equilibrium line for dilute solutions in a packed tower.

of the column is given by $\dot{m}_a(C_{n_1} - C_{n_2})/M_a = KAL(C_n - C_{n_e})_{ave}$ and by the definition that

$$NTU = \frac{C_{n_1} - C_{n_2}}{(C_n - C_{n_e})_{ave}} \quad (11\text{-}81)$$

Then

$$NTU = \frac{KALM_a}{\dot{m}_a} \quad (11\text{-}82)$$

We may also write that $L = NTU(HTU)$ in which HTU is the height of one transfer unit, so that

$$HTU = \frac{\dot{m}_a}{KAM_a} \quad (11\text{-}83)$$

In Eq. (11-80), we define

$$\alpha_A = \frac{H\dot{m}_a M_l}{P\dot{m}_l M_a} \quad (11\text{-}84)$$

and substitute Eq. (11-82), giving

$$NTU = \frac{1}{1 - \alpha_A} \ln\left[\frac{C_{n_1} - HC'_{n_1}/P}{C_{n_2} - HC'_{n_1}/P}(1 - \alpha_A) + \alpha_A\right] \quad (11\text{-}85)$$

Noting now that $C_{n_2} = C_{n_1}(1 - \eta)$, then Eq. (11-85) may be written

$$NTU = \frac{1}{1 - \alpha_A} \ln\left[\frac{C_{n_1} - HC'_{n_1}/P}{C_{n_1}(1 - \eta) - HC'_{n_1}/P}(1 - \alpha_A) + \alpha_A\right] \quad (11\text{-}86)$$

Equation (11-86) may also be solved for η to give

$$\eta = \left(1 - \frac{H}{P}\frac{C'_{n_1}}{C_{n_1}}\right)\frac{e^{(1-\alpha_A)\,\text{NTU}} - 1}{e^{(1-\alpha_A)\,\text{NTU}} - \alpha_A} \qquad (11\text{-}87)$$

EXAMPLE 11-11   Ammonia is to be removed from air in a packed tower. Assume the following data:

$$\dot{m}_a = 10 \text{ kg/s} \qquad\qquad \dot{m}_l = 40 \text{ kg/s}$$
$$C_{mv_1} = 0.10 \text{ kg/m}^3 \qquad C'_{mv_1} = 0$$
$$\eta = 0.95 \qquad\qquad D = 3.0 \text{ m}$$

Compute the NTU required. Assume the liquid is water.

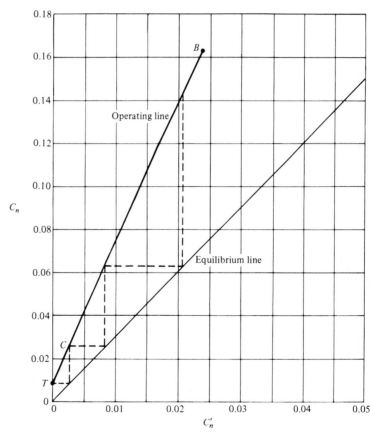

FIGURE 11-10
Operating line and equilibrium line for Example 11-11.

SOLUTION The exit concentration of NH_3 in the airstream is 0.005 kg/m^3. The following equation can be derived for small concentrations in air:

$$C_n = \frac{C_{mv}}{C_{mv} + M_p(\rho - C_{mv})/M_a} = \frac{M_a}{M_p} \frac{C_{mv}}{\rho}$$

From this equation we obtain $C_{n_1} = \frac{29}{15}(0.10/1.185) = 0.163$ and $C_{n_2} = 0.00816$. We note that $C'_{n_1} = 0$, and from Eq. (11-77) we have

$$C'_{n_2} = C'_{n_1} + \frac{\dot{m}_a M_l}{\dot{m}_l M_a}(C_{n_1} - C_{n_2}) = 0 + \frac{10(18)}{40(29)}(0.163 - 0.00816) = 0.0240$$

From Table 11-1, $H = 303{,}000$ N/m^2; and Eq. (11-76) gives

$$C_{n_e} = \frac{303{,}000}{101{,}326} C'_n = 2.99 C'_n$$

This is the equation of the equilibrium line; it and the operating line are shown in Fig. 11-10 (p. 493).

Equation (11-84) gives $\alpha_A = 0.464$, and Eq. (11-85) then gives NTU as

$$\text{NTU} = \frac{1}{1 - 0.464} \ln\left[\frac{0.163 - 2.99(0)}{0.00816 - 2.99(0)}(1 - 0.464) + 0.464\right] = 4.50 \quad Ans. \quad ////$$

Figure 11-10 shows another approximate method to determine the NTU for the tower. Start at operating point T, which represents the top end of the tower, and draw a horizontal line to the equilibrium line. At the point of intersection of this line with the equilibrium line, erect a vertical line which intersects the operating line at point C. Repeat this process until point B is reached, which represents the conditions at the bottom of the tower. The number of such steps is approximately the NTU of the process; in the case of Example 11-11, this graphic method gives NTU $= 3.3$.

For further details on the analysis of packed towers, including the treatment of cases in which Henry's law does not hold, see Treybal [4, chap. 8] or another book on mass transfer.

Height of a Transfer Unit

The height of a transfer (HTU) in a packed tower is a complicated and unknown function of the interfacial surface area in the packing, which itself is complicated and unknown. Empirical methods are necessary to determine this quantity. The data presented here are taken from Treybal [4, pp. 234–239] after conversion to SI units. Once the HTU has been determined, the height of the packing region of the tower is obtained from

$$L = \text{NTU(HTU)} \qquad (11\text{-}88)$$

The HTU is given as the sum of a contribution due to the air and another one due to the liquid, which is expressed as

$$HTU = HTU_a + \alpha_A HTU_l \qquad (11\text{-}89)$$

in which $\alpha_A = H\dot{m}_a/P\dot{m}_l$ as given by Eq. (11-84). The value of HTU_l is obtained from

$$HTU_l = \phi\left(\frac{\dot{m}_l}{A\mu_l}\right)^\delta \sqrt{Sc_l} \qquad (11\text{-}90)$$

in which the subscript l refers to the liquid, as usual. The values of ϕ and δ are given in Table 11-4 for certain types of packing. For the value of HTU_a use

$$HTU_a = \frac{\alpha(\dot{m}_a/A)^\beta}{(\dot{m}_l/A)^\gamma} \sqrt{Sc_a} \qquad (11\text{-}91)$$

Here, the values of α, β, and γ are given in Table 11-5.

EXAMPLE 11-12 Compute the height of the packing in the tower of Example 11-11. The packing consists of 1-in Raschig rings.

SOLUTION The cross-sectional area of the tower is 7.07 m². Using the diffusion coefficients from Table 11-1 and the viscosity values for air and water, the Schmidt numbers are computed to be

$$Sc_a = 0.706 \qquad Sc_l = 440$$

Table 11-4 VALUES OF HTU_l FOR VARIOUS PACKING MATERIALS

Packing	ϕ, $m^{1-\delta}$	δ	Range of $\dot{m}_l/A$, kg/m² · s
Raschig rings:			
$\frac{3}{8}$ in	0.000321	0.46	0.54–20
0.5 in	0.000718	0.35	0.54–20
1 in	0.00235	0.22	0.54–20
1.5 in	0.00261	0.22	0.54–20
2 in	0.00293	0.22	0.54–20
Berl saddles:			
0.5 in	0.00146	0.28	0.54–20
1 in	0.00129	0.28	0.54–20
1.5 in	0.00137	0.28	0.54–20
3-in partition rings (stacked staggered)	0.0171	0.09	4–19
Spiral rings (stacked staggered):			
3-in single spiral	0.00199	0.28	0.54–20
3-in triple spiral	0.00254	0.28	4–19
Drip-point grids (continuous flue):			
No. 6146	0.00357	0.23	4.7–40
No. 6295	0.00153	0.31	3.4–30

SOURCE: R. E. Treybal, "Mass-Transfer Operations," p. 237, McGraw-Hill Book Company, New York, 1955. Used by permission.

The two mass fluxes are $\dot{m}_a/A = 1.41$ kg/m$^2 \cdot$ s and $\dot{m}_l/A = 5.66$ kg/m$^2 \cdot$ s. Table 11-4 gives $\phi = 0.00235$ m$^{1-\delta}$ and $\delta = 0.22$ for 1-in Raschig rings. Table 11-5 gives $\alpha = 0.557$ m(m$^2 \cdot$ s/kg)$^{\beta-\gamma}$, $\beta = 0.32$, and $\gamma = 0.51$; it is true that the air flux exceeds the limit shown in the table, but these are the only data at hand and so we shall use these values with some reservation.

Equations (11-90) and (11-91) then give

$$\text{HTU}_l = 0.00235 \left[\frac{40}{7.07(8.8 \times 10^{-4})} \right]^{0.22} \sqrt{440} = 0.339 \text{ m}$$

$$\text{HTU}_a = \frac{0.557(1.41)^{0.32}\sqrt{0.706}}{(5.66)^{0.51}} = 0.216 \text{ m}$$

Using the value of $\alpha_A = 0.464$, Eq. (11-89) gives

$$\text{HTU} = 0.216 + 0.464(0.339) = 0.373 \text{ m}$$

From Example 11-11, NTU $= 4.50$; Eq. (11-88) gives the height L of the packing as

$$L = 4.50(0.373) = 1.68 \text{ m} \qquad Ans.$$

The total height of the tower will be somewhat greater than this value, to allow for the distribution space at the top and at the bottom and also for any redistribution necessary within the packing region. ////

Table 11-5 VALUES OF HTU$_a$ FOR VARIOUS PACKING MATERIALS

Packing	α^*	β	γ	Range of $\dot{m}_a/A$, kg/m$^2 \cdot$ s	Range of $\dot{m}_l/A$, kg/m$^2 \cdot$ s
Raschig rings:					
$\frac{3}{8}$ in	0.620	0.45	0.47	0.27–0.68	0.68–2.0
1 in	0.608	0.39	0.58	0.27–1.1	0.54–0.68
	0.557	0.32	0.51	0.27–0.81	0.68–6.1
1.5 in	0.830	0.38	0.66	0.27–0.95	0.68–2.0
	0.689	0.38	0.40	0.27–0.95	2.0–6.1
2 in	0.894	0.41	0.45	0.27–1.1	0.68–6.1
Berl saddles:					
0.5 in	0.540	0.30	0.74	0.27–0.95	0.68–2.0
	0.367	0.30	0.24	0.27–0.95	2.0–6.1
1 in	0.461	0.36	0.40	0.27–1.1	0.54–6.1
1.5 in	0.652	0.32	0.45	0.27–1.36	0.54–6.1
3-in partition rings (stacked staggered)	8.33	0.58	1.06	0.2–1.2	4.1–13.6
Spiral rings (stacked staggered):					
3-in single spiral	1.08	0.35	0.29	0.18–0.95	4.1–13.6
3-in triple spiral	1.11	0.38	0.60	0.27–1.36	0.68–4.1
Drip-point grids (continuous flue):					
No. 6146	1.04	0.37	0.39	0.18–1.36	4.1–8.8
No. 6295	0.732	0.17	0.27	0.14–1.36	2.7–15.6

SOURCE: R. E. Treybal, "Mass-Transfer Operations," p. 239, McGraw-Hill Book Company, New York, 1955. Used by permission.
* The units of α are m(m$^2 \cdot$ s/kg)$^{\beta-\gamma}$.

Pressure Drop in Packed Towers

The pressure drop in a packed tower, like the height of a transfer unit, is a quantity which can be evaluated only experimentally. Again referring to Treybal [4, pp. 143–145], the following equation for an irrigated, packed tower is suggested:

$$\Delta P = \frac{Lm}{\rho} \, 10^{n\dot{m}_l/A\rho_l - 8} \left(\frac{\dot{m}_a}{A}\right)^2 \qquad (11\text{-}92)$$

in which m and n are constants for a given packing material. These are listed in Table 11-6, along with a known range of validity.

EXAMPLE 11-13   Compute the pressure drop for the tower of Examples 11-11 and 11-12.

   SOLUTION   From Example 11-12, $\dot{m}_a/A = 1.41$ kg/m$^2 \cdot$ s and $\dot{m}_l/A = 5.66$ kg/m$^2 \cdot$ s. From Table 11-6 we have

$$m = 4.39 \times 10^{10} \text{ m}^{-1} \qquad n = 51.3 \text{ s/m}$$

For a length of 1.68 m, Eq. (11-92) gives for ΔP

$$\Delta P = \frac{1.68(4.39 \times 10^{10})}{1.185} \, 10^{51.3(5.66)/1000 - 8}(1.41)^2 = 2415 \text{ N/m}^2 \qquad Ans.$$

Table 11-6   CONSTANTS IN EQ. (11-92) FOR PRESSURE DROP IN PACKED TOWERS

Packing	$m \times 10^{-10}$, m^{-1}	n, s/m	Range of $\dot{m}_l/A$, kg/m$^2 \cdot$ s	Range of $\Delta P/L$, N/m^3
Raschig rings:				
0.5 in	19.0	85.0	0.41–11.7	0–410
0.75 in	4.50	53.1	2.44–14.6	0–410
1 in	4.39	51.3	0.49–36.6	0–410
1.5 in	1.65	47.0	0.98–24.4	0–410
2 in	1.52	34.8	0.98–29.3	0–410
Berl saddles:				
0.5 in	8.27	40.2	0.41–19.1	0–410
0.75 in	3.30	34.8	0.49–19.5	0–410
1 in	2.19	34.8	0.98–39.1	0–410
1.5 in	1.10	26.6	0.98–29.3	0–410
Intalox saddles:				
1 in	1.70	32.7	3.42–19.5	0–410
1.5 in	0.775	26.6	3.42–19.5	0–410
Drip-point grid tiles:				
No. 6146 continuous flue	0.143	25.3	4.07–23.1	0–80
No. 6146 cross flue	0.167	26.8	0.41–23.7	0–80
No. 6295 continuous flue	0.149	26.5	1.15–17.0	0–80
No. 6295 cross flue	0.196	19.7	1.22–17.0	0–80

SOURCE: R. E. Treybal, "Mass-Transfer Operations," p. 144, McGraw-Hill Book Company, New York, 1955. Used by permission.

This gives $\Delta P/L = 1437$, which is outside the range of validity of m and n; our answer is of little value for this reason.

The power expended in pumping the air through the packing, with this pressure drop, is

$$\dot{W}_a = Q\,\Delta P = \frac{\dot{m}_a}{\rho}\,\Delta P = \frac{10}{1.185}(2415) = 20.4\ \mathrm{kW}$$

The power required to raise the water a height of 1.68 m is

$$\dot{W}_l = \dot{m}_l Lg = 40(1.68)(9.807) = 0.66\ \mathrm{kW} \qquad \text{////}$$

Plate Towers

A plate tower consists of a number of plates, or trays, nested above each other inside a cylindrical shell. The liquid drains over the trays, while the gas flows up through perforations or bubble caps in the trays. In this way, interfacial contact is maintained between the gas and the liquid as the gas bubbles up through the liquid. Such an arrangement of trays, using bubble caps, is shown in Fig. 11-11. The analysis of a plate tower is similar to that of a packed tower, except that instead of a continuous variation of concentration, a number of discrete steps, equal to the number of trays, is employed. Also, there is a certain analogy between the number of transfer units in a packed tower and the number of trays in a plate tower.

Number of Plates Required

Figure 11-12 shows a schematic diagram of a plate tower, which has a total of N_T trays. Assuming that the liquid and gas flow rates are known, as well as the inlet and exit gas concentrations and the inlet liquid concentration, the number of trays required can be computed using the analysis that follows. Each tray is assumed to be partially efficient in changing the concentrations of the gas and liquid streams. If the tray were perfectly efficient, then the concentration in the gas stream leaving the tray would be related to that in the liquid stream leaving the tray by Henry's law, Eq. (11-76). The tray efficiency η_T is defined as

$$\eta_T = \frac{C_{n_{i+1}} - C_{n_i}}{C_{n_{i+1}} - C_{n_e}} = \frac{C_{n_{i+1}} - C_{n_i}}{C_{n_{i+1}} - HC'_{n_{i+1}}/P} \qquad (11\text{-}93)$$

in which Eq. (11-76) has been used. Later we shall see how the tray efficiency can be evaluated.

To seek an equation for the number N_T of trays in the tower, let us first take a material balance of the pollutant around the ith tray, as shown in Fig. 11-12. The result is

$$\frac{\dot{m}_a}{M_a}\left(C_{n_{i+1}} - C_{n_i}\right) = \frac{\dot{m}_l}{M_l}\left(C'_{n_{i+1}} - C'_{n_i}\right) \qquad (11\text{-}94)$$

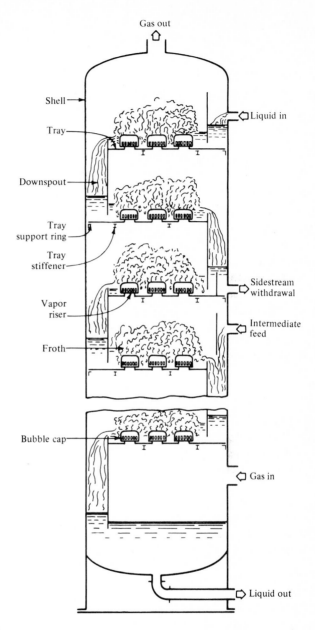

FIGURE 11-11
Bubble-cap plate tower. (*From R. E. Treybal, "Mass-Transfer Operations," p. 111, McGraw-Hill Book Company, New York, 1955. Used by permission.*)

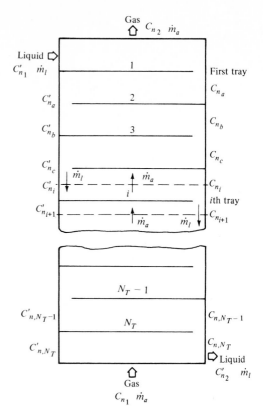

FIGURE 11-12
Schematic diagram of a plate tower
having N_T trays.

Next solve Eq. (11-93) for $C'_{n_{i+1}}$, giving

$$C'_{n_{i+1}} = \frac{P}{H} C_{n_{i+1}} - \frac{P}{H\eta_T} (C_{n_{i+1}} - C_{n_i}) \qquad (11\text{-}95)$$

If we replace i by $i - 1$ in the preceding equation, we obtain a relation valid for each tray except the first:

$$C'_{n_i} = \frac{P}{H} C_{n_i} - \frac{P}{H\eta_T} (C_{n_i} - C_{n_{i-1}}) \qquad (11\text{-}96)$$

Substituting Eqs. (11-95) and (11-96) into Eq. (11-94) and using the definition of α_A given by Eq. (11-84), we obtain

$$(\eta_T \alpha_A + 1 - \eta_T)C_{n_{i+1}} = [(\alpha_A - 1)\eta_T + 2]C_{n_i} - C_{n_{i-1}}$$

If we define $\alpha_B = (\alpha_A - 1)\eta_T$, this equation may be written

$$C_{n_{i+1}} - C_{n_i} = \frac{C_{n_i} - C_{n_{i-1}}}{\alpha_B + 1} \quad (11\text{-}97)$$

Equation (11-97) is not valid for the first tray. When this equation is written for the second to the last tray and the equations are substituted into each other in order, the last equation becomes

$$C_{n, N_T} - C_{n, N_T - 1} = \frac{C_{n_a} - C_{n_2}}{(\alpha_B + 1)^{N_T - 1}} \quad (11\text{-}98)$$

When Eq. (11-95) is substituted into Eq. (11-94) for the first tray and the result is solved for C_{n_a}, we have

$$C_{n_a} = \frac{(\alpha_A \eta_T + 1)C_{n_2} - H\eta_T C'_{n_1}/P}{\alpha_B + 1} \quad (11\text{-}99)$$

Equation (11-99) can be substituted into Eq. (11-98) to give

$$C_{n, N_T} - C_{n, N_T - 1} = \frac{[(\alpha_A \eta_T + 1) - (\alpha_B + 1)]C_{n_2} - H\eta_T C'_{n_1}/P}{(\alpha_B + 1)^{N_T}} \quad (11\text{-}100)$$

Next, a material balance for the pollutant substance is taken around the entire tower, which gives

$$\frac{\dot{m}_a M_l}{\dot{m}_l M_a}(C_{n_1} - C_{n_2}) = C'_{n_2} - C'_{n_1} \quad (11\text{-}101)$$

Note that $C_{n, N_T} = C_{n_1}$, and write Eq. (11-93) for the last tray:

$$C_{n, N_T - 1} = C_{n_1} - \eta_T\left(C_{n_1} - \frac{HC'_{n_2}}{P}\right) \quad (11\text{-}102)$$

These two equations can be combined and the result substituted into Eq. (11-100), then we solve for N_T and simplify, yielding

$$N_T = \frac{-1}{\ln[(\alpha_A - 1)\eta_T + 1]} \ln\left[\frac{C_{n_1} - HC'_{n_1}/P}{C_{n_2} - HC'_{n_1}/P}(1 - \alpha_A) + \alpha_A\right] \quad (11\text{-}103)$$

Since $C_{n_2} = C_{n_1}(1 - \eta)$, the collection efficiency of the entire tower may be brought in to give

$$N_T = \frac{-1}{\ln[(\alpha_A - 1)\eta_T + 1]} \ln\left[\frac{C_{n_1} - HC'_{n_1}/P}{C_{n_1}(1 - \eta) - HC'_{n_1}/P}(1 - \alpha_A) + \alpha_A\right] \quad (11\text{-}104)$$

Equation (11-104) may be solved for η:

$$\eta = 1 - \frac{HC'_{n_1}}{PC_{n_1}} - \frac{(1 - HC'_{n_1}/PC_{n_1})(1 - \alpha_A)}{[(\alpha_A - 1)\eta_T + 1]^{-N_T} - \alpha_A} \quad (11\text{-}105)$$

Next, we must examine the tray efficiency η_T. An empirical equation is given by Treybal [4, pp. 224–227]:

$$\eta_T = 1 - e^{-K} \quad \text{(11-106)}$$

in which K is given by

$$K = \frac{h}{(23.4 + 55.5HM_l/P\rho_l)\mu_l^{0.68}b^{0.33}} \quad \text{(11-107)}$$

where h represents the vertical height between the centers of the slots in the bubble cap and the top of the weir which controls the liquid level on the tray, and b is the width of the slot in the bubble cap. All these quantities are to be expressed in SI units. An example will illustrate the design procedure; we shall not attempt to construct a detailed design of the trays and bubble caps. For that, the reader is referred to Treybal's work or to some other book on mass transfer.

EXAMPLE 11-14 Use the data of Example 11-11 and determine the number of trays required if a plate tower is used. Assume $\eta_T = 0.50$.

SOLUTION From Example 11-11, we have that $C_{n_1} = 0.163$, $C_{n_2} = 0.00816$, $C'_{n_1} = 0$, $C'_{n_2} = 0.0240$, and $\alpha_A = 0.464$. Equation (11-103) gives

$$N_T = \frac{-1}{\ln\left[(0.464 - 1)0.5 + 1\right]} \ln \left[\frac{0.163}{0.00816}(1 - 0.464) + 0.464\right] = 7.74$$

Since we must use an integral number of trays, take $N_T = 8$.   *Ans.*   ////

The pressure drop through the bubble caps and trays in a plate tower is too involved to deal with here. There is a strong dependence on the geometry of the bubble cap, shape of the slots, and height of the liquid above the slots. Even the last quantity is not easily determined since it depends on the shape and size of the weir. At higher rates of gas flow, a froth forms above the liquid, which adds to the pressure drop. There are further problems of flooding and hydrodynamic stability which must be considered in the design of a tower. The interested reader is referred to Treybal [4, pp. 118–131], where a detailed discussion of these matters is presented. For our immediate purposes, let us observe that the pressure drop through a plate tower is comparable to that through a packed tower.

Other types of trays besides bubble-cap designs are used; these include the sieve-tray, or perforated-plate, tower and the grid-tray tower, which are discussed by Treybal [4, pp. 118–131].

REFERENCES

1 Geankoplis, C. J.: "Mass Transport Phenomena," p. 291, Holt, Rinehart and Winston, Inc., New York, 1972.
2 Dean, J. A., ed.: "Lange's Handbook of Chemistry," 11th ed., sec. 10, McGraw-Hill Book Company, New York, 1973.
3 National Research Council: "International Critical Tables," vol. 3, pp. 255–261, McGraw-Hill Book Company, New York, 1928.
4 Treybal, R. E.: "Mass-Transfer Operations," McGraw-Hill Book Company, New York, 1955.

PROBLEMS

11-1 Supply the missing steps in the derivation of Eq. (11-11).
11-2 Supply the missing steps in the derivation of Eq. (11-12).
11-3 Derive Eq. (11-28).
11-4 Derive Eq. (11-29).
11-5 Derive Eq. (11-31).
11-6 Derive Eq. (11-32).
11-7 Derive Eq. (11-33).
11-8 Derive Eq. (11-34).
11-9 Derive exact equations to replace Eqs. (11-32) to (11-34).
11-10 Derive Eq. (11-35).
11-11 Derive Eq. (11-36).
11-12 Derive Eq. (11-37).
11-13 Ammonia is to be scrubbed from a mixture of air and ammonia for which $C_{mv_x} = 0.002$ kg/m^3. The scrubbing fluid is a water-ammonia mixture for which C'_{mv_i} is 0.0001 kg/m^3. If the drop diameter is 0.5 mm and V_∞ is 10 m/s, determine σ and η_d.
11-14 Repeat Prob. 11-13 for values of V_∞ of 1, 5, 20, 30, and 50 m/s.
11-15 Hydrogen cyanide is to be scrubbed from air for which $C_{mv_\infty} = 0.0005$ kg/m^3. The scrubbing water is initially pure and is formed in drops of 0.75 mm diameter, moving at a velocity of 40 m/s. Compute σ and η_d for this case.
11-16 Methyl chloride is to be scrubbed from standard air by using drops of initially pure water. The concentration of CH_3Cl in the air is 0.00025 kg/m^3. It is required that the collection efficiency of the drop shall exceed 0.001 for the first 0.25 s of drop motion. For a drop diameter of 0.5 mm, what drop velocity is required?
11-17 Repeat Prob. 11-16 for a drop diameter of 1.0 mm.
11-18 Derive Eqs. (11-47) and (11-48).
11-19 Derive Eq. (11-50).
11-20 Derive Eq. (11-52).
11-21 Repeat Example 11-4 if the initial concentration of potassium sulfide is 1.0 kg/m^3.
11-22 Use the reaction of Example 11-4 and the value of $\mathscr{D}''$ given there. If the drop diameter is 0.5 mm, its velocity is 10 m/s, and $C_{mv_x} = 0.0005$ kg/m^3, compute and plot curves of η_d as a function of time for values of C''_{mv_i} ranging from 0.01 kg/m^3 to 50 kg/m^3.

11-23 The reaction of Example 11-4 is to be used inside a drop to scrub SO_2 from standard air. If $C_{mv_\infty} = 0.0005$ kg/m^3, $D = 0.75$ mm, and $V_\infty = 20$ m/s, determine the necessary value of C''_{mv_i} in order that the collection efficiency exceed 0.001 for a period of 5.0 s.

11-24 Hydrogen sulfide H_2S mixed in standard air with a concentration $C_{mv_\infty} = 0.001$ kg/m^3 is to be scrubbed by using drops having a diameter of 1.0 mm and a velocity of 20 m/s. The water contains sodium carbonate Na_2CO_3 with an initial concentration of 10 kg/m^3. Assuming $\mathscr{D}''$ is roughly 1.5×10^{-9} m^2/s, what collection efficiency is expected after 2.5 s? The chemical reaction is

$$H_2S + Na_2CO_3 \longrightarrow NaHCO_3 + NaHS$$

11-25 Use the data of Example 11-5, with $Q_s = 1.0$ m^3/s, and compute the overall collection efficiency of the spray chamber for a range of drop diameters. Consider the increase in concentration of the SO_2 in the drop.

11-26 Repeat Prob. 11-25 if the diameter of the spray chamber is 4.0 m, all other data remaining the same.

11-27 Use the data of Example 11-5, with $Q_s = 0.2$ m^3/s, and consider the increase in concentration of the SO_2 in the drop. Compute the overall collection efficiency if the diameter of the spray chamber varies from 1.0 to 4.0 m.

11-28 Compute the power loss for the data of Example 11-5 if $\Delta P_{ns} = 150$ N/m^2. Take $Q_s = 0.01$ m^3/s.

11-29 Repeat Example 11-6, using the data and chemical reaction of Example 11-4.

11-30 Hydrogen sulfide H_2S is to be removed from a stream of 25 m^3/s of standard air for which $C_{mv_\infty} = 0.0075$ kg/m^3 in a spray chamber. The chamber has a diameter of 3.5 m and a length of 10.0 m. Assume a drop diameter of 1.0 mm, and compute the overall collection efficiency for a water flow rate of 0.5 m^3/s. Take into account the increase of concentration of the H_2S in the drop.

11-31 Repeat Prob. 11-30, but assume that a chemical reaction like the one in Prob. 11-24 occurs. Assume $C''_{mv_i} = 1.0$ kg/m^3, and take $\mathscr{D}'' = 1.5 \times 10^{-9}$ m^2/s.

11-32 Repeat Prob. 11-29 if C''_{mv_i} is 1.0 kg/m^3.

11-33 Repeat Example 11-7 if the chemical reaction of Example 11-4 occurs. Assume the initial concentration of potassium sulfide in the water is 5 kg/m^3.

11-34 Use the chemical reaction of Prob. 11-24 to remove H_2S from standard air if $C_{mv_\infty} = 0.002$ kg/m^3. The air flow rate is 10 m^3/s, and the air velocity is 10 m/s. The initial concentration of sodium carbonate in the water is 20 kg/m^3, and the drop diameter is 1.5 mm. A spray scrubber is used with one point of injection of water drops into the airstream; the length of the scrubber is 3.0 m. Compute the water injection rate if the scrubber efficiency is to be 0.99.

11-35 Derive Eq. (11-70).

11-36 Derive Eq. (11-72).

11-37 Derive Eqs. (11-73) and (11-74).

11-38 Rework Example 11-8 if all the data remain the same except that $Q_s = 0.1$ m^3/s and $C''_{mv_i} = 5$ kg/m^3.

11-39 Use the chemical reaction of Prob. 11-24 to remove H_2S from standard air in a cyclone scrubber. Assume $Q = 50$ m^3/s, $Q_s = 0.2$ m^3/s, $C_{mv_0} = 0.005$ kg/m^3, $C''_{mv_i} = 2$ kg/m^3, and that the drop diameter is 1.0 mm. Determine the collection efficiency. Take $R_2 = 1.5$ m and $W = 0.5$ m.

11-40 A cyclone scrubber is to remove SO_2 from a stream of 30 m³/s of standard air for which C_{mv} is initially 0.001 kg/m³. The water drops are 0.5 mm in diameter and contain potassium sulfide with an initial concentration of 10 kg/m³. Take $V_\theta = 40$ m/s, and let $W = R_2/4$. Determine R_2 and compute the flow rate of spray liquid to achieve a collection efficiency of 0.99.

11-41 Design a cyclone scrubber to remove SO_2 from standard air where the initial concentration is 0.003 kg/m³ and the flow rate of air is 300 m³/s. Use the chemical reaction involving potassium sulfide. A collection efficiency of 0.95 is required.

11-42 Repeat Prob. 11-41 if no chemical reaction is to be used.

11-43 A venturi scrubber is to be used to remove SO_2 from a stream of 100 m³/s of standard air without the use of a chemical reaction. The initial concentration of the SO_2 is 0.001 kg/m³, and the inlet velocity is 20 m/s. The throat diameter is one-half the inlet diameter, and the water-drop diameter is 0.25 mm. The length of the throat is 3.0 m. Compute the water flow rate required to collect all of the SO_2.

11-44 Repeat Prob. 11-43 if potassium sulfide is present in the water with an initial concentration of 10 kg/m³.

11-45 Compute the power required in Probs. 11-43 and 11-44. Assume a divergence angle of 10°.

11-46 Derive an equation for the collection efficiency of a venturi scrubber if the value of σ_{cr}, as given by Eq. (11-57), is allowed to increase as the scrubbing process proceeds while σ_{ncr} is assumed constant.

11-47 Repeat Prob. 11-44 using the efficiency expression derived in Prob. 11-46.

11-48 Design a venturi scrubber to remove H_2S from a stream of 50 m³/s of standard air with an initial concentration of 0.002 kg/m³. Use the chemical reaction of Prob. 11-24.

11-49 Repeat Prob. 11-44 if the drop diameter is 0.8 mm.

11-50 Repeat Prob. 11-49 if the throat diameter is one-third the inlet diameter.

11-51 Compute the power required for the venturi scrubbers of Probs. 11-49 and 11-50.

11-52 Derive Eqs. (11-79), (11-80), and (11-85).

11-53 Repeat Example 11-11 using $\dot{m}_l = 20$ kg/s. Also determine the height of the transfer unit and the total height of the packing in the tower.

11-54 In Prob. 11-53, determine the NTU by the graphic procedure shown in Fig. 11-10. Compare with the NTU computed in Prob. 11-53.

11-55 Determine the pressure drop and power requirements for Prob. 11-53.

11-56 A stream of 100 m³/s of standard air containing SO_2 with an initial concentration $C_{mv} = 0.001$ kg/m³ is to be treated in a tower packed with 1-in Berl saddles. The collection efficiency is to be 0.97 with the tower designed for 10 transfer units. The liquid is water. Design a tower to meet these requirements; determine the tower diameter and packing height, the pressure drop, and the power expended in pumping the air and the water.

11-57 An absorption tower is packed with No. 6146 continuous-flue drip-point grid tiles. The tower has a diameter of 1.0 m and a height of packing equal to 8.0 m. The tower is used to treat a stream of 0.75 m³/s of standard air containing SO_2 with C_{mv} initially equal to 0.002 kg/m³. The liquid is initially pure water flowing at the rate of 2.0 kg/s. Determine the NTU and the collection efficiency of the tower.

11-58 Determine the power requirements for the air and water streams in Prob. 11-57.

11-59 Derive Eq. (11-98).

11-60 Derive Eq. (11-103).

11-61 Repeat Example 11-14 if $\dot{m}_l = 20$ kg/s.

11-62 Repeat Prob. 11-56, except that a bubble-cap plate tower is used. Assume the tray efficiency η_T is 0.75.

11-63 Use the data of Prob. 11-61 and compute and plot N_T as a function of η_T.

11-64 A bubble-cap plate tower uses water as the liquid and treats a mixture of ammonia and standard air. The bubble caps and trays are constructed so that the height from the tray to the weir is 8 cm, the slots in the bubble cap are 5 mm wide by 2 cm high with centers located 3 cm above the tray. Compute the tray efficiency.

11-65 Repeat Prob. 11-64 if the air contains SO_2 instead of ammonia.

11-66 A bubble-cap plate tower treating a mixture of standard air and SO_2 has $C_{mv_1} = 0.0025$ kg/m^3 and $C'_{mv_1} = 0$. The flow rates of air and water are 20 and 5 m^3/s, respectively. The bubble caps and trays have the pertinent dimensions of Probs. 11-64 and 11-65. If the tower has 15 trays, what collection efficiency is expected?

ADSORPTION DEVICES

One of the basic phenomena of chemistry is that of adsorption. When a solid medium is surrounded by a gas, the molecules of the solid at the interface tend to strongly attract molecules of the gas. Several layers of gas molecules are held tightly bound to the surface of the solid. The exact state of the adsorbed gas molecules, that is, whether they are in the gaseous or liquid state, is unclear. It is known, however, that energy is given off when adsorption takes place and that the magnitude of this energy is roughly comparable to that given off upon condensation. The adsorption process is reversible, with energy required to effect a release of the adsorbed gas. Thus the adsorption process is exothermic, while its reverse process, desorption, is endothermic. The amount of gas which a substance can adsorb is limited, and this amount decreases with increasing temperature.

Gases are adsorbed preferentially; gases with higher molecular weights and lower boiling temperatures generally are more readily adsorbed. When a mixture of gases surrounds the solid, those gases more readily adsorbed displace those which are adsorbed less easily and a separation effect occurs. For example, organic gases and vapors are more easily adsorbed than air and can be entirely separated from the air under proper conditions.

The solid substance upon which the gas adsorbs is known as the *adsorbent*, while the gas attaching to the adsorbent is known as the *adsorbate;* the gas from which the adsorbate is removed is known as the *solvent*. Although all solids act as adsorbents to some degree, only a few are very effective. Effective adsorbents have a large surface area per unit mass, caused by a highly irregular structure and containing capillary pores which, it is possible, may be filled with condensed gas as adsorption proceeds. Among the numerous commercial adsorbents, including charcoal, silica gel, Fuller's earth, bauxite, and many others, activated carbon is the one widely used for air pollution control work. This chapter will consider the use of activated carbon in the types of equipment commonly used for this purpose.

12-1 PRINCIPLES OF ADSORPTION

All common adsorbents except activated carbon show a strong preference for polar molecules over nonpolar ones. Such materials will adsorb water vapor in preference to organic molecules, which is not desirable. For this reason, activated carbon is normally used and must be used if water vapor is in the airstream. Since activated carbon is not unduly expensive, it is normally chosen. It is finely ground and contained in beds through which the air passes.

Activated carbon is effective in removing virtually all gases and vapors with molecular weights exceeding about 45. Permanent gases are not effectively adsorbed. As adsorption proceeds, a saturation state is eventually reached in which the adsorbent can hold no more of the adsorbate at that temperature; if the temperature is lowered, more can be adsorbed. Once saturation has been reached, the air moves through the bed unchanged. Of course, since heat is evolved in adsorption, the bed will likely operate at an elevated temperature. If further air movement occurs after saturation is reached at the elevated temperature, the cooling effect of the air will lower the temperature, resulting in further adsorption. Normally, in air pollution control work, the initial concentrations of pollutant are so small that the bed will operate at only a slightly elevated temperature.

When the air contains a mixture of pollutants, the carbon shows some preference for one substance over another. The bed will initially become saturated with the mixture of pollutants contained in the air. Upon further air flow, the least attractive adsorbate will be released and replaced by more attractive ones. Eventually, the carbon will be saturated with only one pollutant substance from the mixture. Of course, the carbon will lose its effectiveness as soon as it becomes saturated with the mixture, and further operation would be useless in controlling air pollution.

Equilibrium Conditions

Although the adsorbent bed will not operate in equilibrium and will cease to be useful when equilibrium is reached, the equilibrium behavior of the adsorbent-adsorbate-solvent combination is important as an upper limit of practical performance. This behavior is shown by a series of isotherms on a graph of

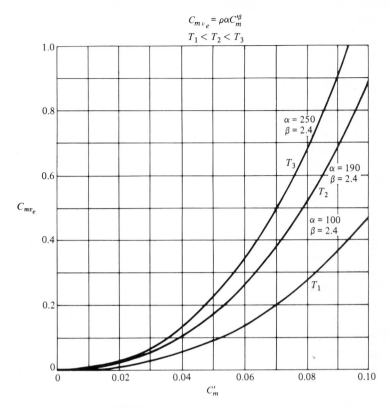

FIGURE 12-1
Equilibrium curves for a pollutant-adsorbent combination.

concentration of the pollutant in the gas stream as a function of concentration of the pollutant attached to the adsorbent. The gas concentration may be taken as either C_m or C_{mv}; the concentration on the adsorbent is given by C'_m, the mass of adsorbate per unit mass of adsorbent. Figure 12-1 shows such a series of three hypothetical isotherms. Note that the higher isotherms represent less of the pollutant attached to the adsorbent and thus occur at higher temperatures. The operating curve will lie above the equilibrium curve since a concentration driving force is required for finite speed of operation.

The curves of Fig. 12-1 are represented by the Freundlich equation

$$C_{mv_e} = \rho \alpha C'^{\beta}_m \qquad (12\text{-}1)$$

in which α and β are constants for the particular substances involved and for the temperature of the isotherm. The subscript e means that C_{mv} is the mass-volume concentration of the pollutant in the gas when the pollutant on the adsorbent is in equilibrium with that in the gas stream. The Freundlich equation is only an approximation to the equilibrium curve for a real substance but will be adequate for our analysis.

Adsorption Kinetics

The speed with which the mass transfers from the solvent to the adsorbent determines, among other things, the thickness of bed required. The mass flow rate of pollutant from a dissolved state in the gas to an attached state on the adsorbent is proportional to the difference in concentration between the actual state in the gas and the equilibrium state corresponding to the concentration on the adsorbent; this relation is given as

$$\dot{m}_p = K\overline{V}(C_{mv} - C_{mv_e}) \qquad (12\text{-}2)$$

The mass flow rate is also shown as being proportional to the volume $\overline{V}$ of the bed. The coefficient K depends on the effective interface area between the grains of adsorbent and the gas and on the film resistance on the gas side of this interface. As a result of the microscopic complexity of this interface, the coefficient K is best determined experimentally for a given adsorbent bed and air flow rate. One realistic value of K is 20 s^{-1}, but the value can vary over a wide range.

Regeneration

After a certain time, the adsorption bed will approach saturation and will no longer be effective in its assigned function. The fate of the adsorbent at this point must be decided in advance in the design of the unit. The recovered pollutant is frequently of little economic value, and the carbon is not inordinately expensive. For small units which need to be replenished only occasionally, the cheapest method may be simply to discard the adsorbent. In other cases, the pollutant is of little value but the carbon must be saved. The carbon can be regenerated by heating it to drive off the vapors under controlled conditions.

Usually steam, but sometimes hot air, is used to heat the adsorbent; this is done in such a way that the effluent steam or airstream is much more highly concentrated than the original polluted airstream. The effluent stream can be cooled and the pollutant condensed and removed by decantation or fractional distillation if necessary. In some cases, the adsorbent bed is replaced and sent back to the vendor for regeneration. A technique sometimes used is to embed a catalyst in the adsorbent; the catalyst transforms the pollutant substance into harmless substances once the bed is raised to a suitable temperature. In other cases, recovery of the pollutant substance is a prime consideration; this is particularly true where a single pollutant is present. An example is the recovery of solvent from a dry-cleaning plant. Another example is the recovery of gasoline vapors from an automobile fuel tank; these vapors are expelled when the tank is filled or when the liquid expands due to an increase in the ambient temperature. The fuel vapors are collected in a charcoal adsorber and then fed to the carburetor when the engine is started.

The regeneration process is similar to the adsorption process except that the concentration driving force must act the other way. The operating line must lie below the equilibrium curve of Fig. 12-1.

12-2 FIXED-BED ADSORBERS

Adsorbers used in the process industries are often arranged in true counterflow operation; the adsorbent flows downward as the gas flows upward. This arrangement poses some problems in that the granular adsorbent flows with some difficulty. In most pollution control activities, this counterflow arrangement is unnecessary and other arrangements are used.

A fixed-bed adsorber operates under transient conditions. It continuously approaches equilibrium conditions until it reaches a state where it no longer performs adequately; it is then removed from service and regenerated. Such units are made with one, two, three, four, or more adsorbent beds. The single-bed unit is the simplest. As shown in Fig. 12-2, this unit consists of an adsorbent bed of granulated activated carbon mounted inside a shell. The polluted airstream enters at the top of the bed, travels downward through the bed, and leaves at the bottom. The gas could be arranged to travel upward, but this arrangement has the disadvantage of requiring smaller gas velocities to protect the integrity of the bed.

When the single-bed adsorber is placed in service, it will remove essentially all of the pollutant. It will continue to do this for some time, but after a while, the pollutant level in the effluent gas stream will start to rise and will continue to rise until it approaches the level in the inlet gas stream. The point at which the effluent pollutant level begins to rise is called the *breakthrough point*, and the point at which the effluent pollutant level is essentially equal to the inlet pollutant level is called the *exhaustion point*. The single-bed adsorber may be operated to the breakthrough point when it must be regenerated. Regeneration is accomplished by stopping the flow of air and then allowing steam to flow through the adsorbent bed. After the bed has been regenerated, it must be allowed to cool before it is again placed in service. Later in this section, we shall determine the time elapsed from the start of collection until the breakthrough point is reached.

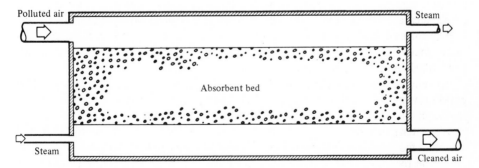

FIGURE 12-2
Single-bed adsorber.

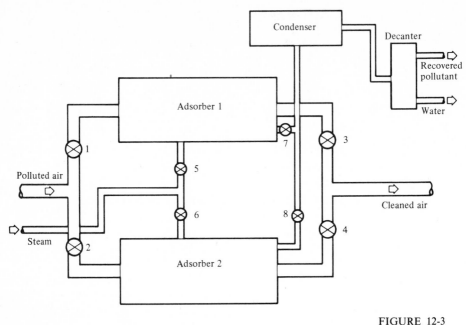

FIGURE 12-3
Two-bed adsorber.

The single-bed adsorber requires that the process which it serves must be shut down during regeneration of the adsorbent. Where shutdown is inconvenient, the use of two or more adsorbent beds can enable the adsorbing unit to operate continuously. A two-bed adsorber is shown in Fig. 12-3. With two beds available, one will be in operation while the other is being regenerated, is cooling, or is waiting its turn to come into service. The different valves shown in the Fig. 12-3 must be properly positioned for each function. For example, if adsorber 1 is in service while adsorber 2 is being regenerated, then valves 1, 3, 6, and 8 will be open while valves 2, 4, 5, and 7 will be closed. After adsorber 2 has been regenerated, while it is cooling or waiting its turn to be used, valves 6 and 8 will be closed also. When adsorber 2 comes on line and adsorber 1 is being regenerated, then valves 2, 4, 5, and 7 will be open and valves 1, 3, 6, and 8 will be closed. The cycle repeats indefinitely.

With one-and two-bed adsorbers, the bed can no longer be used once the breakthrough point has been reached and yet the bed can still adsorb additional pollutant. This feature can be utilized if three or more adsorbant beds are used. In a three-bed arrangement, the bed which has passed its breakthrough point but which has not yet reached exhaustion is the first bed encountered by the polluted air. The air then goes to the last regenerated bed where the remaining pollutant is removed. While this is going on, the third bed is being regenerated and cooled or is standing by. When the second bed reaches

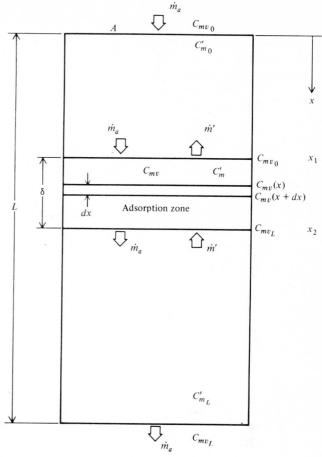

FIGURE 12-4
Section through an adsorption bed showing the moving adsorption zone and an element within this zone.

breakthrough, the beds are cycled so that 1-2-3 becomes 2-3-1. With four beds, two beds can be in service while one bed is being regenerated and one bed is cooling. With additional beds available, one bed or more can be out of commission for repairs.

Transient Behavior of an Adsorbent Bed

Since the practical adsorption devices with which we are concerned operate transiently, let us perform an analysis of a bed which has a concentration changing with time. Figure 12-4 shows a section through an adsorption bed, with the air flowing downward. The air enters with an initial pollutant concentration C_{mv_0}; until the breakthrough point is reached, the pollutant concentration leaving is zero. Since at first the air travels through a region of spent adsorbent, little change in concentration will be obtained. Then the air reaches a zone in which the adsorbent is very active, and the concentration drops rapidly. Finally, in the last zone no change in concentration occurs because the concentration is already essentially zero. This concentration curve is shown in Fig. 12-5.

The adsorption zone moves slowly downward at a velocity V'; let us take its top and bottom positions at any time as x_1 and x_2, respectively. These values of position are functions of time, but for simplicity let us assume that $x_2 - x_1$, which is called δ, is constant; that is, the width of the adsorption zone is constant. Our analysis will consider only the variation of concentration within this adsorption zone. To facilitate the analysis, let us assume that the adsorption zone is fixed in position and that the adsorbent moves upward with velocity V' and mass flow rate $\dot{m}'$. An element of height dx is also shown within the adsorption zone, and it, too, will be fixed, with adsorbent moving into it with mass flow rate $\dot{m}'$.

A pollutant mass balance on the adsorption zone, which is now assumed to be fixed in position, gives

$$\frac{\dot{m}_a}{\rho} C_{mv_0} = \dot{m}' C'_{m_0} = \rho' c A V' C'_{m_0}$$

in which c is the packing density of the adsorbent, defined as the ratio of the mass of a body of adsorbent as packed to the mass of a body of the same size which is solidly packed with the same adsorbent. Using Eq. (12-1) and assuming that $C_{mv_e} = C_{mv_0}$ at the top of the adsorption zone, even though this is not true within the zone, we obtain

$$V' = \frac{\dot{m}_a}{\rho \rho' c A} (\alpha \rho)^{1/\beta} C_{mv_0}{}^{(\beta - 1)/\beta} \qquad (12\text{-}3)$$

Next, write a pollutant mass balance around the top of the adsorption zone between x_1 and x:

$$\frac{\dot{m}_a}{\rho} C_{mv} + \dot{m}'(0) = \dot{m}' C'_m + \dot{m}_a(0)$$

or

$$C'_m = \frac{\dot{m}_a}{\rho \dot{m}'} C_{mv} \qquad (12\text{-}4)$$

A pollutant mass balance written around the element of height dx gives

$$d\dot{m}_p = \dot{m}' \, dC'_m = -\frac{\dot{m}_a}{\rho} \, dC_{mv}$$

Using Eq. (12-2),

$$\frac{\dot{m}_a}{\rho} \, dC_{mv} = -KA(C_{mv} - C_{mv_e}) \, dx \qquad (12\text{-}5)$$

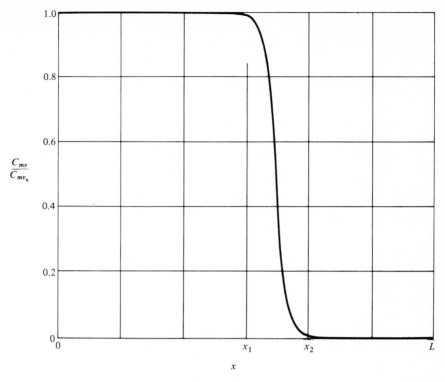

FIGURE 12-5
Variation of pollutant concentration in an adsorption bed.

Substituting Eq. (12-1) into Eq. (12-5) gives

$$\frac{\dot{m}_a}{\rho} dC_{mv} = -KA(C_{mv} - \alpha\rho C_m'^{\beta}) \, dx \qquad (12\text{-}6)$$

Using Eq. (12-4) in Eq. (12-6), we obtain

$$\dot{m}_a \, dC_{mv} = -KA\rho \left[C_{mv} - \alpha\rho \left(\frac{\dot{m}_a}{\rho\dot{m}'}\right)^{\beta} C_{mv}{}^{\beta} \right] dx \qquad (12\text{-}7)$$

Note that $\dot{m}' = A\rho'cV'$; we substitute Eq. (12-3), giving

$$\left(\frac{\dot{m}_a}{\dot{m}'}\right)^{\beta} = \frac{\rho^{\beta}}{\alpha\rho} C_{mv_0}{}^{1-\beta} \qquad (12\text{-}8)$$

Substituting Eq. (12-8) into Eq. (12-7) and integrating between the limits x_1, C_{mv_0} and x_2, 0, we obtain

$$\delta = \frac{\dot{m}_a}{\rho KA} \int_0^{C_{mv_0}} \frac{dC_{mv}}{C_{mv} - C_{mv_0}{}^{1-\beta} C_{mv}{}^{\beta}}$$

Letting $C_{mv}/C_{mv_0} = \varepsilon$, this integral may be written

$$\frac{\rho K A \delta}{\dot{m}_a} = \int_0^1 \frac{d\varepsilon}{\varepsilon(1 - \varepsilon^{\beta - 1})} \qquad (12\text{-}9)$$

The integral in Eq. (12-9) is undefined for the limits of 0 and 1. However, it can be readily evaluated for other limits, such as 0.01 and 0.99. Since these limits give almost the same range of C_{mv} over the adsorption zone, the difference in the final answer will not be significant. The result is

$$\frac{\rho K A \delta}{\dot{m}_a} = 4.595 + \frac{1}{\beta - 1} \ln \frac{1 - (0.01)^{\beta - 1}}{1 - (0.99)^{\beta - 1}} \qquad (12\text{-}10)$$

This equation enables us to determine the thickness δ of the adsorption zone.

An integral similar to Eq. (12-9) but with an indefinite lower limit is

$$x = x_1 + \frac{\dot{m}_a}{\rho K A} \int_{C_{mv}/C_{mv_0}}^{0.99} \frac{d\varepsilon}{\varepsilon(1 - \varepsilon^{\beta - 1})}$$

Upon integration, this equation becomes

$$x = x_1 + \frac{\dot{m}_a}{\rho K A} \left[\ln \frac{0.99 C_{mv_0}}{C_{mv}} + \frac{1}{\beta - 1} \ln \frac{1 - (C_{mv}/C_{mv_0})^{\beta - 1}}{1 - (0.99)^{\beta - 1}} \right] \qquad (12\text{-}11)$$

This is the equation of the concentration distribution through the adsorption zone shown in Fig. 12-5; the curve is calculated for $\beta = 2$.

Next we shall obtain an equation for the breakthrough curve. The breakthrough time t_B is the time at which the bottom point of the adsorption zone x_2 reaches the bottom of the bed. Let us no longer treat the adsorption zone as fixed, but instead let it move downward with velocity V' through the fixed bed. The breakthrough time is given by

$$t_B = \frac{L - \delta}{V'} \qquad (12\text{-}12)$$

This is only an approximation since it assumes that the adsorption zone emerges with full thickness instantaneously at time zero. The exhaustion point occurs at time t_E when the top point x_1 of the adsorption zone reaches the bottom of the bed; that is,

$$t_E = \frac{L}{V'} \qquad (12\text{-}13)$$

The breakthrough curve, which is the curve of C_{mv_L} as a function of time, is obtained from Eq. (12-11) by letting $x = L$ and $x_1 = V't$, giving

$$\ln \frac{C_{mv_0}}{C_{mv_L}} + \frac{1}{\beta - 1} \ln \left[1 - \left(\frac{C_{mv_L}}{C_{mv_0}} \right)^{\beta - 1} \right]$$

$$= \frac{\rho K A}{\dot{m}_a} (L - V't) - \ln 0.99 + \frac{1}{\beta - 1} \ln (1 - 0.99^{\beta - 1}) \qquad (12\text{-}14)$$

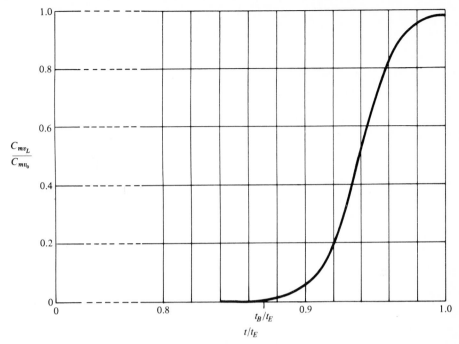

FIGURE 12-6
Breakthrough curve for $\beta = 2$ and for $\rho KAL/\dot{m}_a = 70$.

For the case of $\beta = 2$, the breakthrough curve is shown in Fig. 12-6. The following example illustrates the computation of the breakthrough time.

EXAMPLE 12-1   An adsorbent bed is 0.5 m thick and 10 m² in cross-sectional area. For $\dot{m}_a = 1.5$ kg/s, $K = 30$ s^{-1}, $\alpha = 150$, $\beta = 2.5$, $\rho' = 600$ kg/m³, $c = 0.5$, and $C_{mv_0} = 0.002$ kg/m³, determine the breakthrough time and the exhaustion time for the bed.

SOLUTION   Equation (12-3) gives the speed of the adsorption zone as

$$V' = \frac{1.5}{1.185(600)(0.5)(10)} [150(1.185)]^{1/2.5}(0.002)^{1.5/2.5}$$

$$= 0.0000805 \text{ m/s}$$

Equation (12-10) gives the width δ of the adsorption zone as

$$\delta = \frac{1.5}{1.185(30)(10)} \left[4.595 + \frac{1}{1.5} \ln \frac{1 - (0.01)^{1.5}}{1 - (0.99)^{1.5}} \right] = 0.0312$$

Equations (12-12) and (12-13) give the breakthrough and exhaustion times as

$$t_B = \frac{L - \delta}{V'} = \frac{0.5 - 0.0312}{0.0000805} = 5823 \text{ s} = 1.62 \text{ h} \qquad Ans.$$

$$t_E = \frac{L}{V'} = \frac{0.5}{0.0000805} = 6211 \text{ s} = 1.73 \text{ h} \qquad Ans. \qquad ////$$

Regeneration of an Adsorbent Bed

When the breakthrough point has been reached or, in the case of multiple-bed adsorbers, when the exhaustion point has been reached, the bed must be regenerated. Regeneration can be accomplished with either a cold fluid or a hot fluid. The cold fluid to use is air; the hot fluid can be steam, air, or some other substance. The fluid must be initially free of the pollutant substance in order to completely regenerate the bed. In practice, though, complete regeneration may be unnecessary, so that the fluid could contain some residual amount of pollutant if necessary. With cold regeneration, the bed will be ready for use as soon as regeneration is complete.

For the regeneration process, Eq. (12-3) can be written to apply. The subscript R refers to the flow rate and properties of the regenerating fluid; V'_R is the speed of the desorption zone as it moves upward or downward through the bed. The equation for V'_R is

$$V'_R = \frac{\dot{m}_R}{\rho_R \rho' c A} (\alpha_R \rho_R)^{1/\beta_R} C_{mv_L}{}^{(\beta_R - 1)/\beta_R} \qquad (12\text{-}15)$$

The regeneration time is given by Eq. (12-13) as

$$t_R = \frac{L}{V'_R} \qquad (12\text{-}16)$$

In Eq. (12-15), C_{mv_L} is the concentration in the regeneration fluid at the exit from the bed, given by

$$C_{mv_L} = \alpha_R \rho_R C'_{m_L}{}^{\beta_R} = \alpha_R \rho_R \left(\frac{C_{mv_0}}{\alpha \rho} \right)^{\beta_R / \beta} \qquad (12\text{-}17)$$

Equation (12-17) can be substituted into Eq. (12-15) to yield

$$V'_R = \frac{\dot{m}_R}{\rho' c A} \alpha_R \left(\frac{C_{mv_0}}{\alpha \rho} \right)^{(\beta_R - 1)/\beta} \qquad (12\text{-}18)$$

Substituting Eq. (12-18) into Eq. (12-16) gives

$$t_R = \frac{\rho' c A L}{\dot{m}_R \alpha_R} \left(\frac{\alpha \rho}{C_{mv_0}} \right)^{(\beta_R - 1)/\beta} \qquad (12\text{-}19)$$

The width of the desorbing zone, although it could readily be found, is not of particular importance and will be left alone. The constants α_R and β_R may differ from α and β for the adsorption process, although for cold regeneration they may well be the same.

EXAMPLE 12-2 The adsorbent bed of Example 12-1 is to be regenerated using standard air. The air flow rate is 0.25 kg/s. Assume α_R and β_R are the same as α and β in Example 12-1. Compute the regeneration time t_R.

SOLUTION Equation (12-15) gives V'_R as

$$V'_R = \frac{0.25}{1.185(600)(0.5)(10)}[150(1.185)]^{1/2.5}(0.002)^{1.5/2.5}$$

$$= 0.00001341 \text{ m/s}$$

Equation (12-16) gives the regeneration time t_R:

$$t_R = \frac{0.5}{0.00001341(3600)} = 10.4 \text{ h} \qquad Ans. \qquad ////$$

This example shows why cold regeneration is seldom used. The regeneration time is several times the useful adsorption time, and the result is an airstream enriched in the pollutant. However, by employing several stages in this manner, each stage being considerably smaller than the preceding one, an airstream rich in pollutant can be obtained.

The next example illustrates the use of hot regeneration.

EXAMPLE 12-3 The adsorbent bed of Example 12-1 is to be regenerated using saturated steam at 400 K. The steam flow rate is 0.25 kg/s, and we assume that $K = 200 \text{ s}^{-1}$, $\alpha_R = 1500$, and $\beta_R = 2.5$. Determine the regeneration time t_R.

SOLUTION The density of the steam is found from the steam tables to be 1.37 kg/m³. Equation (12-18) gives for the velocity of the adsorption zone

$$V'_R = \frac{0.25}{600(0.5)(10)}\, 1500\left[\frac{0.002}{150(1.185)}\right]^{1.5/2.5}$$

$$= 0.0001342 \text{ m/s}$$

Equation (12-16) then gives for the regeneration time

$$t_R = \frac{L}{V'_R} = \frac{0.5}{0.0001342(3600)} = 1.04 \text{ h} \qquad Ans. \qquad ////$$

Drying and Cooling an Adsorbent Bed

The process of drying the adsorbent bed can be handled in very much the same way as the adsorption of the pollutant or the process of regeneration. A drying zone is set up within the bed, with a velocity given by

$$V'_D = \frac{\dot{m}_D C_{mv_D}}{\rho_D \rho' c A C'_{m_D}} \quad (12\text{-}20)$$

in which the subscript D refers to the drying fluid, usually air, or to the drying process. The concentration C_{mv_D} is the saturation concentration of steam in air at the initial temperature T_0 of the bed, given by the following equation:

$$C_{mv_D} = \frac{M_w}{R_u T_0} P_w = 0.002165 \frac{P_w}{T_0} \quad (12\text{-}21)$$

in which P_w is the saturation pressure of steam at temperature T_0.

The concentration C'_{m_D} is the mass of liquid water attached to the adsorbent per unit mass of adsorbent at the start of the drying process. This quantity is impossible to obtain except by experiment; however, we can estimate an upper and a lower bound on the value of C'_{m_D}. For the maximum value of C'_{m_D}, we write

$$C'_{m_{D,\,max}} = \frac{\rho_w (V - V'')}{\rho' c V}$$

in which V is the volume of the adsorbent bed and V'' is the volume of the adsorbent material inside the bed, with all internal pores within the adsorbent granules excluded. The latter volume is given by $V'' = \rho' c V / \rho''$, so that

$$C'_{m_{D,\,max}} = \rho_w \left(\frac{1}{\rho' c} - \frac{1}{\rho''} \right) \quad (12\text{-}22)$$

For the minimum value of C'_{m_D}, we write

$$C'_{m_{D,\,min}} = \frac{\rho_w (V' - V'')}{\rho' c V}$$

in which V' is the volume of the adsorbent granules within the bed, including the internal pore volume. Since $V' = cV$, the preceding equation becomes

$$C'_{m_{D,\,min}} = \rho_w \left(\frac{1}{\rho'} - \frac{1}{\rho''} \right) \quad (12\text{-}23)$$

The time required for drying is given by

$$t_D = \frac{L}{V'_D} \quad (12\text{-}24)$$

The cooling process occurs simultaneously with the drying process, but for now, let us treat the cooling as though it occurs by itself. A cooling zone will form which is similar to the adsorption zone of Fig. 12-4. An energy balance around the cooling zone yields the following equation:

$$V'_c = \frac{\dot{m}_c c_{pc}}{\rho' c A c'} \quad (12\text{-}25)$$

in which c_{pc} is the specific heat at constant pressure of the cooling fluid, usually air, and c' is the specific heat of the adsorbent substance. Also, $\dot{m}_c$ is the mass flow rate of the cooling fluid. The cooling time is given by

$$t_c = \frac{L}{V'_c} \quad (12\text{-}26)$$

Since the drying and cooling processes occur at the same time, the combined process will not require the total $t_D + t_c$ given by the sum of Eqs. (12-24) and (12-26). The latent heat of vaporization of the water which is to be removed is a complicating factor, but since the water is at the higher temperature, its vaporization should assist in cooling the bed. At any rate, we shall neglect the effect of the latent heat of vaporization of the water and recommend that the time required for cooling and drying the bed be the maximum of t_D or t_c, given by Eqs. (12-24) or (12-26), respectively.

For carbon, the value of ρ', which is the density of the adsorbent granules with the volume of the internal pores included, is of the order of 750 kg/m^3. The value of ρ'', the density of the carbon if the internal pores are filled in with carbon, is of the order of 2000 kg/m^3. The value of the specific heat c' of the carbon is around 800 J/kg · K.

EXAMPLE 12-4   Compute the drying and cooling time for a bed for which the following data are available:

$$A = 10 \text{ m}^2 \qquad\qquad L = 1 \text{ m}$$
$$\dot{m}_D = 0.2 \text{ m}^3/\text{s} \qquad\quad T = 100°\text{C}$$
$$\rho' = 700 \text{ kg/m}^3 \qquad \rho'' = 2200 \text{ kg/m}^3$$
$$c = 0.7 \qquad\qquad\quad c' = 850 \text{ J/kg K}$$

The drying fluid is air. Assume that C'_{m_D} is the arithmetic average of the minimum and maximum values.

SOLUTION   The value of ρ_w from the steam tables is 958.3 kg/m^3, and that of P_w is 101,326 N/m^2. Equations (12-22) and (12-23) give the maximum and minimum values of C'_{m_D}:

$$C'_{m_{D,\,\text{max}}} = 958.3 \left[\frac{1}{700(0.7)} - \frac{1}{2200} \right] = 1.52$$

$$C'_{m_{D,\,\text{min}}} = 958.3 \left(\frac{1}{700} - \frac{1}{2200} \right) = 0.933$$

The average of these values gives $C'_{m_D} = 1.23$.

Equation (12-21) gives $C_{mv_D} = 0.002165(101326)/373 = 0.588$ kg/m^3. Equation (12-20) gives

$$V'_D = \frac{0.2(0.588)}{0.946(700)(0.7)(10)(1.23)} = 0.00002061 \text{ m/s}$$

in which the density of air at 100°C is 0.946 kg/m^3. The drying time is computed from Eq. (12-24) as

$$t_D = \frac{1.0}{0.00002061(3600)} = 13.5 \text{ h}$$

Equation (12-25) gives for V'_c

$$V'_c = \frac{0.2(1010)}{700(0.7)(10)(850)} = 0.0000485$$

Equation (10-26) then gives the cooling time as $1.0/(0.0000485 \times 3600) = 5.73$ h. Since the drying time is longer, it will be the time required to complete the cooling and drying process.   ////

Pressure Drop through an Adsorbent Bed

The pressure drop through a packed adsorbent bed is a standard problem in chemical engineering, and considerable attention has been devoted to it. Leva [1, 2] has given a generalized empirical equation which correlates the data reasonably well over a wide range. His correlation makes use of the Reynolds number based on the diameter of a granule in the bed, as given by

$$\text{Re} = \frac{V_0 D'}{\nu} = \frac{QD'}{A\nu} \qquad (12\text{-}27)$$

in which D' is the mean diameter of an adsorbent granule, defined as the diameter of a sphere having the same volume as the granule. This is equivalent to D_s of Eq. (4-15). The equation for the pressure drop is

$$\Delta P = \frac{2f\rho Q^2 L c^{3-n}}{A^2 D' \phi^{3-n}(1-c)^3} \qquad (12\text{-}28)$$

In Eq. (12-28), f is the friction factor, c is the packing density as we have used it previously, and L is the thickness of the bed. The quantity ϕ is a factor relating the shape of the granule and must be interpreted carefully for an adsorbent particle. The basic definition of ϕ is the ratio of the surface area of a sphere of diameter D' (having volume equal to that of the particle) to the actual surface area of the particle. This relation may be written as

$$\phi = \pi^{1/3} 6^{2/3} \frac{V'^{2/3}}{A} = 4.836 \frac{V'^{2/3}}{A} \qquad (12\text{-}29)$$

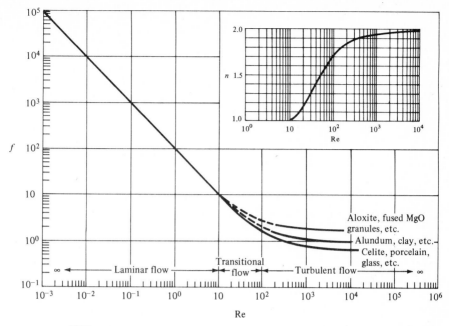

FIGURE 12-7

Curve of friction factor f and exponent n for flow through an adsorbent bed for use in Eqs. (12-25) and (12-27). (*From M. Leva, "Fluidization," p. 49, McGraw-Hill Book Company, New York, 1959. Used by permission.*)

in which V' is the volume of the adsorbent particle including the interior pore structure. The area A is the surface area of the particle without including the surface area of the interior pore structure. In other words, A is the surface area of a solid particle of about the same gross shape as the porous particle. It is unlikely that the interior pore structure will have a noticeable effect on the flow resistance through the bed. The exponent n in Eq. (12-28) is a function of the Reynolds number of the flow through the bed.

For laminar flow, which occurs if Re < 10, Eq. (12-28) may be replaced by

$$\Delta P = \frac{200\mu QLc^2}{AD'^2\phi^2(1-c)^3} \quad (12\text{-}30)$$

There will also be a transition region for $10 < \text{Re} < 100$ and a fully turbulent region for $\text{Re} > 100$. In these regions, the size and shape of the particles affect the friction factor f. Although empirical formulas are available for determining f in the fully turbulent region, these will not be presented here. Figure 12-7 shows curves of f and n as functions of Reynolds number. Although the degree of roughness for adsorbent particles is difficult to determine, it is suggested that one of the curves for rough particles be used.

EXAMPLE 12-5 Compute the pressure drop for the adsorbent bed of Example 12-1. Assume the average diameter of the adsorbent particles to be 0.1 mm. Take $\phi = 0.6$.

SOLUTION Using the data from Example 12-1, the volumetric flow rate of air through the adsorber is $1.5/1.185 = 1.266$ m³/s. Equation (12-27) gives the Reynolds number

$$\text{Re} = \frac{QD'}{Av} = \frac{1.266(10^{-4})}{10(1.55 \times 10^{-5})} = 0.817$$

The flow is in the laminar range. Equation (12-30) applies, giving

$$\Delta P = \frac{200(1.84 \times 10^{-5})(1.266)(0.5)(0.5)^2}{10(10^{-4})^2(0.6)^2(0.5)^3} = 129{,}413 \text{ N/m}^2 \qquad Ans.$$

The power loss for this pressure drop is $1.266(129.4) = 164$ kW.　　　////

Reduction of Pressure Drop

It is seen from Example 12-5 that the pressure drop through an adsorbent bed can easily become excessive. One is then led to ask if such large pressure drops are necessary. Clearly, to reduce the pressure drop through the bed, a larger bed is the most obvious answer. Problem 12-5 shows that the capacity of the bed is proportional to the volume of the bed, while Eq. (12-30) indicates that the pressure drop decreases as the bed area increases or as the bed thickness decreases. These observations suggest that a large thin bed is best. Equation (12-10) shows that the ratio of the height of the adsorption zone to the height of the bed is proportional to the volume of the bed; thus no complications to the analysis will ensue upon making the bed large and thin. There is undoubtedly an optimum set of proportions for the bed geometry. If the bed is too large and thin, the cost of the housing and support structure will be excessive while little saving in power cost will be obtained over that for a thicker bed. No attempt will be made here to systematically optimize the bed proportions, although such an attempt seems reasonable; it is suggested that several alternative designs be tried to see which is most economical.

　　　Various methods are employed in practice to increase the bed area without proportionately increasing the cost of the structure. One of the simplest methods involves the use of a conical bed inside a cylindrical shell. Such an arrangement is discussed by Danielson [3]. Another method for increasing the bed area is to arrange a number of trays inside the shell, each tray supporting an adsorbent bed and with the flow split to pass through each tray; such an arrangement is shown in Fig. 12-8. Other arrangements involve the use of hollow cylindrical cannisters, corrugated cells, and perforated tubes filled with adsorbent. These last three arrangements are discussed by Mantell [4].

　　　The next example shows how increasing the surface area reduces the pressure drop; in fact, the pressure drop in this example is almost negligible, indicating that the area is probably too large.

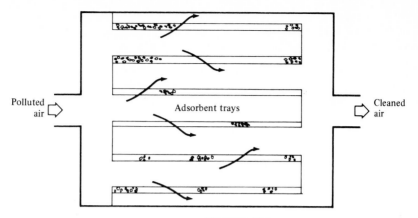

FIGURE 12-8
Adsorber with a number of adsorbent trays.

EXAMPLE 12-6 A stream of 10 m³/s of standard air polluted with a substance for which $C_{mv_0} = 0.003$ kg/m³ passes through an adsorbent bed. The bed has an area of 1000 m² and a thickness of 2 cm. Assume that $\alpha = 200$ and $\beta = 4$ at standard temperature. Assume that $K = 50$ s⁻¹, $\rho' = 650$ kg/m³, $c = 0.4$, $D' = 1$ mm, and $\phi = 0.5$.

(a) Find the breakthrough time. What is the pressure drop and the power required?

(b) The bed is regenerated with a stream of 1.0 m³/s of air at 600 K, at which temperature $\alpha = 820{,}000$ and $\beta = 4$. Determine the regeneration time and the concentration of pollutant leaving with the regeneration air. Compute the pressure drop and power loss.

(c) The bed is cooled with 30 m³/s of standard air. How much time is required, and what will be the pressure drop and the power loss? Assume $c' = 750$ J/kg · K.

SOLUTION (a) Equation (12-3) gives the velocity of the adsorption zone as it passes downward through the bed:

$$V' = \frac{10}{650(0.4)(1000)} [200(1.185)]^{0.25}(0.003)^{0.75} = 1.934 \times 10^{-6} \text{ m/s}$$

Equation (12-10) gives for the height δ of the adsorption zone:

$$\delta = \frac{10}{50(1000)} \left[4.595 + \frac{1}{3} \ln \frac{1 - (0.01)^3}{1 - (0.99)^3} \right] = 0.001153 \text{ m}$$

Then Eq. (12-12) gives for the breakthrough time t_B

$$t_B = \frac{0.02 - 0.001153}{1.934 \times 10^{-6}} = 9745 \text{ s} = 2.71 \text{ h} \qquad Ans.$$

The Reynolds number is computed from Eq. (12-27) as 0.645 for which the flow is laminar. Equation (12-30) applies for the pressure drop and gives

$$\Delta P = \frac{200(1.84 \times 10^{-5})(10)(0.02)(0.4)^2}{1000(10^{-3})^2(0.5)^2(0.6)^3} = 2.181 \text{ N/m}^2 \qquad Ans.$$

The power loss associated with this pressure drop is

$$\dot{W} = Q\,\Delta P = 10(2.181) = 21.8 \text{ W} \qquad Ans.$$

(b) The density of the regenerating air is computed from the perfect-gas law to be 0.588 kg/m³. The viscosity at 600 K is found from Fig. 2-10 to be 3×10^{-5} kg/m s. Equation (12-18) gives for the velocity V'_R of the regenerating zone

$$V'_R = \frac{1.0}{650(0.4)(1000)}\,820{,}000(0.588)\left[\frac{0.003}{200(1.185)}\right]^{0.75} = 0.00394 \text{ m/s}$$

Equation (12-16) gives for the regeneration time

$$t_R = \frac{0.02}{0.000394(3600)} = 0.0141 \text{ h} \qquad Ans.$$

For the concentration of the exit regeneration air, we have from Eq. (12-17)

$$C_{mv_L} = \alpha_R \rho_R \left(\frac{C_{mv_0}}{\alpha\rho}\right)^{\beta_R/\beta}$$

$$= 820{,}000(0.588)\left[\frac{0.003}{200(1.185)}\right]^{1.0}$$

$$= 6.10 \text{ kg/m}^3 \qquad Ans.$$

A value for the exit concentration can be obtained alternatively by equating the amount of pollutant in the gas which leaves the bed to that which was in the bed initially, giving

$$Q_R C_{mv_L} t_R = C'_{m_L}\rho'cAL = \rho'cAL\left(\frac{C_{mv_0}}{\alpha\rho}\right)^{1/\beta}$$

$$C_{mv_L} = \frac{650(0.4)(1000)(0.02)}{1.0(0.0141)(3600)}\left[\frac{0.003}{200(1.185)}\right]^{0.25} = 6.11 \text{ kg/m}^3$$

The agreement is virtually exact. The pressure drop and power loss are computed in the same manner as in (a) to be 0.355 N/m² and 0.355 W, respectively.

(c) Equation (12-25) gives the velocity of the cooling zone as

$$V'_c = \frac{30(1.185)(1005)}{650(0.4)(1000)(750)} = 0.000183 \text{ m/s}$$

The cooling time is calculated to be

$$t_c = \frac{0.02}{0.000183(3600)} = 0.03032 \text{ h} \qquad Ans.$$

The pressure drop and power loss are computed to be 6.54 N/m² and 196 W.

////

The use of air at 600 K to regenerate the adsorbent in the preceding example would not be acceptable if the adsorbent were carbon owing to the possibility of combustion.

12-3 MOVING-BED ADSORBERS

With a fixed-bed adsorber, it is necessary to periodically remove an adsorber from service for regeneration, drying, and cooling. This process can be done automatically if the bed is arranged to move in the right way. Adsorbers where the bed moves from the polluted fluid to the regeneration fluid to the drying and cooling fluid are called *moving-bed* adsorbers. Although there are several possible arrangements, the most common type of moving-bed adsorber has a cylindrical bed which slowly rotates about its axis. A cross-sectional view of such an adsorber is shown in Fig. 12-9. In this arrangement the bed rotates about its axis with angular velocity ω while the frame and partitions are stationary.

Three sections are shown in Fig. 12-9: an adsorption section, a regeneration section, and a drying and cooling section. The bed has length H in the direction normal to the plane of the paper. The three fluids enter the respective annular sections outside the bed from inlets at the ends of the unit. The fluids pass through the bed to the annular sections inside the bed and then out the ends of the unit. As the bed revolves past the radial partitions, an element of the bed will pass from the adsorption section to the regeneration section; another element will pass from the regeneration section to the drying and cooling section; and a third element will pass into the adsorption section. In this way, the bed becomes continuously saturated, is regenerated, and is then dried and cooled in preparation for another adsorption cycle.

The analysis and design of a moving-bed adsorber is easily done, using the equations developed in Sec. 12-2. Letting R be the radius to the midpoint of the bed, the bed area is given by

$$A = RH\theta \qquad (12\text{-}31)$$

The pertinent equations of Sec. 12-2 can be written in the following form: Equation (12-12) becomes

$$t_B = \frac{AL/Q - A\delta/Q}{AV'/Q} \qquad (12\text{-}32)$$

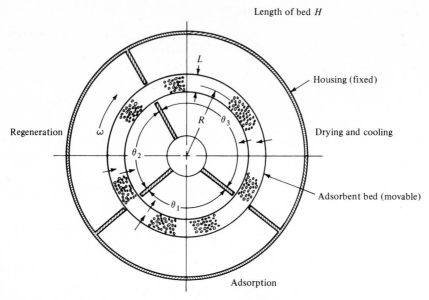

FIGURE 12-9
Cross section of a rotating-bed adsorber.

with Eqs. (12-3) and (12-10) written as

$$\frac{AV'}{Q} = \frac{1}{\rho' c} (\alpha\rho)^{1/\beta} C_{mv_0}{}^{(\beta-1)/\beta} \qquad (12\text{-}33)$$

$$\frac{A\delta}{Q} = \frac{1}{K} \left[4.595 + \frac{1}{\beta-1} \ln \frac{1-(0.01)^{\beta-1}}{1-(0.99)^{\beta-1}} \right] \qquad (12\text{-}34)$$

Equation (12-19) may be written as

$$t_R = \frac{A_R L}{Q_R} \frac{\rho' c}{\rho_R \alpha_R} \left(\frac{\alpha\rho}{C_{mv_0}} \right)^{(\beta_R-1)/\beta} \qquad (12\text{-}35)$$

and Eq. (12-24) for the drying process may be written as

$$t_D = \frac{A_D L}{Q_D} \frac{\rho' c C'_{m_D}}{C_{mv_D}} \qquad (12\text{-}36)$$

For the cooling process, Eq. (12-26) may be written

$$t_c = \frac{A_c L}{Q_c} \frac{\rho' c c'}{\rho_c c_{pc}} \qquad (12\text{-}37)$$

In Eqs. (12-32) to (12-37) the terms on the right are constants, except for the quantities AL/Q, $A_R L/Q_R$ and either $A_D L/Q_D$ or $A_c L/Q_c$, depending on whether t_D or t_c is larger. Now Eq. (12-31) shows that

$$A = RH\theta_1 \qquad A_R = RH\theta_2 \qquad A_D = RH\theta_3 \qquad \text{or} \qquad A_c = RH\theta_3$$

Using the angular velocity ω of the bed, we may also write

$$t_B = \frac{\theta_1}{\omega} \qquad t_R = \frac{\theta_2}{\omega} \qquad t_D = \frac{\theta_3}{\omega} \qquad \text{or} \qquad t_c = \frac{\theta_3}{\omega}$$

Substituting these equations into Eqs. (12-35) to (12-37) and (12-32) gives

$$\frac{\omega RHL}{Q_R} = \frac{\rho_R \alpha_R}{\rho' c}\left(\frac{C_{mv_0}}{\alpha\rho}\right)^{(\beta_R - 1)/\beta} \tag{12-38}$$

$$\frac{\omega RHL}{Q_D} = \frac{C_{mv_D}}{\rho' c C'_{m_D}} \qquad \text{or} \qquad \frac{\omega RHL}{Q_c} = \frac{\rho_c c_{pc}}{\rho' cc'} \tag{12-39}$$

$$\frac{\theta_1}{\omega} = \frac{(RHL/Q)\theta_1 - A\delta/Q}{AV'/Q} \tag{12-40}$$

Let us solve Eq. (12-40) for $\omega RHL/Q$:

$$\frac{\omega RHL}{Q} = \frac{AV'}{Q} + \frac{\omega}{\theta_1}\frac{A\delta}{Q} \tag{12-41}$$

We may then write for Q_D and Q_R

$$Q_D = \frac{(\omega RHL/Q)Q}{(\omega RHL/Q_D)} \qquad \text{or} \qquad Q_c = \frac{(\omega RHL/Q)Q}{(\omega RHL/Q_c)} \tag{12-42}$$

$$Q_R = \frac{(\omega RHL/Q)Q}{(\omega RHL/Q_R)} \tag{12-43}$$

We may also write

$$RHL = \left(\frac{\omega RHL}{Q}\right)\frac{Q}{\omega} \tag{12-44}$$

The design procedure is to first select ω and θ_1 (if the first choice turns out unsatisfactorily, a second choice can be made and the calculations repeated). The angles θ_2 and θ_3 can be selected arbitrarily as long as $\theta_1 + \theta_2 + \theta_3 \leq 2\pi$. Next solve for $\omega RHL/Q$ and for Q_D and Q_R using Eqs. (12-41) to (12-43). Then solve for RHL from Eq. (12-44), and after selecting two of these quantities, solve for the third. The quantities in this design procedure which can be arbitrarily selected can of course be chosen to meet other criteria, such as those involving pressure drop through the bed, proportions of the adsorber, or limiting values of the adsorber dimensions. The following example illustrates the design procedure.

EXAMPLE 12-7 A movable-bed adsorber is to handle a stream of polluted air for which the following data are given:

$Q = 10 \text{ m}^3/\text{s}$ (standard air, $C_{mv_0} = 0.005 \text{ kg/m}^3$) $c' = 770 \text{ J/kg} \cdot \text{K}$
$L = 10 \text{ cm}$ $\alpha = 400$
$R = 1.0 \text{ m}$ $\beta = 3.5$
$\rho' = 600 \text{ kg/m}^3$ $\alpha_R = 3000$
$\rho'' = 2000 \text{ kg/m}^3$ $\beta_R = 3.5$
$c = 0.3$ $K = 75 \text{ s}^{-1}$

Assume the regeneration fluid is steam at 100°C, the drying and cooling fluid is standard air, and design the unit.

SOLUTION The density of liquid water at 100°C is 958 kg/m³, and the density of the steam is 0.598 kg/m³. Equation (12-33) gives

$$\frac{AV'}{Q} = \frac{1}{600(0.3)}[400(1.185)]^{1/3.5}(0.005)^{2.5/3.5} = 0.000734$$

Equation (12-34) gives

$$\frac{A\delta}{Q} = \frac{1}{75}\left[4.595 + \frac{1}{2.5}\ln\frac{1-(0.01)^{2.5}}{1-(0.99)^{2.5}}\right] = 0.08098 \text{ s}$$

Equations (12-22) and (12-23) give

$$C'_{m_{D,\text{max}}} = 958\left[\frac{1}{600(0.3)} - \frac{1}{2000}\right] = 4.845$$

$$C'_{m_{D,\text{min}}} = 958\left(\frac{1}{600} - \frac{1}{2000}\right) = 1.118$$

Then, $C'_{m_D} = 2.98$, which is the average of these two values. Equation (12-38) gives

$$\frac{\omega RHL}{Q_R} = \frac{0.598(3000)}{600(0.3)}\left[\frac{0.005}{400(1.185)}\right]^{2.5/3.5} = 0.00278 \text{ s}$$

Equation (12-21) gives $C_{mv_D} = 0.588 \text{ kg/m}^3$; Eq. (12-39) gives

$$\frac{\omega RHL}{Q_D} = \frac{0.588}{600(0.3)(2.98)} = 0.001096 \text{ s}$$

It can be verified that t_D exceeds t_c, so that the drying process controls.
     Let us now take $\theta_1 = \pi$ and $\omega = 4.0$ rev/h = 0.00698 rad/s. Equation (12-41) gives

$$\frac{\omega RHL}{Q} = 0.000734 + \frac{0.00698}{\pi}(0.08098) = 0.000914$$

Equations (12-42) and (12-43) give

$$Q_D = \frac{0.000914(10)}{0.001096} = 8.34 \text{ m}^3/\text{s} \qquad \textit{Ans.}$$

$$Q_R = \frac{0.000914(10)}{0.00278} = 3.29 \text{ m}^3/\text{s} \qquad \textit{Ans.}$$

Equation (12-44) gives

$$RHL = 0.000914 \frac{10}{0.00698} = 1.31 \text{ m}^3$$

Since $R = 1.0$ m and $L = 0.1$ m, then this gives $H = 13.1$ m. We may select θ_2 and θ_3 as 60 and 120°, respectively.   ////

REFERENCES

1 Leva, M.: Fluid Flow Through Packed Beds, *Chem. Eng.*, vol. 56, no. 5, pp. 115–117, May 1949.
2 Leva, M.: "Fluidization," pp. 45–61, McGraw-Hill Book Company, New York, 1959.
3 Danielson, J. A., Ed.: Air Pollution Engineering Manual, pp. 197–198, *Public Health Service Publ. No. 999-AP-40*, National Center for Air Pollution Control, Cincinnati, Ohio, 1967. (Available from the U.S. Government Printing Office, Washington, D.C.)
4 Mantell, C. L.: "Adsorption," pp. 262–265, McGraw-Hill Book Company, New York, 1951.

PROBLEMS

12-1 Derive Eq. (12-3).

12-2 Obtain the integral of Eq. (12-9) prior to substitution of the limits. Show that the integral becomes infinite at the limits of 0 and 1.

12-3 Evaluate the integral in Prob. 12-2 between the limits of a and b, where a is slightly greater than 0 and b is slightly less than 1. Plot the left side of Eq. (12-9) as a function of a if b is taken as $1 - a$; use a log plot and let a range from 0.1 to 0.00001. Use $\beta = 3$.

12-4 Using the data of Example 12-1, what area of bed is required if the breakthrough time is to be lengthened to 6 h?

12-5 Show that with all other parameters fixed, the exhaustion time is directly proportional to the volume of the bed.

12-6 A bed of adsorbent material for which $\alpha = 250$ and $\beta = 4$ is to remove a pollutant from a stream of 10 m³/s of standard air in which $C_{mv_0} = 0.001$ kg/m³. The bed area is 20 m², and its depth is 2.0 m. Assume that $K = 40$ s⁻¹, $\rho' = 700$ kg/m³, and $c = 0.7$. Compute the breakthrough time and the exhaustion time.

12-7 How thick must the adsorbent bed in Prob. 12-6 be in order for the breakthrough time to be 8 h? What will be the exhaustion time in this case?

12-8 The parameter α in Eq. (12-1) varies strongly with temperature. It appears that a representation of the form

$$\alpha = k_1 e^{T/k_2}$$

is not unreasonable, at least over a moderate temperature range. Assume that the value of α is known at temperature T_1 (call this value α_1), and assume that the temperature difference needed to double the value of α is ΔT_2. Show that k_1 and k_2 can be evaluated from the following equations:

$$k_2 = 1.443 \Delta T_2$$
$$k_1 = \alpha_1 e^{-T_1/k_2}$$

Assume the value of α in Example 12-1 occurs when $T_1 = 300$ K. Plot a curve of α as a function of temperature over the range from 300 to 700 K. Assume $\Delta T_2 = 25°C$.

12-9 Consider the adsorbent bed of Prob. 12-6. Determine the regeneration time if 1.0 m^3/s of standard air is used to regenerate the bed. Use the values of α and β given in Prob. 12-6.

12-10 Rework Example 12-3 over a range of temperatures from 300 to 700 K. Use the equation for α developed in Prob. 12-8. Plot the regeneration time as a function of steam temperature T.

12-11 Consider the adsorbent bed of Prob. 12-6. Determine the regeneration time if 1.0 kg/s of steam at 373 K is used to regenerate the bed. Assume $\alpha_R = 2000$, $\beta_R = 4$, and $K = 50$ s^{-1}.

12-12 Derive Eq. (12-25).

12-13 Consider the bed of Probs. 12-6 and 12-11. Assume $\rho'' = 1900$ kg/m^3 and $c' = 810$ J/kg·K. Let the bed be cooled and dried with a stream of air at 25°C flowing at the rate of 25 m^3/s. Compute the cooling and drying time required.

12-14 For the regeneration process of Example 12-2, determine the pressure drop and the power expended if the adsorbent particle diameter is 0.1 mm and the shape factor ϕ is 0.6.

12-15 For the regeneration process of Example 12-3, determine the pressure drop and the power expended if the adsorbent particle diameter is 0.1 mm and the shape factor ϕ is 0.6.

12-16 Repeat Prob. 12-15 for the drying and cooling process of Example 12-4.

12-17 Compute the pressure drop and the power expended in overcoming this pressure drop for the adsorbent bed of Prob. 12-6. Use $D' = 0.2$ mm and $\phi = 0.5$.

12-18 Repeat Prob. 12-17 but for the regeneration process of Prob. 12-11.

12-19 Repeat Prob. 12-17 but for the drying and cooling process given in Prob. 12-13.

12-20 Take the adsorbent bed of Examples 12-1 and 12-5. Allow the diameter of the adsorbent particles to vary over the range from 0.05 to 5 mm. Keeping the remaining data the same, compute and plot the pressure drop as a function of D'.

12-21 Use the data of Example 12-1 and determine the area of the bed required for the pressure drop to be 10,000 N/m^2. Recompute the breakthrough time for this new area.

12-22 Rework Example 12-6 for an area of 200 m^2 and a bed thickness of 10 cm, the other data remaining the same.

12-23 Write Eq. (12-19) for the regeneration time in terms of the volume of the bed.

12-24 Derive an equation similar to Eq. (12-18) for the case in which the regeneration fluid has an initial concentration C_{mv_i} of pollutant when it enters the bed.

12-25 Use the results of Prob. 12-24 to estimate the regeneration time of the bed in Example 12-6. Assume that the hot air enters the bed with a concentration C_{mv_i} which is in equilibrium with an adsorbent concentration equal to 35 percent of C'_{m_L}. Compute the values of C_{mv_i} and C_{mv_L}. What collection efficiency can be obtained on the adsorption cycle if the bed is regenerated to this extent?

12-26 When the adsorbent bed of Example 12-6 has been fully saturated, what mass of pollutant does the bed contain? Assuming a latent heat of adsorption of 3×10^6 J/kg, how much energy is evolved during the adsorption process? What rise in bed temperature does this entail in the absence of cooling? What flow rate of air is required to cool the bed if a 10°C rise in temperature of the cooling air is allowed and if the cooling process continues for 1.0 h?

12-27 Determine the pressure drop and the power expended in overcoming this pressure drop for the moving-bed adsorber of Example 12-7 if $D' = 1.0$ mm and $\phi = 0.7$.

12-28 A stream of 2.0 m³/s of air containing a pollutant with $C_{mv_0} = 0.0025$ kg/m³ is to be treated in a moving-bed adsorber. Design an adsorber to do the job if the following data are known: $\rho' = 700$ kg/m³, $\rho'' = 1900$ kg/m³, $c = 0.5$, $c' = 740$ J/kg · K, $\alpha = 300$, $\beta = 4.5$, $\alpha_R = 5000$, $\beta_R = 4$, and $K = 40$ s^{-1}. The regeneration fluid is steam at 100°C. Make any necessary assumptions.

12-29 If $D' = 0.6$ mm and $\phi = 0.67$, determine the pressure drop and the power required for your design of Prob. 12-28.

12-30 Use the data of Probs. 12-28 and 12-29. Design a moving-bed adsorber for which the power required to overcome the pressure drop through the bed does not exceed 10 kW.

12-31 A stream of 30 m³/s of polluted air for which $C_{mv_0} = 0.003$ kg/m³ is to be cleaned with a moving-bed adsorber. The adsorbent bed is to be regenerated using air at 475 K. The power required to overcome the pressure drop through the bed must not exceed 50 kW. Design an adsorber if the following data are given: $\rho' = 710$ kg/m³, $\rho'' = 2200$ kg/m³, $c = 0.45$, $c' = 750$ J/kg · K, $\alpha = 250$, $\beta = 4$, $\alpha_R = 15,000$, $\beta_R = 4$, $K = 70$ s^{-1}, $D' = 0.8$ mm, and $\phi = 0.75$.

13

COMBUSTION DEVICES

Throughout the process industries a wide variety of odoriferous substances is produced, and these would be discharged into the atmosphere if suitable control measures were not used. Most of these substances, particularly those having the most objectionable odors, are organic compounds. They are distinguishable by the fact that their odor is detectable when the substance is present in very small concentrations in the ambient air. Often these substances are not harmful in the concentrations in which they are likely to occur in the atmosphere, except for the effects of their odor. But since their odor is very noticeable and ever present, much pressure is brought to bear on the polluter to control the emissions of these odoriferous compounds. These organic compounds can be in either gaseous or aerosol form.

Some examples of operations in which organic compounds are generated include paper pulping, coffee roasting, meat processing, painting, varnish cooking, petroleum refining, and chemical processing. Residents of many communities will gladly point out the operation of greatest concern to them.

Owing to the extremely low concentration threshold for odor detection, a high collection efficiency is normally required for any acceptable control device. Collection of the offending substance is not necessary; its destruction or its transformation to unoffensive substances is all that is required. Collection can best be accomplished by adsorption or condensation, while destruction is best done by combustion.

Fortunately, organic compounds can be oxidized to carbon dioxide, water vapor, and in some cases an additional oxide such as sulfur dioxide, nitrogen dioxide, or a metallic oxide. The carbon dioxide and water vapor are harmless, while the other oxides and other products of oxidation of the original compound remain a problem unless they can be easily separated from the air; even so, the problem will be much less severe than the original one.

In addition to control of odors, there are cases in which it is desirable to remove organic gases or aerosols prior to removal of particulates in order to protect the particle-collection device. An example is a blast furnace or foundry cupola in which an afterburner is employed before the baghouse in order to prevent clogging of the filter pores by condensed organic vapors.

The control of organic compounds by combustion is done in an afterburner or furnace. Where an afterburner is employed, it is generally constructed for the particular polluted airstream, although it is practical to use a central afterburner into which a number of polluted streams are fed. An existing furnace can also serve as a combustion chamber for destruction of pollutants. In this case, however, the furnace must serve two functions and cannot be removed from service for one function without disrupting the other function as well. In this chapter, we shall consider the design of afterburners in detail.

Where a pollutant is to be destroyed by combustion, the combustion must be complete; otherwise, intermediate products of combustion will form, which may prove worse than the original substance. To achieve complete combustion requires an excess of oxygen, a sufficiently high temperature, sufficiently long residence time at this temperature, and high degree of turbulence to achieve intimate mixing of pollutant and oxygen (the three T's of combustion: temperature, time, and turbulence). To achieve the high temperature, the afterburner normally must be separately fueled, so that the fuel must also be mixed with the oxygen. And, of course, the fuel must also burn completely without forming any pollutants.

The use of afterburners and combustion chambers is costly in fuel usage. There are ways in which fuel consumption can be reduced, and we shall examine these in some detail. At best, a considerable amount of fuel is required.

Catalytic afterburners are also used in which the chemical reactions involved are facilitated by the use of a catalyst. These afterburners can operate at lower temperatures than combustion chambers, with a consequent saving in fuel. We shall examine the catalytic afterburner in this chapter.

13-1 CHEMISTRY AND THERMODYNAMICS OF COMBUSTION

Combustion is an exothermic chemical reaction, which means that thermal energy is liberated when the reaction occurs. Such a reaction can be expressed by the usual chemical reaction equation, which places the reactants on the left and the products on the right. For instance, the combustion of carbon in oxygen is represented by the following equation:

$$C + O_2 \longrightarrow CO_2 \qquad (13\text{-}1)$$

For the combustion of hydrogen in oxygen, the equation is

$$H_2 + \tfrac{1}{2}O_2 \longrightarrow H_2O \qquad (13\text{-}2)$$

and for sulfur,

$$S + O_2 \longrightarrow SO_2 \qquad (13\text{-}3)$$

Combustion usually takes place in air, which consists of oxygen, nitrogen, and certain other substances that can be lumped in with the nitrogen. The proportions in air are such that C_v is 0.21 for oxygen and 0.79 for nitrogen, which means that for every mole of oxygen present there are $0.79/0.21 = 3.76$ mol of nitrogen present. The nitrogen can be carried along in the combustion equation. In this way, Eqs. (13-1) to (13-3) may be written for combustion in air:

$$C + O_2 + 3.76N_2 \longrightarrow CO_2 + 3.76N_2 \qquad (13\text{-}4)$$
$$H_2 + \tfrac{1}{2}O_2 + 1.88N_2 \longrightarrow H_2O + 1.88N_2 \qquad (13\text{-}5)$$
$$S + O_2 + 3.76N_2 \longrightarrow SO_2 + 3.76N_2 \qquad (13\text{-}6)$$

Equations (13-4) to (13-6) are written for complete combustion in theoretical, or stoichiometric, air, which can be achieved only if every fuel molecule mates perfectly with an oxygen molecule. Such a condition requires an infinite time at elevated temperature and perfect mixing. We do not expect this to occur in practical combustion, so that Eqs. (13-4) to (13-6) are not practical as they stand, only theoretical. To modify these equations to fit a practical situation, we can allow for incomplete combustion, which means that a wide variety of products may form, each of which represents partial oxidation of fuel molecules. We can also provide excess air to the combustion process, which, with reasonable time and turbulence, will allow virtually complete combustion to occur. An example of partial combustion in stoichiometric air is

$$C + O_2 + 3.76N_2 \longrightarrow 0.8CO_2 + 0.2CO + 0.1O_2 + 3.76N_2 \qquad (13\text{-}7)$$

An example of complete combustion in excess air is

$$H_2 + O_2 + 3.76N_2 \longrightarrow H_2O + \tfrac{1}{2}O_2 + 3.76N_2 \qquad (13\text{-}8)$$

In Eq. (13-8) the amount of air used is twice the theoretical amount of Eq. (13-5) and is described as 200 percent stoichiometric air, or 100 percent excess air.

Note that Eqs. (13-7) and (13-8) are written in terms of number of moles of each species, which is proportional to the number of molecules of the species since each mole contains the same number of molecules (Avogadro's number, 6.02×10^{26} molecules/kg-mol). The number of atoms of each element is conserved in a reaction (the number of moles is not conserved). This conservation forms the basis for balancing the equations, as has been done in each of the preceding equations; the procedure for balancing the equation should readily be recalled from basic chemistry. A chemical equation can also be written on the basis of mass in which each coefficient represents the mass of that compound in the reaction. However, this is not conventionally done, and we shall not do it here.

We may also write the combustion equation for typical organic pollutants. For the complete combustion of diethyl ketone in stoichiometric air,

$$C_5H_{10}O + 7O_2 + 26.3N_2 \longrightarrow 5CO_2 + 5H_2O + 26.3N_2 \qquad (13\text{-}9)$$

For the complete combustion of methyl mercaptan in stoichiometric air,

$$CH_4S + 3O_2 + 11.3N_2 \longrightarrow CO_2 + 2H_2O + SO_2 + 11.3N_2 \qquad (13\text{-}10)$$

Note the undesirable product SO_2.

An important parameter in designing a combustion chamber is the air-fuel ratio defined as the mass of air to the mass of fuel in the reaction. Since $m = nM$, we may multiply the number of moles of air by its molecular weight and divide the result by the product of the number of moles of fuel times its molecular weight. Thus

$$AF = \frac{28.97 n_a}{M_f n_f} \qquad (13\text{-}11)$$

in which n_a is the sum of the number of moles of oxygen and that of nitrogen.

EXAMPLE 13-1 Methane (CH_4) is to be burned in 150 percent stoichiometric air. Write and balance the combustion equation, and determine the air-fuel ratio required.

SOLUTION First, we write the combustion equation with the unknown molar coefficients written in terms of symbols. The equation for 1 mol of methane is

$$CH_4 + xO_2 + 3.76xN_2 \longrightarrow yCO_2 + zH_2O + 3.76xN_2$$

Now we shall balance the equation for stoichiometric air. A carbon balance taken first shows that there is 1 atom of carbon on the left, so that $y = 1$. An H_2 balance shows that there are 2 molecules of hydrogen on the left, and hence $z = 2$. An O_2 balance gives $x = y + z/2 = 1 + 1 = 2$. Then the equation may be written

$$CH_4 + 2O_2 + 7.52N_2 \longrightarrow CO_2 + 2H_2O + 7.52N_2$$

To write the equation for 150 percent stoichiometric air, we increase the air by 50 percent and carry the excess to the right, giving

$$CH_4 + 3O_2 + 11.3N_2 \longrightarrow CO_2 + 2H_2O + O_2 + 11.3N_2$$

The air-fuel ratio AF is obtained from Eq. (13-11):

$$AF = \frac{28.97(3 + 11.3)}{16(1)} = 25.9 \qquad Ans.$$

EXAMPLE 13-2 Determine the stoichiometric air-fuel ratio for the combustion of formaldehyde.

SOLUTION The balanced stoichiometric equation for combustion of formaldehyde in air is

$$CH_2O + O_2 + 3.76N_2 \longrightarrow CO_2 + H_2O + 3.76N_2$$

Equation (13-11) gives the air-fuel ratio:

$$AF = \frac{28.97(1 + 3.76)}{30(1)} = 4.60 \qquad Ans. \qquad ////$$

Heat of Combustion

The product of an exothermic reaction, such as CO_2, is in a lower energy state than the reactants C and O_2 which form it if the pressure and temperature of the product and reactants are the same. This means that if a reaction occurs at room temperature and atmospheric pressure, for example, the energy of the product is less than the combined energies of the reactants. We thus speak of a change of energy and an associated change of enthalpy of the reaction; for an exothermic reaction these changes of energy and enthalpy are negative. By heat of combustion we usually mean the negative of the change in enthalpy during the reaction at constant pressure and temperature, which is the meaning we shall use here. The heat of combustion differs from one fuel to another; Table 13-1 lists the heats of combustion for several common fuels.

Adiabatic Combustion Temperature

The thermal energy of the reactants and products of the reaction is determined by the pressure and temperature at which these species exist. We have mentioned that when the pressure and temperature are the same, the reactants are in a higher energy state than the products. If the reaction occurs adiabatically, that is, with no heat or work transfers, the total energy must be conserved in the reaction. To compensate for the reduced energy state of the products, the thermal energy of the products will be elevated. Thus, the products of an adiabatic combustion reaction will emerge at a high temperature. Our concern is with a constant-pressure reaction. Where no chemical reaction is involved, the reactants and products are heated sensibly, with the change in enthalpy related to the energy transfer as in Eq. (2-105).

Let us consider an adiabatic combustion reaction which begins with reactants supplied to the reaction chamber at state 1, as shown in Fig. 13-1. The products will emerge from the chamber at state 2. We assume that the change of enthalpy of the reaction at room temperature is known. In Fig. 13-2 a process is shown which connects states 1 and 2 and which allows us to use the change in enthalpy at room temperature. For the adiabatic process 1-2, the overall change in enthalpy is zero, so that state 2 lies at the same enthalpy as state 1, as shown in Fig. 13-2.

We shall be very interested in the temperature T_2 with which the products leave the reactor; this temperature is called the *adiabatic combustion temperature*. We observe that $H_2 = H_1$ and follow the process 1-3-4-2 in Fig. 13-2, writing

$$H_2 - H_1 = (H_2 - H_4) + (H_4 - H_3) + (H_3 - H_1) = 0$$

Now $H_4 - H_3 = \Delta H_R$, $H_2 - H_4 = m_{pr}(h_2 - h_4)$, and $H_3 - H_1 = m_r(h_3 - h_1)$, where m_{pr} and m_r are the masses of the products and reactants, respectively. Note that $m_{pr} = m_r$. For the enthalpies h_1, h_2, h_3, and h_4, take the ordinary values based on the assumption of a perfect gas having variable specific heats. with zero value of enthalpy assumed to be at some convenient datum temperature such as 0°C. The preceding equation may be written

$$h_2 = h_1 + (h_4 - h_3) - \frac{\Delta H_R}{m_r}$$

Noting that $m_r = m_a + m_f = m_f(1 + AF)$, this equation may be written

$$h_2 = h_1 + (h_4 - h_3) + \frac{-\Delta H_R/m_f}{1 + AF} \qquad (13\text{-}12)$$

Table 13-1 HEATS OF COMBUSTION OF COMMON FUELS

Substance	Formula	Molecular weight, M	Heat of combustion,* $-\Delta H_R/m_f$, MJ/kg
Hydrogen	H_2	2.016	120.9
Carbon	C	12.0	32.79
Sulfur	S	32.0	9.28
Carbon monoxide	CO	28.0	10.11
Methane	CH_4	16.0	50.14
Ethane	C_2H_6	30.1	47.60
Propane	C_3H_8	44.1	46.45
Butane	C_4H_{10}	58.1	45.88
Octane (*l*)	C_8H_{18}	114.2	44.55
Methyl alcohol (*l*)	CH_3OH	32.0	19.97
Benzene (*l*)	C_6H_6	78.1	40.23

* These numbers are based on the fuel in the gaseous state except where indicated otherwise; in all cases the water formed by combustion is in the gaseous state.

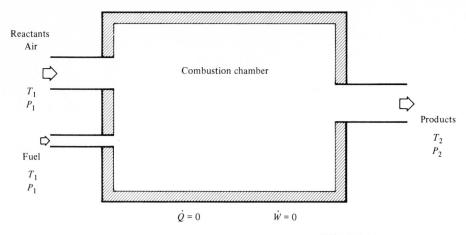

FIGURE 13-1
Adiabatic combustion reactor.

In most combustion reactions of interest, the reactants and the products will be mostly nitrogen, with the specific enthalpies of the other constituents differing slightly from that of nitrogen. A fairly good approximation is obtained if the specific enthalpy of air is used throughout for both the reactants and the products. If this approximation is made, then $h_4 = h_3$ and Eq. (13-12) becomes

$$h_2 = h_1 + \frac{-\Delta H_R/m_f}{1 + AF} \qquad (13\text{-}13)$$

Figure 13-3 shows the variation of specific enthalpy h with temperature for air. The value of $-\Delta H_R/m_f$ is taken from Table 13-1.

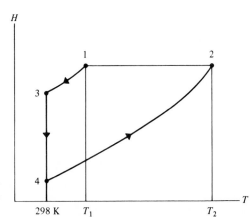

FIGURE 13-2
Enthalpy-temperature diagram for a combustion process. The direct process 1-2 is replaced by the process 1-3-4-2.

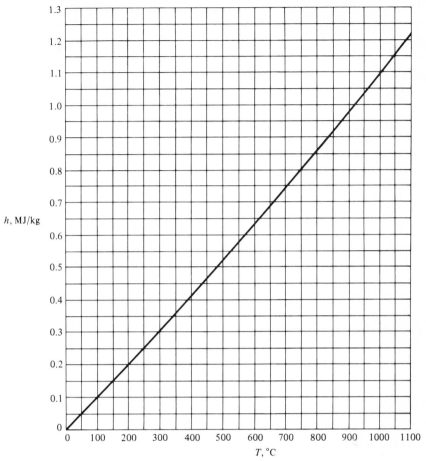

FIGURE 13-3
Specific enthalpy of air as a function of temperature. Datum temperature is 0°C.

EXAMPLE 13-3   Determine the adiabatic combustion temperature for methane burned in 200 percent excess air. The inlet temperature is 50°C.

SOLUTION   From Example 13-1 we have an air-fuel ratio of 25.9 for 150 percent stoichiometric air; in the present example we are dealing with 300 percent stoichiometric air, so that the air-fuel ratio is twice the preceding value, or 51.8. Table 13-1 shows that for methane $-\Delta H_R/m_f = 50.14$ MJ/kg, while Fig. 13-3 gives $h_1 = 0.05$ MJ/kg at 50°C. Equation (13-13) can be solved for h_2, giving

$$h_2 = 0.05 + \frac{50.14}{1 + 51.8} = 0.9996 \text{ MJ/kg}$$

Referring back to Fig. 13-3 shows that $T_2 = 920°C$.   *Ans.*   ////

EXAMPLE 13-4   With what amount of excess air must methane be burned to produce an adiabatic combustion temperature of 700°C if the inlet temperature is 150°C?

SOLUTION   Figure 13-3 gives $h_1 = 0.15$ MJ/kg at 150°C and $h_2 = 0.745$ MJ/kg at 700°C. From Table 13-1, $-\Delta H_R/m_f = 50.14$ MJ/kg. Equation (13-13) can be solved for AF to give

$$AF = \frac{-\Delta H_R/m_f}{h_2 - h_1} - 1 = \frac{50.14}{0.745 - 0.15} - 1 = 83.3$$

From Example 13-1, we may easily compute AF_s, the air-fuel ratio for combustion in stoichiometric air, as

$$AF_s = \frac{25.9}{1.5} = 17.3$$

Our air-fuel ratio of 83.3 corresponds to $83.3/17.3 = 4.82$, or 482 percent of stoichiometric air, giving an excess air of 382 percent.   *Ans.*   ////

Additional Equations

If a mixture of fuels is burned, and the heat of combustion of each constituent of the mixture is available, the heat of combustion of the mixture can be obtained from the following equation:

$$\frac{-\Delta H_R}{m_f} = \frac{m_1}{m_f}\left(\frac{-\Delta H_R}{m_f}\right)_1 + \frac{m_2}{m_f}\left(\frac{-\Delta H_R}{m_f}\right)_2 + \cdots \qquad (13\text{-}14)$$

in which the subscripts 1, 2, ..., refer to the different constituents of the fuel mixture.

An equation similar to Eq. (13-14) can be obtained to account for the heating value of the pollutant substances in the stream. By adjusting the heating value in this way, the equations previously derived remain valid even when this extra heat release is taken into account. The equation is

$$\frac{-\Delta H_R}{m_f} = \left(\frac{-\Delta H_R}{m_f}\right)_s + C_{m_p}\left(\frac{-\Delta H_R}{m_f}\right)_p AF = \left(\frac{-\Delta H_R}{m_f}\right)_s + \frac{C_{m v_p}}{\rho_a}\left(\frac{-\Delta H_R}{m_f}\right)_p AF$$
$$(13\text{-}14a)$$

in which $(-\Delta H_R/m_f)_f$ is the heat of combustion of the fuel and $(-\Delta H_R/m_f)_p$ is the heat of combustion of the pollutant substances.

Equations similar to Eqs. (13-12) and (13-13) can be derived for the case in which heat is transferred from the combustion chamber at the rate $\dot{Q}$. These equations are

$$h_2 = h_1 + (h_4 - h_3) + \frac{-\Delta H_R/m_f}{1 + AF} - \frac{\dot{Q}}{\dot{m}_a(1 + 1/AF)} \qquad (13\text{-}15)$$

$$h_2 = h_1 + \frac{-\Delta H_R/m_f}{1 + AF} - \frac{\dot{Q}}{\dot{m}_a(1 + 1/AF)} \qquad (13\text{-}16)$$

13-2 COMBUSTION-CHAMBER DESIGN

In this section we shall consider the rough design of a combustion chamber for the combustion of waste gases. For complete combustion of these gases, as we have mentioned, the gases must be maintained at a sufficiently high temperature for a sufficiently long time in the presence of adequate turbulence. The required temperature ranges from 375 to 825°C; some pollutant gases will be adequately destroyed at the lower temperatures, while virtually all organic pollutant gases will be destroyed at the higher temperatures if the other factors are suitable. An adequate residence time is of the order of 0.2 to 0.5 s. An adequate degree of turbulence will generally occur if the gas velocity is in the range 4.5 to 7.5 m/s; in addition, turbulence can be promoted if the air and fuel are injected into the combustion chamber tangentially. These values will give an idea of the range of parameters necessary for the design of the combustion chamber.

One of the factors not yet considered is the loss of heat from the chamber to the surroundings by radiation and convection. Heat loss can be minimized by insulating the combustion chamber, which should be done, not only to conserve energy, but for safety as well. This heat loss can be computed using the standard methods of heat-transfer analysis. We shall not make such computations in this book; the reader is referred to standard heat-transfer texts for the necessary equations. Once the heat-transfer rate has been computed, Eq. (13-15) or (13-16) can be used to find the exit-gas temperature. For preliminary design purposes, ordinarily a simple estimate of the heat loss will be enough, and a loss of 10 percent is suggested in the absence of more specific data based on experience with units similar to the one being designed. This heat loss can be handled very neatly by increasing the required adiabatic combustion temperature an amount equal to 10 percent of the difference between the adiabatic combustion temperature and the inlet temperature.

The volume of the combustion chamber must be sufficient to provide the needed residence time for the flow rate of air and fuel, or products of the combustion. We may write

$$V = AVt = Qt \qquad (13\text{-}17)$$

Let us base the volumetric flow rate Q on the gas at its highest temperature and write

$$V = \frac{\dot{m}}{\rho_2} t = \frac{\rho_1 Q_1 + \dot{m}_f}{\rho_2} t \qquad (13\text{-}18)$$

In order to provide good turbulence, the velocity of the product gases should be in the range 4.5 to 7.5 m/s, mentioned earlier. The cross-sectional area of the chamber is then given by

$$A = \frac{Q_2}{V} = \frac{\dot{m}}{\rho_2 V} \qquad (13\text{-}19)$$

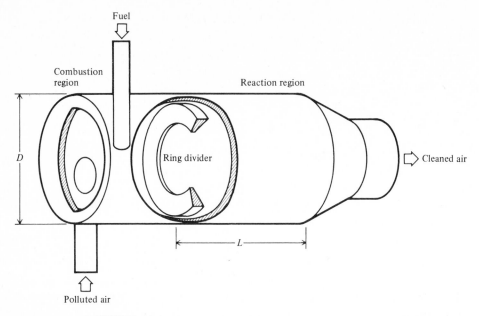

Fuel

Combustion region

Reaction region

D

Ring divider

Cleaned air

L

Polluted air

FIGURE 13-4
Partial cross section of a combustion chamber showing the tangential entries for the polluted air and fuel streams.

Figure 13-4 shows the shape and burner arrangement of a typical combustion chamber used for destruction of pollutant gases, although most actual combustion chambers have additional burners spaced around the periphery of the chamber. The chamber is lined with refractory material on the inside and covered with insulation on the outside. Further details of the construction and operation of afterburners are given by Danielson [1, pp. 171–178]. The afterburner can be oriented either horizontally or vertically as needed.

The length L of the reaction chamber shown in Fig. 13-4 can be readily obtained from Eqs. (13-18) and (13-19). We may write

$$L = \frac{V}{A} \quad (13\text{-}20)$$

This is the length of the reaction region only; the length of the combustion region to the left of the ring divider must be added, as well as the lengths of the stack cone and supports at the other end.

EXAMPLE 13-5 A combustion chamber is to treat 10 m³/s of standard air carrying a pollutant with $C_{mv} = 0.005 \text{ kg/m}^3$. The fuel is methane with $AF = 51.8$ (see Example 13-3). Take a chamber velocity (axial component) of 5 m/s and a residence time of 0.5 s, and determine the length and diameter of a combustion chamber similar to that of Fig. 13-4.

SOLUTION   The low value of C_{mv} of the pollutant means that it will contribute negligibly to the heat of combustion. Let us assume here that the process is adiabatic, and with this assumption the exit temperature is 920°C, as computed in Example 13-3. The exit density ρ_2 is computed as 0.296 kg/m^3 at this temperature. The mass flow rate in the reaction section is

$$\dot{m} = \dot{m}_a\left(1 + \frac{1}{AF}\right) = \rho_1 Q_1\left(1 + \frac{1}{AF}\right) = 1.185(10)\left(1 + \frac{1}{51.8}\right) = 12.08 \text{ kg/s}$$

Equation (13-18) gives the volume of the reaction chamber:

$$V = \frac{\dot{m}}{\rho_2}t = \frac{12.08(0.5)}{0.296} = 20.4 \text{ m}^3$$

Equation (13-19) gives for the cross-sectional area of the reaction chamber

$$A = \frac{12.08}{0.296(5)} = 8.16 \text{ m}^2$$

which corresponds to a diameter of 3.22 m. The length of the reaction chamber is given by Eq. (13-20) as

$$L = \frac{20.4}{8.16} = 2.50 \text{ m} \qquad Ans.$$

The combustion region would have a length of 1.0 to 1.5 m, giving a total length of 3.5 to 4.0 m exclusive of the exhaust cone.   ////

13-3   ARRANGEMENT FOR ENERGY CONSERVATION

It is apparent that the simple combustion-chamber arrangement of Sec. 13-2 is wasteful of energy; not only that, it discharges high-temperature gases into the atmosphere, which is particularly bad when energy supplies are inadequate. Fortunately, the energy consumption of the typical afterburner can be greatly reduced by the use of additional equipment or by modifications in the design of the unit.

The simplest procedure is to utilize the hot exhaust gases leaving the afterburner as a source of heat. To do this, a heat exchanger is placed in the exhaust stream following the afterburner; a second fluid is heated in the exchanger and is used for some needed purpose. This procedure works well provided there is a need for a heated fluid which is matched to the heat available in the exhaust gases in quantity, period of availability, and location. Unfortunately, such a match occurs only occasionally; too often, the need for heat is too far removed from the afterburner, requires too much or too little heat, or operates on a different schedule than the afterburner.

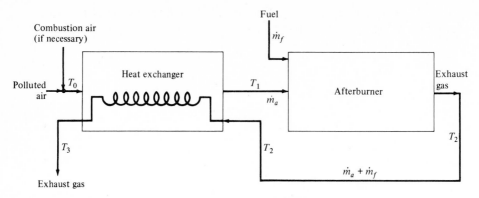

FIGURE 13-5
Regenerative arrangement for an afterburner.

A better arrangement is the use of regenerative heat recovery, as shown in Fig. 13-5. The hot exhaust gases are used to heat the incoming airstream. The polluted airstream is mixed with any additional air required for the combustion of fuel in the afterburner. This mixture is then heated in the exchanger and then goes to the afterburner, where it is heated further by the combustion of fuel and where the pollutant is destroyed. The exhaust gases then pass through the heat exchanger and are discharged at a temperature somewhat above that of the inlet polluted stream. If this temperature is sufficiently above the ambient value, further heat recovery for some other purpose can be effected with a second heat exchanger.

Analysis of the Regeneration System

To analyze the system of Fig. 13-5, let us first define the heat-exchanger effectiveness ε:

$$\varepsilon = \frac{h_1 - h_0}{h_2 - h_0} \quad (13\text{-}21)$$

which is in accord with the standard definition of effectiveness. We shall not discuss heat-exchanger theory in this book; the reader is referred to a suitable heat-transfer text, such as Parker, Boggs, and Blick [2], for effectiveness data for particular types and designs of heat exchangers.

Next, we take an energy balance around the heat exchanger, which gives

$$\dot{m}_a(h_1 - h_0) = (\dot{m}_a + \dot{m}_f)(h_2 - h_3)$$

Solving this equation for h_3 and introducing the air-fuel ratio $AF = \dot{m}_a/\dot{m}_f$ gives

$$h_3 = h_2 - \frac{AF}{AF + 1}(h_1 - h_0) \quad (13\text{-}22)$$

Next, we take Eq. (12-13), solve it for h_1, and substitute the resulting equation into Eq. (13-21). Solving for ε gives

$$\varepsilon = 1 - \frac{-\Delta H_R / m_f}{(h_2 - h_0)(AF + 1)} \qquad (13\text{-}23)$$

Solving Eq. (13-23) for AF, we have

$$AF = \frac{-\Delta H_R / m_f}{(h_2 - h_0)(1 - \varepsilon)} - 1 \qquad (13\text{-}24)$$

We substitute Eqs. (13-21) and (13-24) into Eq. (13-22), giving

$$h_3 = h_2 - \varepsilon\left[1 - \frac{(h_2 - h_0)(1 - \varepsilon)}{-\Delta H_R / m_f}\right](h_2 - h_0) \qquad (13\text{-}25)$$

The enthalpies can be read from the temperatures and vice versa from Fig. 13-3. We see from Eq. (13-24) that the air-fuel ratio can be made as large as desired, which means that the fuel consumption can be made as small as desired by using a heat exchanger whose effectiveness approaches unity. It should be pointed out, however, that for this application, where the flow rates and specific heats of the two fluids are nearly equal, a practical heat-exchanger effectiveness lies in the range 0.6 to 0.7.

EXAMPLE 13-6 In a regenerative arrangement such as that of Fig. 13-5, an afterburner handles 10 m³/s of standard air which contains a pollutant. Assume no extra combustion air is required. The afterburner is fueled with methane, and the exhaust-gas temperature T_2 must be 900°C. Assume a heat-exchanger effectiveness of 0.65, and compute the air-fuel ratio and the temperature T_3 of the exhaust gas leaving the heat exchanger.

SOLUTION From Table 13-1 and Fig. 13-3 we have

$$\frac{-\Delta H_R}{m_f} = 50.14 \qquad h_0 = 0.025 \qquad h_2 = 0.975 \text{ MJ/kg}$$

Equation (13-24) gives

$$AF = \frac{50.14}{(0.975 - 0.025)(1 - 0.65)} - 1 = 149.8 \qquad Ans.$$

From Eq. (13-25) the enthalpy of the exhaust gas leaving the heat exchanger is

$$h_3 = 0.975 - 0.65\left[1 - \frac{0.95(0.35)}{50.14}\right]0.95 = 0.362 \text{ MJ/kg}$$

Then Fig. 13-3 gives the temperature $T_3 = 355°C.$ Ans. ////

13-4 FLAMMABLE MIXTURES AND FLARES

One important aspect of the design of combustion equipment for the destruction of pollutants is safety from explosion and improper combustion. Most industrial polluted airstreams are not combustible, and combustion occurs only after the addition of fuel, which can be done under properly controlled conditions. A number of streams, however, are combustible or even explosive, and these must be handled with great care. Examples include certain effluent streams from petroleum refineries and blast furnaces.

Flammable mixtures can be handled in one of two ways: They may be destroyed in special burners designed to safely burn such mixtures—a flare is commonly used for this purpose—or they may be diluted with air to a point where the mixture is no longer flammable. Diluted mixtures can be safely handled in an afterburner of the type already discussed. A special case is that of a mixture too rich in fuel to burn; such a mixture becomes flammable with a certain degree of dilution. If such a mixture is to be diluted and destroyed, care must be exercised in both the dilution and combustion processes. A better plan for a fuel-rich mixture is to use it for fuel.

Limits of Flammability

A mixture of a fuel and air will burn (or perhaps explode) if certain conditions are met. These conditions relate to the relative concentrations of fuel and oxygen in the mixture, the way in which the mixture is confined, the velocity of the mixture, and the presence of some source of ignition. Burning can occur when the concentration of fuel in oxygen, or in air, lies in a range between lower and upper flammability limits. These flammability limits are influenced by the nature of confinement of the mixture. Table 13-2 shows the limits of flammability for certain gases and vapors in air; these values are for a burner having a large diameter and directed vertically upward. More complete data are presented in Ref. [3].

The shape and size of the combustion chamber or other confinement device has a very great influence on combustion, not only on the flammability limits, but also on whether combustion can occur at all. A small-diameter tube will not allow combustion to occur if the walls of the tube are cooled owing to the quenching effect of the walls. Various types of fittings inside the tube can have the same effect; flame arrestors are often made of perforated plates which operate on this quenching principle. Orientation of the combustion chamber is important, too, since a flame will propagate vertically upward more readily than in other directions. Except in the case of explosions, flames propagate with a certain finite velocity, and it is possible to prevent the spread of a flame along a duct by making the duct velocity sufficiently large.

For safety in handling potentially flammable mixtures, it is best to dilute the mixture to a point below its lower flammability limit. To allow for accidental variations in the mixture proportions and in the flammability limit,

Table 13-2 FLAMMABILITY LIMITS FOR GASES AND VAPORS IN AIR*

Substance	Lower flammability limit†	Upper flammability limit†
Ammonia	0.15	0.28
Hydrazine	0.047	1.00
Hydrogen	0.04	0.75
Hydrogen cyanide	0.06	0.41
Hydrogen sulfide	0.043	0.45
Carbon monoxide	0.125	0.74
Cyanogen	0.06	0.32
Methane	0.053	0.14
Ethane	0.030	0.125
Propane	0.022	0.095
Butane	0.019	0.085
Hexane	0.012	0.075
Heptane	0.012	0.067
Isooctane	0.011	0.060
Ethylene	0.031	0.32
Acetylene	0.025	0.81
Benzene	0.014	0.071
Toluene	0.014	0.067
Naphthalene	0.009	0.059
Cyclopropane	0.024	0.104
Cyclohexane	0.013	0.080
Water gas	0.070	0.72
Carbureted water gas	0.055	0.36
Natural gas	0.038–0.065	0.13–0.17
Gasoline	0.014	0.076
Naphtha	0.008	0.05
Kerosine	0.007	0.05
Coal gas	0.053	0.32
Coke-oven gas	0.044	0.34
Blast-furnace gas	0.35	0.74
Producer gas	0.17	0.70
Oil gas	0.047	0.33
Methyl alcohol	0.073	0.36
Ethyl alcohol	0.043	0.19
n-Propyl alcohol	0.021	0.135
n-Butyl alcohol	0.014	0.112
Methyl ether	0.034	0.18
Ethyl ether	0.019	0.48
Acetaldehyde	0.041	0.55
Acetone	0.03	0.11
Methyl ethyl ketone	0.018	0.10
Cyclohexanone	0.011	
Methylamine	0.049	0.207
Ethylamine	0.035	0.140
Methyl chloride	0.107	0.174
Methyl bromide	0.135	0.145
Vinyl chloride	0.040	0.220
Dichloroethylene	0.097	0.128
Dimethyl sulfide	0.022	0.197
Ethyl mercaptan	0.028	0.180

SOURCE: These values are selected from H. F. Coward and G. W. Jones, "Limits of Flammability of Gases and Vapors," tables 44 and 45, pp. 130–134, *Bull.* 503, Bureau of Mines, U.S. Government Printing Office, 1952.

* These values are for upward flame propagation in large-diameter combustion chambers.

† Volumetric concentration C_v of the flammable substance in air.

it is customary to dilute the mixture to a fuel concentration equal to 0.2 to 0.25 times the lower flammability limit as estimated or as determined by calculation. Other methods of control, such as flame arrestors or maintenance of a high gas velocity, are not considered safe enough to avoid the need for dilution to a point well below the flammability limit.

In most industrial processes, as mentioned previously, the mixture is normally diluted well below the lower flammability limit without any additional provision being made. However, explosive or inflammable mixtures can occur at special times even in such systems. For example, when the process has been shut off for a period of time, flammable substances can accumulate in the ductwork and with little or no air circulation can build up to explosive concentrations. In such a case, when normal operation is resumed, an explosion can occur when this mixture moves into the ignited afterburner. To minimize the danger of this occurring, the ductwork and equipment should be purged with air before turning on any source of ignition.

Flares

The most common method of destroying fuel gases when it is not deemed feasible to salvage them is by the use of flares. A flare is essentially a burner open to the atmosphere that is sufficiently removed from personnel and other structures so as not to cause damage. Most flares are elevated, located from 5 to 100 m above ground level. However, some flares, particularly emergency flares, are located at ground level in a well-protected area. Figure 13-6 shows a sketch of a flare burner which could be placed in any suitable location.

Ross [4] gives the following empirical equation for the required height of an elevated flare:

$$H = \frac{\sqrt{\dot{Q}}}{e} - 20D \qquad (13\text{-}26)$$

in which H is the height of the flare above surrounding objects, $\dot{Q}$ is the energy release rate of the flare in watts, D is the diameter of the flare, and e is a constant. The recommended values of e are $550 \sqrt{W}/m$ for continuous exposure of structures or equipment or for intermittent exposure of personnel for periods up to 20 min, and $320 \sqrt{W}/m$ for continuous exposure of personnel.

What are the advantages of using flares instead of combustion chambers for destruction of waste fuels? One advantage is simplicity and low cost. However, the main advantage relates to safety in the type of application for which flares are best suited. There is an inherent danger of explosion when an enclosed chamber is used with a flammable fuel mixture. No such danger exists with the flare since it is not enclosed, although there is a certain danger associated with the piping leading to the flare; this danger would exist in any case. Flares are best used for intermittent flows or for continuous flows of low-quality fuel

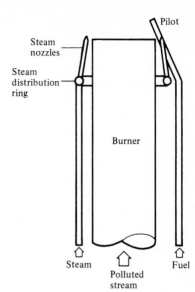

FIGURE 13-6
Sketch of a flare burner. This type of
burner is normally mounted several
meters above ground.

mixtures since otherwise it would be preferable to salvage the fuel. If the heating
value of the fuel mixture is sufficiently low, it may be necessary to add additional
fuel to the mixture before it can be flared. In such a case, a combustion
chamber may be preferable to a flare.

The primary disadvantage of the flare, aside from its wasteful nature,
is that it tends to discharge a considerable quantity of pollutants into the
atmosphere. Combustion in the open air is difficult to control to the extent
that all hydrocarbons are completely burned. As a result, flares tend to be
smoky. The smoke can be almost entirely eliminated by injection of steam into
the flare. Figure 13-6 shows a ring of steam nozzles by which steam is injected
into the combustion zone. One effect of the steam is to retard the formation
of long-chain hydrocarbons and inhibit polymerization of the hydrocarbons.
This effect is particularly striking with regard to the unsaturated hydrocarbons
in the mixture. An equation for the amount of steam required as a function of
the fraction by mass of the unsaturated hydrocarbons in the mixture is, for
smokeless operation

$$\frac{m_s}{m_f} = 0.1 + 4.15\frac{m_{unsat}}{m_f} \quad (13\text{-}27)$$

Equation (13-27) is written from a graph presented by Danielson [1, p. 596]; m_s
is the mass of steam required, m_f is the mass of fuel mixture burned, and
m_{unsat} is the mass of unsaturated hydrocarbons in mass m_f of fuel.

13-5 CATALYTIC AFTERBURNERS

A modification to the simple regenerative afterburner system of Fig. 13-5 involves the addition of a catalytic unit. The catalytic unit is inserted in order to facilitate the reaction between the pollutants and the oxygen; it is not intended to affect the combustion reaction. The effect of the catalyst is to lower the temperature required for the complete destruction of the pollutants. This saves fuel but at the expense of the catalytic unit as well as of periodic regeneration or replacement of the catalyst as it deteriorates with use. The catalytic unit is inserted after the combustion chamber, as shown in Fig. 13-7.

The action of the catalyst is to speed up the reaction at a given temperature or to lower the temperature required for a given reaction speed. The same heat of reaction occurs when a catalyst is used, and the adiabatic reaction temperature is the same whether a catalyst is present or not. The temperature of the gases increases across the catalytic unit, but this increase of temperature is normally small since the pollutant concentration is usually small. The analysis of the unit of Fig. 13-7 is very similar to that of the unit in Fig. 13-5. An example will illustrate the procedure.

EXAMPLE 13-7   The afterburner of Example 13-6, using the catalytic arrangement of Fig. 13-7, requires a temperature T_2 of 600°C. With the other data remaining the same, compute the AF ratio and the exhaust temperature T_3.

SOLUTION   At temperature $T_2 = 600$°C, the value of h_2 from Fig. 13-3 is 0.63 MJ/kg. As we did previously, we shall ignore the heat of reaction of the pollutant and use Eq. (13-24) for the AF ratio:

$$AF = \frac{50.14}{(0.63 - 0.025)(0.35)} - 1 = 236 \qquad Ans.$$

Equation (13-25) gives h_3 as

$$h_3 = 0.63 - 0.65 \left[1 - \frac{(0.63 - 0.025)(0.35)}{50.14} \right] (0.63 - 0.025) = 0.238 \text{ MJ/kg}$$

Figure 13-3 then gives $T_3 = 230$°C.   Ans.   ////

EXAMPLE 13-8   Repeat Example 13-7, but take into account the heat of reaction of the pollutant. Assume that $C_{mv} = 0.003$ kg/m^3 and $-\Delta H_R/m_{fp} = 20$ MJ/kg for the pollutant.

SOLUTION   Equation (13-14a) gives

$$\frac{-\Delta H_R}{m_f} = 50.14 + \frac{0.003(20)}{1.185} AF = 50.14 + 0.0506 AF$$

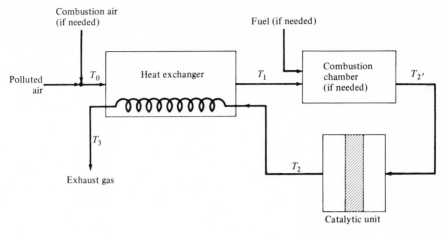

FIGURE 13-7
Catalytic combustion system.

Substituting this into Eq. (13-24), we have

$$AF = \frac{50.14 + 0.0506AF}{(0.63 - 0.025)(0.35)} = 235.8 + 0.239AF$$

Thus

$$AF = \frac{235.8}{1 - 0.239} = 310 \qquad Ans.$$

The temperature of the exhaust gas leaving the heat exchanger is the same as in Example 13-7. Thus $T_3 = 230°C$.   *Ans.*   ////

Comparing Examples 13-7 and 13-8 shows that the assumption of negligible heat of reaction of the pollutants is questionable. However, it is a conservative assumption and may provide a decent margin for error if this heat of reaction is neglected.

The Catalytic Unit

The catalytic unit consists of a catalyst bed located inside a shell or container. Suitable catalysts include platinum and platinum alloys, copper chromite, copper oxides, manganese, nickel, chromium, and cobalt. These catalysts should be arranged in such a way as to provide the maximum possible surface area in contact with the gas. The arrangement of a catalyst bed is somewhat similar to the arrangement of an adsorption bed. The catalyst is coated onto suitable elements such as metal ribbons, ceramic rods, or alumina pellets. These elements are then packed into the catalyst bed.

We shall determine the volume of the catalyst bed required for complete reaction. Our analysis will necessarily be superficial. Let us assume that the residence time t for the gas within the catalyst is known or can be estimated; in reality, the computation of this residence time is quite complex, and reference should be made to the literature on chemical kinetics and catalysis. For a volumetric flow rate Q of the gas entering the bed and a packing density c of the elements within the bed, the volume V of the bed is given by

$$V = \frac{Qt}{1-c} \qquad (13\text{-}28)$$

The velocity of the gas within the bed is given by

$$V = \frac{Q}{A(1-c)} \qquad (13\text{-}29)$$

in which A is the cross-sectional area of the bed.

EXAMPLE 13-9   For the catalytic afterburner of Example 13-7, determine a suitable size of the catalyst bed if $c = 0.5$ and if the required residence time t is 0.25 s.

SOLUTION   At a temperature T_2 of 600°C the density of air is calculated to be 0.404 kg/m^3. The volumetric rate is 10 m^3/s at standard conditions; at this elevated temperature it is

$$Q_2 = \frac{\rho Q}{\rho_2} = \frac{1.185(10)}{0.404} = 29.3 \text{ m}^3/\text{s}$$

Equation (13-28) then gives the volume of the catalyst bed as

$$V = \frac{29.3(0.25)}{1 - 0.5} = 14.7 \text{ m}^3$$

This volume can be obtained with a square bed 0.5 m thick and 5.4 m on each of the other two sides.   ////

REFERENCES

1   Danielson, J. A., ed.: Air Pollution Engineering Manual, *Public Health Service Publ. No. 999-AP-40*, National Center for Air Pollution Control, Cincinnati, Ohio, 1967. (Available from the U.S. Government Printing Office, Washington, D.C.)
2   Parker, J. D., J. H. Boggs, and E. F. Blick: "Introduction to Fluid Mechanics and Heat Transfer," pp. 401–409, Addison-Wesley Publishing Company, Inc., Reading, Mass., 1969.
3   Coward, H. F., and G. W. Jones: Limits of Flammability of Gases and Vapors, *Bull. 503*, Bureau of Mines, U.S. Government Printing Office, Washington, D.C., 1952.
4   Ross, R. D.: "Air Pollution and Industry," p. 327, Van Nostrand Reinhold Company, New York, 1972.

PROBLEMS

13-1 Derive Eq. (13-12).

13-2 Compute the stoichiometric air-fuel ratio for the combustion of butane in air.

13-3 Compute the stoichiometric air-fuel ratio for the combustion of methyl alcohol in air.

13-4 Acetylene (C_2H_2) is to be burned using 30 percent excess air. What air-fuel ratio is required?

13-5 Octane is burned with an air-fuel ratio of 25. What excess air is present?

13-6 Hexanone is to be burned in air. What air-fuel ratio is required for stoichiometric combustion?

13-7 A stream of polluted air contains amyl acetate with a concentration $C_{mv} = 0.01$ kg/m^3. When the pollutant is completely burned, what change in concentration C_{mv} of oxygen results?

13-8 A stream contains methyl ether, oxygen with a concentration of $C_v = 0.05$, and the remainder nitrogen and other inert substances. The stream is heated to a sufficiently high temperature that all the methyl ether burns. What maximum concentration C_{mv} of the ether could have been present?

13-9 Propane is to be burned with 300 percent stoichiometric air. What is the adiabatic combustion temperature? The inlet temperature is 20°C.

13-10 A coal consists of 75 percent carbon, 16 percent hydrogen, and the remainder inert substances measured by mass. Write the chemical equation for the stoichiometric combustion of this coal in air.

13-11 The coal of Prob. 13-10 is to be burned in 300 percent stoichiometric air. What will be the adiabatic combustion temperature? Assume an inlet temperature of 50°C.

13-12 The coal in Prob. 13-10 is to be burned so as to produce an adiabatic combustion temperature of 1000°C from an inlet temperature of 25°C. What air-fuel ratio is required, and what will be the percentage of excess air?

13-13 What is the adiabatic combustion temperature for burning methyl alcohol in 150 percent excess air? The inlet temperature is 10°C.

13-14 Butane is to be burned in air so as to produce an adiabatic combustion temperature of 800°C. What air-fuel ratio is required? Assume an inlet temperature of 25°C.

13-15 A fuel contains carbon monoxide, methane, and hydrogen in the following proportions by volume:

CO	40%
CH$_4$	35%
H$_2$	25%

If this fuel is burned in 200 percent excess air, what is the adiabatic combustion temperature? The inlet temperature is 50°C.

13-16 Suppose the fuel of Prob. 13-15 is to be diluted with nitrogen until it reaches a composition which is just capable of producing an adiabatic combustion temperature of 900°C when burned with the same amount of air. What volume of nitrogen must be mixed with each unit volume of the original fuel to achieve this degree of dilution? Assume an inlet temperature of 50°C.

13-17 The combustion process of Prob. 13-12 is to be used to treat a stream of 25 m^3/s of polluted air. If the necessary residence time is 0.25 s and the axial velocity in the reaction region is to be 10 m/s, find the dimensions of a suitable combustion chamber.

13-18 Use the fuel of Prob. 13-15 with the air-fuel ratio of that problem to fire a combustion chamber. Take $V = 7.5$ m/s and $t = 0.35$ s. A stream of 2.0 m³/s of polluted air is to be treated. Find the dimensions of the combustion chamber.

13-19 A stream of 5.0 m³/s of polluted air at 400°C is to be treated in an afterburner fueled with methane. The exit temperature must be at least 750°C, the axial velocity in the reaction chamber must be 7 m/s, and the required residence time is 0.4 s. There will be a heat loss by convection and radiation from the burner which can be assumed equal to 10 percent of the heat liberated in the combustion chamber. Determine suitable dimensions for the afterburner.

13-20 An existing furnace has a volume of 50 m³ and handles 30 m³/s of air. It is proposed to use this boiler to treat a stream of 10 m³/s of polluted air in addition to its present duty. What will be the residence time of the polluted air in this boiler? Assume the temperature following combustion is 1000°C.

13-21 For Example 13-6, compute the rate at which fuel is used, in kilograms per hour. Also compute the fuel rate if no heat exchange is used and compare the two values. What percent saving results?

13-22 A stream of 5.0 m³/s of nitrogen, carbon dioxide, and a pollutant gas (assume that the properties are the same as for air) enters the heat exchanger of the arrangement of Fig. 13-5 at 275°C and 1.0 atm pressure. Combustion air must be added before the stream enters the heat exchanger; assume standard air is available for this. Natural gas, for which $-\Delta H_R/m_f = 55$ MJ/kg, is used as the fuel in the afterburner. If $T_2 = 700°C$ and $\varepsilon = 0.6$, determine the amount of fuel required per hour and the temperature at which the exhaust gas leaves the heat exchanger.

13-23 The afterburner of Prob. 13-19 is to be used with a heat exchanger in a regenerative arrangement. The heat-exchanger effectiveness is 0.5. What will be the new fuel-consumption rate, and what saving in fuel consumption will result? Also, compute the temperature at which the exhaust gas leaves the heat exchanger.

13-24 The combustion chamber of Prob. 13-18, using the fuel of Prob. 13-15, is to be used with a heat exchanger in a regenerative arrangement. Assume the fuel costs $.10/kg. If the unit operates 2000 h/year, compute the annual fuel cost without the heat exchanger. A heat exchanger with some value of effectiveness is to be installed. A tabulation of heat-exchanger effectiveness and cost for the size of heat exchanger involved shows that

Effectiveness	Cost
0.4	$10,000
0.55	20,000
0.65	30,000
0.70	40,000
0.75	50,000
0.77	60,000
0.78	70,000

Assume the annual capitalization rate is 0.15. What heat-exchanger effectiveness results in the least annual cost? What is that cost, and what is the corresponding fuel-consumption rate?

13-25 A truckload of gasoline is to be delivered into a storage tank. During the delivery, which takes 1.5 h, 35 m³ of gasoline is transferred into the tank. While the tank is being filled, a mixture of gasoline vapor and air is exhausted from the tank; this mixture originally occupied the space in the tank above the liquid. For purposes of safety, we must assume the mixture is explosive, while for design purposes, let us assume that the mixture has the properties of pure gasoline vapor. The mixture leaving the tank is to be conveyed to an afterburner through a pipe after diluting with air so that the mixture is no longer flammable. Assuming uniform flow of gasoline during delivery, what flow rate of air must be used to dilute the gasoline vapor?

13-26 A fuel gas for which the heating value is 25 MJ/kg and for which the average molecular weight is 32.5 is to be burned in a flare. The flow rate of the gas is 0.25 m^3/s measured at 3 atm pressure and 25°C. Assuming that personnel will not be exposed to the flare for periods exceeding 20 min, how high should the flare be erected above the ground? The diameter of the flare is 0.2 m.

13-27 The fuel gas of Prob. 13-26 has a mass ratio of unsaturated hydrocarbons of 0.30. How much steam should be supplied to the flare to prevent the formation of smoke?

13-28 The fuel tank of an automobile is filled with 0.06 m^3 of gasoline during a period of 2.0 min. The temperature in the tank is 35°C, and during the filling process gasoline vapor is expelled from the tank. At what rate is the gasoline vapor expelled? If this vapor is to be diluted to a safe concentration and piped to a flare, what flow rate of dilution air is required?

13-29 The exhaust from an automobile engine is at 1000°C and 1 atm pressure and flows at the rate of 0.25 m^3/s. It has the following volumetric composition by percentage:

Constituent	Volumetric percentage
CO_2	13.2
O_2	0.3
CO	2.2
H_2	0.4
CH_4	0.3
N_2	83.6

A thermal reactor is to be installed in the exhaust system. This exhaust mixture, which has some combustible constituents present, is to be diluted with enough additional air at standard conditions to provide twice the theoretical oxygen required for complete combustion of the exhaust gas.

(a) What flow rate of air must be added to the exhaust gas to provide the needed oxygen for combustion?

(b) What will be the exhaust-gas temperature after dilution?

(c) What is the heating value of the exhaust gas?

(d) If the reaction occurs adiabatically, what will be the temperature of the exhaust gas leaving the reactor?

(e) Determine a suitable diameter and length of the reactor.

13-30 For Example 13-7, compute the rate at which fuel is used in kilograms per hour. Compare with the corresponding value from Prob. 13-21.

13-31 Suppose that the pollutant in Example 13-7 has a concentration of $C_{mv} = 0.005$ kg/m^3 and that this concentration is not diminished in the combustion chamber. Suppose further that the pollutant, if pure, would have a heating value of 30 MJ/kg. Determine the temperature rise $T_2 - T_{2'}$ across the catalyst.

13-32 A catalytic afterburner using the arrangement of Fig. 13-7 is to treat 25 m³/s of standard air polluted with a substance in a concentration of $C_{mv} = 0.003$ kg/m³. The temperature $T_2 = 700°C$, and the heat-exchanger effectiveness $\varepsilon = 0.75$. The fuel of Prob. 13-15 is to be used to fire the combustor. Determine the rate of consumption of the fuel. Assume the heat of combustion of the pollutant is 25 MJ/kg.

13-33 Determine the size of a catalyst bed suitable for the catalytic afterburner of Prob. 13-32. Assume $t = 0.2$ s and $c = 0.6$.

13-34 Assume that the catalyst bed of Example 13-9 consists of pellets 3.0 mm in diameter. Determine the pressure drop across the bed and the power expended in overcoming this pressure drop.

13-35 The exhaust gas from an automotive engine flows at 0.3 m³/s at a temperature of 900°C. It is to pass through a catalyst unit with a required residence time of 0.1 s and a packing density of 0.3. Compute the volume of the catalyst bed required, and suggest a suitable size for the bed. Compute the pressure drop across the bed you have selected if the bed consists of pellets 1.0 mm in diameter.

14

CONDENSATION DEVICES

Often the best way to recover an organic gaseous pollutant, where the substance is of value, is by the use of condensation. Usually, condensation alone will not remove enough of the pollutant to constitute an effective pollution control measure unless a refrigerated fluid is used in the condenser. Even so, a condenser using room-temperature water as the cooling fluid can be a valuable preliminary device before employing an adsorbing unit or an afterburner, for example. A condenser using chilled water or a refrigerant at a low temperature can be an effective pollution control device by itself.

In this chapter we shall consider the use of condensers as either partial or complete pollution control devices. We shall not go deeply into the analysis or design of heat exchangers since any good heat-transfer book covers this subject very well. We shall, however, consider those aspects of heat-exchanger theory and practice which are of particular concern to the types of mixtures involved in air pollution studies. We begin with a study of the physical properties of pollutants.

14-1 THERMODYNAMIC PROPERTIES OF POLLUTANTS

Let us consider certain thermodynamic properties of a number of substances for which condensation is a possible method of control. The property of primary interest is the relation of vapor pressure to temperature for saturated vapor. Of secondary interest is the variation of the latent heat of condensation h_{fg} as a function of temperature. Table 14-1 lists data from which the saturation pressure and latent heat can be estimated at a specified temperature for several pollutant gases. Data for other substances can be found in the sources listed at the end of Table 14-1 as well as in other sources. We shall examine how the data of this table may be used at a specified temperature.

The suggested equation for vapor pressure of the pollutant is the Frost-Kalkwarf-Thodos equation given by Reid and Sherwood [1, p. 124]. This equation appears in terms of the critical pressure P_c and critical temperature T_c and also a constant B, which can be evaluated if the vapor pressure and temperature are known at one other point. The equation, which is implicit in the vapor pressure P_p, is

$$\ln P_p = \ln P_c - 2.303B\left(\frac{1}{T} - \frac{1}{T_c}\right) + \left(2.67 - \frac{1.8B}{T_c}\right)\ln\frac{T}{T_c} + 0.422\left(\frac{T_c^2}{P_c T^2}P_p - 1\right)$$

(14-1)

Equation (14-1) may be solved for B, giving

$$B = \frac{\ln P_c + 2.67\ln T_b/T_c + 0.422(T_c^2/P_c T_b^2 - 1)}{2.303(1/T_b - 1/T_c) + (1.8/T_c)\ln T_b/T_c}$$

(14-2)

in which the temperature T is taken as T_b, the normal boiling temperature at a pressure of 1.0 atm. A different temperature could be used if the pressure at that temperature were known, resulting in a slightly different form to Eq. (14-2); we shall not write the more general form of Eq. (14-2), since it is the normal boiling temperature that is usually known. Table 14-1 lists the normal boiling

Table 14-1 PROPERTIES OF SOME POLLUTANT GASES

Gas	h_{fg_b}, kJ/kg	T_b, K	T_c, K	P_c, atm	B, K
Acetaldehyde	570	293	461	54.7	2182
Acetic acid	405	391	594	57.1	3384
Acetone	520	330	510	47.2	2473
Allyl alcohol	684	370	545	55.5	3582
Ammonia	1375	239	405	111.3	1802
n-Amyl alcohol	503	411	583	37.4	3990
Benzene	394	353	562	48.3	2387
Bromine	194	331	584	102	2136
n-Butane	386	272	425	37.7	1765
n-Butyl alcohol	592	391	563	43.6	3786
Carbon disulfide	352	319	552	78	1975
Carbon tetrachloride	194	350	556	45.0	2313
Chloroform	247	334	536	54	2301
Cyclohexane	358	354	553	40.2	2369
n-Decane	252	447	618	20.8	3707

Table 14-1 PROPERTIES OF SOME POLLUTANT GASES (*Continued*)

Gas	$h_{f_{g_b}}$, kJ/kg	T_b, K	T_c, K	P_c, atm	B, K
Diethylamine	381	329	496	36.6	2432
Diethyl ketone	380	376	561	36.9	2912
Ethane	489	184	305	48.2	1073
Ethyl acetate	434	305	523	38.0	2730
Ethyl alcohol	855	351	516	63.0	3590
Ethylamine	611	290	456	55.5	2178
Ethyl benzene	339	409	617	35.6	2976
Ethyl chloride	388	285	460	52	1892
Ethylene	483	169	283	50.5	971
1,2 Ethylene chloride	324	355	561	53	2568
Ethyl ether	351	308	467	35.9	2204
Ethyl formate	407	327	508	46.8	2393
Freon 22	232	232	370	48.5	1562
n-Heptane	320	371	540	27.0	2755
n-Hexane	337	342	507	29.3	2439
Hydrogen bromide	218	207	363	84	1265
Hydrogen chloride	443	188	324	81.5	1200
Hydrogen cyanide	880	299	457	53.2	2458
Hydrogen fluoride	1562	188	461	64	368
Hydrogen sulfide	548	213	373	88.9	1336
Isobutyl acetate	309	391	561	31.4	3334
Isobutyl alcohol	578	381	548	42.4	3665
Isopropyl alcohol	667	355	508	47.0	3640
Methane	510	111	190	45.4	562
Methyl alcohol	1100	338	512	79.9	3345
Methyl chloride	428	249	416	65.9	1618
Methyl ethyl ketone	444	353	535	41.0	2688
Methyl formate	470	305	487	59.2	2223
Naphthalene	316	491	748	40.0	3621
n-Octane	306	399	569	24.5	3098
n-Pentane	357	309	470	33.4	2108
Propane	426	231	370	41.9	1432
Propionic acid	414	414	612	53	3880
n-Propyl acetate	336	375	549	33.2	3000
n-Propyl alcohol	688	371	537	51.0	3739
Propylene	438	225	365	45.6	1383
Pyridine	450	388	620	55.6	2748
Sulfur dioxide	389	263	431	77.9	1947
Sulfur trioxide	533	318	491	83.8	2980
Toluene	363	384	592	40.6	2719
Trichloroethylene	240	360	544	49.5	3005
Water	2257	373	647	218.3	3233
Xylene	347	417	630	36.8	3063

SOURCE: J. M. Coulson and J. F. Richardson, "Chemical Engineering," vol. 3, pp. 502–503, Pergamon Press, Oxford, 1971; J. A. Dean, "Lange's Handbook of Chemistry," 11th ed., tables 4-1, 7-4, and 9-4, McGraw-Hill Book Company, New York, 1973; R. C. Reid and T. K. Sherwood, "The Properties of Gases and Liquids, Their Estimation and Correlation," 2d ed., pp. 571–584, McGraw-Hill Book Company, New York, 1966; and J. H. Perry, ed., "Chemical Engineers' Handbook," 4th ed., tables 3-7, 3-9, 3-169, and 3-171, McGraw-Hill Book Company, New York, 1963. Used by permission. The values of B were calculated from T_b, T_c, and P_c.

temperature T_b as well as the critical temperature and pressure T_c and P_c. Note that P_c is expressed in atmospheres in Eq. (14-2).

The latent heat of condensation h_{fg} is listed in Table 14-1 at the normal boiling temperature. At other temperatures, the latent heat can be estimated from the following equation, also taken from Reid and Sherwood [1, p. 148]:

$$h_{fg} = h_{fg_b}\left(\frac{T_c - T}{T_c - T_b}\right)^n \qquad (14\text{-}3)$$

In this equation, h_{fg_b} is the latent heat at the normal boiling point, listed in Table 14-1, while the exponent n can have various values depending on conditions. A value of $n = 0.38$ is suggested as a suitable average, giving

$$h_{fg} = h_{fg_b}\left(\frac{T_c - T}{T_c - T_b}\right)^{0.38} \qquad (14\text{-}4)$$

An example shows how the values of P_p and h_{fg} are computed. Since Eq. (14-1) is implicit in both pressure and temperature, it is necessary to solve it by trial and error or by an iterative procedure. This process is readily accomplished using a programmable electronic calculator and can be solved without great difficulty using other computational aids.

EXAMPLE 14-1   Compute the pressure and latent heat of condensation of saturated octane vapor at 20°C.

SOLUTION   The values from Table 14-1 are $T_b = 399$ K, $T_c = 569$ K, $P_c = 24.5$ atm, $B = 3098$ K, and $h_{fg_b} = 306$ kJ/kg. Equation (14-1) gives

$$\ln P_p = \ln 24.5 - 2.303(3098)\left(\frac{1}{293} - \frac{1}{569}\right) + \left[2.67 - \frac{1.8(3098)}{569}\right]\ln\frac{293}{569}$$

$$+ 0.422\left[\frac{(569)^2}{24.5(293)^2}P_p - 1\right]$$

$$= 0.06496P_p - 4.302$$

The two solutions to this equation are $P_p = 0.0135$ atm and $P_p = 142.6$ atm. We shall take the smaller value as correct; thus $P_p = 0.0135$ atm $= 1368$ N/m^2. *Ans.*

For the latent heat of condensation, Eq. (14-4) gives

$$h_{fg} = 306\left(\frac{569 - 293}{569 - 399}\right)^{0.38} = 368 \text{ kJ/kg} \qquad Ans. \qquad ////$$

Change in Concentration of Pollutant

As a way to assess the effectiveness of a condensation process, let us derive expressions for the change in concentration of the pollutant in a mixture as the mixture is cooled and the pollutant condenses. Using the definition of $C_{mv} = m_p/V$ and using the perfect-gas law, then

$$C_{mv} = \frac{M_p P_p}{R_u T} \qquad (14\text{-}5)$$

If we write Eq. (14-5) for states 1 and 2 and form the ratio, we have

$$C_{mv_2} = C_{mv_1} \frac{P_{p_2} T_1}{P_{p_1} T_2} \qquad (14\text{-}6)$$

In a similar way we obtain, in terms of C_v,

$$C_{v_2} = C_{v_1} \frac{P_{p_2}}{P_{p_1}} \qquad (14\text{-}7)$$

We would normally apply Eqs. (14-6) and (14-7) by letting state 1 be the initial state and state 2 be the state following condensation. However, P_{p_1} would be known directly only if the pollutant were saturated in the initial state, which is not the usual condition. These equations are useful, but next we shall derive expressions for collection efficiency in terms of C_{m_1}, C_{mv_1}, and C_{v_1} rather than P_{p_1}.

Collection Efficiency for a Condensation Process

The collection efficiency for the removal of a gaseous pollutant is defined as

$$\eta = 1 - \frac{\dot{m}_{p_2}}{\dot{m}_{p_1}} \qquad (14\text{-}8)$$

Equation (14-8) can be written in terms of C_{mv_1} and P_{p_2} as

$$\eta = \frac{P}{P - P_{p_2}} \left(1 - \frac{M_p P_{p_2}}{R_u T_1 C_{mv_1}} \right) \qquad (14\text{-}9)$$

Using Eq. (4-24), Eq. (14-9) can also be written as

$$\eta = \frac{P - P_{p_2}/C_{v_1}}{P - P_{p_2}} \qquad (14\text{-}10)$$

Equation (14-8) may also be written as

$$\eta = 1 - \frac{1 - C_{m_1}}{C_{m_1}} \frac{M_p P_{p_2}}{M_a (P - P_{p_2})} \qquad (14\text{-}11)$$

If the concentration of the pollutant is small at both the entrance and the exit from the condenser (states 1 and 2), then Eqs. (14-9) to (14-11) become

$$\eta = 1 - \frac{M_p P_{p_2}}{R_u T_1 C_{mv_1}} \qquad (14\text{-}12)$$

$$\eta = 1 - \frac{P_{p_2}}{P C_{v_1}} \qquad (14\text{-}13)$$

$$\eta = 1 - \frac{M_p P_{p_2}}{M_a P C_{m_1}}\,(1 - C_{m_1}) \qquad (14\text{-}14)$$

EXAMPLE 14-2   A mixture of octane vapor in air has $C_{mv} = 0.5$ kg/m^3 initially. It is cooled to 20°C, as in Example 14-1. What collection efficiency results, assuming that the octane is saturated in the final state and that all of the octane that condenses is collected? Assume $T_1 = 75°C$.

SOLUTION   From Example 14-1, $P_{p_2} = 1368$ N/m^2. Equation (14-9) gives

$$\eta = \frac{101{,}326}{101{,}326 - 1368}\left[1 - \frac{114(1368)}{8314(348)(0.5)}\right] = 0.9044 \qquad Ans.$$

Equation (14-12) gives $\eta = 0.892$ for comparison. ////

14-2   DIRECT-CONTACT CONDENSERS

In a direct-contact condenser a stream of water or other cooling liquid is brought into direct contact with the vapor to be condensed. This process is conducted in a chamber especially designed for the purpose; one such chamber is shown in Fig. 14-1. The liquid stream $\dot{m}_m$ leaving the chamber contains the original cooling liquid plus the condensed substances. The gaseous stream $\dot{m}_2$ leaving the chamber contains the noncondensable gases and such condensable vapors as did not condense; it is reasonable to assume that the vapors in the leaving gas stream are saturated. It is then the temperature of the leaving gas stream which determines the collection efficiency of the condenser, as determined by the equations of Sec. 14-1.

A variety of arrangements of direct-contact condenser is possible; the one shown in Fig. 14-1 is a counterflow arrangement. Another common arrangement is the simple spray chamber. The cyclone scrubber and the venturi scrubber also function well as direct-contact condensers.

Equations for the Flow Rate of Cooling Fluid

An energy balance may be taken around the entire condenser of Fig. 14-1, giving

$$\dot{H}_1 + \dot{H}_w = \dot{H}_2 + \dot{H}_m \qquad (14\text{-}15)$$

Let us break these enthalpy rates into terms due to the air, pollutant, and steam portions. For generality, we shall allow the incoming stream to contain

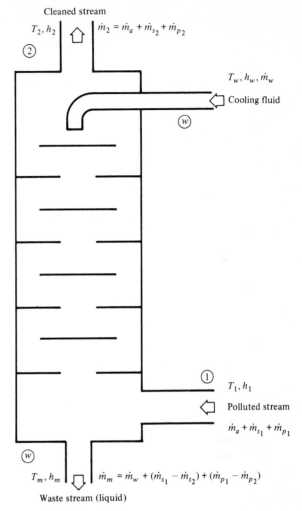

Cleaned stream

T_2, h_2 ⇧ $\dot{m}_2 = \dot{m}_a + \dot{m}_{s_2} + \dot{m}_{p_2}$

②

$T_w, h_w, \dot{m}_w$

⇦ Cooling fluid

ⓦ

① T_1, h_1

⇦ Polluted stream

$\dot{m}_a + \dot{m}_{s_1} + \dot{m}_{p_1}$

ⓦ

T_m, h_m ⇩ $\dot{m}_m = \dot{m}_w + (\dot{m}_{s_1} - \dot{m}_{s_2}) + (\dot{m}_{p_1} - \dot{m}_{p_2})$

Waste stream (liquid)

FIGURE 14-1
Direct-contact condenser used as a pollution-control device.

air, steam, and one or more pollutant vapors. After rearrangement, Eq. (14-15) can be written

$$(\dot{H}_{a1} - \dot{H}_{a2}) + (\dot{H}_{s1} - \dot{H}_{s2} - \dot{H}_{sm}) + (\dot{H}_{p1} - \dot{H}_{p2} - \dot{H}_{pm}) = \dot{H}_{wm} - \dot{H}_w \quad (14\text{-}16)$$

The enthalpy changes in Eq. (14-16) can be written

$$\dot{H}_{a1} - \dot{H}_{a2} = \dot{m}_a(h_{a1} - h_{a2}) = C_{m_{a1}} \dot{m}_1(h_{a1} - h_{a2})$$

$$\dot{H}_{wm} - \dot{H}_w = \dot{m}_w(h_{wm} - h_w) = \dot{m}_w c_w(T_m - T_w)$$

$$\dot{H}_{s1} - \dot{H}_{s2} - \dot{H}_{sm} = C_{m_{s1}} \dot{m}_1 h_{s1} - C_{m_{s2}} \dot{m}_2 h_{s2} - (C_{m_{s1}} \dot{m}_1 - C_{m_{s2}} \dot{m}_2) h_{sm}$$

$$\dot{H}_{p1} - \dot{H}_{p2} - \dot{H}_{pm} = C_{m_{p1}} \dot{m}_1 h_{p1} - C_{m_{p2}} \dot{m}_2 h_{p2} - (C_{m_{p1}} \dot{m}_1 - C_{m_{p2}} \dot{m}_2) h_{pm}$$

For the specific enthalpies in the preceding equations

$$h_{p1} - h_{pm} = c_{p_p}(T_1 - T_m)$$
$$h_{p2} - h_{pm} = c_{p_p}(T_2 - T_m)$$

The enthalpies h_{s1}, h_{sm}, and h_{s2} are obtained from the steam tables approximately at the saturated conditions:

$$h_{s1} = h_g(T_1) = h_{g_1}$$
$$h_{sm} = h_f(T_m) = h_{f_m}$$
$$h_{s2} = h_g(T_2) = h_{g_2}$$

When all the preceding equations for enthalpy are substituted into Eq. (14-16) and the resulting equation is solved for $\dot{m}_w$, we obtain

$$
\dot{m}_w = \frac{1}{c_w(T_m - T_w)} \{ \dot{m}_1 [C_{m_{a1}}(h_{a1} - h_{a2}) + C_{m_{s1}}(h_{g_1} - h_{f_m})
$$
$$
+ C_{m_{p1}} c_{p_p}(T_1 - T_m) + C_{m_{p1}} h_{fg_p}]
$$
$$
- \dot{m}_2 [C_{m_{s2}}(h_{g_2} - h_{f_m}) - C_{m_{p2}} c_{p_p}(T_m - T_2) + C_{m_{p2}} h_{fg_p}] \} \qquad (14\text{-}17)
$$

In Eq. (14-17), c_w is the specific heat of the cooling water and c_{p_p} is the specific heat of the pollutant vapor. If the temperature difference $T_1 - T_2$ is not too large, we may write for the air component

$$h_{a1} - h_{a2} = c_{p_a}(T_1 - T_2) \qquad (14\text{-}18)$$

in which c_{p_a} is the specific heat of the air.

In applying Eq. (14-17), we shall assume that normally $\dot{m}_1$, $C_{m_{a1}}$, $C_{m_{s1}}$, $C_{m_{p1}}$, T_1, T_2, T_w, T_m, and η are known. In order to use the equation as it stands, the quantities $\dot{m}_2$, $C_{m_{a2}}$, $C_{m_{s2}}$, $C_{m_{p2}}$, and the various enthalpies and specific heats must be known as well. The following equation for $\dot{m}_{s2}$, the mass flow rate of steam leaving the condenser, may be derived readily enough:

$$\dot{m}_{s2} = \frac{\rho_{g_2} \dot{m}_1 [C_{m_{a1}}/M_a + (1 - \eta)C_{m_{p1}}/M_p]}{P/R_u T_2 - \rho_{g_2}/M_s} \qquad (14\text{-}19)$$

For the mass flow rates of air and pollutant leaving the condenser

$$\dot{m}_a = C_{m_{a1}} \dot{m}_1 \qquad (14\text{-}20)$$
$$\dot{m}_{p2} = (1 - \eta)C_{m_{p1}} \dot{m}_1 \qquad (14\text{-}21)$$

Then for the mass flow rate leaving

$$\dot{m}_2 = [C_{m_{a1}} + (1 - \eta)C_{m_{p1}}]\dot{m}_1 + \dot{m}_{s2} \qquad (14\text{-}22)$$

For the three leaving mass concentrations

$$C_{m_{s2}} = \frac{\dot{m}_{s2}}{\dot{m}_2} \tag{14-23}$$

$$C_{m_{p2}} = \frac{\dot{m}_{p2}}{\dot{m}_2} = \frac{(1 - \eta)C_{m_{p1}}\dot{m}_1}{\dot{m}_2} \tag{14-24}$$

$$C_{m_{a2}} = 1 - C_{m_{s2}} - C_{m_{p2}} \tag{14-25}$$

It is necessary to evaluate Eqs. (14-19) to (14-24) prior to evaluating Eq. (14-17).

Steam-Pollutant Mixtures

If the polluted stream consists only of steam and pollutant, with no air present, as will occur from time to time in industrial practice, then Eqs. (14-17) to (14-25) may be simplified to the following equations:

$$\dot{m}_w = \frac{1}{c_w(T_m - T_w)} \{\dot{m}_1[C_{m_{s1}}(h_{g_1} - h_{f_m}) + C_{m_{p1}}c_{p_p}(T_1 - T_m) + C_{m_{p1}}h_{fg_p}]$$
$$- \dot{m}_2[C_{m_{s2}}(h_{g_2} - h_{f_m}) - C_{m_{p2}}c_{p_p}(T_m - T_2) + C_{m_{p2}}h_{fg_p}]\} \tag{14-26}$$

$$\dot{m}_{s2} = \frac{\rho_{g_2}\dot{m}_1(1 - \eta)C_{m_{p1}}}{M_p(P/R_u T_2 - \rho_{g_2}/M_s)} \tag{14-27}$$

$$\dot{m}_{p2} = (1 - \eta)C_{m_{p1}}\dot{m}_1 \tag{14-28}$$

$$\dot{m}_2 = (1 - \eta)C_{m_{p1}}\dot{m}_1 + \dot{m}_{s2} \tag{14-29}$$

$$C_{m_{s2}} = \frac{\dot{m}_{s2}}{\dot{m}_2} \tag{14-30}$$

$$C_{m_{p2}} = \frac{\dot{m}_{p2}}{\dot{m}_2} = \frac{(1 - \eta)C_{m_{p1}}\dot{m}_1}{\dot{m}_2} \tag{14-31}$$

It should be noted that T_2 must be equal to or greater than T_w; the closeness of the approach of T_2 to T_w is a measure of the effectiveness of the condenser. We shall make no attempt to analyze the effectiveness of an exchanger here; instead, both temperatures T_w and T_2 will be specified in working problems.

EXAMPLE 14-3   A mixture of steam and octane vapor is initially at 120°C with $C_{m_s} = 0.9$. It is cooled and condensed in a direct-contact condenser, leaving at 20°C. The flow rate of steam-pollutant mixture is 1.0 kg/s, while cooling water enters at 15°C and leaves at 60°C. Compute the cooling-water flow rate.

SOLUTION Example 14-1 gives $P_{p_2} = 1368$ N/m². Equation (14-11) gives for the collection efficiency

$$\eta = 1 - \frac{0.9}{0.1} \frac{114(1368)}{18(101,326 - 1368)} = 0.780$$

From the steam tables, we have $h_{g_1} = 2.706$ MJ/kg, $h_{f_m} = 0.251$ MJ/kg, $h_{g_2} = 2.538$ MJ/kg, and $\rho_{g_2} = 0.0173$ kg/m³. Equations (14-27) to (14-31) give

$$\dot{m}_{s2} = \frac{0.0173(1.0)(1 - 0.78)(0.1)}{114[101,326/(8314)(293) - 0.0173/18]} = 0.0000822 \text{ kg/s}$$

$$\dot{m}_{p2} = (1 - 0.78)(0.1)(1.0) = 0.022 \text{ kg/s}$$

$$\dot{m}_2 = \dot{m}_{s2} + \dot{m}_{p2} = 0.0000822 + 0.022 = 0.0221 \text{ kg/s}$$

$$C_{m_{s2}} = \frac{0.0000822}{0.0221} = 0.00372$$

$$C_{m_{p2}} = \frac{0.022}{0.0221} = 0.9963$$

Using $c_w = 4190$ J/kg K, $c_{p_p} = 1589$ J/kg K from Table 2-1, and $h_{fg_p} = 368$ kJ/kg from Example 14-1, Eq. (14-26) gives

$$\dot{m}_w = \frac{1}{4190(60 - 15)} \{1.0[0.9(2.706 - 0.251)(10^6) + 0.1(1589)(120 - 60)$$

$$+ 0.1(368)(10^3)] - 0.0221[0.00372(2.538 - 0.251)(10^6)$$

$$- 0.0063(1589)(60 - 20) + 0.9963(368)(10^3)]\} = 11.9 \text{ kg/s} \qquad Ans. \qquad ////$$

The results of Example 14-3 indicate that the steam could be regarded as the pollutant and that the condenser is a very efficient steam remover.

Air-Pollutant Mixtures

If no steam is present, then Eqs. (14-17) to (14-25) reduce to the following:

$$\dot{m}_w = \frac{1}{c_w(T_m - T_w)} \{\dot{m}_1[C_{m_{a1}}c_{p_a}(T_1 - T_2) + C_{m_{p1}}c_{p_p}(T_1 - T_m) + C_{m_{p1}}h_{fg_p}]$$

$$+ \dot{m}_2 C_{m_{p2}}[c_{p_p}(T_m - T_2) - h_{fg_p}]\} \tag{14-32}$$

$$\dot{m}_a = C_{m_{a1}}\dot{m}_1 = (1 - C_{m_{p1}})\dot{m}_1 \tag{14-33}$$

$$\dot{m}_{p2} = (1 - \eta)C_{m_{p1}}\dot{m}_1 \tag{14-34}$$

$$\dot{m}_2 = \dot{m}_a + \dot{m}_{p2} = (1 - \eta C_{m_{p1}})\dot{m}_1 \tag{14-35}$$

$$C_{m_{p2}} = \frac{\dot{m}_{p2}}{\dot{m}_2} = \frac{(1 - \eta)C_{m_{p1}}}{1 - \eta C_{m_{p1}}} \tag{14-36}$$

$$C_{m_{a2}} = 1 - C_{m_{p2}} = \frac{1 - C_{m_{p1}}}{1 - \eta C_{m_{p1}}} \tag{14-37}$$

In Eq. (14-32), Eq. (14-18) was used for the change in enthalpy of the air.

EXAMPLE 14-4 Repeat Example 14-3 if the mixture is of air and octane vapor, with $C_{m_p} = 0.1$ initially.

SOLUTION Using the previous data, Eq. (14-11) gives for the collection efficiency

$$\eta = 1 - \frac{0.9}{0.1} \frac{114(1368)}{29(101,326 - 1368)} = 0.516$$

Equations (14-35) and (14-36) give

$$\dot{m}_2 = [1 - 0.516(0.1)](1.0) = 0.9484 \text{ kg/s}$$

$$C_{m_{p2}} = \frac{(1 - 0.516)(0.1)}{1 - 0.516(0.1)} = 0.05103$$

Using $c_{p_a} = 1005$ J/kg K for air, Eq. (14-32) yields

$$\dot{m}_w = \frac{1}{4190(60 - 15)} \{1.0[0.9(1005)(120 - 20) + 0.1(1589)(120 - 60)$$

$$+ 0.1(368,000)] + 0.9484(0.05103)[1589(60 - 20) - 368,000]\}$$

$$= 0.647 \text{ kg/s} \quad Ans. \qquad\qquad ////$$

Dilute Air-Pollutant Mixtures

If the initial concentration of pollutant vapor in the air is small, say, for $C_{m_{p1}} < 0.01$, then various simplifications can be made in Eqs. (14-32) to (14-37). The resulting equations are

$$\dot{m}_2 = \dot{m}_1 \qquad\qquad (14\text{-}38)$$

$$C_{m_{p2}} = (1 - \eta)C_{m_{p1}} \qquad\qquad (14\text{-}39)$$

$$\dot{m}_w = \frac{\dot{m}_1}{c_w(T_m - T_w)} \{c_{p_a}(T_1 - T_2) + C_{m_{p1}}c_{p_p}[(T_1 - T_2) - \eta(T_m - T_2)] + \eta C_{m_{p1}}h_{fg_p}\}$$

$$(14\text{-}40)$$

One precaution to take in dealing with dilute mixtures is to be sure that enough of the pollutant is present initially so that the pollutant vapor is saturated leaving the condenser. If not, then theoretically there will be no condensation and the use of a condenser will only cool the mixture. This precaution applies to any condenser, not just to a direct-contact condenser. This situation will occur if the efficiency, as computed by Eqs. (14-9) to (14-14), is negative; such a value suggests that vapor must be added to the air to saturate it. Should this problem arise, either the gas will have to be cooled to a lower temperature or the use of a condenser must be abandoned.

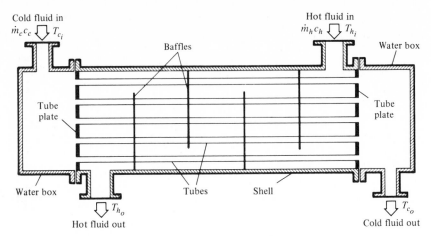

FIGURE 14-2
Counterflow shell-and-tube heat exchanger having one shell pass and one tube pass.

14-3 SURFACE HEAT EXCHANGERS

A surface heat exchanger is a device which promotes the transfer of heat from one fluid to the other but in which the two fluids are separated by a wall so that no mixing of the fluids occurs. The simplest arrangement of a surface heat exchanger is the counterflow exchanger shown in Fig. 14-2. In the arrangement shown, the hot fluid flows through the shell around the tubes while the cold fluid flows through the tubes. The heat transfer takes place across the wall of the tubes from the hot fluid to the cold fluid. It is equally practicable to direct the hot fluid through the tubes and the cold fluid in the shell. The same exchanger could be used for parallel flow by reversing the direction of one of the fluids, but counterflow is preferred. The baffle plates direct the path of the shell fluid so that it makes contact with all of the tubes. This type of exchanger is also referred to as a *shell-and-tube heat exchanger*.

In a shell-and-tube heat exchanger, either or both fluids can make several passes back and forth along the length of the unit. Figure 14-3 shows such an exchanger having one shell pass and two tube passes, which is designated as a *1-2 exchanger*. A number of different arrangements are possible, such as two shell and two tube passes (2-2), two shell and four tube passes (2-4), to mention two very common arrangements. Another very common type of exchanger is the cross-flow arrangement in which the fluids move at right angles to each other. For details of the design and arrangement of various types of heat exchangers, the reader should consult books on heat transfer, such as Parker, Boggs, and Blick [2, Chap. 13] or Kreith [3], or more specialized books on heat exchangers, such as Fraas and Ozisik [4] or Kays and London [5].

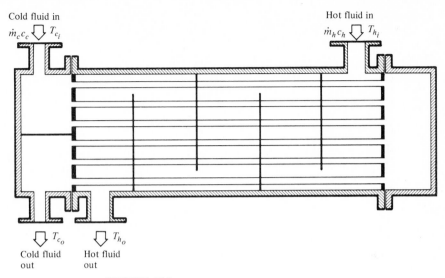

Cold fluid in
$\dot{m}_c c_c \;⬇\; T_{c_i}$

Hot fluid in
$\dot{m}_h c_h \;⬇\; T_{h_i}$

⬇ T_{c_o} ⬇ T_{h_o}

Cold fluid   Hot fluid
out   out

FIGURE 14-3
Shell-and-tube heat exchanger having one shell pass and two tube passes.

Overall Heat-Transfer Coefficient

In computing the heat-transfer rate across a wall, such as the cylindrical wall of a tube, it is convenient to define an overall heat-transfer coefficient U. If the coefficient U_o is based on the outside surface area of the tube, for which the radius is r_o, then U_o is related to the film coefficients h_i and h_0 on the inside and outside surfaces, respectively, by the following equation:

$$\frac{1}{U_o} = \frac{1}{h_o} + \frac{r_o}{r_i}\frac{1}{h_i} + \frac{r_o}{k}\ln\frac{r_o}{r_i} + \frac{1}{h_{f_i}} + \frac{1}{h_{f_o}} \qquad (14\text{-}41)$$

in which r_i is the inner radius of the tube, k is the thermal conductivity of the tube wall, and h_{f_i} and h_{f_o} are fouling factors on the inside and outside walls, respectively. For usual applications, k is large and r_o/r_i is close to unity, so that Eq. (14-41) becomes

$$\frac{1}{U} = \frac{1}{h_o} + \frac{1}{h_i} + \frac{1}{h_{f_i}} + \frac{1}{h_{f_o}} \qquad (14\text{-}42)$$

If the fouling factors can be ignored,

$$\frac{1}{U} = \frac{1}{h_o} + \frac{1}{h_i} \qquad (14\text{-}43)$$

For the film coefficient h_i on the inside surface of the tube, Eq. (2-115) can often be used. This same equation can also be used in many cases for the flow outside the tubes where the flow is predominantly along the tubes and where the hydraulic diameter based on the passage around the tube bundle is used. For flow normal to a single tube, the following equation is frequently used [2, pp. 254–263]:

$$h_o = C \frac{k}{D} Re_D{}^n Pr^{1/3} \qquad (14\text{-}44)$$

in which suggested values of C and n are listed in Table 14-2. Data for the film coefficient due to the cross flow of a fluid over a bank of tubes are also presented by Parker, Boggs, and Blick [2, p. 254–263] but will not be repeated here.

Mean Temperature Difference

The heat-transfer rate between the hot and cold fluid is given by

$$\dot{Q} = UA\,\Delta T_m \qquad (14\text{-}45)$$

in which U and A must be based on the same surface area, such as the outside surface area of the tube. The mean temperature difference ΔT_m in Eq. (14-45) is a weighted mean of the temperature difference between the hot and cold fluid at each point of the wall; the weighting function is the surface area of the exchanger. For the counterflow heat exchanger of Fig. 14-2, the mean temperature difference is the logarithmic mean of the temperature differences at each end of the heat exchanger:

$$\Delta T_{m_{cf}} = \frac{(T_{h_i} - T_{c_o}) - (T_{h_o} - T_{c_i})}{\ln\,(T_{h_i} - T_{c_o})/(T_{h_o} - T_{c_i})} \qquad (14\text{-}46)$$

Equation (14-46) is also valid for a parallel-flow heat exchanger, although $\Delta T_{m_{pf}}$ has a different value from $\Delta T_{m_{cf}}$ owing to the different values for the temperatures of the hot and cold fluids at the inlet and outlet.

Table 14-2 VALUES OF C AND n FOR EQ. (14-44)

Re_D	C	n
1–4	0.989	0.330
4–40	0.911	0.385
40–4,000	0.683	0.466
4,000–40,000	0.193	0.618
40,000–250,000	0.0266	0.805

SOURCE: J. D. Parker, J. H. Boggs, and E. F. Blick, "Introduction to Fluid Mechanics and Heat Transfer," p. 259, Addison-Wesley Publishing Company, Inc., Reading, Mass., 1969.

For other heat-exchanger arrangements, the mean temperature difference is defined in terms of that for counterflow:

$$\Delta T_m = F_t \, \Delta T_{m_{cf}} \qquad (14\text{-}47)$$

in which F_t is a complicated equation, usually presented graphically, of the two quantities P and R:

$$P = \frac{T_{c_o} - T_{c_i}}{T_{h_i} - T_{c_i}} \qquad (14\text{-}48)$$

$$R = \frac{T_{h_i} - T_{h_o}}{T_{c_o} - T_{c_i}} = \frac{\dot{m}_c \, c_c}{\dot{m}_h \, c_h} \qquad (14\text{-}49)$$

Figure 14-4 shows the factor F_t as a function of P and R for the 1-2 heat exchanger of Fig. 14-3. For other types of heat exchangers, Refs. [3] to [6] present F_t or similar data. Substituting Eq. (14-47) into Eq. (14-45) gives for the heat-transfer rate

$$\dot{Q} = U A F_t \, \Delta T_{m_{cf}} \qquad (14\text{-}50)$$

EXAMPLE 14-5   A 1-2 heat exchanger is to cool a stream of 1.0 m³/s of air at 100 to 30°C using a stream of water entering at 20 and leaving at 40°C. The water velocity inside the pipes is 1.0 m/s; the film coefficient on the outside tube surface is 500 W/m² · K. The tubes are 1.0 cm inside diameter with a wall thickness of 1.0 mm. The thermal conductivity of the tube-wall material is 175 W/m · K; there is a fouling factor of $h_{f_i} = 1000$ W/m² · K on the inside surface, and no fouling factor on the outside surface ($h_{f_o} = \infty$).
(a) Compute the overall heat-transfer coefficient.
(b) How many tubes are required and what length of tube is required?

SOLUTION   (a) Use the average temperature of 30°C for the water; its properties of interest are: $\mu = 8 \times 10^{-4}$ kg/m · s, $\rho = 996$ kg/m³, $c = 4190$ J/kg · K, and $k = 0.643$ W/m · K, taken from Chap. 2. The Reynolds number for flow inside the tubes is Re = 12,450, and the Prandtl number of the water is Pr = $c\mu/k$ = 4190(8 × 10⁻⁴)/0.643 = 5.21. Equation (2-115) gives the film co-efficient on the inside surface as

$$h_i = 0.023 \frac{0.643}{0.01} (12{,}450)^{0.8} (5.21)^{1/3} = 4842 \text{ W/m}^2 \cdot \text{K}$$

Using the inside and outside tube radii of 0.005 and 0.006 m, respectively, Eq. (14-41) gives for the overall heat-transfer coefficient based on the outside surface area of the tube

$$\frac{1}{U_o} = \frac{1}{500} + \frac{0.006}{0.005} \frac{1}{4842} + \frac{0.006}{175} \ln \frac{0.006}{0.005} + \frac{1}{1000} + \frac{1}{\infty} = 0.003254$$

$$U_o = 307.3 \text{ W/m}^2 \cdot \text{K}$$

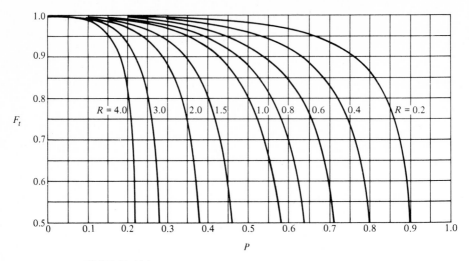

FIGURE 14-4
Correction factor F_t for a heat exchanger having one shell pass and two tube passes (1-2 heat exchanger).

(b) Next, we shall proceed to evaluate the number of tubes required. The density of the air is computed to be 0.947 kg/m³ at the entering temperature, giving a mass flow rate of 0.947(1.0) = 0.947 kg/s. At the average air temperature the specific heat c_h is 1010 J/kg · K. The heat-transfer rate is

$$\dot{Q} = \dot{m}_h c_h(T_{h_i} - T_{h_o}) = \dot{m}_c c_c(T_{c_o} - T_{c_i})$$
$$= 0.947(1010)(100 - 30)$$
$$= 66{,}919 \text{ W}$$

For the mass flow rate of the water

$$\dot{m}_c = \frac{\dot{Q}}{c_c(T_{c_o} - T_{c_i})} = \frac{66{,}919}{4190(40 - 20)} = 0.799 \text{ kg/s}$$

The volumetric flow rate of water is 0.799/996 = 0.0008022 m³/s. Using the inlet and outlet temperatures $T_{c_o} = 40°C$, $T_{c_i} = 20°C$, $T_{h_i} = 100°C$, and $T_{h_o} = 30°C$, we have from Eqs. (14-48) and (14-49)

$$P = \frac{40 - 20}{100 - 20} = 0.25 \qquad R = \frac{100 - 30}{40 - 20} = 3.5$$

Figure 14-4 gives $F_t = 0.5$. Equation (14-46) gives the mean temperature difference for counterflow:

$$\Delta T_{m_{cf}} = \frac{(100 - 40) - (30 - 20)}{\ln(60/10)} = 27.9°C$$

We may now solve for the surface area of the tubes using Eq. (14-50):

$$A = \frac{\dot{Q}}{U_o F_t \Delta T_{m_{cf}}} = \frac{66,919}{307.3(0.5)(27.9)} = 15.6 \text{ m}^2$$

We may also compute the total cross-sectional area of the tubes in one pass:

$$A_c = \frac{Q_c}{V} = \frac{0.0008022}{1.0} = 0.0008022 \text{ m}^2$$

We now have the data necessary to compute the number and length of the tubes. For the cross-sectional area of a single tube, we have

$$A_T = \pi r_i^2 = \pi(0.005)^2 = 0.00007854 \text{ m}^2$$

Then the number of tubes in each pass is

$$N_T = \frac{A_c}{A_T} = \frac{0.0008022}{0.00007854} = 10.2$$

Since a fractional tube is impossible, use 11 tubes.   *Ans.*

The surface area of a single tube is $2\pi r_o L = 0.0377L$. The tube length is

$$L = \frac{A}{2N_T(0.0377)} = \frac{15.6}{2(11)(0.0377)} = 37.6 \text{ m} \qquad Ans.$$

The great length of tubes required in this example suggests that 6 to 10 tube passes would be more appropriate in this case.   ////

14-4 CONDENSATION OF STEAM-POLLUTANT MIXTURES ON SURFACE CONDENSERS

To deal with steam-pollutant mixtures, some additional properties of water vapor are needed. Figure 14-5 shows the variation with temperature of the specific heat at constant pressure, the dynamic viscosity, and the thermal conductivity for steam at very low pressures. These values may be used with reasonable confidence up to a pressure or partial pressure of the water vapor of around 1 atm. For pressures substantially above atmospheric, more accurate values should be sought.

We can determine the mass flow rate leaving the condenser from Eqs. (14-27), (14-28), and (14-11), noting that $\dot{m}_2 = \dot{m}_{s2} + \dot{m}_{p2}$:

$$\dot{m}_2 = (1 - C_{m_{p1}})\dot{m}_1 \frac{M_p P_{p_2}}{M_s(P - P_{p_2})}\left[1 + \frac{\rho_{g_2}}{M_p(P/R_u T_2 - \rho_{g_2}/M_s)}\right] \qquad (14\text{-}51)$$

Using the equation

$$Q_2 = \frac{\dot{m}_{s2}}{\rho_{g_2}} + \frac{\dot{m}_{p2}}{\rho_{p_2}}$$

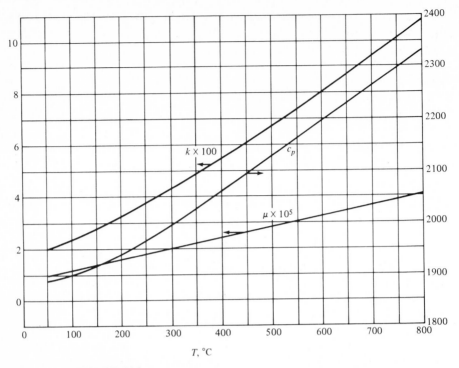

FIGURE 14-5
Specific heat at constant pressure (J/kg · K), dynamic viscosity (kg/m · s), and thermal conductivity (W/m · K) for steam at low pressure.

the perfect-gas law for ρ_{p_2}, and Eqs. (14-27), (14-28), and (14-11), we may derive the volumetric flow rate leaving the condenser:

$$Q_2 = \frac{(1 - C_{m_{p1}})\dot{m}_1 P_{p_2} R_u T_2}{M_s P(P - P_{p_2})} \frac{2 - \rho_{g_2} R_u T_2/M_s P}{1 - \rho_{g_2} R_u T_2/M_s P} \quad (14\text{-}52)$$

An energy balance on the steam side of the condenser will give the following equation for the heat transfer from the steam-pollutant mixture across the tube walls of the exchanger and to the cooling fluid

$$\dot{Q} = \dot{m}_1 C_{m_{p1}} \left\{ \frac{1 - C_{m_{p1}}}{C_{m_{p1}}} (h_{g_1} - h_{f_2}) + c_{p_p}(T_1 - T_2) + \eta h_{fg_p} \right.$$

$$\left. - \frac{\rho_{g_2}(1 - \eta)}{M_p(P/R_u T_2 - \rho_{g_2}/M_s)} [(h_{g_1} - h_{f_2}) - c_{p_s}(T_1 - T_2)] \right\} \quad (14\text{-}53)$$

in which η is given by Eqs. (14-9) to (14-11).

The film coefficients for use in determining the overall heat-transfer coefficient are difficult to determine with good accuracy from existing correlations for mixtures of two condensing vapors in practical condensers. Also, theoretical predictions for the mean temperature difference are inadequate when two phases condense at separate temperatures. We cannot deal with these two difficulties in this book; the reader who must have accurate values of film coefficient and mean temperature difference is referred to the literature or to the laboratory. For our immediate needs, we shall assume that U can be estimated and that ΔT_m, or F_t, can be evaluated from existing charts and used with acceptable accuracy.

EXAMPLE 14-6   A mixture of steam and isobutyl alcohol vapor having $C_{m_{p1}} = 0.25$ flows at the rate of 2.5 kg/s at 450 K and 1 atm pressure. It enters a surface condenser, and leaves at 30°C. Compute the leaving mass and volume flow rates and the heat transferred to the cooling water.

SOLUTION   Using the method of Example 14-1, the vapor pressure of the pollutant (the alcohol) at the leaving temperature is $P_{p_2} = 0.028$ atm $= 2837$ N/m². The following property values are obtained from steam tables and from the tables and graphs in this book:

$$h_{fg_p} = 578 \text{ kJ/kg} \qquad h_{g_1} = 2.78 \text{ MJ/kg}$$
$$c_{p_p} = 1503 \text{ J/kg} \cdot \text{K} \qquad h_{f_2} = 0.126 \text{ MJ/kg}$$
$$c_{p_s} = 1930 \text{ J/kg} \cdot \text{K} \qquad \rho_{g_2} = 0.0304 \text{ kg/m}^3$$

Equation (14-51) gives for the leaving mass flow rate $\dot{m}_2$

$$\dot{m}_2 = (1 - 0.25)2.5 \frac{74.1(0.028)}{18(1 - 0.028)}\left\{1 + \frac{0.0304}{74.1[101,326/8314(303) - 0.0304/18]}\right\}$$
$$= 0.3002 \text{ kg/s} \qquad Ans.$$

Equation (14-52) gives the leaving volumetric flow rate:

$$\frac{\rho_{g_2} R_u T_2}{M_s P} = \frac{0.0304(8314)(303)}{18(101,326)} = 0.0420$$
$$Q_2 = \frac{(1 - 0.25)(2.5)(0.0280)(8314)(303)}{18(101,326)(1 - 0.028)} \frac{2 - 0.0420}{1 - 0.0420}$$
$$= 0.1525 \text{ m}^3/\text{s} \qquad Ans.$$

The collection efficiency is computed from Eq. (14-11):

$$\eta = 1 - \frac{0.75}{0.25}\frac{74.1(0.0280)}{18(1 - 0.028)} = 0.6442$$

Then the heat-transfer rate is obtained from Eq. (14-53):

$$\dot{Q} = 2.5(0.25)\left\{\frac{0.75}{0.25}(2.78 - 0.126)(10^6) + 1503(450 - 303)\right.$$

$$+ 0.6442(578,000) - \frac{0.0304(0.3558)}{74.1[101,326/(8314)(303) - 0.0304/18]}$$

$$\left. \times [(2.78 - 0.126)(10^6) - 1930(450 - 303)]\right\}$$

$$= 5.34 \text{ MJ/s} \quad Ans. \qquad\qquad ////$$

EXAMPLE 14-7   Using the data of Example 14-6, suppose $U = 300 \text{ W/m}^2 \cdot \text{K}$, that the cooling water flows at the rate of 20 kg/s, and that the cooling water enters at 25°C. What condenser surface area is required? Assume $F_t = 0.6$.

SOLUTION   Note that $T_{h_i} = 450 \text{ K} = 177°C$, $T_{h_o} = 30°C$, $T_{c_i} = 25°C$, and that

$$T_{c_o} = T_{c_i} + \frac{\dot{Q}}{\dot{m}_c c_c} = 25 + \frac{5.34 \times 10^6}{20(4190)} = 88.7°C$$

Equation (14-46) gives the mean temperature difference for counterflow:

$$\Delta T_{m_{cf}} = \frac{(177 - 88.7) - (30 - 25)}{\ln 88.3/5} = 29.0°C$$

Then solving Eq. (14-50) for the heat-exchanger surface area,

$$A = \frac{\dot{Q}}{U F_t \Delta T_{m_{cf}}} = \frac{5.34 \times 10^6}{300(0.6)(29.0)} = 1023 \text{ m}^2 \quad Ans. \qquad ////$$

14-5   CONDENSATION OF AIR-POLLUTANT MIXTURES ON SURFACE CONDENSERS

The equations which are equivalent to Eqs. (14-51) to (14-53), dealing with a mixture of air and a pollutant substance, with no steam present, are

$$\dot{m}_2 = (1 - \eta C_{m_{p1}})\dot{m}_1 \qquad\qquad (14\text{-}54)$$

$$Q_2 = \frac{\dot{m}_1 R_u T_2}{P}\left[\frac{1 - C_{m_{p1}}}{M_a} + \frac{(1 - \eta)C_{m_{p1}}}{M_p}\right] \qquad\qquad (14\text{-}55)$$

$$\dot{Q} = \dot{m}_1[C_{m_{p1}}c_{p_p}(T_1 - T_2) + \eta C_{m_{p1}}h_{fg_p} + (1 - C_{m_{p1}})c_{p_a}(T_1 - T_2)] \qquad (14\text{-}56)$$

Note that Eq. (14-54) is the same as Eq. (14-35). The collection efficiency is given by Eqs. (14-9) to (14-11).

EXAMPLE 14-8   Repeat Example 14-6 but let the mixture be of air and isobutyl alcohol, the other data remaining the same.

SOLUTION   Using the data and the calculated value of P_{p_2} from Example 14-6, Eq. (14-11) gives for the collection efficiency

$$\eta = 1 - \frac{1 - 0.25}{0.25} \frac{74.1(0.028)}{29(1 - 0.028)} = 0.221$$

Equations (14-54) to (14-56) give for $\dot{m}_2$, Q_2, and $\dot{Q}$, respectively

$$\dot{m}_2 = [1 - 0.221(0.25)](2.5) = 2.36 \text{ kg/s} \qquad Ans.$$

$$Q_2 = \frac{2.5(8314)(303)}{101,326} \left[\frac{1 - 0.25}{29} + \frac{(1 - 0.221)(0.25)}{74.1} \right] = 1.77 \text{ m}^3/\text{s} \qquad Ans.$$

$$\dot{Q} = 2.5[0.25(1503)(450 - 303) + 0.221(0.25)(578,000)$$
$$+ (1 - 0.25)(1005)(450 - 303)] = 0.495 \text{ MJ/s} \qquad Ans. \qquad ////$$

14-6   REGENERATIVE SYSTEMS FOR ENERGY CONSERVATION

Condensation systems have a considerable field of usefulness in the economic recovery of pollutants; for example, solvent vapors can often be recovered by this means. For the control of emissions, however, our analysis to this point indicates that there are severe difficulties to be overcome. To achieve a high collection efficiency, particularly with dilute mixtures, it is usually necessary to cool the mixture to very low temperatures, possibly as low as $-100°C$. This procedure requires a large expenditure of energy and requires that the air be emitted at a very low temperature, which can in itself cause problems. The use of a regenerative system, as shown in Fig. 14-6, can ease this problem considerably. We would be tempted to ignore condensation systems for the removal of gaseous pollutants or at least to downgrade their importance, except that no other known method is very satisfactory either. Let us explore the system of Fig. 14-6 to see if it may prove to have some practical value. Note that the cooling fluid must be at a temperature below that to which the mixture is cooled, which means that it must be refrigerated, so that even a small amount of heat transfer will be very expensive.

To analyze the system of Fig. 14-6, let us first take an energy balance around condenser 2. We shall assume that stream 3-4 has the properties of pure air and that its flow rate is $\dot{m}_a$. This assumption introduces some error, but if the collection efficiency is high, the error will be small. To simplify the analysis, we shall introduce several approximations of this sort. Let us designate the heat transfer $\dot{Q}$ as that which is transferred to the cooling fluid in condenser 2. Then

$$\dot{Q} = \dot{m}_w c_w (T_{w_2} - T_{w_1}) \qquad (14\text{-}57)$$

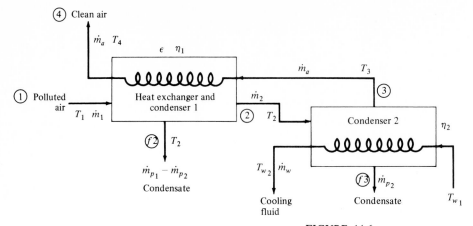

FIGURE 14-6
Regenerative condensation system.

Let us use the following equations:

$$h_{a_2} - h_{a_3} = c_{p_a}(T_2 - T_3)$$
$$h_{p_2} - h_{p_3} = c_{p_p}(T_2 - T_3)$$
$$h_{p_3} - h_{f_3} = h_{fg_{p3}}$$
$$\dot{m}_a = (1 - C_{m_{p1}})\dot{m}_1$$
$$\dot{m}_{p1} = C_{m_{p1}}\dot{m}_1$$
$$\dot{m}_{p2} = (1 - \eta_1)C_{m_{p1}}\dot{m}_1$$

The energy balance may be written as

$$\dot{m}_a h_{a_2} + \dot{m}_{p2} h_{p_2} - \dot{m}_a h_{a_3} - \dot{m}_{p2} h_{f3} = \dot{m}_w c_w (T_{w_2} - T_{w_1})$$

which simplifies to

$$\dot{Q} = (1 - C_{m_{p1}})\dot{m}_1 c_{p_a}(T_2 - T_3) + (1 - \eta_1)C_{m_{p1}}\dot{m}_1[c_{p_p}(T_2 - T_3) + h_{fg_{p3}}] \qquad (14\text{-}58)$$

An energy balance around heat exchanger and condenser 1 gives

$$\dot{m}_a h_{a_1} + \dot{m}_{p1} h_{p1} - \dot{m}_a h_{a_2} - \dot{m}_{p2} h_{p_2} + \dot{m}_a h_{a_3} - \dot{m}_a h_{a_4} - (\dot{m}_{p1} - \dot{m}_{p2})h_{f2} = 0$$

This equation may be written in a manner similar to Eq. (14-58) and the result solved for T_2 to give

$$T_2 = \frac{(1 - C_{m_{p1}})c_{p_a}(T_1 + T_3 - T_4) + C_{m_{p1}}(c_{p_p} T_1 + \eta_1 h_{fg_{p2}})}{c_{p_a}(1 - C_{m_{p1}}) + c_{p_p} C_{m_{p1}}} \qquad (14\text{-}59)$$

We define the effectiveness of heat exchanger and condenser 1 as

$$\varepsilon = \frac{T_4 - T_3}{T_1 - T_3} \qquad (14\text{-}60)$$

and solve for T_4:

$$T_4 = T_3 + \varepsilon(T_1 - T_3) \quad (14\text{-}61)$$

Substituting Eq. (14-61) into Eq. (14-59),

$$T_2 = \frac{(1 - C_{m_{p1}})c_{p_a}[T_1 - \varepsilon(T_1 - T_3)] + C_{m_{p1}}(c_{p_p}T_1 + \eta_1 h_{fg_{p2}})}{c_{p_a}(1 - C_{m_{p1}}) + c_{p_p}C_{m_{p1}}} \quad (14\text{-}62)$$

We shall need some of the equations already derived for use in conjunction with the preceding equations. From Eq. (14-11)

$$\eta_1 = 1 - \frac{1 - C_{m_{p1}}}{C_{m_{p1}}} \frac{M_p P_{p_2}}{M_a(P - P_{p_2})} \quad (14\text{-}63)$$

$$\eta = 1 - \frac{1 - C_{m_{p1}}}{C_{m_{p1}}} \frac{M_p P_{p_3}}{M_a(P - P_{p_3})} \quad (14\text{-}64)$$

When Eq. (14-64) is solved for P_{p_3},

$$P_{p_3} = \frac{(1 - \eta)P}{(1 - \eta) + (1 - C_{m_{p1}})M_p/C_{m_{p1}}M_a} \quad (14\text{-}65)$$

We shall need to use Eq. (14-1); when solved to facilitate the trial-and-error calculation of T_3, it becomes

$$\frac{2.303B}{T_3} - \frac{0.422T_c^2 P_{p_3}}{P_c} \frac{1}{T_3^2} - \left(2.67 - \frac{1.8B}{T_c}\right) \ln T_3$$

$$+ \ln \frac{P_{p_3}}{P_c} + \left(2.67 - \frac{1.8B}{T_c}\right) \ln T_c - \frac{2.303B}{T_c} + 0.422 = 0 \quad (14\text{-}66)$$

Equation (14-4) will also be needed:

$$h_{fg} = h_{fg_b}\left(\frac{T_c - T}{T_c - T_b}\right)^{0.38} \quad (14\text{-}4)$$

Let us summarize the calculation procedure. In addition to the properties of the air and of the pollutant, we shall assume that initially the values of T_1, $C_{m_{p1}}$, $\dot{m}_1$, ε, and η are known. The procedure is as follows:

1 Solve for P_{p_3} from Eq. (14-65).
2 Solve for T_3 by trial and error or iteratively, using Eq. (14-66).
3 Solve for T_2, using Eqs. (14-62), (14-63), (14-1), and (14-4), by a trial-and-error procedure or by an iterative procedure.
4 Solve for T_4 from Eq. (14-61).
5 Solve for η_1 from Eq. (14-63) and for $h_{fg_{p3}}$ from Eq. (14-4) if these have not already been obtained in step 3.

6 Solve for the heat-transfer rate $\dot{Q}$ from Eq. (14-58).

7 Determine $\dot{m}_w$, T_{w_1} and T_{w_2} either arbitrarily or by using the heat-exchanger relations of Sec. 14-3. Note that $T_{w_1} < T_3$ and that Eq. (14-57) must be satisfied.

Let us examine this procedure by means of an example.

EXAMPLE 14-9 A mixture of octane vapor and air is initially saturated at 35°C. It is to be collected using a regenerative system with an overall collection efficiency of 0.99. Assume $\varepsilon = 0.85$ and that the mass flow rate $\dot{m}_1$ is 0.01 kg/s. Compute T_3 and the necessary heat-transfer rate $\dot{Q}$.

SOLUTION The following data are obtained from Tables 14-1 and 2-1:

$$h_{fg_b} = 306{,}000 \text{ J/kg} \qquad B = 3098 \text{ K}$$
$$T_b = 399 \text{ K} \qquad M_p = 114.2$$
$$T_c = 569 \text{ K} \qquad c_{p_p} = 1589 \text{ J/kg} \cdot \text{K}$$
$$P_c = 24.5 \text{ atm} \qquad c_{p_a} = 1005 \text{ J/kg} \cdot \text{K}$$

Using the preceding data, Eq. (14-1) becomes

(a) $$\ln P_p - 5577 \frac{P_p}{T^2} + \frac{7135}{T} + 7.13 \ln \frac{T}{569} - 15.32 = 0$$

At $T_1 = 308$ K, P_{p_1} is evaluated from this equation as

$$P_{p_1} = 0.0311 \text{ atm} = 3151 \text{ N/m}^2$$

The following equation is useful to evaluate $C_{m_{p1}}$:

$$C_m = \frac{P_p}{(1 - M_a/M_p)P_p + M_a P/M_p} \qquad (14\text{-}67)$$

which gives for $C_{m_{p1}}$

$$C_{m_{p1}} = \frac{0.0311}{(1 - 29/114.2)(0.0311) + 29(1)/114.2} = 0.1122$$

For an overall collection efficiency of 0.99, Eq. (14-65) gives for P_{p_3}

$$P_{p_3} = \frac{0.01(1)}{0.01 + (1 - 0.1122)(114.2)/0.1122(29)} = 0.0003208 \text{ atm}$$

We may use Eq. (a) to evaluate T_3, using the preceding value of P_{p_3}; the result of the trial-and-error calculation is

$$T_3 = 242 \text{ K} = -31°\text{C} \qquad Ans.$$

Upon substitution into Eq. (14-62), the simplified result is

(b) $$T_2 = 261.2 + 0.0001048\eta_1 h_{fg_{p2}}$$

Equation (*a*) can be used again, this time to evaluate P_{p_2} from T_2. Then, upon substitution of known values, Eq. (14-63) becomes

(*c*)
$$\eta_1 = 1 - 31.16\left(\frac{P_{p_2}}{1 - P_{p_2}}\right)$$

Equation (14-4) becomes

(*d*)
$$h_{fg_{p2}} = 43,466(569 - T_2)^{0.38}$$

We may solve Eqs. (*a*) to (*d*) iteratively in the following manner: First, assume reasonable values for η_1 and $h_{fg_{p2}}$. Calculate T_2 using Eq. (*b*); then calculate P_{p_2} by trial and error from Eq. (*a*). Now compute η_1 and $h_{fg_{p2}}$ from Eqs. (*c*) and (*d*). Repeat this process until a single value of T_2 results. Following this procedure, the resulting values obtained are

$$T_2 = 287.8 \text{ K}$$
$$P_{p_2} = 0.00995 \text{ atm} = 1008 \text{ N/m}^2$$
$$\eta_1 = 0.687$$
$$h_{fg_{p2}} = 370,486 \text{ J/kg}$$

Replacing T_2 by T_3 in Eq. (*d*), we may calculate $h_{fg_{p3}} = 392,354$ J/kg. Equation (14-61) gives $T_4 = 25°C$. Equation (14-58) gives for the heat-transfer rate $\dot{Q}$

$$\dot{Q} = (1 - 0.1122)(0.01)(1005)(287.8 - 242)$$
$$+ (1 - 0.687)(0.1122)(0.01)[1589(287.8 - 242) + 392,354]$$
$$= 572 \text{ W} \quad \textit{Ans.} \quad ////$$

The computations in Example 14-9, though lengthy, can be accomplished readily by using a programmable electronic calculator or computer terminal. The data on which the example is based could arise in considering the control of hydrocarbon emissions into the atmosphere during the filling of fuel tanks at a service station. The required power output seems modest enough that a condenser could be seriously considered as a control measure for this application. The next example shows how much power is required.

EXAMPLE 14-10 Suppose the heat-transfer rate of Example 14-9 is to be absorbed by a coolant which enters at $T'_3 = -36°C$. The coolant is supplied at this temperature by a refrigerator which rejects heat at 25°C. If the refrigerator has a coefficient of performance equal to 0.4 times that of a Carnot cycle operating between these temperatures, what work input to the refrigerator is required?

SOLUTION   The coefficient of performance of a refrigerator is defined as

$$cop = \frac{\dot{Q}}{\dot{W}}$$

ih which $\dot{Q}$ is the heat-transfer rate to the evaporator or to the cooling fluid and $\dot{W}$ is the work input to the compressor. For an ideal Carnot cycle receiving heat at temperature T'_3 and rejecting heat at temperature T_0, the coefficient of performance is given by

$$cop_{\text{Carnot}} = \frac{T'_3}{T_0 - T'_3}$$

We then obtain

$$cop = 0.4cop_{\text{Carnot}} = \frac{0.4(237)}{25 - (-36)} = 1.56$$

The work input to the compressor of the refrigerator is then

$$\dot{W} = \frac{\dot{Q}}{cop} = \frac{572}{1.56} = 368 \text{ W} \qquad Ans. \qquad ////$$

REFERENCES

1   Reid, R. C., and T. K. Sherwood: "The Properties of Gases and Liquids, Their Estimation and Correlation," 2d ed., McGraw-Hill Book Company, New York, 1966.
2   Parker, J. D., J. H. Boggs, and E. F. Blick: "Introduction to Fluid Mechanics and Heat Transfer," Addison-Wesley Publishing Company, Inc., Reading, Mass., 1969.
3   Kreith, F.: "Principles of Heat Transfer," 2d ed., chap. 11, International Textbook Company, Scranton, Pa., 1965
4   Fraas, A. P., and M. N. Ozisik: "Heat Exchanger Design," John Wiley & Sons, Inc., New York, 1965.
5   Kays, W. M., and A. L. London: "Compact Heat Exchangers," McGraw-Hill Book Company, New York, 1958.

PROBLEMS

14-1   Derive Eq. (14-2) from Eq. (14-1).
14-2   Check the value of B for octane by direct calculation using Eq. (14-2).
14-3   Derive Eqs. (14-5) and (14-6).
14-4   Derive Eq. (14-7).
14-5   Derive Eqs. (14-9) to (14-14).
14-6   Compute and plot curves of saturated vapor pressure and latent heat of condensation for octane vapor over a range of temperature from -20 to $70°C$.
14-7   Compute the saturated vapor pressure and the latent heat of condensation for methyl alcohol at $5°C$.

14-8 At what temperature is the vapor pressure of methyl alcohol equal to 2000 N/m^2?

14-9 Using the data of Table 14-1, compute and plot the vapor pressure of water as a function of temperature over the range from 0 to 100°C. Compare with values taken from the steam tables.

14-10 A mixture of naphthalene in air is initally at 100°C, at which state the naphthalene is saturated. The mixture is passed through a condenser and emerges as a saturated mixture at 20°C. What collection efficiency is expected?

14-11 Sulfur dioxide in air at 120°C is to be collected by a condenser; the initial concentration C_{mv_1} is 0.07 kg/m^3. Compute the temperature to which the mixture must be cooled in order for the collection efficiency to be 0.98.

14-12 What must be the final temperature in Example 14-2 if the collection efficiency is to be 0.99?

14-13 Derive Eq. (14-17).

14-14 Derive Eqs. (14-19) to (14-25).

14-15 Carbon tetrachloride is present in air at 40°C with an initial concentration of 0.005 kg/m^3. To what temperature must the mixture be cooled in order to have a collection efficiency of 0.95?

14-16 A mixture of steam and propyl alcohol for which $C_{m_{s1}} = 0.6$ is initially at 100°C, with the steam at saturated vapor. This mixture is to be cooled in a direct-contact condenser to a temperature T_2 of 20°C, using water which enters at 15°C and leaves at 40°C. How much cooling water is required per kilogram of mixture?

14-17 A mixture of air and diethyl ketone at 200°C with $C_{mv_{p1}} = 0.15$ kg/m^3 is to be condensed in a direct-contact condenser to 30°C using water which enters at 20°C and leaves at 80°C. The flow rate of the mixture is 1.0 m^3/s at the entrance to the condenser. What flow rate of water is required?

14-18 Standard air flows normally to a 3.0-cm-diameter tube with a velocity of 5.0 m/s. What value of the film coefficient is expected?

14-19 Standard air flows through a tube having an inside diameter of 5.0 cm. If the air velocity is 10 m/s, what is the film coefficient on the inside surface of the tube?

14-20 A 1-2 heat exchanger handles a stream of 5.0 m^3/s of air at 300°C on the outside of the tubes and is cooled by a stream of water entering at 20°C and flowing at the rate of 3.5 kg/s. The air leaves at 70°C, and the overall heat-transfer coefficient is 300 W/m^2 K. If the outside tube diameter is 2.0 cm, how many tubes are required and of what length shall they be?

14-21 Derive Eq. (14-53).

14-22 A mixture of steam and ethyl formate at 100°C, with $C_{m_{p1}} = 0.4$, is to be collected in a surface condenser with a collection efficiency of 0.95.

(a) To what temperature must the mixture be cooled?

(b) Compute the leaving mass flow rate of mixture and the heat transferred from the mixture during condensation.

(c) Assume that the cooling fluid has the properties of water, that it leaves the exchanger 10°C higher than it enters, that $U = 500$ W/m^2 K, and that $F_t = 0.7$. Determine the flow rate of the water and the heat-exchanger surface area required. Assume that the cooling water enters the condenser at a temperature 5°C below the temperature of the leaving mixture. Calculate the quantities specified for a flow rate of 1.0 kg/s. Note that for a portion of the condensation process, the condensate is in the form of ice.

14-23 Suppose that the condenser of Example 14-8 is cooled by a stream of water which enters at 25°C and leaves at 75°C. If $U = 200$ W/m$^2 \cdot$ K and $F_t = 0.6$, compute the water flow rate and the heat-exchanger surface area required.

14-24 Determine the heat-transfer rate for condensation of the octane-air mixture of Example 14-2. Assume that the flow rate of the mixture is 1.0 kg/s.

14-25 Derive Eqs. (14-58), (14-59), and (14-62).

14-26 A service station is filling four automobile fuel tanks simultaneously at the rate of 40 l/min into each tank. As the liquid fuel enters the fuel tank, an equal volume of air-fuel mixture is exhausted from the tank. In order to control emissions into the atmosphere, this air-fuel mixture is collected and passed into a regenerative condenser system such as the one in Fig. 14-6. On a hot day, let us assume that the air leaves the fuel tanks at 50°C saturated with octane vapor. Compute the mass flow rate $\dot{m}_1$ of the mixture entering the condenser, and also compute the concentration $C_{m_{p1}}$.

14-27 For the mixture of air and octane vapor of Prob. 14-26, compute the temperature to which the air must be cooled and the heat-transfer rate required for this cooling if a collection efficiency of 0.97 is required. Assume $\varepsilon = 0.80$.

14-28 The sulfur dioxide-air mixture of Prob. 14-11 is to be collected in a regenerative condenser arrangement such as the one shown in Fig. 14-6. Assume that the effectiveness of the heat exchanger is 0.85, and compute the heat-transfer rate required for a flow rate of the mixture of 1000 m^3/s.

14-29 A saturated mixture of air and carbon tetrachloride at 60°C is condensed in a regenerative condenser arrangement to a low temperature of -10°C. Compute the collection efficiency of the condenser arrangement, and compute the heat-transfer per kilogram of carbon tetrachloride collected. Assume $\varepsilon = 0.85$.

DERIVATION OF THE DRAG EQUATION FOR CREEPING FLOW AROUND A SPHERE

Because of its great importance to the study of particle motion, we shall develop in detail in this appendix the equation for the drag force acting on a sphere in creeping motion in a fluid. Our starting point in this derivation will be the Navier-Stokes equations, which are available in many fluid-dynamics textbooks. We consider the flow around a sphere such that at some distance away from the sphere the flow is uniform in the positive x direction, as indicated in Fig. A1-1. This free-stream velocity is designated V_∞. The derivation will be conducted in spherical coordinates; by symmetry, u_ϕ will be zero and all variations in the ϕ direction will likewise be zero.

Basic Differential Equations

The differential equations of motion are taken from Ref. [1]. The continuity equation is

$$\frac{1}{r^2}\frac{\partial}{\partial r}(r^2 u_r) + \frac{1}{r\sin\theta}\frac{\partial}{\partial \theta}(u_\theta \sin\theta) + \frac{1}{r\sin\theta}\frac{\partial u_\phi}{\partial \phi} = 0 \qquad \text{(A1-1)}$$

Equation (A1-1) simplifies to

$$\frac{\partial u_r}{\partial r} + \frac{2u_r}{r} + \frac{1}{r}\frac{\partial u_\theta}{\partial \theta} + \frac{\cot\theta}{r}u_\theta = 0 \qquad \text{(A1-2)}$$

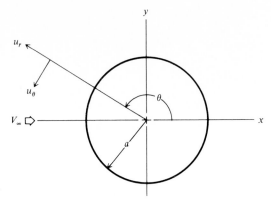

FIGURE A1-1
Flow around a sphere.

The Navier-Stokes equations in the r and θ directions, also taken from Ref. [1], are

$$\frac{\partial u_r}{\partial t} + u_r \frac{\partial u_r}{\partial r} + \frac{u_\theta}{r} \frac{\partial u_r}{\partial \theta} + \frac{u_\phi}{r \sin \theta} \frac{\partial u_r}{\partial \phi} - \frac{u_\theta{}^2 + u_\phi{}^2}{r}$$

$$= -\frac{1}{\rho} \frac{\partial P}{\partial r} + \nu\left(\nabla^2 u_r - \frac{2}{r^2} u_r - \frac{2}{r^2} \frac{\partial u_\theta}{\partial \theta} - \frac{2}{r^2} u_\theta \cot \theta - \frac{2}{r^2 \sin \theta} \frac{\partial u_\phi}{\partial \phi}\right) + \frac{R}{\rho} \qquad \text{(A1-3)}$$

$$\frac{\partial u_\theta}{\partial t} + u_r \frac{\partial u_\theta}{\partial r} + \frac{u_\theta}{r} \frac{\partial u_\theta}{\partial \theta} + \frac{u_\phi}{r \sin \theta} \frac{\partial u_\theta}{\partial \phi} + \frac{u_r u_\theta}{r} - \frac{u_\phi{}^2 \cot \theta}{r}$$

$$= -\frac{1}{\rho r} \frac{\partial P}{\partial \theta} + \nu\left(\nabla^2 u_\theta + \frac{2}{r^2} \frac{\partial u_r}{\partial \theta} - \frac{u_\theta}{r^2 \sin^2 \theta} - \frac{2 \cos \theta}{r^2 \sin^2 \theta} \frac{\partial u_\phi}{\partial \phi}\right) + \frac{\Theta}{\rho} \qquad \text{(A1-4)}$$

For creeping flow, we may neglect the inertia terms (the left sides of Eqs. (A1-3) and (A1-4)) and the body forces (R and Θ in these equations). When we also replace $\nabla^2 u_r$ and $\nabla^2 u_\theta$ by their definitions, the Navier-Stokes equations become

$$\frac{1}{\mu} \frac{\partial P}{\partial r} = \frac{\partial^2 u_r}{\partial r^2} + \frac{2}{r} \frac{\partial u_r}{\partial r} + \frac{1}{r^2} \frac{\partial^2 u_r}{\partial \theta^2} + \frac{\cot \theta}{r^2} \frac{\partial u_r}{\partial \theta} - \frac{2u_r}{r^2} - \frac{2}{r^2} \frac{\partial u_\theta}{\partial \theta} - \frac{2}{r^2} (\cot \theta) u_\theta \qquad \text{(A1-5)}$$

$$\frac{1}{\mu} \frac{\partial P}{\partial \theta} = r \frac{\partial^2 u_\theta}{\partial r^2} + 2 \frac{\partial u_\theta}{\partial r} + \frac{1}{r} \frac{\partial^2 u_\theta}{\partial \theta^2} + \frac{\cot \theta}{r} \frac{\partial u_\theta}{\partial \theta} + \frac{2}{r} \frac{\partial u_r}{\partial \theta} - \frac{u_\theta}{r \sin^2 \theta} \qquad \text{(A1-6)}$$

The basic assumption in creeping flow is that only the viscous and pressure effects are significant, and hence the inertia and body forces are neglected. This effectively limits the validity of the solution to very small Reynolds numbers. When Eq. (A1-5) is differentiated with respect to θ, Eq. (A1-6) with respect to r, and the resulting equations subtracted, the following equation results, which no longer contains the pressure P:

$$\frac{\partial^3 u_r}{\partial \theta \, \partial r^2} - \frac{1}{r} \frac{\partial^3 u_\theta}{\partial r \, \partial \theta^2} + \frac{1}{r^2} \frac{\partial^3 u_r}{\partial \theta^3} - r \frac{\partial^3 u_\theta}{\partial r^3} + \frac{\cot \theta}{r^2} \frac{\partial^2 u_r}{\partial \theta^2} - 3 \frac{\partial^2 u_\theta}{\partial r^2} - \frac{1}{r^2} \frac{\partial^2 u_\theta}{\partial \theta^2}$$

$$- \frac{\cot \theta}{r} \frac{\partial^2 u_\theta}{\partial r \, \partial \theta} - \frac{1}{r^2 \sin^2 \theta} \frac{\partial u_r}{\partial \theta} - \frac{\cot \theta}{r^2} \frac{\partial u_\theta}{\partial \theta} + \frac{1}{r \sin^2 \theta} \frac{\partial u_\theta}{\partial r} + \frac{u_\theta}{r^2 \sin^2 \theta} = 0 \qquad \text{(A1-7)}$$

Equation (A1-7) can be simplified further if the notion of stream function is employed. The stream function is a single function of the coordinates from which the velocity components can be obtained by differentiation; it also has the property that the continuity equation is satisfied automatically. Denoting the stream function by ψ, the velocity components are related to it by

$$u_r = \frac{1}{r^2 \sin \theta} \frac{\partial \psi}{\partial \theta} \qquad \text{(A1-8)}$$

$$u_\theta = -\frac{1}{r \sin \theta} \frac{\partial \psi}{\partial r} \qquad \text{(A1-9)}$$

When all the derivatives of u_r and u_θ present in Eq. (A1-7) are expressed in terms of the stream function ψ by using Eqs. (A1-8) and (A1-9), and when these are substituted into Eq. (A1-7) and the results simplified as much as possible, the final equation is

$$\frac{\partial^4 \psi}{\partial r^4} + \frac{2}{r^2} \frac{\partial^4 \psi}{\partial r^2 \partial \theta^2} + \frac{1}{r^4} \frac{\partial^4 \psi}{\partial \theta^4} - \frac{2 \cot \theta}{r^4} \frac{\partial^3 \psi}{\partial \theta^3} - \frac{4}{r^3} \frac{\partial^3 \psi}{\partial r \partial \theta^2} - \frac{2 \cot \theta}{r^2} \frac{\partial^3 \psi}{\partial \theta \partial r^2}$$

$$+ \frac{5}{r^4} \frac{\partial^2 \psi}{\partial \theta^2} + \frac{3}{r^4 \sin^2 \theta} \frac{\partial^2 \psi}{\partial \theta^2} + \frac{4 \cot \theta}{r^3} \frac{\partial^2 \psi}{\partial r \partial \theta} - \frac{6 \cot \theta}{r^4} \frac{\partial \psi}{\partial \theta} - \frac{3 \cot \theta}{r^4 \sin^2 \theta} \frac{\partial \psi}{\partial \theta} = 0 \qquad \text{(A1-10)}$$

Although it does not enter into the current derivation, Eq. (A1-10) can be expressed in functional form as

$$\left(\frac{\partial^2}{\partial r^2} + \frac{1}{r^2} \frac{\partial^2}{\partial \theta^2} - \frac{\cot \theta}{r^2} \frac{\partial}{\partial \theta} \right)^2 \psi = 0 \qquad \text{(A1-11)}$$

The boundary conditions which apply to this problem state that the velocity is zero at the surface of the sphere and that the velocity approaches the free-stream value at a large distance from the surface of the sphere. Thus when $r = a$, $u_r = u_\theta = 0$, or

$$\frac{\partial \psi}{\partial \theta} = \frac{\partial \psi}{\partial r} = 0 \qquad \text{(A1-12)}$$

and when $r \to \infty$, $u_r \to V_\infty \cos \theta$ and $u_\theta \to -V_\infty \sin \theta$, or

$$\frac{\partial \psi}{\partial \theta} \to V_\infty r^2 \sin \theta \cos \theta \qquad \text{(A1-13)}$$

$$\frac{\partial \psi}{\partial r} \to V_\infty r \sin^2 \theta \qquad \text{(A1-14)}$$

Equations (A1-13) and (A1-14) integrate to

$$\psi = \tfrac{1}{2} V_\infty r^2 \sin^2 \theta \qquad \text{(A1-15)}$$

except for an additive constant, which is of no consequence. Equation (A1-15) is the stream function for uniform flow.

Let us try a solution to Eq. (A1-10) of the form

$$\psi = \tfrac{1}{2} V_\infty f(r) \sin^2 \theta \qquad \text{(A1-16)}$$

This equation automatically satisfies the boundary condition (A1-13); it will also satisfy Eq. (A1-12) if

$$f(a) = f'(a) = 0 \qquad \text{(A1-17)}$$

while Eq. (A1-14) requires that

$$f(r) \to r^2 \qquad \text{as } r \to \infty \qquad \text{(A1-18)}$$

When Eq. (A1-16) is substituted into Eq. (A1-10) and the results simplified, the following linear ordinary differential equation results:

$$r^4 f^{iv} - 4r^2 f'' + 8rf' - 8f = 0 \qquad \text{(A1-19)}$$

The solution to this equation is

$$f(r) = \frac{A}{r} + Br + Cr^2 + Dr^4$$

When Eqs. (A1-17) and (A1-18) are applied, the result is

$$f(r) = \frac{a^3}{2r} - \frac{3ar}{2} + r^2 \qquad \text{(A1-20)}$$

The stream function then follows from Eq. (A1-16) and is

$$\psi = \frac{1}{4} V_\infty \left(\frac{a^3}{r} - 3ar + 2r^2 \right) \sin^2 \theta \qquad \text{(A1-21)}$$

The velocity components in spherical coordinates then follow from Eqs. (A1-8) and (A1-9):

$$u_r = \frac{1}{2} V_\infty \left(\frac{a^3}{r^3} - \frac{3a}{r} + 2 \right) \cos \theta \qquad \text{(A1-22)}$$

$$u_\theta = \frac{1}{4} V_\infty \left(\frac{a^3}{r^3} + \frac{3a}{r} - 4 \right) \sin \theta \qquad \text{(A1-23)}$$

The velocity components can also be expressed in rectangular cartesian coordinates by means of the transformation equations:

$$u = u_r \cos \theta - u_\theta \sin \theta$$
$$v = u_r \sin \theta + u_\theta \cos \theta$$

giving

$$u = \frac{1}{2} V_\infty \left(\frac{a^3}{r^3} - \frac{3a}{r} + 2 \right) \cos^2 \theta$$

$$- \frac{1}{4} V_\infty \left(\frac{a^3}{r^3} + \frac{3a}{r} - 4 \right) \sin^2 \theta \qquad \text{(A1-24)}$$

$$v = V_\infty \left(\frac{3a^3}{4r^3} - \frac{3a}{4r} \right) \sin \theta \cos \theta \qquad \text{(A1-25)}$$

Note that u and v both become zero when $r = a$, that u approaches V_∞ when $r \to \infty$ for any θ, and that $v \to 0$ as $r \to \infty$.

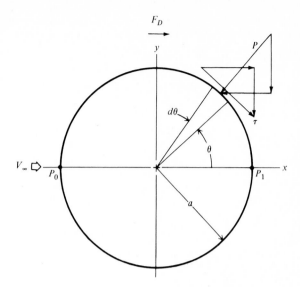

FIGURE A1-2
Forces acting on a sphere.

When Eqs. (A1-22) and (A1-23) are used to evaluate Eq. (A1-6) for $\partial P/\partial\theta$ at the surface of the sphere, $r = a$, the value is

$$\left(\frac{\partial P}{\partial\theta}\right)_{r=a} = \frac{3\mu V_\infty}{2a}\sin\theta \qquad \text{(A1-26)}$$

Let us now determine the pressure distribution around the sphere, relative to the upstream stagnation pressure P_0 located at the point where $\theta = \pi$. Figure A1-2 indicates this stagnation pressure. The pressure distribution on the surface of the sphere is obtained from the formula

$$P(a, \theta) = P_0 + \int_\pi^\theta \left(\frac{\partial P}{\partial\theta}\right)_{r=a} d\theta \qquad \text{(A1-27)}$$

Upon substituting Eq. (A1-26), Eq. (A1-27) integrates to

$$P(a, \theta) = P_0 + \frac{3\mu V_\infty}{2a}\int_\pi^\theta \sin\theta \, d\theta$$

$$= P_0 - \frac{3\mu V_\infty}{2a}(1 + \cos\theta) \qquad \text{(A1-28)}$$

and when $\theta = 0$, the pressure at the rear stagnation point P_1 is given by

$$P_1 = P(a, 0) = P_0 - \frac{3\mu V_\infty}{a} \qquad \text{(A1-29)}$$

The shear stress at the surface of the sphere is given by

$$\tau = -\mu\left(\frac{du_\theta}{dr}\right)_{r=a} = \frac{3\mu V_\infty}{2a}\sin\theta \qquad \text{(A1-30)}$$

Next, we shall compute the drag force acting on the sphere. Take the element of angle $d\theta$ shown in Fig. A1-2, whose area is $2\pi a^2 \sin\theta\, d\theta$. The component of net force acting in the x direction is given by

$$dF_D = -P(a, \theta)\cos\theta\, dA + \tau\sin\theta\, dA$$

Substituting Eq. (A1-28) for $P(a, \theta)$ and substituting for dA, give, upon integration,

$$F_D = 2\pi a^2 \left\{ -P_0 \int_0^\pi \sin\theta\cos\theta\, d\theta + \frac{3\mu V_\infty}{2a} \int_0^\pi \sin\theta\cos\theta\, d\theta \right.$$

$$\left. + \frac{3\mu V_\infty}{2a} \int_0^\pi \sin\theta\cos\theta\, d\theta + \frac{3\mu V_\infty}{2a} \int_0^\pi \sin^3\theta\, d\theta \right\}$$

The first two integrals are zero, the third integral has the value $\frac{2}{3}$, and the fourth integral has the value $\frac{4}{3}$, giving

$$F_D = 6\pi\mu V_\infty a = 3\pi\mu V_\infty d \qquad \text{(A1-31)}$$

REFERENCE

1   Parker, J. D., J. H. Boggs, and E. F. Blick: "Introduction to Fluid Mechanics and Heat Transfer," pp. 77 and 91, Addison-Wesley Publishing Company, Inc., Reading, Mass. 1969.

The error function is defined as

$$\mathrm{erf}\, x = \frac{2}{\sqrt{\pi}} \int_0^x e^{-z^2}\, dz \qquad \text{(A2-1)}$$

As x approaches infinity, the error function approaches unity. The complementary error function is defined as

$$\mathrm{erfc}\, x = 1 - \mathrm{erf}\, x \qquad \text{(A2-2)}$$

which may also be written as

$$\mathrm{erfc}\, x = \frac{2}{\sqrt{\pi}} \int_x^\infty e^{-z^2}\, dz \qquad \text{(A2-3)}$$

For negative arguments, it can be easily seen from Eq. (A2-1) that

$$\mathrm{erf}\, (-x) = -\mathrm{erf}\, x \qquad \text{(A2-4)}$$

The complementary error function for negative arguments gives

$$\mathrm{erfc}\, (-x) = 1 + \mathrm{erf}\, x = 2 - \mathrm{erfc}\, x \qquad \text{(A2-5)}$$

A series representation for the error function is

$$\mathrm{erf}\, x = \frac{2}{\sqrt{\pi}} \sum_{i=0}^{\infty} \frac{(-1)^i x^{2i+1}}{i!\,(2i+1)} \qquad \text{(A2-6)}$$

The error function can be readily evaluated for any value of the argument x by means of a programmable electronic desk calculator. The ith term of the series may be written as

$$T_i = \frac{(-1)^i x^{2i+1}}{i!\,(2i+1)}$$

The $i + 1$ term can be written as

$$T_{i+1} = \frac{(-1)^{i+1} x^{2i+3}}{(i+1)!\,(2i+3)} = -\frac{x^2(2i+1)}{(i+1)(2i+3)}\,T_i \quad \text{(A2-7)}$$

Using the second part of Eq. (A2-7), the series can be programmed to automatically form the required sum. The data in Table A2-1 were computed in this way.

Table A2-1   erf x, erfc x, and erfc $(-x)$

x	erf x	erfc x	erfc $(-x)$
0	0	1.0000	1.0000
0.01	0.01128	0.9887	1.0113
0.02	0.02256	0.9774	1.0226
0.03	0.03384	0.9662	1.0338
0.04	0.04511	0.9549	1.0451
0.05	0.05637	0.9436	1.0564
0.06	0.06762	0.9324	1.0676
0.07	0.07886	0.9211	1.0789
0.08	0.09008	0.9099	1.0901
0.09	0.1013	0.8987	1.1013
0.10	0.1125	0.8875	1.1125
0.11	0.1236	0.8764	1.1236
0.12	0.1348	0.8652	1.1348
0.13	0.1459	0.8541	1.1459
0.14	0.1569	0.8431	1.1569
0.15	0.1680	0.8320	1.1680
0.16	0.1790	0.8210	1.1790
0.17	0.1900	0.8100	1.1900
0.18	0.2009	0.7991	1.2009
0.19	0.2118	0.7882	1.2118
0.20	0.2227	0.7773	1.2227
0.21	0.2335	0.7665	1.2335
0.22	0.2443	0.7557	1.2443
0.23	0.2550	0.7450	1.2550
0.24	0.2657	0.7343	1.2657
0.25	0.2763	0.7237	1.2763
0.26	0.2869	0.7131	1.2869
0.27	0.2974	0.7026	1.2974
0.28	0.3079	0.6921	1.3079
0.29	0.3183	0.6817	1.3183
0.30	0.3286	0.6714	1.3286
0.31	0.3389	0.6611	1.3389
0.32	0.3491	0.6509	1.3491
0.33	0.3593	0.6407	1.3593

(continued)

Table A2-1   (*continued*)

x	erf x	erfc x	erfc $(-x)$
0.34	0.3694	0.6306	1.3694
0.35	0.3794	0.6206	1.3794
0.36	0.3893	0.6107	1.3893
0.37	0.3992	0.6008	1.3992
0.38	0.4090	0.5910	1.4090
0.39	0.4187	0.5813	1.4187
0.40	0.4284	0.5716	1.4284
0.41	0.4380	0.5620	1.4380
0.42	0.4475	0.5525	1.4475
0.43	0.4569	0.5431	1.4569
0.44	0.4662	0.5338	1.4662
0.45	0.4755	0.5245	1.4755
0.46	0.4847	0.5153	1.4847
0.47	0.4937	0.5063	1.4937
0.48	0.5027	0.4973	1.5027
0.49	0.5117	0.4883	1.5117
0.50	0.5205	0.4795	1.5205
0.51	0.5292	0.4708	1.5292
0.52	0.5379	0.4621	1.5379
0.53	0.5465	0.4535	1.5465
0.54	0.5549	0.4451	1.5549
0.55	0.5633	0.4367	1.5633
0.56	0.5716	0.4384	1.5716
0.57	0.5798	0.4201	1.5798
0.58	0.5879	0.4121	1.5879
0.59	0.5959	0.4041	1.5959
0.60	0.6039	0.3961	1.6039
0.61	0.6117	0.3883	1.6117
0.62	0.6194	0.3806	1.6194
0.63	0.6270	0.3730	1.6270
0.64	0.6346	0.3654	1.6346
0.65	0.6420	0.3580	1.6420
0.66	0.6494	0.3506	1.6494
0.67	0.6566	0.3434	1.6566
0.68	0.6638	0.3362	1.6638
0.69	0.6708	0.3292	1.6708
0.70	0.6778	0.3222	1.6778
0.71	0.6847	0.3153	1.6847
0.72	0.6914	0.3086	1.6914
0.73	0.6981	0.3019	1.6981
0.74	0.7047	0.2953	1.7047
0.75	0.7112	0.2888	1.7112
0.76	0.7175	0.2825	1.7175
0.77	0.7238	0.2762	1.7238
0.78	0.7300	0.2700	1.7300
0.79	0.7361	0.2639	1.7361
0.80	0.7421	0.2579	1.7421
0.81	0.7480	0.2520	1.7480
0.82	0.7538	0.2462	1.7538
0.83	0.7595	0.2405	1.7595
0.84	0.7651	0.2349	1.7651
0.85	0.7707	0.2293	1.7707

(*continued*)

Table A2-1 (*continued*)

x	erf x	erfc x	erfc $(-x)$
0.86	0.7761	0.2239	1.7761
0.87	0.7814	0.2186	1.7814
0.88	0.7867	0.2133	1.7867
0.89	0.7918	0.2082	1.7918
0.90	0.7969	0.2031	1.7969
0.91	0.8019	0.1981	1.8019
0.92	0.8068	0.1932	1.8068
0.93	0.8116	0.1884	1.8116
0.94	0.8163	0.1837	1.8163
0.95	0.8209	0.1791	1.8209
0.96	0.8254	0.1746	1.8254
0.97	0.8299	0.1701	1.8299
0.98	0.8342	0.1658	1.8342
0.99	0.8385	0.1615	1.8385
1.00	0.8427	0.1573	1.8427
1.01	0.8468	0.1532	1.8468
1.02	0.8508	0.1492	1.8508
1.03	0.8548	0.1452	1.8548
1.04	0.8586	0.1414	1.8586
1.05	0.8624	0.1376	1.8624
1.06	0.8661	0.1339	1.8661
1.07	0.8698	0.1302	1.8698
1.08	0.8733	0.1267	1.8733
1.09	0.8768	0.1232	1.8768
1.10	0.8802	0.1198	1.8802
1.11	0.8835	0.1165	1.8835
1.12	0.8868	0.1132	1.8868
1.13	0.8900	0.1100	1.8900
1.14	0.8931	0.1069	1.8931
1.15	0.8961	0.1039	1.8961
1.16	0.8991	0.1009	1.8991
1.17	0.9020	0.09800	1.9020
1.18	0.9048	0.09516	1.9048
1.19	0.9076	0.09239	1.9076
1.20	0.9103	0.08968	1.9103
1.21	0.9130	0.08704	1.9130
1.22	0.9155	0.08447	1.9155
1.23	0.9181	0.08195	1.9181
1.24	0.9205	0.07949	1.9205
1.25	0.9229	0.07710	1.9229
1.26	0.9252	0.07476	1.9252
1.27	0.9275	0.07249	1.9275
1.28	0.9297	0.07027	1.9297
1.29	0.9319	0.06810	1.9319
1.30	0.9340	0.06599	1.9340
1.31	0.9361	0.06394	1.9361
1.32	0.9381	0.06193	1.9381
1.33	0.9400	0.05998	1.9400
1.34	0.9419	0.05809	1.9419
1.35	0.9438	0.05624	1.9438
1.36	0.9456	0.05444	1.9456
1.37	0.9473	0.05269	1.9473

(*continued*)

Table A2-1 (*continued*)

x	erf x	erfc x	erfc $(-x)$
1.38	0.9490	0.05098	1.9490
1.39	0.9507	0.04933	1.9507
1.40	0.9523	0.04771	1.9523
1.41	0.9539	0.04615	1.9539
1.42	0.9554	0.04462	1.9554
1.43	0.9569	0.04314	1.9569
1.44	0.9583	0.04170	1.9583
1.45	0.9597	0.04030	1.9597
1.46	0.9611	0.03895	1.9611
1.47	0.9624	0.03763	1.9624
1.48	0.9637	0.03635	1.9637
1.49	0.9649	0.03510	1.9649
1.50	0.9661	0.03389	1.9661
1.51	0.9673	0.03272	1.9673
1.52	0.9684	0.03159	1.9684
1.53	0.9695	0.03048	1.9695
1.54	0.9706	0.02941	1.9706
1.55	0.9716	0.02838	1.9716
1.56	0.9726	0.02737	1.9726
1.57	0.9736	0.02640	1.9736
1.58	0.9745	0.02545	1.9745
1.59	0.9755	0.02454	1.9755
1.60	0.9763	0.02365	1.9763
1.61	0.9772	0.02279	1.9772
1.62	0.9780	0.02196	1.9780
1.63	0.9788	0.02116	1.9788
1.64	0.9796	0.02038	1.9796
1.65	0.9804	0.01962	1.9804
1.66	0.9811	0.01889	1.9811
1.67	0.9818	0.01819	1.9818
1.68	0.9825	0.01751	1.9825
1.69	0.9832	0.01685	1.9832
1.70	0.9838	0.01621	1.9838
1.71	0.9844	0.01559	1.9844
1.72	0.9850	0.01500	1.9850
1.73	0.9856	0.01442	1.9856
1.74	0.9861	0.01387	1.9861
1.75	0.9867	0.01333	1.9867
1.76	0.9872	0.01281	1.9872
1.77	0.9877	0.01231	1.9877
1.78	0.9882	0.01183	1.9882
1.79	0.9886	0.01136	1.9886
1.80	0.9891	0.01091	1.9891
1.81	0.9895	0.01048	1.9895
1.82	0.9899	0.01006	1.9899
1.83	0.9903	0.00965	1.9903
1.84	0.9907	0.00926	1.9907
1.85	0.9911	0.00889	1.9911
1.86	0.9915	0.00853	1.9915
1.87	0.9918	0.00818	1.9918
1.88	0.9922	0.00784	1.9922
1.89	0.9925	0.00752	1.9925

(*continued*)

Table A2-1   (*continued*)

x	erf x	erfc x	erfc $(-x)$
1.90	0.9928	0.00721	1.9928
1.91	0.9931	0.00691	1.9931
1.92	0.9934	0.00662	1.9934
1.93	0.9937	0.00634	1.9937
1.94	0.9939	0.00608	1.9939
1.95	0.9942	0.00582	1.9942
1.96	0.9944	0.00557	1.9944
1.97	0.9947	0.00534	1.9947
1.98	0.9949	0.00511	1.9949
1.99	0.9951	0.00489	1.9951
2.00	0.9953	0.00468	1.9953

A	area, m^2
A_c	area of control surface, m^2
A_p	projected frontal area, m^2
A_{ss}	specific surface, m^2/kg
AF	air-fuel ratio
a	radius, m
b	bypass factor
b	cylinder radius around filter, m
C	concentration
C	constant
C	Cunningham correction factor
C_m	concentration by mass
C_{ma}	mass per unit area, kg/m^2
C_{mv}	mass-volume concentration, mass per unit volume, kg/m^3
C_n	mole fraction
C_{na}	number of particles per unit area, m^{-2}
C_{nv}	number of particles per unit volume, m^{-3}
C_{ppm}	concentration in parts per million by volume
C_v	concentration by volume, mole fraction
c	speed of sound, m/s
c	flow coefficient
c	packing density of filter

c	specific heat of liquid or solid, J/kg · K
c_d	discharge coefficient
c_p	specific heat at constant pressure, J/kg · K
c_v	specific heat at constant volume, J/kg · K
D	diameter, m or mm
D_h	hydraulic diameter, m
$\mathscr{D}$	diffusion coefficient, m²/s
d	particle diameter, m or μm
$d_1 \cdots d_6$	mean particle diameters, m or μm
$\bar{d}$	mean diameter, m or μm
E	electric field strength, V/m
$\dot{E}_f$	flow-pressure-energy flow rate, W
e	roughness height, m
F	force, N
F_D	drag force, N
F_t	correction factor for mean temperature difference
f	friction factor, roughness factor
f	make-up air fraction
g	acceleration of gravity, m/s²
H	height, m
H	Henry's law constant, N/m²
$\dot{H}$	enthalpy flow rate, W
HTU	height of a transfer unit, m
h	film coefficient, W/m² · K
h	specific enthalpy, J/kg
h_m	mass-transfer coefficient, m/s
i, j, k	unit vectors
K	coefficient
K	ion mobility, m²/V · s
K_1	filter constant, kg/m² · s
K_2	filter constant, s⁻¹
K_L	loss coefficient
Ku	Kuwabara parameter
k	ratio of specific heat, c_p/c_v
k	thermal conductivity, W/m · K
L	length, m
l	distance, m
M	molecular weight
M	momentum, kg · m/s
M	Mach number
$\dot{M}$	momentum flow rate, kg · m/s²
m	mass, kg
$\dot{m}$	mass flow rate, kg/s
N	number of cyclones
N	number of ions
N	total number of particles
N	rotative speed, rev/min (rpm)
$N(d)$	number of particles with diameter $\leq d$

N_T	number of trays in a plate tower
N_T	number of tubes in a condenser
NTU	number of transfer units
n	exponent
n	number
n	number of drops encountered
n	number of filter elements
n	number of compartments
n	number of moles
n	number of trays in a settling chamber
$n(d)$	rate of change of number of particles with diameter, m^{-1}
P	generalized property
P	pressure, N/m^2
P	pulvation distance, m
Pr	Prandtl number
p	perimeter, m
Q	volumetric flow rate, m^3/s
Q	heat transfer, J
$\dot{Q}$	heat transfer rate, W
q	charge, C
q_v	charge density, C/m^3
R	body force per unit volume in the r direction, N/m^3
R	gas constant, $J/kg \cdot K$
R	radius, m
R	rate of discharge of pollutant into a conditioned space, m^3/s or kg/s
R_u	universal gas constant, $J/kg\text{-mol} \cdot K$
Re	Reynolds number
r, θ	polar coordinates, m and rad
S	cost, money
Sc	Schmidt number
Sh	Sherwood number
SND	static-no-delivery condition
s	coordinate in the direction of particle motion, m
s	number of stages
T	temperature, K
t	time, s
U	overall heat-transfer coefficient, $W/m^2 \cdot K$
$\dot{U}$	internal energy flow rate, J/s
u	specific internal energy, J/kg
u, v, w	velocity components, m/s
$\bar{u}$	mean molecular velocity, m/s
V	velocity, m/s
V_c	control velocity, m/s
V_t	terminal velocity, m/s
$\tilde{V}$	voltage, V
$\mathcal{V}$	volume, m^3
$|V|$	speed, m/s
v	specific volume, m^3/kg

W	width, m
W	work, J
$\dot{W}$	power, W
w	weight, N
X	distance from hood to control surface, m
x, y, z	coordinates, m
Y	compressibility factor
α	constant in Freundlich equation
α	exit angle, radians
α	solubility
β	blade angle, radians
β	constant in Freundlich equation
β	skewness
β	solubility
γ	specific weight, N/m^3
δ	boundary-layer thickness, m
δ	deviation, radians
δ	solubility
δ	width of adsorption zone, m
η	collection efficiency
η_A	collection efficiency based on surface area
η_d	collection efficiency for an individual drop
η_m	collection efficiency based on mass
η_n	collection efficiency based on number of particles
$\eta(d)$	collection efficiency of particles having a diameter d
Θ	body force per unit volume in the θ direction, N/m^3
κ	dielectric constant
λ	molecular mean free path, $kg/m \cdot s$
μ	dynamic viscosity, $kg/m \cdot s$
v	kinematic viscosity, m^2/s
ρ	density, kg/m^3
σ	attachment coefficient
σ	standard deviation
σ	surface tension, N/m
τ	shear stress, N/m^2
ϕ	shape factor for an adsorbent bed
ψ	stream function, m^2/s
∞	infinity

Subscripts

A	area
a	air
ave	average
c	circular, collected, collector, compartment, control surface
cf	counterflow
d	diffusion, drop, dust
e	electron, end, equivalent, exit
f	face, filter, fluid, fouling, fuel

g	gas
i	inlet, inside, interception and inertial impaction
j	jet
l	liquid, lower
m	mass, maximum, mean
mb	mean bulk
me	median
mo	modal
mw	mean wall
max	maximum
min	minimum
n	number
ns	no spray
o	outlet, output, outside, passed through
p	particle, pollutant
pf	parallel flow
pr	product
R	reaction, regenerating fluid
r	reactant, rectangular, return
s	saturation, spray, static, stoichiometric, supply, surface
sc	spray chamber
ss	specific surface
t	terminal
tr	throat
w	at the wall, water
∞	free-stream value

APPENDIX 4

FREQUENTLY USED CONSTANTS IN SI UNITS

Standard atmospheric pressure	P_{atm}	101,326 N/m^2
Standard gravitational acceleration	g	9.8067 m/s^2
Universal gas constant	R_u	8314.3 J/kg-mol $\cdot$ K
Electrical permittivity constant	ε_0	8.85×10^{-12} C/V $\cdot$ m
Electron charge	q_e	1.60×10^{-19} C
Boltzmann's constant	k	1.38×10^{-23} J/K

APPENDIX 5

PROPERTIES OF AIR AT STANDARD CONDITIONS IN SI UNITS*

Molecular weight	M	28.97
Gas constant	R	287 J/kg · K
Specific heat at constant pressure	c_p	1005 J/kg · K
Specific heat at constant volume	c_v	718 J/kg · K
Density	ρ	1.185 kg/m^3
Dynamic viscosity	μ	1.84 × 10^{-5} kg/m · s
Kinematic viscosity	v	1.55 × 10^{-5} m^2/s
Thermal conductivity	k	0.0257 W/m · K
Ratio of specific heats, c_p/c_v	k	1.3997
Prandtl number	Pr	0.720

* Measured at 101,326 N/m^2 pressure and 298 K temperature.

APPENDIX 6

PROPERTIES OF SATURATED WATER AT 298 K IN SI UNITS

Molecular weight	M	18.02
Gas constant	R	461.4 J/kg · K
Density	ρ	999 kg/m^3
Specific heat	c	4181 J/kg · K
Dynamic viscosity	μ	9.239×10^{-4} kg/m · s
Kinematic viscosity	v	9.248×10^{-7} m^2/s
Prandtl number	Pr	6.395
Thermal conductivity	k	0.604 W/m · K

LIST OF SELECTED
TABLES AND FIGURES

Tables

Figures

ANSWERS TO PROBLEMS

Chapter 1

1-4 Precipitator, $10,660/year

1-6 $c = \dfrac{r}{1 - e^{-rt_f}}$; 0.0859

Chapter 2

2-2 $\rho = 1.08 \text{ kg/m}^3$
$c_p = 1006 \text{ J/kg} \cdot \text{K}$
$c_v = 718 \text{ J/kg} \cdot \text{K}$
$k = 1.402$
$\mu = 1.98 \times 10^{-5} \text{ kg/m} \cdot \text{s}$
$v = 1.9 \times 10^{-5} \text{ m}^2/\text{s}$

2-6 $\rho = 1.1614 \text{ kg/m}^3$
$c_p = 1005.05 \text{ J/kg} \cdot \text{K}$
$c_v = 717.04 \text{ J/kg} \cdot \text{K}$
$k = 1.4017$
$\mu = 1.84 \times 10^{-5} \text{ kg/m} \cdot \text{s}$
$v = 1.584 \times 10^{-5} \text{ m}^2/\text{s}$

2-4 $\rho = 0.0705 \text{ lb}_m/\text{ft}^3$
$c_p = 0.240 \text{ Btu/lb}_m \cdot {}^\circ\text{F}$
$c_v = 0.171 \text{ Btu/lb}_m \cdot {}^\circ\text{F}$
$\mu = 1.263 \times 10^{-5} \text{ lb}_m/\text{ft} \cdot \text{s}$
$v = 1.722 \times 10^{-4} \text{ ft}^2/\text{s}$

2-8 $C_{v_p} = 1.205 \times 10^{-8}$
$\rho = 1.205 \text{ kg/m}^3$
$c_p = 1004.99 \text{ J/kg} \cdot \text{K}$
$c_v = 718.00 \text{ J/kg} \cdot \text{K}$
$\mu = 1.84 \times 10^{-5} \text{ kg/m} \cdot \text{s}$
$v = 1.53 \times 10^{-5} \text{ m}^2/\text{s}$

2-10 $\rho = 998$ kg/m^3
  $\mu = 9.9 \times 10^{-4}$ kg/m $\cdot$ s
  $v = 0.992 \times 10^{-6}$ m^2/s
  $\sigma = 0.0728$ N/m
  $e = 4.368$ J/kg
2-14 24.78 m
2-20 153.3 N/m^2

2-32 $C_D = 24$ for Re $= 1.0$

2-36 $\dot{W} = 10,130$ kW; 15,585 kW electrical
2-40 $T_2 = 300°$C
  $\dot{Q} = 171$ kW
2-48 $D_2 = 0.51$ mm
  $P_2 = 54,035$ N/m^2

2-12 $V_{ave} = 4.95$ m/s
    Re $= 150.000$
    $P_1 - P_2 = 6.028 \times 10^6$ N/m^2
2-16 $\dfrac{D_h}{D} = 0.707$ for $\dfrac{D_o}{D_i} = 3.0$

2-29 $C_D = 24$
  $F_D = 4.24 \times 10^{-10}$ N
2-34 (a) $V_{ave} = 12.7$ m/s
    $V_{max} = 25.4$ m/s
    (b) $\dot{M} = 1685$ kg $\cdot$ m/s^2
    (c) $\dot{E}_k = 16,088$ J/s
    (d) Re $= 1.27 \times 10^6$
    (e) $Q = 0.000181$ m^3/s
2-38 99 m/s
2-46 $Q = 7.134$ m^3/s
  $V_1 = 25.2$ m/s

Chapter 3

3-10 $V_t = 0.740$ cm/s
  Re$_t = 0.00477$
    $t = 2.26$ ms
    $s = 11.45$ μm
3-14 (a) 0.044 m/s
   (b) 0.111 m/s
   (c) 0.370 m/s
   (d) 0.837 m/s
3-20 $s_{max} = 0.249$ m   for $d = 50$ μm
3-26 1.21 m

3-12 $V_t = 7.53$ m/s
  Re$_t = 486$
    $t = 1.42$ s
    $s = 7.50$ m
3-18 1.088 m

3-22 0.978 m

Chapter 4

4-2   300 m^2/kg

4-10 0.271 mg
4-20 $C_m = 0.0066$
  $C_v = 6.995 \times 10^{-4}$
  $C_{mv} = 0.007867$ kg/m^3
4-24 $C_m = 0.000202$
  $C_v = 0.000075$
  $C_{mv} = 0.000239$ kg/m^3

4-6 $d_6 = 0.3316 \dfrac{\rho V_t^2}{\rho_p g}$

4-13 100.6 m^2/kg
4-22 $C_m = 0.0474$
  $C_v = 0.000062$
  $C_{mv} = 0.0589$ kg/m^3

4-30 $\bar{u} = \ln \dfrac{\bar{d}}{\sqrt{1 + \sigma^2/\bar{d}^2}}$

  $\sigma_u = \left[\ln \left(1 + \dfrac{\sigma^2}{\bar{d}^2} \right) \right]^{1/2}$

4-33 $d_m = 29.6\ \mu m$
$d_{mo} = 28.8\ \mu m$
$\sigma_m = 1.180$
$\bar{u} = -10.43$
$\sigma_u = 0.1655$
$\beta = 3.98\ \mu m$

4-35 $C_m = 0.1837$
$C_v = 0.0001451$
$C_{mv} = 0.2177\ kg/m^3$

4-44 157,890 kg/s; 8310 kg/s

4-56 $\eta < \eta_1$
4-60 0.57
4-64 $C_r = 235\ \mu g/m^3$
4-67 $Q_s = 17,550\ m^3/s$
$C_{ppm_s} = 0.5$
4-72 14.4 m^3/s; 15.0 m^3/s

4-34 2177 kg/s

4-38 $C_v = \dfrac{\pi}{6}\ Ne^{3\bar{u}}e^{9\sigma_u{}^2/2}$

4-51 $\eta_m = 0.9726$
$\eta_A = 0.9625$
$\eta_n = 0.9231$

4-58 $\eta = \eta_1\eta_3 + \eta_2\eta_3 - \eta_1\eta_2\eta_3$
4-63 0.850
4-66 0.964
4-71 $Q_s = 17,547\ m^3/s$
$C_s = 0.5\ ppm$

Chapter 5

5-2 0.419 m^3/s
5-6 0.515 m
5-10 1.92 m^3/s
5-16 (a) 1.3 cm
(b) $Q = 0.0030\ m^3/s$
$D = 2.0$ cm
5-20 $A_c = 2L(2y + h)$

5-26 $a = 3.53$ m
$V = 9.44$ m/s
5-30 38.41 N/m^2
5-34 11.27 N/m^2
5-38 20.8 by 69.2 cm
5-42 5.44 m
5-48 $Q_2 = 150\ m^3/s$
$\Delta P_2 = 6750\ N/m^2$
5-52 $N_2 = 1800$ rpm
$\Delta P_2 = 1868\ N/m^2$
5-56 153.8 kW

5-65 $\eta = 0.43$
$\dot{W} = 3.46$ kW

5-4 $C_m = 3.11 \times 10^{-5}$; 4.86 m
5-8 14.1 m^3/s
5-14 $Q = 4.58\ m^3/s$ for $V_f = 15$ m/s
5-18 $A_c = 6yL$

5-22 $X = 70$ cm
$Q_c = 1.84\ m^3/s$
5-28 $D_c = 12$ cm
$V = 9.2$ m/s
5-32 3.84 N/m^2
5-36 11.85 N/m^2
5-40 41.8 cm
5-47 3600 rpm
5-50 2170 rpm

5-54 8.65 kW

5-58 $\dot{W}_1 = 10.22$ kW
$\dot{W}_2 = 14.77$ kW
$163/year increased cost
$\dot{W}_3 = 2.55$ kW
$72/year savings
5-66 $.0745/hr; 9237 h/year

5-67 $Q_j = 1.21 \text{ m}^3/\text{s}$
    $\dot{W}_j = 41.5 \text{ kW}$
    $\text{SND} = 6673 \text{ N/m}^2$
    $Q_{1_{max}} = 10.6 \text{ m}^3/\text{s}$

Chapter 6

6-2 (a) Laminar
    (b) $\eta = 0.500$ for $d = 3.0 \ \mu\text{m}$
6-6 Turbulent
    $\eta = 0.377$ for 40-μm particles
6-16 $n = 8$
    $L = W = 5.03 \text{ m}$
    $H = 2.4 \text{ m}$
6-22 7.5 h

6-4 $n = 10$
    $L = 42.2 \text{ m}$
6-12 (a) Turbulent
    (b) $\eta = 0.901$ for $d = 25 \ \mu\text{m}$
6-18 $n = 2$
    $L = W = 0.102 \text{ m}$

6-26 1.000

Chapter 7

7-2 $\eta = 0.282$ for $d = 25 \ \mu\text{m}$
7-6 $\eta_A = 0.822$
    $\eta_n = 0.711$
7-12 $r_2 = 0.818 \text{ m}$
    $W = r_1 = 0.409 \text{ m}$
7-22 $\eta = 0.962$ for $d = 5.0 \ \mu\text{m}$

7-4 $\eta = 0.780$ for $d = 60 \ \mu\text{m}$
7-8 39.1 rad or 6.22 turns

7-18 $D_2 = 2.60 \text{ m}$
    $L = 29.2 \text{ m}$
7-24 $D_2 = 8.4 \text{ m}$ $L = 33.6 \text{ m}$
    $L_1 = L_2 = 16.8 \text{ m}$ $H = 4.2 \text{ m}$
    $D_e = 4.2 \text{ m}$ $B = 2.1 \text{ m}$
    $D_d = 2.1 \text{ m}$ $L_3 = 1.05 \text{ m}$
7-28 $\eta = 0.83$ for $d = 10 \ \mu\text{m}$

7-26 $D_2 = 0.115 \text{ m}$ $\alpha_2 = 60°$
    $L_1 = L_2 = 0.231 \text{ m}$ $D_e = 0.057 \text{ m}$
    $D_d = 0.029 \text{ m}$ $L_3 = 0.014 \text{ m}$
7-34 $\Delta P = 70{,}533 \text{ N/m}^2$
    $\dot{W} = 7.05 \text{ MW}$
7-38 $\Delta P = 147{,}826 \text{ N/m}^2$
    $\dot{W} = 444 \text{ kW}$
7-43 $D_2 = 17 \text{ cm}$
    $r_1 = 4.31 \text{ cm}$
    $L = 19.0 \text{ m}$
    $N = 8679$
    $322{,}850/\text{year power cost};$
    $743{,}048/\text{year total cost}$

7-36 $\Delta P = 189{,}600 \text{ N/m}^2$
    $\dot{W} = 19.0 \text{ MW}$
7-40 $\Delta P = 2{,}973{,}000 \text{ N/m}^2$
    $\dot{W} = 157 \text{ kW}$

Chapter 8

8-2 1.28 m

8-6 $\tilde{V} = -140{,}000 \text{ V}$

    at $\dfrac{i}{L} = -0.1257 \text{ mA/m}$

8-4 $\tilde{V} = 64{,}125 \text{ V}$ and $E = 10^6 \text{ V/m}$
    at $y = 5 \text{ cm}$
8-9 $D_0 = 3.90 \text{ mm}$
    $E_0 = -3.023 \times 10^6 \text{ V/m}$

8-15 $t_0 = 0.101$ ms
    $q_{ps} = 0.695$ fC
8-18 $t_0 = 1.005 \times 10^{-5}$ s
    $q_{ps} = 0.0111$ fC   for $d = 0.4$ μm
8-24 $L = 30.8$ m
8-30 $L = 3.56$ m
    $\tilde{V}_w = -124{,}636$ V
8-34 $L = 32.0$ m
    $r_w = 0.00328$ m
    $\tilde{V}_{w_0} = -161{,}739$ V
    $\tilde{V}_{w_L} = -154{,}956$ V
8-38 $W = 0.10$ m
    $H = 1.0$ m
    $L = 15.6$ m
    $N = 2$
8-42 $L_2 = 3.01$ m
    $\tilde{V}_2 = \tilde{V}_w = 57{,}000$ V
8-48 $\eta = 1.0$

8-16 $q_{ps} = 11.9$ fC   at $r = 10$ cm

8-22 $q_p = 4.04 \times 10^{-4}$ fC   for $t = 0.1$ s

8-26 0.35 m
8-32 -0.115 mA

8-36 $L = 2.84$ m
    $W = 8.8$ m
    $H = 3.0$ m

8-40 $L = 3.08$ m

8-44 $L_2 = 1.98$ m
    $N_c = 85$ for $r_c = 0.10$ m
8-50 $L_2 = 14.1$ m   for $\tilde{V}_2 = 200{,}000$ V

Chapter 9

9-10 0.1354
9-18 $\eta_{dd} = 0.000111$ for $V_\infty = 20$ m/s

9-22 0.00108
9-26 0.723
9-32 3.113 kW

9-36 2.05×10^5
9-40 0.0419
9-44 116 by 58 cm; 7.22 kW
9-48 $\eta = 0.920$ for $d = 5.0$ μm

9-52 $R_2 = 2.58$ m
    $W = 0.645$ m
    $Q_s = 0.202$ m³/s
    $\Delta P = 7522$ N/m²
    $\dot{W} = 188$ kW
9-56 $Q_s = 0.0143$ m³/s   for $D = 1.0$ mm
9-60 (a) 0.0157 m³/s; 215 kW
    (b) 0.0235 m³/s; 49.7 kW
    (c) 0.0133 m³/s; 1587 kW

9-12 0.7566
9-20 $D = 0.1$ mm
    $V_\infty = 19.8$ cm/s
9-24 0.632; 0.99995
9-30 145.3 kW
9-34 (a) $\rho_d = 1002$ kg/m³
    (b) $\eta = 0.270$ for $d = 0.4$ μm
9-38 0.5 mm approximately
9-42 11.61 kW
9-46 $\dot{W} = 50$ kW   for $d = 0.8$ μm
9-50 $\eta = 1.0$
    $Q_s = 0.277$ m³/s
9-54 $Q_s = 0.0285$ m³/s

9-58 $\dot{W} = 40$ kW   for $D = 1.0$ mm

Chapter 10

10-6 0.972

10-10 4.17

10-8 Re $= 0.258$ and $\eta_{fi} = 0.0235$
    for $V_\infty = 1.0$ m/s
10-12 $\eta_{fd} = 0.0149$ for $D = 5.0$ μm

10-14 $\eta_{fi} = 0.00475$
$\quad \eta_{fd} = 0.00222$
$\quad \eta_{f} = 0.00696$
10-18 5.94 mm
10-24 2.43 by 2.43 m
10-28 72,398 N/m^2; 72.4 kW
10-36 $\dot{W} = 772$ kW $\quad$ for $t = 5.0$ h
10-40 75 kW

10-44 \$20,000/year for cotton;
$\quad$ \$18,000/year for nylon
10-50 51.3 min

10-16 0.0207

10-22 7.23 mm
10-26 12,188 N/m^2
10-30 786 N/m^2; 1.57 kW
10-38 13.9 h
10-42 $K_1 = 52,500$ N · s/m^3
$\quad K_2 = 101,310$ s^{-1}
10-48 8.46 min

Chapter 11

11-14 $\eta_d = 0.140$ $\quad$ for $V_\infty = 20$ m/s
11-22 $\eta_d = 0.000566$ $\quad$ for
$\quad C''_{mv_i} = 0.1$ kg/m^3 and $t = 1.0$ s
11-26 $\eta = 0.356$ $\quad$ for $D = 1.0$ mm
11-30 0.000107
11-34 0.289 m^3/s
11-40 $R_2 = 2.45$ m
$\quad Q_s = 0.116$ m^3/s

11-44 1.94 m^3/s
11-54 12.2 graphically

11-58 $\dot{W}_a = 107$ kW
$\quad \dot{W}_l = 157$ W
11-64 0.728

11-16 $V_\infty < 142.8$ m/s
11-24 0.0888

11-28 3.38 kW
11-32 $\eta = 0.498$ $\quad$ for $Q_s = 0.1$ m^3/s
11-38 $\eta = 0.0879$
11-42 $R_2 = 20$ m
$\quad W = 5$ m
$\quad L = 40$ m
$\quad D = 1.0$ mm
$\quad Q_s = 110$ m^3/s
11-50 $Q_s = 1.027$ m^3/s
11-56 $D = 23.1$ m
$\quad L = 4.21$ m
$\quad \Delta P = 56.8$ N/m^2
$\quad \dot{W}_a = 5.68$ kW
$\quad \dot{W}_l = 129.7$ kW
11-62 $Q_l = 3.44$ m^3/s; 10 trays; other data
$\quad$ as in Prob. 11-56
11-66 0.9978

Chapter 12

12-4 35.4 m^2

12-8 $\alpha = 2398$ $\quad$ for $T = 400$ K
12-14 21,569 N/m^2; 4.55 kW
12-18 734,925 N/m^2; 1230 kW
12-22 (a) $t_B = 2.71$ h $\qquad \dot{W} = 545$ W
$\quad$ (b) $t_R = 0.0141$ h $\qquad \dot{W} = 8.89$ W
$\quad$ (c) $t_c = 0.0303$ h $\qquad \dot{W} = 4.91$ kW

12-6 $t_B = 22.5$ h
$\quad t_E = 23.3$ h
12-10 $t_R = 10.35$ h $\quad$ for $T = 300$ K
12-16 17,276 N/m^2; 2.92 kW
12-20 $\Delta P = 5176$ N/m^2 $\quad$ for $D' = 0.5$ mm
12-26 $m'_p = 310.2$ kg
$\quad E = 9.31 \times 10^8$ J
$\quad \dot{m}_c = 25.7$ kg/s
$\quad Q_c = 21.7$ m^3/s

12-27 $\Delta P = 64.0$ N/m^2 for the adsorption
      $\dot{W} = 0.64$ kW   section
      $\Delta P = 79.4$ N/m^2 for the drying
      $\dot{W} = 0.662$ kW   section

12-30 $L = 0.08$ m
      $R = 0.5$ m
      $H = 3.8$ m
      $\Delta P = 2020$ N/m^2 for the adsorp-
      $\dot{W} = 4.04$ kW   tion section
      $\Delta P = 579$ N/m^2 for the drying
      $\dot{W} = 0.43$ kW   section
      $\dot{W} = 4.47$ kW total

Chapter 13

13-2   154.
13-6   11.72
13-10 $AF_s = 14.09$
13-14 $AF = 54.28$
13-18 $D = 1.19$ m
      $L = 2.62$ m
13-22 $\dot{m}_f = 47.9$ kg/h
      $T_3 = 430°C$
13-26 $H = 5.07$ m

13-30 $\dot{m}_f = 181$ kg/h; 36.5 percent savings
13-34 57.0 N/m^2; 1.67 kW

13-4   17.2
13-8   0.0314 kg/m^3
13-12 $AF = 40.22$; 185.5 percent
13-16 2.21
13-20 0.293 s

13-24 0.75 effectiveness; \$23,590/year
      $\dot{m}_f = 160,896$ kg/year
13-28 $\dot{m}_f = 0.00226$ kg/s
      $Q_a = 0.142$ m^3/s
13-32 $\dot{m}_f = 476$ kg/h

Chapter 14

14-2   3098 K

14-8 $-6°C$
14-12 $-13°C$
14-18 43.1 W/m$^2 \cdot$ K
14-22 (a) $-40°C$
      (b) $\dot{m}_2 = 0.020$ kg/s
           $\dot{Q} = 2.08$ MJ/kg
      (c) $A = 151$ m^2/(kg/s)
14-26 $\dot{m}_1 = 0.00348$ kg/s
      $C_{m_{p1}} = 0.216$

14-6 $P_p = 382$ N/m^2 and
      $h_{fg} = 378$ kJ/kg   at $T = 0°C$
14-10 0.991
14-16 $m_w = 17.05$ kg/kg
14-20 12 tubes, 33.4 m long
14-24 151,042 J/s

14-28 68.7 MW

INDEX